Renewable Energy, Second Edition

A First Course

Renewable Energy, Second Edition

A First Course

Robert Ehrlich
Harold A. Geller

CRC Press is an imprint of the
Taylor & Francis Group, an **informa** business

CRC Press
Taylor & Francis Group
6000 Broken Sound Parkway NW, Suite 300
Boca Raton, FL 33487-2742

Printed on acid-free paper

International Standard Book Number-13: 978-1-4987-3695-4 (Paperback)

Library of Congress Cataloging-in-Publication Data

Names: Ehrlich, Robert, 1938- author. | Geller, Harold, 1954- author.
Title: Renewable energy : a first course / Robert Ehrlich, Harold A. Geller.
Description: Second edition. | Boca Raton : Taylor & Francis, CRC Press, 2017. | Includes bibliographical references and index.
Identifiers: LCCN 2017018670| ISBN 9781498736954 (pbk. : alk. paper) | ISBN 9781138297388 (hardback : alk. paper)
Subjects: LCSH: Renewable energy sources.
Classification: LCC TJ808 .E34 2017 | DDC 333.79/4--dc23
LC record available at https://lccn.loc.gov/2017018670

Visit the Taylor & Francis Web site at
http://www.taylorandfrancis.com

and the CRC Press Web site at
http://www.crcpress.com

Printed and bound in Great Britain by
TJ International Ltd, Padstow, Cornwall

For Lochlan, Meridian, Richard and Eugenio

Brief Contents

Contents

Preface to the First Edition

If you are a student in one of the sciences or engineering who has taken a few introductory courses in physics and calculus, you will find this book useful, because it covers a variety of technologies in renewable energy and explains the basic principles. It avoids, if at all possible, technical jargon and mathematically advanced approaches found in many books on the subject. It is also not overly long, unlike many other books on renewable energy, and its 14 chapters should easily fit within a standard semester, at least in most schools in the United States. I personally find the sheer weight of some textbooks intimidating, so hopefully, this book will not fall into that category. Most importantly, until about four years ago, I was in your shoes. No, I am not a young faculty member, but I am relatively new to the energy field. In fact, I am pretty much at the end of my teaching career. Four years ago, I wanted to find something meaningful to which I could devote the remainder of my career, and renewable energy certainly seemed a good fit for me. Until that point, my teaching and research had been entirely or almost entirely in the field of physics. Thus, four years ago, I was a relative "newbie" in the field of renewable energy, and I had to figure out a lot of things for myself. I thus still remember the kinds of issues that confuse students in this subject and how to explain these to them as clearly as possible.

Renewable energy assumes great significance for the future of the world, given the environmental issues that are related to the ways we generate most of our energy and the central place that energy occupies in our society. Proper energy choices need to be made in order to avoid an environmental disaster, severe energy shortages, and even social chaos or war. These proper choices are not obvious, and certainly, it is not as simple as saying "Let us stop using fossil fuels and nuclear power now in order to save the environment!" Making wise decisions involves sound consideration of all the implications and a thorough look at economic, environmental, technical, political, and other perspectives, weighing relative costs and benefits for a host of possible technologies. Thus, even if you are not planning a career in this field, the book should help you make more intelligent choices as a citizen and consumer.

The book, despite its title, does include three chapters on nonrenewable energy: one on fossil fuels and two on nuclear, the first focused on the science and the second on the technology. This is an important addition, because renewable energy needs to be compared with the other primary ways in which we now produce energy in order to better evaluate its advantages and shortcomings. Moreover, these other technologies will probably remain for some time to come, even if some nations such as Germany have opted to phase out both their nuclear- and coal-fired electricity-generating plants. Some observers believe that it is realistic to move entirely toward renewable energy (such as solar, wind, biofuels, geothermal, and hydropower) by 2030, while most would probably put the date further into the future. The book also includes four overarching topics that go beyond any specific type of energy, namely, energy conservation, energy storage, energy transmission, and energy policy. The energy field is a continually changing one, and so it is important to keep up with the latest advances. This book provides up-to-date information, although it will inevitably require some revisions in a few years. I hope you enjoy it—let me know if you do, and certainly, let me know if you find any errors or ambiguities.

RE

Preface to the Second Edition

The first edition of any textbook is bound to include some number of typographical errors. This revision gives a full update on the original text, and it fixes any problems or mistakes of its predecessor. As before, it is intended for undergraduate students in physics or engineering who have taken an introductory course in physics and calculus. It not only mainly focuses on renewable energy, but also addresses nonrenewables such as fossil fuels and nuclear technology. We continue with a physics orientation, while also addressing important conservation, economic, and public policy issues. The emphasis is on a nontechnical presentation, avoiding advanced math while teaching fundamental analytical skills with wide application.

We have appreciated the feedback of current adopters in developing this second edition. The addition of two new chapters on population growth and energy use and probability and risk constitutes a major addition. We have tried throughout to maintain a readable, lively text with healthy use of anecdotes, history, illustrations, and sidebar topics.

Problem sets have been updated and expanded. Instructors may access some additional test bank questions and answers through the publisher's website at http://www.crcpress.com (search for the page for the book and click on *Downloads/Updates*).

Acknowledgments

We would like to thank the students in our classes for giving us useful feedback on the book and the two reviewers contacted by the publisher, Professor Michael Ogilvie (Washington University in St. Louis) and Professor John Smedley (Bates College), for their review of the entire manuscript. We are especially grateful to Lu Han, senior editor at CRC Press/Taylor & Francis Group, for his suggestions.

Authors

Robert Ehrlich is a professor of physics emeritus at George Mason University, Fairfax, Virginia. He earned his BS in physics from Brooklyn College and his PhD from Columbia University. He is a fellow of the American Physical Society. He formerly chaired the physics departments at George Mason University and The State University of New York at New Paltz and has taught physics for nearly four decades.

Dr. Ehrlich is an elementary particle physicist and has worked in a number of other areas. He has authored or edited 20 books and about 100 journal articles. His current scholarly interests include renewable energy and the existence of faster-than-light particles.

Harold A. Geller is an associate professor of physics and astronomy at George Mason University, Fairfax, Virginia. He earned his BS from the University of the State of New York, Albany, and his MA in astronomy and informatics and his doctorate in education from George Mason University. Dr. Geller has been teaching physics and astronomy for over a quarter century. He has been the associate chair of the Department of Physics and Astronomy, manager of the Washington Operations for the Consortium for International Earth Science Information Networks, program manager at Science Applications International Corporation, and doctoral fellow of the State Council of Higher Education for Virginia. He has authored or edited seven books and has published about 85 papers in education, astrobiology, astrophysics, and biochemistry. His current scholarly interests include energy and the environment, the search for life in the universe, and the exploration of space.

Introduction

CONTENTS

1.1 WHY ANOTHER BOOK ON ENERGY?

The idea for this book arose as a result of one of the author's first time teaching a course on renewable energy. The course was not Energy 101, but it was intended for students who had completed an introductory physics sequence and taken a few courses in calculus. Most available books were either too elementary or too advanced, and the handful of books at the right level seemed too focused on technicalities that obscured the basic ideas. In addition, many of those texts lacked the desired informal writing style, with even some occasional touches of humor that can enhance readability. Moreover, any course focused on renewable energy must also cover nonrenewable energy (fossil fuels and nuclear, specifically), because only then could useful contrasts be drawn. Renewable energy is a multidisciplinary subject that goes well beyond physics, although it is fair to say that this book has a physics orientation. Physicists do have a certain way of looking at the world that is different from other scientists and from engineers. They want to understand how things work and strip things down to their fundamentals. It is no accident that many new technologies, from the laser, to the computed tomography scanner, to the atomic bomb, were invented and developed by physicists, while their refinement is often done by engineers.

1.2 WHY IS ENERGY SO IMPORTANT TO SOCIETY?

Those of us who are fortunate to live in the developed world often take for granted the availability of abundant sources of energy, and we do not fully appreciate the difficult life faced by half of the population of the world, who substitute their own labor or that of domestic animals for the machines and devices that are so common in the developed world. A brief taste of what life is like without access to abundant energy sources is provided at those times when the power goes out. But while survival during such brief interludes may not be in question (except in special circumstances), try to imagine what life would be like if the power were to go out for a period of, say, 6 months. Not having cell phones, television, Internet, or radio might be the least of your problems, especially if the extended power failure occurred during a cold winter when food was not available, and your "taking up farming" was a complete joke, even if you had the knowledge, tools, and land to do so. As much as some of us might imagine the pleasures of a simple preindustrial lifestyle without all the trappings of our high-technology society, the reality would likely be quite different if we were suddenly plunged into a world without electricity. It is likely that a large fraction of the population would not survive 6 months. The idea of a prolonged failure of the power grid in many nations simultaneously is not just some outlandish science fiction prospect and could occur as a result of a large solar flare directed at the planet. The last one that was large enough to pose a threat of catastrophic damage was apparently the Carrington event, which occurred in 1859 before our electrified civilization existed, but it did cause telegraph systems all over North America and Europe to fail.

1.3 EXACTLY WHAT IS ENERGY?

In elementary school, many of us learned that "energy is the ability to do work" and that "it cannot be created or destroyed" (conservation of energy). But these memorized and parroted phrases are not always easy to apply to real situations. For example, suppose you had a hand-cranked or pedal-driven electric generator that was connected to a light bulb. Do you think it would be just as hard to turn the generator if the light bulb were unscrewed from its socket or replaced by one of lower wattage? Most people (even some engineering students) who were asked this question answer yes and are often surprised to find on doing the experiment that the answer is no—the generator is easier to turn with the bulb removed or replaced by one of lower wattage. This of course must be the case by conservation of energy, since it is the mechanical energy of your turning the crank that is being converted into electrical energy, which is absent when the light bulb is unscrewed. Were the handle on the generator just as easy to turn regardless of whether a bulb is being lit or how brightly it glows, then it would be just as easy for a generator to supply electric power to a city of a million people as one having only a thousand! Incidentally, you can probably forget about supplying all your own power by using a pedal-powered generator, since even an avid cyclist would be able to supply at most only a few percent of what the average American consumes.

Aside from misunderstanding what the law of energy conservation implies about specific situations, there are also some interesting and subtle complexities to the law itself. Richard Feynman was one of the great physicists of the twentieth century who made many important discoveries, including the field of quantum electrodynamics, which he coinvented with Julian Schwinger. Feynman was both a very colorful person and a gifted teacher, who came up with novel ways to look at the world. He understood that the concept of energy and its conservation was more complex and abstract than many other physical quantities such as electric charge where the conservation law involves a single number—the net amount of charge. With energy, however, we have the problem that it comes in a wide variety of forms, including kinetic, potential, heat, light, electrical, magnetic, and nuclear, which can be converted into one another. To keep track of the net amount of energy and to recognize that it is conserved involve some more complicated "bookkeeping," for example, knowing how many units of heat energy (calories) are equivalent to how many units of mechanical energy (joules).

HOW MANY JOULES IS EQUAL TO 1 CALORIE?

The calorie is the amount of heat needed to raise 1 g of water by 1°C. But since this amount slightly depends on temperature, one sometimes sees slightly different values quoted for the conversion factor commonly taken to be 4.1868 J/cal.

In presenting the concept of energy and the law of its conservation, Feynman made up a story of a little boy playing with 28 indestructible blocks (Feynman, 1985). Each day, the boy's mother returns home and sees that there are in fact 28 blocks, until one day, she notices that only 27 are present. The observant mother notices one block lying in the backyard and realizes that her son must have thrown it out the window. Clearly, the number of blocks (like energy) is "conserved" only in a closed system, in which no blocks or energy enters or leaves. In the future, she is more careful not to leave the window open. Another day when the mother returns, she finds that only 25 blocks are present, and she concludes that the missing three blocks must be hidden somewhere—but where?

The boy seeking to make his mother's task harder does not allow her to open a box in which blocks might be hidden. However, the clever mother finds that when she weighs the box, it is heavier than it was when empty by exactly three times the weight of one block, and she draws the obvious conclusion. The game between the mother and the child continues day after day, with the child finding more ingenious places to hide the blocks. One day, for example, he hides several under the dirty water in the sink, but the mother notices that the level of the water has risen by an amount equivalent to the volume of two blocks. Notice that the mother never sees any hidden blocks, but can infer how many are hidden in different places by making careful observations, and now that the windows are closed, she always finds the total number to be conserved. If the mother is so inclined, she might write her finding in terms of the equation for the "conservation of blocks":

$$\begin{aligned}&\text{Number of visible blocks} + \text{number hidden in box}\\&\quad + \text{number hidden in sink} + \ldots = 28,\end{aligned}$$

where each of the numbers of hidden blocks had to be inferred from careful measurements, and the three dots suggest any number of other possible hiding places.

Energy conservation is similar to the story with the blocks in that when you take into account all the forms of energy (all the block hiding places), the total amount works out to be a constant. But remember that in order to conclude that the number of blocks was conserved, the mother needed to know exactly how much excess weight in the box, how much rise in dishwater level, etc., corresponded to one block. Exactly the same applies to energy conservation. If we want to see if energy is conserved in some process involving motion and heat, we need to know exactly how many units of heat (calories) are equivalent to each unit of mechanical energy (joules). In fact, this was how the self-taught physicist James Prescott Joule proved that heat was a form of energy. Should we ever find a physical situation in which energy appears not to be conserved, there are only four possible conclusions. See if you can figure out what they are before reading any further.

1.4 MIGHT THERE BE SOME NEW FORMS OF ENERGY NOT YET KNOWN?

Feynman's story of the boy and his blocks is an appropriate analogy to humanity's discovery of new forms of energy that are often well hidden and found only when energy conservation seems to be violated. A century ago, for example, who would have dreamed that vast stores of energy exist inside the nucleus of all atoms and might actually be released? Even after the discovery of the atomic nucleus, three decades elapsed before scientists realized that the vast energy the nucleus contained might be harnessed. Finding a new form of energy is of course an exceptionally rare event, and the last time it occurred was in fact with nuclear energy.

It remains conceivable that there exists some as-yet undiscovered forms of energy, but all existing claims for it are unconvincing. The likelihood is that in any situation where energy seems not to be conserved, either the system is not closed or else we simply have not accounted for all the known forms of energy properly. Likewise, those who believe in energy fields surrounding the human body that are not detectable by instruments, but which can be manipulated by skilled hand-waving "therapeutic touch" practitioners, are deluding themselves. The idea that living organisms operate based on special energy fields different from the normal electromagnetic fields measureable by instruments is essentially the discredited nineteenth-century belief known as "vitalism." This theory holds that there exists some type of energy innate in living structures or a vital force peculiar to life itself.

FOUR POSSIBLE CONCLUSIONS IF ENERGY APPEARS NOT TO BE CONSERVED

- We are not dealing with a closed system—energy in one form or another is entering or leaving the system.
- Energy stays within the system but is in some form we neglected to consider (possibly because we did not know that it existed).
- We have made an error in our measurements.
- We have discovered an example of the violation of the law of conservation of energy.

Figure 1.1 Therapeutic touch practitioner (*left*) attempting to sense which of her two hands was in the presence of the young experimenter's hand hidden from her view on the right. (Courtesy of the Skeptics Society, Altadena, CA.)

For most physicists, the last possibility is considered sufficiently unthinkable, so that when it seems to be occurring, it prompts proposals for highly radical alternatives—the neutrino, for example, to account for the "missing" energy in the case of the phenomenon known as beta decay—see Chapter 3.

In one clever experiment designed and conducted by a sixth-grade student, and published in a prestigious medical journal, practitioners of therapeutic touch were unable to perceive any energy fields where they should have been able to. In fact, they guessed correctly only 44% of the time, i.e., less than chance (Rosa et al., 1998). Needless to say, believers in such nonsense are unlikely to find much of interest in this book (Figure 1.1).

1.5 WHAT ARE THE UNITS OF ENERGY?

The fact that energy exists in many forms is part of the reason why there are so many different units for this quantity—for example, calories and British thermal units (BTUs) are typically used for heat; Joules, ergs, and foot-pounds for mechanical energy; kilowatt-hours for electrical energy; and million electron volts (MeV) for nuclear energy. However, since all these units describe the same fundamental entity, there must be conversion factors relating them all. To make matters more even confusing,

Table 1.1 Some Units of Energy

Name	Definition
Joule (J)	Work done by a 1 N force acting through 1 m (also 1 W s)
Erg	Work done by a 1 dyne force acting through 1 cm
Calorie (cal)	Heat needed to raise 1 g of water by 1°C
BTU	Heat needed to raise 1 lb of water by 1°F
Kilowatt-hour (kWh)	Energy of 1 kW of power flowing for 1 h
Quad	A quadrillion (10^{15}) BTU
Therm	100,000 BTU
Electron volt (eV)	Energy gain of an electron moved through a 1 V potential difference
Megaton (Mt)	Energy released when a million tons of trinitrotoluene explodes
Foot-pound	Work done by a 1 lb force acting through 1 ft

Note: A calorie associated with food is actually 1000 cal by the aforementioned definition or a kilocalorie (kcal). Sometimes 1 kcal is written as 1 Cal (capitalized C). Readers should be familiar with some of the more important conversion factors.

there are a whole host of separate units for the quantity power, which refers to the rate at which energy is produced or consumed, i.e.,

$$p = \frac{dE}{dt} = \dot{E} \quad \text{or} \quad E = \int p\,dt. \tag{1.1}$$

Note that a dot over any quantity is used as shorthand for its time derivative. Many power and energy units unfortunately sound similar, e.g., kilowatts are power, whereas kilowatt-hour (abbreviated kWh) is energy (Table 1.1).

DO YOU PAY FOR POWER OR ENERGY?

Electric power plants are rated according to the electric power they produce in megawatts (MW), but for the most part, they charge residential customers merely for the total energy they consume in kilowatt-hours and not the rate at which they use it, or the time of day you use it. The situation is often very different for large consumers, where these factors are taken into account. Moreover, in order to smooth out their demand, some electric utilities actually do allow residential customers to pay a special rate if their usage tends to be very uniform, and in another plan, they bill for very different rates for on-peak and off-peak usage. These special pricing options aside, the utility company charges you the same price to supply you with 100 kWh of energy, whether you use it to light a 100 W bulb for 1000 h or a 200 W bulb for 500 h.

1.6 LAWS OF THERMODYNAMICS

The law of conservation of energy is also known as the first law of thermodynamics, and as we have noted, it has never been observed to be violated. Essentially, as applied to energy, the first law says that "you cannot get something from nothing." The second law, however, is the more interesting one, and it says that "you cannot even break even." Although the second law has many forms, the most common one concerns the generation of mechanical work W from heat Q_C, where the subscript C stands for heat of combustion. In general, we may define the energy efficiency of any process as

$$e \equiv \frac{E_{useful}}{E_{input}} = \frac{W}{Q_C} = \frac{\dot{W}}{\dot{Q}_C}. \quad (1.2)$$

The last equality in Equation 1.2 reminds us that the equation for efficiency applies equally well to power as to energy. By the first law, the maximum possible value of the efficiency would be 1.0 or 100%. However, the second law places a much more stringent limit on its value. For a process in which fuel combustion takes place at a temperature T_C and heat is expelled to the environment at ambient temperature T_a, the efficiency in general, defined as the useful work output divided by the heat input, cannot exceed the Carnot efficiency:

$$e_C = 1 - \frac{T_a}{T_C}, \quad (1.3)$$

where both temperatures must be in kelvin. This limitation is a direct consequence of the second law of thermodynamics, which states that heat energy spontaneously always flows from high temperatures to low temperatures. The Carnot efficiency would hold only for ideal processes that can take place in either direction equally, which do not exist in the real world, except as limiting cases. For example, were you to take a movie of any real process, such as an isolated swinging pendulum slowing down gradually, there would be no doubt when the movie was run backward or forward. Time-reversible, ideal processes would require that the net entropy S remain constant, where a small change in entropy can be defined in terms of the heat flow dQ at some particular temperature T as

$$dS = \frac{dQ}{T}. \quad (1.4)$$

Thus, an alternative definition of the second law of thermodynamics is that for any real process, $dS > 0$.

1.6.1 Example 1: Calculating Energy When Power Varies in Time

Suppose during a test of a nuclear reactor, its power level is ramped up from zero to its rated power of 1000 MW over a 2 h period, and then after running at full power for 6 h, it is ramped back down to zero over a 2 h period. Calculate the total energy generated by the reactor during those 10 h.

PERPETUAL-MOTION MACHINES

Over the course of history, many inventors have come up with ideas for devices known as perpetual-motion machines, which generate either energy from nothing (and violate the first law of thermodynamics or the law of conservation of energy) or violate the second law of thermodynamics. In the latter case, the useful work they produce, while less than the heat they consume, exceeds the amount dictated by the Carnot limit. None of these machines have ever worked, although patent applications for them have become so common that the US Patent and Trademark Office (USPTO) has made an official policy of refusing to grant patents for perpetual-motion machines without a working model. In fact, it is interesting that the USPTO has granted quite a few patents for such devices—even some in recent years. However, it is also important to note that granting of a patent does not mean that the invention actually works, only that the patent examiner could not figure out why it would not.

Solution

We shall assume here that during the time that the power is ramped up and down, it varies linearly, so that the power the reactor generates varies accordingly during the 10 h test as shown in Figure 1.2.

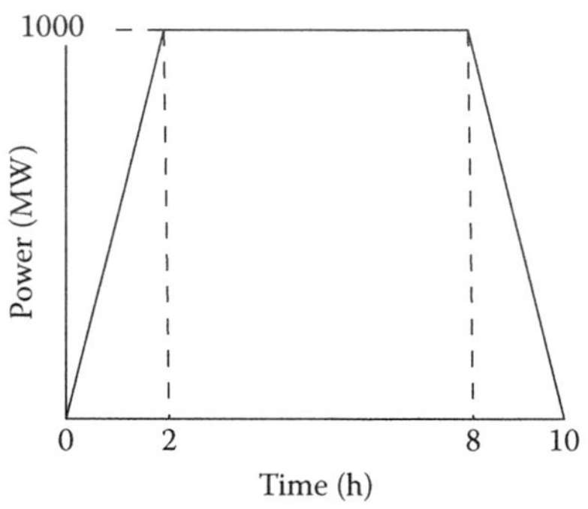

Based on Equation 1.1, and the definition of the integral as the area under the power–time curve, the energy must be equal to the area of the trapezoid in Figure 1.2 or 8000 MWh (Table 1.2).

Figure 1.2 Power profile of nuclear reactor during the 10 h test.

Table 1.2 Some Common Prefixes Used to Designate Various Powers of 10

Prefix	Definition
Terra (T)	10^{12}
Giga (G)	10^{9}
Mega (M)	10^{6}
Kilo (k)	10^{3}
Milli (m)	10^{-3}
Micro (μ)	10^{-6}
Nano (n)	10^{-9}
Pico (p)	10^{-12}

1.7 WHAT IS AN ENERGY SOURCE?

Some energy sources are stores (repositories) of energy, typically chemical or nuclear, that can be liberated for useful purposes. Other energy sources are flows of energy through the natural environment that is present in varying degrees at particular times and places. An example of the first type of source might be coal, oil, or uranium, while wind or solar energy would be examples of the second type of source. Consider the question of electricity—is it an energy source or not? Electricity does exist in the natural environment in the extreme form of lightning, and therefore, it can be considered to fall into the second category. In fact, lightning could be considered an energy source, since the electric charge from a lightning strike could be captured and stored (in a capacitor) and then later released for useful purposes. Anyone watching a storm is likely to marvel at the awesome power of a lightning bolt, which is indeed prodigious—typically about 1 TW (10^{12} W). This amount is equal to the power output of a thousand 1000 MW nuclear reactors—more than what exists in the entire world! Such a comparison may prompt the thought: Great! Why not harness lightning as an energy source? The problem is not figuring out how to capture the lightning, but rather that while the power is very high, the energy lightning contains is quite small, since a lightning bolt lasts for such a short time—around 30 μs = 3×10^{-5} s; so by Equation 1.1, the energy contained is around $10^{12} \times 3 \times 10^{-5} = 3 \times 10^{7}$ J = 30 MJ. Thirty million joules may sound impressive, but suppose we designed a "lightning catcher" that managed to capture say 10% of this energy. It would be sufficient to only light a 100 W light bulb for a time, $t = E/p = 3 \times 10^{6}$ J/100 W = 3000s, which is just under an hour—hardly a useful energy source, considering the likely expense involved.

What about electricity that humans create—can it be thought of as an energy source? Hardly! Any electricity that we create requires energy input of an amount that is greater than that of the electricity itself, since some energy will always be lost to the environment as heat. Thus, human-created electricity, whether it be from batteries, generators, or solar panels, is not an energy source itself, but merely the product of whatever energy source that created it. In the case of a generator, it would be whatever gave rise to the mechanical energy forcing it to turn, while in the case of a solar panel, it would be the energy in the sunlight incident on the panel.

1.8 WHAT EXACTLY IS THE WORLD'S ENERGY PROBLEM?

All sources of energy have some environmental impact, but as you are aware, the impacts of different sources considerably vary. The energy sources people worry the most about are fossil fuels (coal, oil, and gas) as well as nuclear, while the renewable ("green") energy sources are considered much more benign—even though they too have some harmful impacts. Moreover, the environmental impact of fossil fuel and nuclear energy usage has gotten worse over time, as the human population has grown and the energy usage per capita has also grown—an inevitable consequence of the rise in living standards worldwide. This is not to say that higher per capita wealth invariably requires higher per capita energy usage, but the two are strongly correlated. People are well aware of the harmful environmental impacts of fossil fuel and nuclear plants based on dramatic events, reported in the news of oil spills, coal mine disasters, and nuclear meltdowns such as that at Fukushima, Japan. Other impacts involving air, water, and land pollution may be ongoing and less dramatic but may cost many more lives over the long term.

1.8.1 Climate Change

The long-term environmental impact raising perhaps the greatest level of concern among many people is that of global climate change or global warming associated with the increasing level of greenhouse gases put into the Earth's atmosphere from a variety of causes, but most notably the burning of fossil fuels. The basic science behind the greenhouse effect is solid. There is no debate among scientists concerning whether

(1) atmospheric greenhouse gas levels have been significantly rising over time due to human actions and (2) these rising emissions are responsible for some degree of climate change—which most climate scientists consider the predominant cause.

Periodically, the Intergovernmental Panel on Climate Change (IPCC), an international collaboration of hundreds of climate scientists, issues reports summarizing the state of the science behind climate change. The most recent comprehensive assessment *The IPCC Fifth Assessment Report* (issued in 2014) states that "it is extremely likely that human influence has been the dominant cause of the observed warming since the mid-20th Century." Finally, a widely cited survey found that 97.4% of active researchers in climate science believe that "human activity is a significant contributing factor in changing mean global temperatures" (Doran and Kendall Zimmerman, 2009). Some have used the results of this survey to draw the conclusion that the issue of human-caused global warming is therefore entirely settled among climate scientists, which is perhaps a bit of an overstatement. Agreeing that the human-caused component of climate change is "significant," i.e., not trivial, is not at all the same as agreeing that it is the only cause. More importantly, issues in science are never decided on the basis of a majority vote, but on the merits of the arguments. Nevertheless, surveys of the general public on the issue of global warming sharply contrast with those of climate scientists, with far smaller percentages of people believing that human actions are primarily responsible. Chapter 9 discusses the topic of climate change in much greater depth, and Chapter 14 discusses why levels of climate change skepticism have risen so significantly in the United States and suggests a way forward to bridge the divide.

1.8.2 Is Human Population Growth the Root Cause of Our Energy and Environmental Problem?

The Reverend Thomas Robert Malthus, who lived from 1766 to 1834, was an economist noted for his highly influential writings on demography and the growth in human population. Like many other economists, he was also a pessimist about human nature. Writing at a time when the impact of the industrial revolution had begun to fuel a growth in human population that had been static for many centuries, Malthus realized that "the power of population is indefinitely greater than the power of the Earth to produce subsistence for man." Advances in technology undreamed of by Malthus have led part of humanity to live in a manner to which kings of his day might aspire, and they allowed the numbers of humans to reach levels far in excess of what then existed. Malthus, being pessimistic regarding the future progress of humanity, believed that throughout history, through wars, epidemics, or famines abetted by population, pressures would always lead to a substantial fraction of humanity to live in misery. To Malthus's list of scourges of famine, war, and disease, modern-day observers might add drastic climate change; pollution; species loss; and shortages in natural resources, energy, and water—all of which are exacerbated by overpopulation (Figure 1.3).

Although the growth in the human population has significantly slowed in recent decades, it is unclear if it has happened in time to avert catastrophe, with some observers maintaining that the Earth has already far too many humans to have a long-term future that is sustainable. Currently, half of humanity survives on less than $2.50/day, and the gap between the developed and developing worlds may widen rather than narrow because of demographic trends. Even though the populations in many developed nations have begun to decline, demographers foresee an inevitable increase in population throughout the first half of this century given the high fertility of previous generations and the numbers of future parents who are already alive (even if their fertility is relatively lower), with the largest increases coming in regions where poverty is endemic.

One of the prominent twentieth-century environmentalists who foresaw disaster stemming from overpopulation was the biologist Paul Ehrlich (no relation), whose famous and controversial 1968 book, *The Population Bomb*, began with the dramatic and explicit statement: "The battle to feed all of humanity is over. In the 1970s hundreds of millions of people will starve to death in spite of any crash programs embarked upon now. At this late date nothing can prevent a substantial increase in the world death rate..." (Ehrlich, 1968). Of course, while many drought- or war-induced famines have occurred, none has been on the scale and time frame suggested by Ehrlich. Yet the concern over an eventual day of reckoning unabatedly continues among many environmentalists who believe that the Earth is well past its carrying capacity, in terms of the maximum human population it can support.

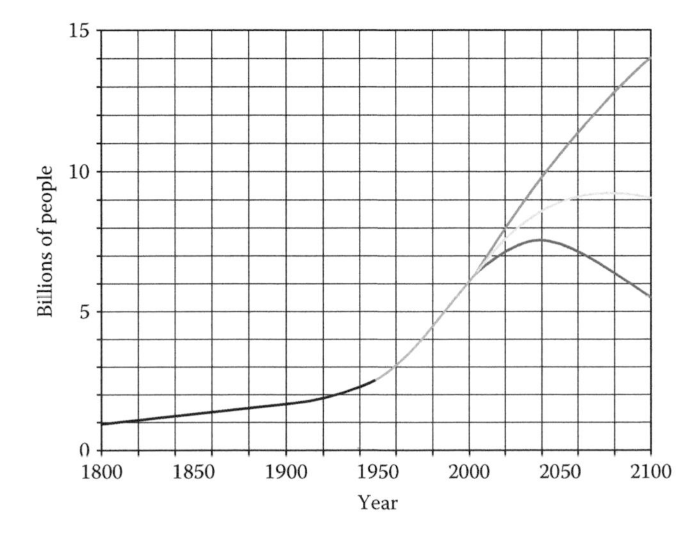

Figure 1.3 World population growth since 1800 based on United Nations (UN) 2010 projections and US Census Bureau historical estimates. The two curves show the high and low estimates beyond 2010 for the population growth according to the UN.

If the Earth is indeed already 50% beyond its capacity as some environmentalists such as Paul Gilding believe, then improvements in energy efficiency might do little to solve the root cause of humanity's problem, namely, too many people. Given that demographers tell us that the population will continue to rise by roughly another 50% by around 2050, with an ever-larger percentage living in poverty, the old scourges of epidemics, famine, and war and the new ones of climate change, species loss, and resource shortages might well cause mass suffering and death on an unimaginable scale. Surprisingly, Gilding himself believes in a possible happier ending to the story. Just as the imminent prospect of a hanging does wonders to concentrate the mind, Gilding thinks that when the coming "Great Disruption" does arrive, we will finally act like grown-ups and take the concerted drastic actions required "at a scale and speed we can barely imagine today, completely transforming our economy, including our energy and transport industries in just a few short decades." Let us hope he is right and Malthus and Paul Ehrlich are wrong.

1.8.3 How Much Time Do We Have?

The question of how quickly the world needs to move away from fossil fuels is, of course, a matter of considerable debate, depending on how serious the threat of climate change is viewed. If it is likely to be as catastrophic as some citizens and scientists believe, with the possibility of a "tipping point" if the global average temperature should rise by 2°C, then we would have almost no margin for error and need to take urgent action. As noted earlier, some environmentalists believe that it is already too late to forestall disaster.

Quite apart from climate change and the other environmental issues connected with fossil fuels, there are many other reasons the world needs to transition away from these energy sources, most importantly, that we do not have a choice. None of them can be considered renewable, and all will gradually be running out—some sooner than others. It is believed, for example, the world has perhaps a 40-year supply of oil left, and that "peak oil" production is probably occurring about the time you are reading this, meaning that depending on economic conditions, oil should become increasingly scarce in years to come. Thus, shifting away from fossil fuels (oil in particular) is a matter of assuring an adequate energy supply, as well as promoting national (and global security) and economic well-being—especially for nations such as Japan that depends so heavily on foreign sources.

1.9 HOW IS GREEN OR RENEWABLE ENERGY DEFINED?

We have already used the term *renewable energy*, so it might be worthwhile to define it and delineate its properties. One definition is that energy is considered renewable if it comes from natural resources. Many of these renewable sources are driven by the sun, including wind, hydropower, ocean waves, biomass from photosynthesis, and, of course, direct solar energy. Hydropower is solar driven because solar heating is what drives the planet's water cycle. Several other types of renewable energy are the tides (mainly due to the moon, not the sun) and geothermal power from the Earth's hot interior. The magnitude of the amount of renewable energy sources available at the surface of the Earth is in total truly astounding. The numbers given in Figure 1.4 are on a per capita basis, so if you wanted to find the actual totals for the planet, just multiply by the world's population—about 7 billion. They have been expressed on a per capita basis because they can then be easily compared to the per capita power used on a worldwide basis, 2.4 kW. (The figure for the United States is four times as great or about 10 kW.) As shown in Figure 1.4, the influx of solar radiation dwarfs all the other flows, and it is about 5000 times the power now used by humans worldwide. One consequence of this fact is that if we could collect solar energy with 100% efficiency, it would be necessary to cover only a mere 1/5000 of the surface of the planet with solar collectors to generate all the energy currently used in the world.

One further type of renewable energy not derived from natural resources involves converting the wastes of human civilization into energy—which is done at some landfills that use garbage to create methane gas from which they then create electricity. There are five key properties that renewable energy sources share that make them very desirable, and there are also some drawbacks to some of them (Table 1.3).

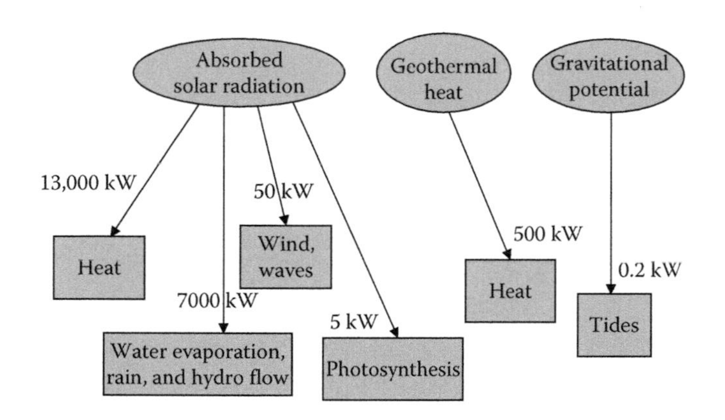

Figure 1.4 Per capita power influxes from renewable sources accessible at the Earth's surface. In the case of geothermal, however, the power would require drilling wells to be accessed.

Table 1.3 Desirable Properties and Drawbacks of Renewable Energy Sources

Desirable Properties	Drawbacks
Virtually inexhaustible	Some highly intermittent in time
Intrinsically nonpolluting	May be distant from populations
Sustainable	Very dilute (large footprint)
Fuel is free	Upfront costs involved
Ideal for off-grid use and distributed power	May be more costly (ignoring extrinsic costs) May involve some degree of environmental issues

The concept of sustainability essentially means that their usage in no way compromises the needs of future generations' need for energy, since nothing is being "used up." Some of the renewable sources satisfy these conditions better than others. For example, geothermal energy, while it is present everywhere, is much more accessible in some places than others, depending on how deep underground you need to go to access high temperatures, but it is also much less intermittent in time than most of the others. Wind (much more than solar) is highly dependent on location, since in many areas, the wind speed is insufficient to make it a viable alternative. Thus, in some sense, we can talk about "prospecting" for renewable sources (finding the best places for particular ones), just as we talk about prospecting for mineral resources. It is interesting, however, that a nation's policies may count for more than the amount of the resource available. Germany, for example, the world's leader in solar energy, is not noted for many sunny days!

1.10 WHY HAS RENEWABLE ENERGY AND CONSERVATION BEEN NEGLECTED UNTIL FAIRLY RECENTLY?

It is a bit misleading to think of humans' use of renewable energy being especially recent, since some renewable sources have been with us for millennia, including wind (to propel sailing ships and windmills), biomass/solar (growing food and lumber), and hydropower. Nevertheless, there has clearly been a relatively recent effort to move toward the greater usage of renewable sources, which currently account for a very small fraction of society's total energy use, at least in most nations. There are many reasons aside from simple inertia why moving away from fossil fuels and toward renewable energy has and will continue to be a challenge. First, the awareness of the environmental problems associated with fossil fuels has come very gradually, and views on the seriousness of the threat posed by climate change considerably vary. Moreover, in times of economic uncertainty, long-term environmental issues can easily take a backseat to more immediate concerns, especially for homeowners (Figure 1.5).

Second, compared to fossil fuels, there are problems with renewable sources, which may be very dispersed, intermittent, and expensive—although the cost differential widely varies and often fails to take into account what economists refer to as *externalities*, i.e., costs incurred by society as a whole or the environment. The intermittency poses special problems if the renewable source is used to generate electricity at large central power plants connected to the grid. One can cope with this problem using various energy storage methods and upgrades to the electric power grid, but, of course, both have costs. Cost, in fact, is perhaps the biggest problem with some renewable sources, especially upfront costs. While the fuel may be free, many renewable sources have in the past not been cost-competitive compared to fossil fuels, although this is rapidly changing and does not apply to renewable sources across the board.

Figure 1.5 Solar homes near Boston, Massachusetts.

As Table 1.4 shows, some of the renewable sources, including geothermal and biomass, and especially hydropower and onshore wind, compare quite favorably in terms of cost of electric power generation. The low values of the capacity for some renewable sources (especially wind and solar), attributable to their intermittent nature, do however represent a serious drawback.

Energy conservation in this section title is, of course, being used in a sense other than the law of energy conservation. Here, it refers to using less of energy and using it more efficiently. Conservation can be thought of as an *energy source* in a sense that it lessens the need for more generating capacity.

Table 1.4 Costs of Generating Electric Power in 2013 Dollars per Megawatt Installed

Source	$/MWh	Capacity (%)
Geothermal	47.8	92.0
Gas (comb cycle)	72.6	87.0
Wind	73.6	36.0
Hydro	83.5	54.0
Coal	95.1	85.0
Adv nuclear	95.2	90.0
Biomass	100.5	83.0
Solar PV	125.3	25.0
Coal with CCS	144.4	85.0
Wind (offshore)	196.9	38.0
Solar thermal	239.7	20.0

Source: EIA (Energy Information Administration), *Annual Energy Outlook 2011*, Energy Information Administration, Washington, DC, 2011.

Note: The *capacity* refers to the average power actually generated as a percentage of the maximum rated power for that source. Renewable (green) sources are shown as italicized. CCS, carbon capture and storage.

There is considerable opportunity for energy conservation to make a major difference given the amount of energy wasted in various sectors of the economy, especially in the United States. Some types of energy conservation such as upgrading the insulation of your home do involve upfront costs, but many do not and instead involve simple behavioral changes, such as carpooling or turning down your home thermostat. As we shall see in Chapter 12, even when upfront costs are involved, the payback on the initial investment can be enormous, for example, in the case of replacing incandescent light bulbs with light-emitting diodes or upgrading poor insulation.

1.11 DOES ENERGY EFFICIENCY REALLY MATTER?

The question posed in the section title is not intended to be provocative, because there are situations where energy efficiency (usually very worthwhile) does not matter. It is always important, for example, to look at overall efficiencies and not merely the efficiency of one part of a process. Thus, the process of heating water using an electric hot water heater is 100% efficient ($e = 1.0$), because all the electrical energy is used to produce heat, but this fact is irrelevant since it ignores the energy inefficiency inherent in producing electricity at the power plant and delivering it to your home. In fact, for this reason gas-fired hot water heaters are a significant improvement over electric ones on an overall efficiency basis. Another case where energy efficiency may be irrelevant involves any renewable energy source (where the fuel is free and abundant). The following example will clarify this point.

1.11.1 Example 2: Which Solar Panels Are Superior?

Suppose that 10 type A solar panels produced enough power for your electricity needs, had a lifetime of 30 years, cost only $1000, but had an efficiency of only 5%. Five type B panels cost $5000, but they had an efficiency of 10%, and lasted only 15 not 30 years. Which panels should you buy assuming the installation costs were the same for both types of panels?

Solution

Obviously, the more efficient panels would take up only half the area on your roof than the type A panels, but who cares if they both met your needs. The cost over a 30-year period would be $1,000 for the type A panels, but $10,000 for the more efficient type B panels that produced the same amount of power (since they last only half as long), so clearly you would opt for the less efficient choice in this case. As a general rule, as long as the fuel is free, and there are no differences in labor or maintenance costs, your primary consideration would almost always be based on cost per unit energy generated over some fixed period—usually the lifetime of the longer-lived alternative.

1.12 WHICH RENEWABLE ENERGY SOURCES HOLD THE GREATEST PROMISE?

Each of the renewable energy sources is best for a given location depending on its availability. It is difficult to say which renewable energy source is likely to hold the greatest promise in the future, since a technological breakthrough could elevate one of the sources considered to have limited application, e.g., geothermal, to the first tier. In the past, two sources have generated the greatest amounts of power, namely, hydropower (3.4%) and biomass (10%), mainly used for heating, with all the other renewable sources constituting about 3% of the final energy consumed. Although there is considerable room for expansion of hydropower in the developing nations, its expansion in the developed world will probably be less significant, given that the best sites have already been used. Biofuels will likely continue to be important, especially as a transportation fuel as an alternative to electric vehicles. On a worldwide (average) basis, however, the two sources likely to have the greatest impact in the future are wind and solar power. Wind power is already economically viable for centralized power generation, and photovoltaic (PV) solar cells may soon be at cost parity with conventional sources and is expected to reach it around 2020 for coal-fired generating plants.

The growth in installed photovoltaic (IPV) solar panels for electric power generation by both central power plants and individuals has been phenomenal, increasing 1 million percent since 1975. As shown in Figure 1.6, due to their declining cost as a result of technological improvements, the growth has been roughly consistent with being an exponential function—the trend line in Figure 1.6, indicating a constant percentage annual growth of 24.7%, is described by

$$IPV = 4.92\exp(0.247t), \tag{1.5}$$

where t is the year minus 1975 and IPV represents the amount of IPV solar cells in megawatts.

"THE RULE OF 70"

According to this rule, the doubling time in years for any quantity that grows by a fixed percentage p each year can be found to be approximately $70/p$ years. You can easily verify this rule by starting with $df/dt = pf$ and integrating to find $f = f_0e^{pt}$. Finally, just solve for t that gives $f = 0.5f_0$ and you obtain $t = 69.3/p \approx 70/p$. The rule of 70 works equally well for a quantity that decreases at a fixed percentage each year if we wish to estimate the halving time.

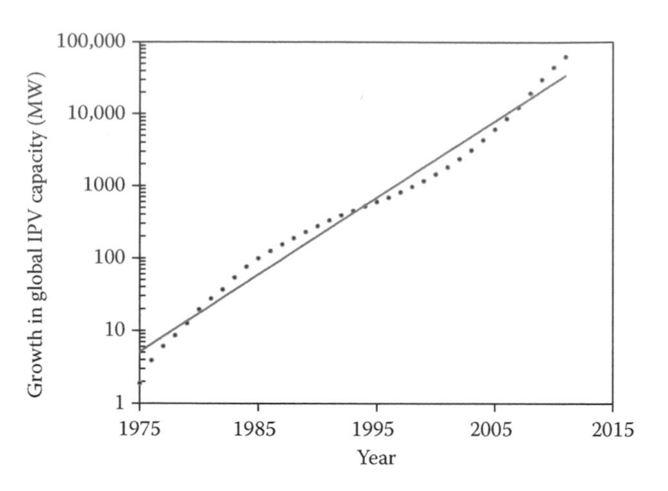

Figure 1.6 Growth in global IPV capacity in megawatts. (Courtesy of Earth Policy Institute, Washington, DC. Data compiled by Earth Policy Institute, with 1975–1979 data from Worldwatch Institute, *Signposts 2004*, CD-ROM, Worldwatch Institute, Washington, DC, 2004; 1980–2000 from Worldwatch Institute, *Vital Signs 2007–2008*, Worldwatch Institute, Washington, DC, 2008, p. 39; 2001–2006 from Prometheus Institute and Greentech Media, 25th annual data collection results: PV production explodes in 2008, *PVNews*, 28(4), 15–18, April 2009.)

As of 2015, the IPV supplies only about 0.06% of the world's total energy. As an exercise, we estimate using the "rule of 70" that with a 24.7% annual growth rate, the amount of IPV doubles every 70/24.7 ≈ 2.5 years. To increase from 0.06% to 100% requires an increase by a factor of 100/0.06 ≈ 1700, which requires between 10 and 11 doublings, or between 25 and 27.5 years. Thus, as of 2012, solar PV would be a major proportion of the world's energy mix by the year 2040, if its exponential growth were to continue.

At the same time that solar panel installations have been exponentially growing, their costs have been steadily declining. In fact, an interesting empirical relation has also been discovered between the cost of PV power and the cumulative amount deployed that holds true over the entire time period since 1975 (Handleman, 2008):

$$IPV \approx \frac{31{,}900}{C^3}, \tag{1.6}$$

where C is the cost in dollars per watt.

Thus, according to this relation, exponentially declining costs are associated with exponentially rising cumulative PV deployment (Figure 1.6).

Although wind and solar may each be the better choice for a particular location, there is some basis for considering solar to be the better source on an overall average basis (Table 1.5).

Table 1.5 Advantages of Solar over Wind Power

Potential: Solar is suited to a much greater range of geographic locations than wind.
Expense: The best location for wind is offshore, which is very expensive to exploit.
Maintenance: Solar requires less maintenance than wind and is easier to install.
Distributed power: Solar is usually more suited to distributed power usage by individuals.
Diversity: Solar has three different ways it can be pursued, including PV, solar thermal, and solar chimneys, which are discussed in various subsequent chapters.

1.13 WHO ARE THE WORLD LEADERS IN RENEWABLE ENERGY?

Three nations—China, the United States, and Germany—rank number 1, 2, and 3 in the world in terms of renewable energy usage. Together, China and the United States have invested half the world's total toward developing renewable energy, but it also needs to be said that they account for half the world's CO_2 emissions, which is not too surprising as they are the two largest economies.

Germany on a per capita basis would rank far ahead of China who is number one in absolute terms, with the United States number two. There are nations that could be considered even "greener" than Germany in terms of the percentage of their energy from renewable sources—Norway and Iceland, for example. However, in such cases, the extensive renewable usage is largely a fortunate accident of geography: Norway generates 99% of its electricity using hydropower, while Iceland gets 100% from renewable sources—both hydro- and geothermal.

- *Germany*: There are few nations (like Germany) whose commitment to "going green" is so strong that they made a commitment to embracing green energy long before it approached economic parity with conventional sources. The Germans have supported green energy through national policies that subsidize its deployment and by removing unwise subsidies for conventional sources, including coal. Germany remains the number one nation in IPV capacity, and in 2011, following the Fukushima accident, it has decided to phase out its nuclear power plants. Germany may serve as a test case for just how fast a nation can move toward renewable sources without harming its economy or paying an excessive price for its energy. Of course, when comparisons are made between the costs of various energy sources, difficult-to-quantify environmental costs are often not factored in to the usual calculation, so the German approach may make considerable sense. However, some observers worry that if Germany abandons nuclear too quickly and is forced to import power from neighboring countries to make up for any energy shortfall, the Germans are simply exporting any environmental impact, and they may even exacerbate the problem of climate change, since nuclear power has no CO_2 emissions.

- *China*: Unlike Germany, whose leaders could possibly be accused of putting emotion ahead of reasoned analysis and paying too much attention to public opinion, China's leadership certainly falls at the other end of the spectrum. Of course, having an authoritarian system does make it easier to engage in long-term planning and execution, unhindered by serious opposition from either the public or an opposition political party—a case in point being the Three Gorges Dam and power plant that displaced over a million people from their homes and did considerable environmental damage. China's ability to forge ahead in the renewable energy area has also been greatly assisted by the government subsidies, which include tax breaks, low interest loans, and free land for factories, which has led to some American solar manufacturers to relocate there. In some cases, government subsidies may be less motivated by promoting renewable energy domestically than increasing the nation's exports, since 95% of China's solar panels are made for export. The Chinese have several other advantages allowing them to become the world leaders in renewable energy, including an abundant pool of scientific and engineering talent, an immense pool of relatively cheap labor, and a near monopoly (96%) on the world supply of rare earth elements. These elements, such as dysprosium, neodymium, terbium, europium, yttrium, and indium, are considered to be of critical importance to clean energy technologies.

  Despite China's commitment to renewable energy it is even more strongly committed to increasing its energy-generating capacity generally, including fossil fuel and nuclear power, and it has been building several new coal-fired generating plant each week with plans to do so for years to come. While China's new coal plants may incorporate pollution abatement technology, on average, its plants are more polluting than those in the West, and air pollution (as well as coal miner deaths) represents serious problems—much as it did in Western nations in years past. The Chinese government very likely cares about the environment, but it probably cares more for building its economy, increasing its citizens' living standards, and, more importantly, becoming a leading power on the world scene.
- *United States*: Although renewable energy still constitutes a tiny fraction of the nation's energy usage, the United States appears to be committed to expanding it, and it is second only to China in the magnitude of its investment. Additionally, according to public opinion polls, many citizens support renewable energy, even if they may be skeptical about human-caused climate change. Unfortunately, many policies that could lead to greater usage of renewable energy, such as a "renewable energy standard (RES)" requiring utilities to generate a certain fraction of their power from renewable sources, exist only at the state and not the federal level, although some states such as California are quite generous in their support, and even states such as Texas, noted for its conservative political outlook, appears very receptive toward

wind power. At the federal government, the political gridlock of a divided government has stymied actions on advancing renewable energy, apart from those mandated by the executive branch during democratic administrations. Even worse, continuing subsidies for energy from fossil fuel and nuclear energy continue to be significantly greater in the United States than those for renewable energy, with the bulk of the subsidies being in the form of tax breaks (Shahan, 2011). In one positive development, the federal government has committed to raising the mileage standard in new automobiles over a period—an important way of achieving greater energy efficiency in the transportation sector.

1.14 WHAT IS OUR LIKELY ENERGY FUTURE?

Given that the world population continues to grow, and many developing nations have a growing appetite for a better living standard, it is virtually inevitable that the demand for energy will grow during the coming decades. The mix of energy sources contributing to that growth is much less certain—especially if it is long term. One such projection is shown in Figure 1.7 made by the German Advisory Council on Global Change through the year 2300.

There are several interesting aspects to the projections in Figure 1.7. The first is that even though renewable sources are expected to provide a greater share of the world's energy, the council foresees little major redistribution of the mix through 2030 and some presence of the three fossil fuels through the entire coming century, with coal—the most environmentally harmful source—cutting back the most. The most interesting projection, however, is that by far, the dominant renewable source, especially after

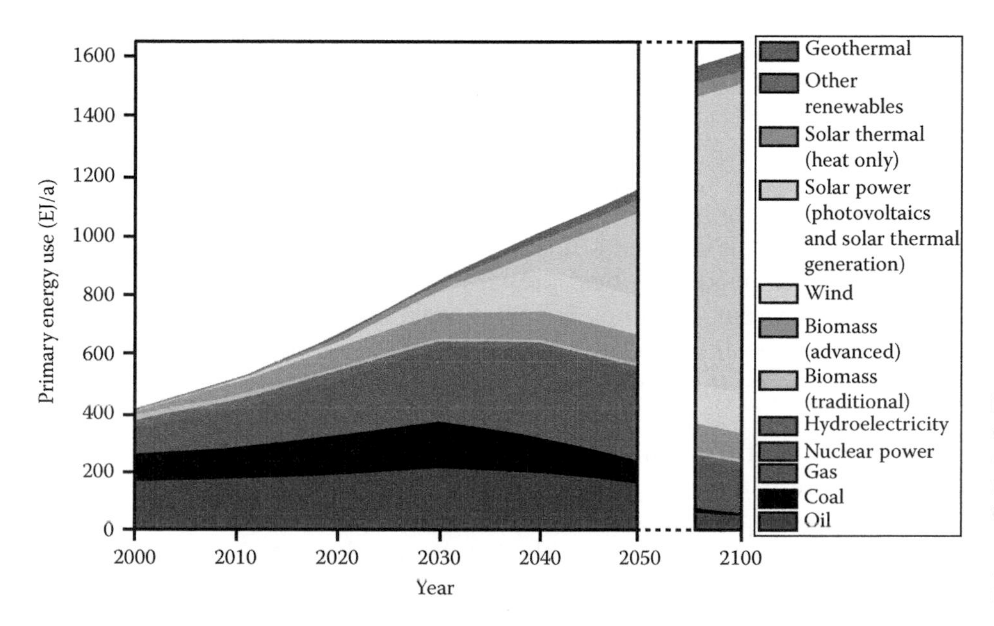

Figure 1.7 Transforming the global energy mix: The exemplary path until 2050/2100. (From WBGU (German Advisory Council on Global Change), *World in Transition: Towards Sustainable Energy Systems. Summary for Policy-Makers*, WBGU, Berlin, 2003. With permission.)

2050, will be solar. Are these projections realistic? Lacking a crystal ball, no one can say, but the existing exponential growth of solar (starting from a very tiny base) offers some justification.

AN INAPPROPRIATE TOPIC?

Some instructors may believe that it is inappropriate to have a section dealing with jobs and careers in a textbook. If you happen to be one of them, please be sure to tell your students that they "are not responsible for the material in this next section, and that it will not be covered on any exams."

1.14.1 What Is Projected for Future Employment in the Renewable Energy Field?

Making projections for future employment can be very hazardous, depending as it does on future human actions and the unknown evolution of the global economy. In fact, projecting the likely employment needs 20 years in the future may have almost as much uncertainty as projecting the likely mean global temperatures a century from now, which of course also greatly depend on human actions and the global economy! Nevertheless, given the very strong past growth in both solar and wind powers, which is likely to continue if costs continue to fall, it is reasonable to imagine that the growth might continue on its present trajectory for the next decade or two.

According to the Energy Information Administration data on the growth of solar energy, strong job growth is expected, reaching as many as 290,000 new jobs by 2030. Another study by the American Solar Energy Society looking at the entire field of renewable energy concluded that by 2030, in the United States alone, some 1.3 million direct and indirect jobs could be created under a business-as-usual scenario, and 7.9 million under a scenario with strong national policies favoring renewable energy, including targets, standards, and invigorated research and development (ASES, 2006).

It is natural for any student thinking about going into the field of renewable energy to wonder what kinds of jobs might be available and what sort of education is needed to best prepare for them. A search on a website advertising current openings in US companies in the renewable energy field came up with the results shown in Table 1.6, with the numbers in parenthesis being the numbers of jobs listed.

Clearly, many of the kinds of jobs listed would require a 4-year degree, and most are in specific areas of study with engineering clearly topping the list, but the various subfields of business also being very important. Although *science* is far down on the list, it must be noted that the website advertises corporate opportunities and would not include opportunities

Table 1.6 Number of Job Openings Listed by http://www.careerbuilder.com in Renewable Energy in the United States on September 22, 2015

Management (135)	Marketing (36)
Engineering (119)	Business development (34)
Sales (105)	General business (28)
Skilled labor trades (20)	Construction (27)
Entry level (58)	Strategy planning (24)
Customer service (54)	Administration (20)
Accounting (49)	Professional services (20)
Finance (49)	Manufacturing (20)
Installation maintenance (49)	Design (18)
Information technology (42)	Quality control (18)
Consultancy (38)	Science (13)

Note: No listings with fewer than 13 openings are included, and a few vague titles have been omitted.

in basic research available at universities, colleges, research institutes, and national laboratories in the renewable energy field. These are not only certainly less numerous, but also have fewer people seeking them. The categories listed in Table 1.6 might apply equally well to work in just about any field, so it might be more relevant to list the kinds of work areas specifically related to renewable energy that one might want to seek to work in. Here is a very partial list in alphabetic order:

- Basic research, consulting, consumer education, designing new materials
- Designing smart grid, energy auditing, energy education, environmental impacts
- Environmental abatement, green buildings, solar panel design, and calibration
- Wind turbine design, testing and maintenance, fluid dynamics simulations
- Wind farm management, wind resource assessment, windsmith

How might one prepare for a career in the renewable energy field? It would be useful to have a few courses in renewable energy or perhaps a minor in the subject, such as the one at George Mason University. A minor is perhaps a better preparation than a degree specifically focusing on energy, since many job listings tend to seek people having conventional academic backgrounds with degrees in engineering, business, or science.

1.15 COMPLEXITIES IN CHARTING THE BEST COURSE FOR THE FUTURE

As noted earlier, it is imperative that over time, the world will move away from fossil fuels, but the degree of urgency for doing so depends on one's views with regard to the possibility of a catastrophic climate change and, in particular, the need to avoid a tipping point in the climate system. Even if one is committed to moving toward renewable on a long-term basis, there remains the serious question of what to do in the interim, bearing in mind that some fossil fuels are more environmentally harmful than others and that in an era of economic uncertainty, we need to be cognizant of economic costs as well as environmental benefits. Other controversial matters include the long-term role of nuclear power and whether carbon sequestration could enable coal to become a clean energy source. Perhaps most controversial is the notion as to whether some form of geoengineering, i.e., manipulating the Earth's climate to counteract rising CO_2 levels, might be worthwhile or whether the dangers are simply unacceptable. These issues will be fully explored in subsequent chapters, especially Chapters 4 and 14.

As one example of the complexities facing us in trying to plan the best way forward, consider our continued reliance on natural gas. There are many possible positions one might take on this issue, and four of them are sketched out in the following; which of them is the best course depends to a large extent on your assumptions and a mix of environmental and economic issues and the weight you assign to each:

1. Phase out use of natural gas as well as all other fossil fuels as quickly as possible
2. Pursue new natural gas discoveries and use it for power generation instead of coal
3. Pursue new natural gas discoveries and use it for transportation instead of petroleum
4. Pursue new natural gas discoveries and use it for both transportation and power generation

Here, for example, is the argument for option 3. Natural gas emits significantly less pollutants as well as greenhouse gases than coal, Even though there are environmental problems with natural gas extraction involving fracking, they should be manageable if adequate precautions are taken, and its overall environmental impact is significantly less than coal. Due to new discoveries of natural gas, its price has considerably dropped, and the amount available in the United States has roughly doubled in the last decade. Currently, even though natural gas is the least expensive way to generate electricity, it has in the past tended to be used mostly for power plants to supply extra power during periods of peak demand, because such plants can be ramped up or down in power much faster than coal or nuclear plants.

This property will become increasingly important as more intermittent renewable sources such as wind and solar are used. In fact, few other energy sources besides natural gas have this desirable property, so a plausible argument can be made for not extending its power generation usage beyond supplying power at times of greatest demand, lest the natural gas reserves be used up too quickly. In contrast, the transportation usage of natural gas (as a replacement of petroleum) may be more crucial, because the alternatives are less clear. It may be true that alternatives to petroleum exist in the transportation sector, including all electric vehicles, but they could face an uncertain market acceptance, unless their range (on a full charge) significantly improves.

Notice how in making the argument to use natural gas mainly for a transportation fuel rather than increasing its use in power generation, we have discussed a mix of environmental and economic concerns and, most importantly, a weighing of the alternatives in both the power generation and transportation sectors. It might be worthwhile for you to reflect on what a similar argument might consist of for some other alternative.

HOW CAN LESS BE MORE?

One illustration of the counterintuitive consequences that can occur when fossil fuel sources are replaced by renewables was done in a test conducted by the Bentek Energy Company (Bentek, 2010). In the test, wind turbines offset a certain fraction of the power supplied by a coal plant, and the coal plant needed to have its power output changed to compensate for the variability of the wind generators. One might imagine the use of wind to replace some of the power from the coal plant would have resulted in a reduction of CO_2 emissions (for the same total power output), but exactly the opposite occurred since coal plants that are cycled up and down on a short timescale are much less efficient. As we have seen, natural gas plants do not suffer from this drawback, and had they been used instead in conjunction with the wind turbine emissions would have been reduced.

1.15.1 Example 3: How the Usage of Wind Power to Offset Coal-Fired Plants Can Generate More Emissions, Not Less

Suppose that a certain fraction of the power produced by a 500 MW coal plant is offset by wind power. Assume that when the coal plant runs at its constant rated power, it has an efficiency of 35%, but when it needs to be ramped up and down to compensate for the wind power variations, its efficiency is reduced according to $e = 0.35 - 0.00001\,p^2$, where p is the amount of wind power. Find the percentage increase in emissions that results when 90 MW of the 500 MW is generated by wind power instead of coal.

Solution

In order to generate the full 500 MW by itself, the coal plant requires 500/0.35 = 1429 MW of heat flow from the coal. If the wind power is 90 MW, the efficiency of the coal plant is reduced to $e = 0.35 - 0.00001(90)^3 = 0.269$, and the heat flow required to generate (500 – 90) = 410 MW is therefore 410/(0.269) = 1524 MW. The percentage increase in emissions is the same as the percentage increase in the heat flow to the coal plant, i.e., 6.7%.

1.16 SUMMARY

This chapter discusses some background topics on energy. It goes on to discuss the nature of renewable energy, the world's energy–environment problem, and the need to transition away from fossil fuel energy sources with their finite supply and harmful environmental impact—climate change being just one of many. The chapter concludes with a section on employment in the renewable energy field.

PROBLEMS

General comments on problems: The following comments refer to the problems that follow each chapter, including this one. Some of the problems in this book may require your ability to make rough estimates, while in other cases, it is expected that you will be able to locate missing data on the web. However, do not use the web as a substitute to doing calculations, although it is fine to perhaps use it to confirm your answers. Be sure to check that the results of all calculations are reasonable. A number of problems mention using Excel to do a calculation. However, if you are more familiar with other tools such as Basic or Mathematica, feel free to use those instead. In a number of problems, hints are given. Be sure to try to figure out the relevance of any hints. The answers to selected problems are provided at the back of the book.

1. Compare the direct costs to the consumer of using a succession of ten 100 W incandescent light bulbs with an efficiency to visible light of 5%, a lifetime of 1,000 h, and a price of 50 cents with one compact fluorescent lamp giving the same illumination at 22% efficiency, a lifetime of 10,000 h, and a price of $3. Assume a price of electricity of 10 cents per kWh.
2. How many kilowatt-hours would a 1000 MW nuclear power plant generate in a year?
3. Consider a nuclear power plant whose power level is ramped up from zero to a maximum 1000 MW and then back down to zero over a 10 h period. Assume that the power level varies as a quadratic function of time during those 10 h. Write an expression for the power as a function of time, and then find the total energy generated by the plant during the 10 h period.
4. The United States generates and uses about 71 quads of energy each year, and its renewable sources generate about 40 GW. If the

renewable sources are generating power about a third of the time, what fraction of its energy usage is based on renewable sources?

5. Based on Equation 1.6, by what factor would the total amount of PV solar panels increase if their costs decreased by 30%?
6. Prove that Equation 1.5 implies a 24.7% annual growth rate.
7. If Equations 1.5 and 1.6 continue to hold, at what date would the cost of IPV reach 50 cents/W?
8. Do you think the trend described by Equation 1.5 is the cause or the effect of that suggested by Equation 1.6? Discuss.
9. If the trend illustrated in Figure 1.6 were continued in the future, when would solar cells be able to meet humanity's present energy needs by itself?
10. How large would a square of side L need to be so that if it were covered by 10% efficient solar cells in the middle of the Sahara desert, the power generated would be enough to satisfy the world's present energy needs? Assume that the incident solar radiation striking each square meter of the Earth's surface is approximately 1000 W.
11. Using the data in Example 3, find the amount of wind power that could be used with a 500 MW coal-fired plant that would result in the least amount of emissions.
12. Although typically, electricity customers are charged based merely on the total number of kilowatt-hours they consume, some utilities have payment plans designed to encourage customers to shift their energy use to off-peak times. Suppose that a utility charges most customers a flat 7.9 cents/kWh under their standard plan, but under a special time-of-use plan, it charges 3 cents/kWh for off-peak times (between 10 p.m. and 11 a.m. on weekdays) and 16 cents/kWh at other times. If a customer consumes electricity at the same rate at all times, which plan should he or she sign up for?
13. Figure 1.7 shows solar PV reaching 200 EJ/year that was installed before 2050. Quantitatively compare that projection with the historical trend illustrated in Figure 1.6—note the different units.
14. What is the efficiency of a Carnot engine if the $T_c = T_h$?
15. What must the T_c temperature be in order to obtain a 100% efficient Carnot engine?

REFERENCES

ASES (Asia-Pacific Student Entrepreneurship Society) (2006) *Green Jobs: Towards Decent Work in a Sustainable, Low-Carbon World*. A 2008 report by the United Nations, United Nations, Geneva, Switzerland, http://www.everblue.edu/renewable-energy-training/solar-and-wind-energy-jobs (Accessed Fall 2011).

Doran, P. T., and M. Kendall Zimmerman (2009) Direct examination of the scientific consensus on climate change, *EOS*, 90(3), 22.

Ehrlich, P. R. (1968) *The Population Bomb*, Ballantine Books, New York.

EIA (Energy Information Administration) (2011) *Annual Energy Outlook 2011*, EIA, Washington, DC.

Feynman, R. (1985) *The Character of Physical Law*, MIT Press, Cambridge, MA. http://www.scribd.com/doc/32653291/The-Character-of-Physical-Law-Richard-Feynman (Accessed Fall 2011).

Handleman, C. (2008) An experience curve based model for the projection of PV module costs and its policy implications, Heliotronics, Hingham, MA. http://www.heliotronics.com/papers/PV-Breakeven.pdf (Accessed on May 29, 2008).

Prometheus Institute and Greentech Media (2009) 25th annual data collection results: PV production explodes in 2008, *PVNews*, 28(4), 15–18, April 2009.

Rosa, L., E. Rosa, L. Sarner, and S. Barrett (1998) A close look at therapeutic touch, *JAMA*, 279, 1005–1010.

Shahan, Z. (2011) *Wind Power Subsidies Don't Compare to Fossil Fuel & Nuclear Subsidies*, http://cleantechnica.com/2011/06/20/wind-power-subsidies-dont-compare-to-fossil-fuel-nuclear-subsidies/?utm_source=feedburner&utm_medium=feed&utm_campaign=Feed%3A+IM-cleantechnica+%28CleanTechnica%29 (Accessed Fall 2011).

WGBU (German Advisory Council on Global Change) (2003) *World in Transition: Towards Sustainable Energy Systems. Summary for Policy-Makers*, WBGU, Berlin.

Worldwatch Institute (2004) *Signposts 2004*, CD-ROM, Worldwatch Institute, Washington, DC.

Worldwatch Institute (2008) *Vital Signs 2007–2008*, Worldwatch Institute, Washington, DC, p. 39.

Fossil Fuels

Chapter

2.1 INTRODUCTION

Most fossil fuels, which include coal, oil, and natural gas, were formed from the remains of ancient life over the course of tens to hundreds of millions of years—hence, the adjective *fossil*. The one exception to this rule is believed to be methane, the main component of natural gas, which has both abiogenic and biogenic origins and may form in much shorter time spans. Some ancient life decomposed in the presence of oxygen and would not have become fossil fuels, because the original stored chemical energy would be released during the oxidation process. The beginnings of the process by which fossil fuels are formed are still going on today in the oceans, and swampy areas, especially peat bogs—*peat* being the term for partially carbonized decomposed organic matter. While the beginnings of the process of future fossil fuel formation may be going on now, the rate of their formation is dwarfed by the rate at which humans have been using existing fossil fuels made over tens to hundreds of millions of years to power the industrialized society, and hence, fossil fuels are a nonrenewable resource.

While no one can be certain exactly how much new coal, oil, and gas will be discovered in the future, reasonable estimates can be made about future discoveries. Thus, we can say that humanity's fossil fuel era that we now find ourselves in is a short blip on a very long timescale, meaning that the world consumption of fossil fuels looks qualitatively like that shown in Figure 2.1.

Clearly, the fossil fuel era is bound to end in a matter of at most a century or two, due to the finite amount of remaining reserves. However, as you might suspect, waiting until fossil fuels begin to run out before making the transition away from them would be an utter disaster for the planet for a host of reasons, climate change being only one among many (Figure 2.2). Today, fossil fuels account for totally 85% of the world's primary energy usage, with nuclear and hydropower comprising 8% and 3% and the renewable sources of geothermal, solar, tidal, wind, and wood waste amounting to a bit over 1% collectively. An obvious question is what has made fossil fuels so attractive as an energy source in the past as well as today and why is it so difficult to move away from them despite the mounting evidence of the environmental problems they pose. The answer has primarily to do with the enormous store of energy they contain. For example, coal, oil, and gas have at least 200 times the energy per kilogram that is stored in a lead acid car battery, and unlike the car battery, the energy was put there courtesy of Mother Nature. Fossil fuels represent highly concentrated stores of energy compared to the much more dilute concentrations typical of renewable sources, such as wind and solar, and usually, they can be more easily and cheaply collected, stored, shipped, and used where and when desired than most renewable energy sources. Of course, the low cost and high energy density of fossil fuels is not the whole story behind our addiction to them, with simple inertia and power politics also playing important roles.

CONTENTS

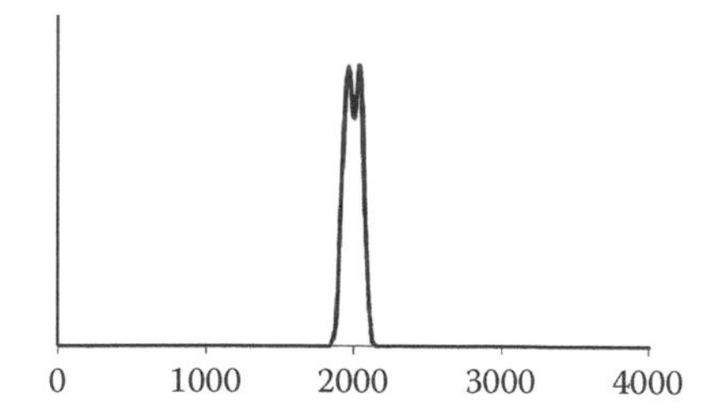

Figure 2.1 Annual consumption of fossil fuels by humans versus year anno Domini. The decline from the peak is a consequence of the three fossil fuels running out in the future. The double peak reflects the fact that coal began to be abundantly utilized before oil.

Figure 2.2 Although coal is largely responsible for rising atmospheric CO_2 levels, there are also many negative impacts in the shorter term that are of much concern, as illustrated in this photo of the 1968 Farmington coal mine disaster that killed 78 miners in West Virginia. (From Wikipedia, http://en.wikipedia.org/wiki/Coal_mining, July 21, 2017.)

HOW MUCH IS THERE?

A major issue about the three fossil fuels that greatly affects how long they will last is the estimates of the remaining amount in each case. Experts in this field classify mineral deposits into two categories, *reserves* and *resources*, based on their geologic certainty and economic value. Thus, while resources are potentially valuable and may eventually be economically extracted, reserves are known to be valuable and economically extractable using present technology. Reserves are often further subdivided into categories of proven, probable, and possible based on the degree of certainty that they can be economically extracted using existing technology—*proven* meaning >90% and *probable* meaning 50%, for example. One can obtain very different estimates for a particular fossil fuel depending on which word one uses. Thus, while the United States has been estimated to have 22 billion barrels of oil remaining as part of its proven reserves, it has more than 10 times this amount (274 billion barrels) as part of its resources, and there is a roughly comparable amount in the Canadian tar sands. However, note that the proven reserves in the United States can significantly vary with time, so, for example, it increased by 39% from 2008 to 2011, as new extraction technologies came into play.

2.1.1 Carbon Cycle

The element carbon is an essential component of all fossil fuels and the ancient (and modern) life from which they arose. In fact, as any science fiction enthusiast is aware, we (and other life on Earth) are carbon-based life forms. In fact, roughly half the dry weight of most living organisms consists of carbon. The carbon cycle describes the host of biogeochemical processes by which carbon is exchanged between a multiplicity of reservoirs on, above, and inside the Earth. These reservoirs on or near the Earth's surface include the atmosphere (where the carbon is mostly CO_2), the biosphere, the oceans, and sediments, which include fossil fuel deposits. The largest of these reservoirs by far is the oceans, and the greatest component there is the deep ocean part (38,000 Gton), which does not rapidly exchange carbon with the upper layers or the atmosphere. Of the reservoirs in the Earth's crust the fossil fuel deposits are the largest, while for aboveground terrestrial carbon, the largest component (86%) is stored in forests. There are many pathways by which carbon can enter or leave the Earth's atmosphere. These include the decay of animal and plant matter, fossil fuel combustion, production of cement, and volcanic eruptions. As the planet warms due to increased atmospheric CO_2, even more CO_2 enters the atmosphere since the equilibrium concentration of dissolved CO_2 in the upper layers of the oceans becomes less.

2.2 COAL

The energy believed to be present in the world's coal supply dwarfs all other fossil fuels combined, and it has been estimated at 2.9×10^{20} kJ, most of which is not economically exploitable at present. Coal has been used by humans as a heating fuel for at least 4000 years, but the earliest known European usage dates back only 1000 years. It was coal that powered the steam engines during the industrial revolution, beginning in the eighteenth century, which arguably would not have taken place without it. The extensive use of coal beginning with the industrial revolution left its imprint on the planet in terms of a significant increase in the atmospheric concentration of CO_2 (the primary greenhouse gas) whose rise started at that time. Studies of bubbles trapped in Antarctic ice cores show that atmospheric CO_2 levels were around 260–280 ppm (parts per million) prior to the industrial revolution, but the rise since then has been very rapid—especially during the last half century. In fact, present atmospheric CO_2 levels have risen to a higher point than has occurred during the last 400,000 years, and the rise since the industrial revolution began is almost entirely human caused—largely due to coal-burning power plants.

Humans are still very dependent on coal—less so for heating than in the past, but more for electricity generation and various industrial processes. Today, the roughly 41% of electricity worldwide is generated by coal-fired power plants, which, together with automobiles using petroleum derivatives, have been the main sources of rising atmospheric CO_2 levels. The problem of climate change associated with the greenhouse effect and the human contribution to it will be discussed at length in another chapter. Suffice to say here that the basic physics behind the greenhouse effect is unquestionable, and the extent of the human contribution to climate change is considerable.

2.2.1 Composition of Coal

Coal is a combustible sedimentary rock. It differs from other kinds of rocks, which are generally made of minerals and, hence, are inorganic by definition. Coal, however, is mostly carbon made primarily from plant material and is therefore organic. While carbon may be its primary component, it does contain minor amounts of hydrocarbons, such as methane, and inorganic mineral material that are considered impurities. Coal does not have a specific chemical composition, because the precise mixture of sulfur, oxygen, hydrogen, nitrogen, and other elements comprising it varies according to the particular rank or grade of coal and even within a grade. For example, for anthracite, the highest and hardest rank of coal, its composition includes 0–3.75% hydrogen, 0–2.5% oxygen, and up to around 1.6% sulfur. Although the number of coal ranks depends on the classification system, one system that is widely used is based on the four grades listed in Table 2.1.

The order of the rows (starting from the top) goes from the lowest rank of coal to the highest. As can be seen from Table 2.1, higher rank coals have both a higher percentage of carbon and higher energy content per unit mass. Higher ranks since they are more carbon rich tend to have less hydrogen, oxygen, and sulfur, and they tend to have lower percentages of volatiles, which are substances that are driven off when the coal is heated

Table 2.1 Four Basic Ranks of Coal Based on the American Standards Association

Rank	Carbon Content (%)	Energy Content (BTU/lb)
Lignite	<46	5,500–8,300
Subbituminous	46–60	8,300–11,000
Bituminous	46–86	11,000–13,500
Anthracite	86–98	13,500–15,600

Note: Lignite, the lowest rank, is also known as brown coal.

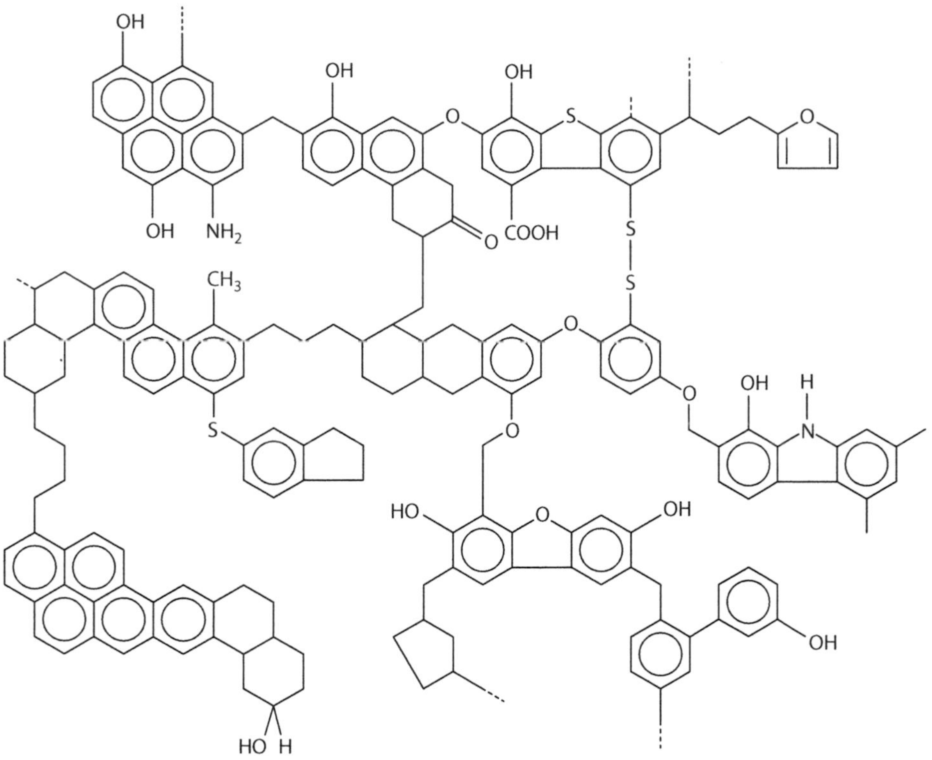

Figure 2.3 Example of the complex molecules found in coal.

to a high temperature in the absence of air. The percentage of volatiles for a given sample of coal is not calculated on the basis of its chemistry, but is found from direct measurements after the coal is subject to some standardized temperature over a period. Although there are some simple compounds in coal, most of the molecules in coal tend to be very massive and complex, since the plant fibers that they originated from are often long. These molecules, many of which even lack names, vary from one piece of coal to another and within a piece. One such nameless molecule is depicted in Figure 2.3.

Figure 2.3 uses the symbols familiar to organic chemists. Thus, for example, the hexagons shown with single lines comprise the rings of six carbon atoms, while those with double lines that strongly resemble a picture of a nut represent benzene rings (six linked carbons having hydrogen atoms always attached to the vertices).

2.2.2 Example 1: Energy Content of Coal

An empirically determined formula for the energy content of coal based on the elemental abundances of carbon, hydrogen, oxygen, and sulfur is

$$E = 337\mathrm{C} + 1442(\mathrm{H} - \mathrm{O}/8) + 93\mathrm{S}, \tag{2.1}$$

where E is in units of kilojoules per kilogram, and the symbols stand for the mass percentages of the elements C, H, O, and S. Use Equation 2.1

and the information provided earlier about anthracite, i.e., H = 0–3.75%, O = 0–2.5%, and S = 1%, to estimate the highest value, lowest value, and average value of the energy content of anthracite assuming that no elements besides C, H, O, and S are present.

Solution

Based on the values of the constants in Equation 2.1, the maximum energy density requires H to be as high as possible and O is as low as possible, and the minimum energy requires the opposite. Thus, using the data from Table 2.1, we have

$$E_{max} = 337(95.25) + 1442(3.75 - 0/8) + 93(1) = 36,700 \text{ kJ/kg}$$
$$= 15,800 \text{ Btu/lb},$$

$$E_{min} = 337(96.5) + 1442(0 - 2.5/8) + 93(1) = 32,200 = \text{kJ/kg}$$
$$= 13,800 \text{ Btu/lb}.$$

As a check, we note that these values are fairly close to those provided in Table 2.1 for anthracite.

2.2.3 Formation of Coal

According to geologists, all ranks or grades of coal were formed through the same process starting with dead plant matter. In most times and places, when plants die they either decompose or are consumed by fire, i.e., the material is oxidized. However, on rare occasions and in specific places, especially in swampy areas, deposited plant matter can accumulate in a layer and be preserved from decay and fire by the absence of oxygen. Swamps are ideal places for such matter to gradually build up in the water, because of its anoxic nature (relative absence of oxygen). Actually, some decay may occur if small amounts of oxygen are present, but as long as the rate of decay is less than the rate of deposition, the debris layer will grow in thickness over time. Usually the growth in thickness will be extremely slow. Estimates are the accumulation of a 10 m thick layer (perhaps eventually leading to a 1 m thick coal seam) might take thousands of years.

Although some coal has been reported as being found in rocks as old as two billion years, and some as young as two million years, the large majority of the world's coal began its formation during what is aptly known as the Carboniferous Period, which lasted during the period of 359–299 million years ago. Conditions were especially suitable then because the sea level was high and the forests tended to be in enormous tropical coastal swamps that were flooded by the seas. After a thick layer of organic debris has been laid down, the inflow of the sea over coastal swamps, or the retreat of the sea and the influx of stream deposits over the swamps,

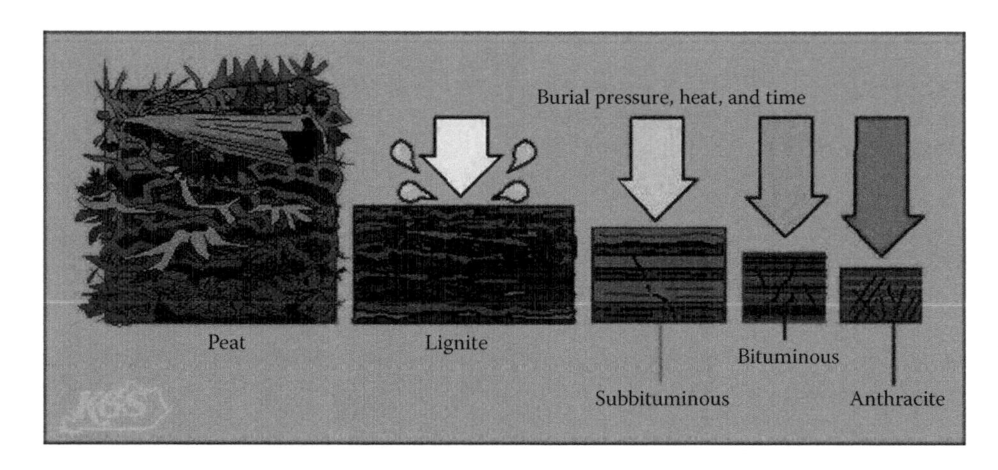

Figure 2.4 Coal formation process. (Courtesy of Kentucky Geological Survey, Jim Cobb, State Geologist, http://www.uky.edu/KGS/coal/coalkinds.htm.)

would cause a layer of sand and mud to be deposited over it. With the rise and fall of sea level, alternating layers of dead organic matter was sandwiched between the sand and mud. Over the course of time, the sediment would turn to rock, and the process of coal formation (coalification) would occur with the application of pressure and heat causing volatiles to be driven off, the layer of organic material to become more compact, and its carbon concentration to increase. According to this scenario, the different ranks of coal, peat → lignite → subbituminous → bituminous → anthracite, form a time sequence, although in any given location, the process might have gone only partway given insufficient time, pressure, or heat (Figure 2.4).

Thus, based on the sequence of steps just described, we see that the various ranks of coal from lowest to highest are not simply arbitrary types defined based on their carbon content, but rather steps in an evolutionary time sequence. The evidence for this theory of coal formation is threefold: (1) coal is always found in seams or strata, (2) the seams sometimes contain actual plant fossils, and (3) successively higher grades of coal tend to be found at greater depths. Incidentally, while anthracite was listed last, sometimes the sequence of increasingly pure carbon content can continue on to graphite—which is essentially pure carbon. Furthermore, it is even possible for pure graphite to become diamonds, but that requires pressures in excess of 14,500 atm = 1.45×10^9 Pa, which would occur only at very great depths.

2.2.4 Resource Base

Coal is the most abundant of the fossil fuels, and around 50 nations have commercially operating mines. Nevertheless, about 85% of the recoverable reserves of coal in the world can be found in these nine nations: the United States (22.6%), Russia (14.4%), China (12.6%), Australia (8.9%), India (7.0%), Germany (4.7%), Ukraine (3.9%), Kazhakstan (3.9%), and South Africa (3.5%). Given the special circumstances leading to coal formation previously described, it is not surprising that coal deposits are often highly localized and that different types of coal are found in different places (Figure 2.5).

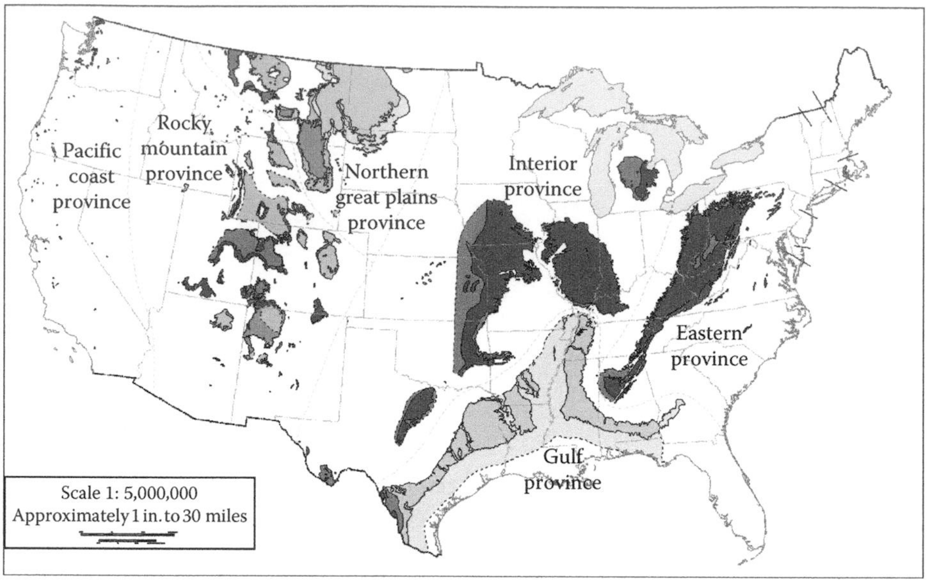

Figure 2.5 Locations of US coal deposits with color coding to designate ranks of deposits: light gray and dark gray being bituminous, light and dark yellow being lignite. For the actual color, see the wikipedia website. (From Wikipedia, http://en.wikipedia.org/wiki/Coal_mining_in_the_United_States.)

The United States has about 240 years worth of coal remaining at its present rate of consumption. On a worldwide basis, the known reserves are sufficient to last about a century at the present rate of consumption. Regrettably, that rate has been steadily rising, owing to the rapidly rising coal production rate in China, which now produces (and uses) roughly half the world's coal output. China's main use for coal is in connection with electricity generation, which fuels 69% of all its electric power, although coal has industrial uses and is sometimes used for home heating there. On average, the Chinese have been building several new coal-fired plants per week and will be doing so for years to come. As a result, in 2006, China has surpassed the United States as the nation emitting the most CO_2 annually. While America may welcome having relinquished the number one spot in this regard, it still far exceeds China in its per capita annual CO_2 emissions (Figure 2.6).

About 60% of the new plants China is building have advanced technology that is highly efficient and limit emissions other than CO_2 more effectively.

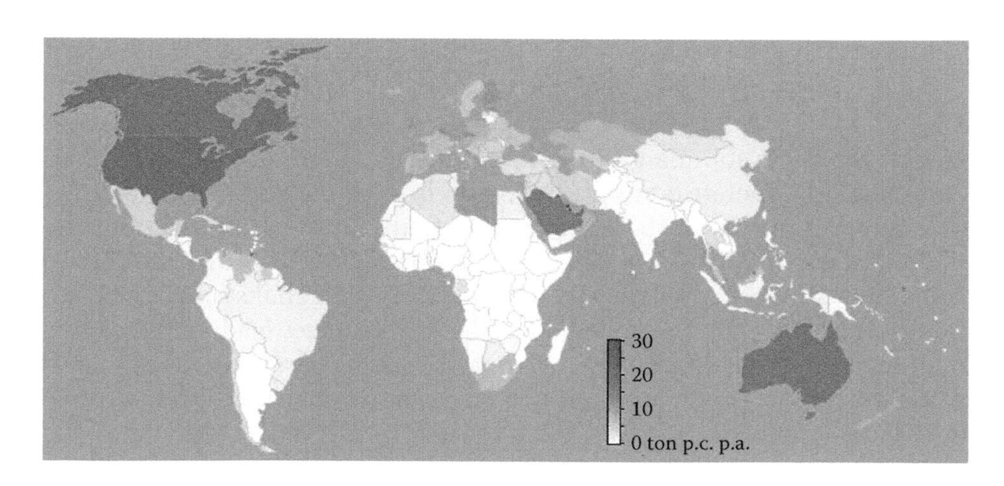

Figure 2.6 Dark shade-coded indicator of per capita CO_2 emissions per country in units of tons of CO_2 per person per year. (Courtesy of Wikimedia Foundation, San Francisco, California, http://en.wikipedia.org/wiki/File:CO2_per_capita_per_country.png.)

To the extent that power companies retire an older, more polluting plant for each new one built, the net result would be a reduction in non-CO_2 emissions. Although introducing more efficient coal-fired plants could also mean less CO_2 emitted per megawatt generated, any such gain is more than outweighed by the annual increase in Chinese coal consumption, which is rising at about 9% per year.

2.2.5 Electricity Generation from Coal

All three fossil fuels can be used for electricity generation, but petroleum is mostly used in other sectors (petrochemicals and transportation fuels), and coal tends to be the dominant fossil fuel source for electricity generation in many countries. There are many possible reasons, however, why a nation might wish to use natural gas or even oil to generate electric power instead of coal, even though coal in the past has been the cheaper alternative—ignoring external (environmental) costs. These reasons include concern for the environment and human health, lack of abundant domestic coal reserves, and greater ease of transport of oil and gas through existing pipelines or nearby ports.

Figure 2.7 shows the basic process for converting the heat from burning coal to make electricity. After coal has been pulverized and delivered to a combustion chamber, it is burned, and the heat boils water creating high-pressure steam that drives a turbine connected to an electric generator. By the second law of thermodynamics, inevitably, a fraction of the heat of combustion is rejected to the environment either up the chimney or in the cooling water that is used to condense the steam back to water, which is why coal plants, like all heat-generating plants, are often near lakes or rivers. Rivers can also serve for barge transport to bring the coal to the plant, although rail cars are also often used. A large coal-fired plant would typically require a train of about 100 cars (over a mile long) or 10,000 tons of coal to be delivered each day.

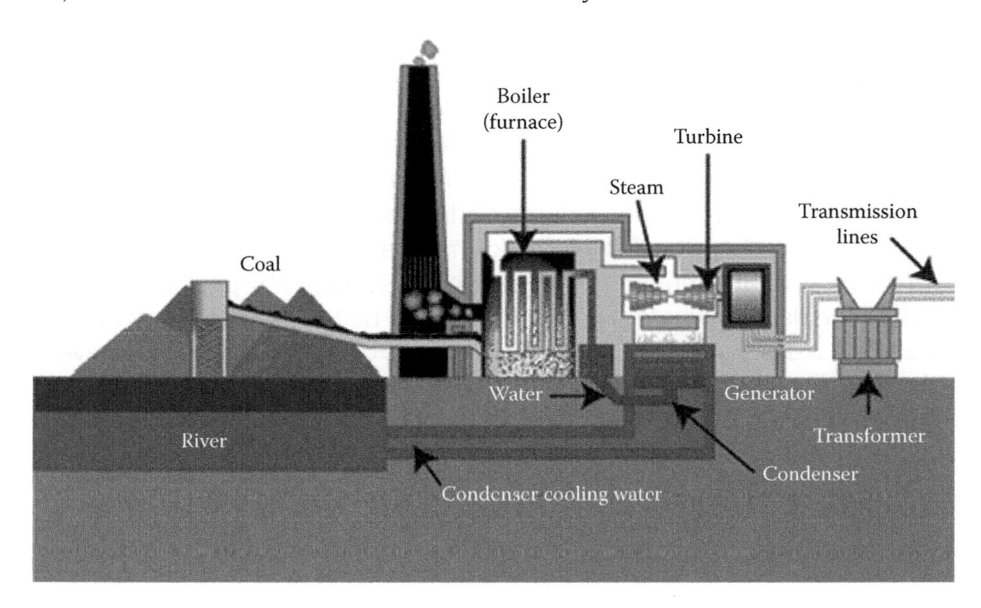

Figure 2.7 Basic components of a coal-fired power plant.

It would be desirable if the heat expelled to the environment were as small as possible, so that more heat could be converted to mechanical energy in the turbine, but there is the fundamental limit imposed by Carnot's theorem.

2.2.5.1 Example 2: Efficiency of a coal-fired power plant Given that coal ignites at around 450°C, how does the 33% efficiency of a coal-fired power plant compare with the highest possible efficiency dictated by the Carnot limit?

$$e_C = 1 - \frac{T_a}{T_C}. \tag{2.2}$$

Solution

Using Equation 2.2 with $T_C = 450°C = 723$ K and $T_a \approx 300$ K, we find $e_C = 1 - (300/723) = 0.59$, i.e., 59%, almost twice as great as the average coal plant efficiency.

2.2.5.2 Rankine cycle All fossil fuel power plants use heat engines to convert heat into mechanical work, and worldwide, about 80% of them rely on the Rankine cycle to accomplish this. An example of a Rankine cycle is depicted in Figure 2.8 using the variables temperature (T) and entropy (S), instead of the usual P–V (pressure–volume) variables to which some readers may be more accustomed. One advantage of using a T–S plot rather than a P–V plot is that in this case, the ideal Carnot cycle of two isothermal and two adiabatic curves can be represented as a simple rectangle. Another point to note is that the closed loop areas defined by $\oint T\,dS$ and $\oint P\,dV$ are identical and represent the work W done in one cycle. Moreover, in both representations by conservation of energy, $W = Q_{in} - Q_{out}$.

WHERE DO DIAMONDS FORM?

Given the known mass and radius of the Earth, we can easily calculate its average density to be 5500 kg/m^3. Although the core of the Earth is made of iron and has a higher density than this, since the core occupies a small fraction of the total volume of the Earth, the average density will be quite typical of the Earth's rocky mantle. From the equation giving pressure as a function of depth beneath the surface $P = \rho g y$, we can calculate the depth y where pure graphite could be turned into diamonds as $y = P/\rho g = 1.45 \times 10^9/(5{,}500 \times 9.8) = 26{,}900$ m = 269 km. However, despite diamonds being the end of an evolutionary sequence involving carbon, geologists do not believe that they formed from coal that was carried downward to sufficient depth. Instead, the generally held belief is that diamond deposits were formed deep in the mantle and delivered to the surface by volcanic eruptions. The accepted value for their formation depth is between 150 and 200 km—a bit lower than our estimate.

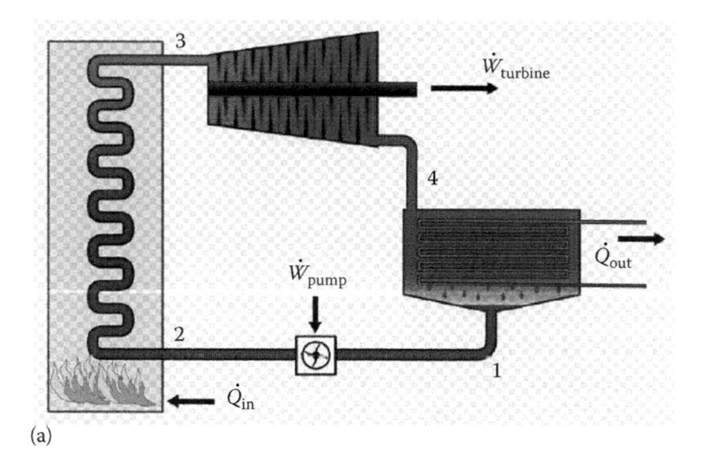

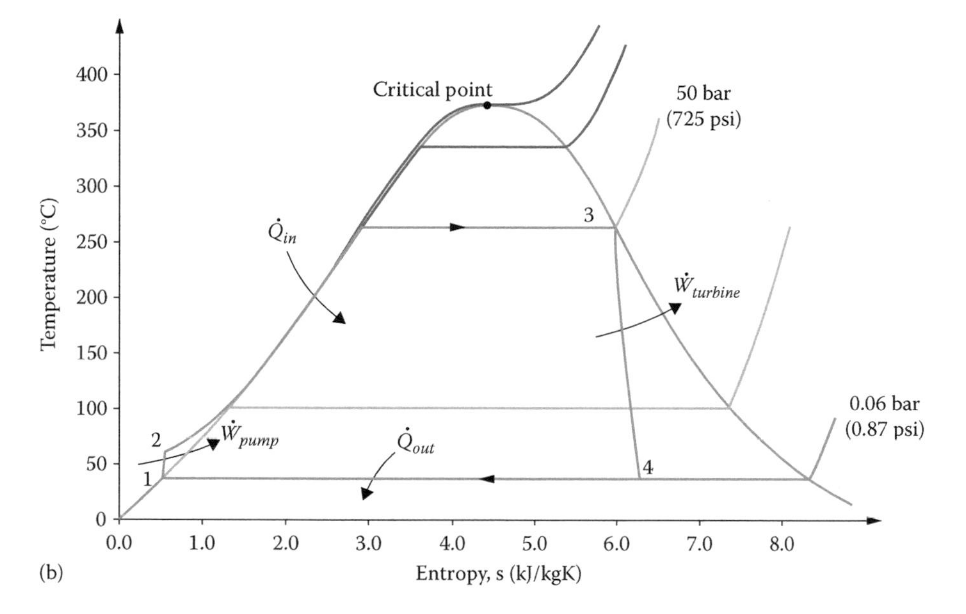

Figure 2.8 Two representations of the Rankine cycle (a) by function and (b) on a temperature–entropy diagram. Example of the Rankine cycle showing for which steps in cycles 1 → 2, 2 → 3, 3 → 4, and 4 → 1 heat is added and removed and positive or negative work is done.

The working fluid in the Rankine cycle is water either in the liquid or vapor phase, with the concave down curve in Figure 2.8 showing the boundary between the two phases in the *T–S* plane. The point labeled the critical point at the maximum of the curve is the maximum temperature for which water could exist for the liquid phase. The steps in the cycles 1 → 2, 2 → 3, 3 → 4, and 4 → 1 represent a time sequence of particular processes taking place in the steam boiler. In order to better understand how the steps in the cycle correspond to physical processes, consider step 2 → 3 as an example. Here we see in the figure that water under high pressure enters the boiler and has its temperature raised from about 60°C to 260°C (mostly following the part of the curve on the left) until it becomes high enough to boil at the start of the horizontal section of the path. As we traverse that section, more and more of the water is converted to steam at constant temperature. Once we reach point 3, all the liquid water has been converted to steam. For an ideal Rankine cycle, steps 1 → 2 and 3 → 4 would be vertical lines, and the cycle would approximate a Carnot cycle, the main difference being that the first part of the cycle bears no resemblance to a vertical straight line—an adiabatic path on a *T–S* diagram.

Given the formula for the Carnot efficiency for a heat engine (Equation 2.2), the two ways to raise the theoretical maximum efficiency are to lower the ambient temperature (unfeasible) or raise the maximum temperature of the working fluid. For plants operating on a Rankine cycle, this can be accomplished by adding an extra step to the cycle $3 \rightarrow 3'$ (not shown in Figure 2.8) that goes up along the 50 bar isobar to a temperature above the critical temperature (generating supercritical steam) before the cycle is finally completed ($3' \rightarrow 4$). This has the effect of increasing the area inside the closed curve and, hence, the efficiency of the process. The supercritical Rankine cycle was made possible in part by the advent of alloys that could tolerate the high temperatures and pressures needed. Supercritical plants around the world (still in the minority) operate at water temperatures of 540°C and 3500 psi and have efficiencies exceeding 45%.

2.2.6 Conversion of Coal to a Transportation Fuel

Apart from steam-powered locomotives powered by solid coal, transportation fuels are generally either liquids or gases. A gaseous fuel *syngas* (short for synthetic gas, which is a mixture of carbon monoxide and hydrogen) can be produced from coal by heating it under high pressure in the presence of water vapor. The syngas reaction known as coal gasification is

$$\text{Coal} + O_2 + H_2O \rightarrow H_2 + CO. \tag{2.3}$$

Although syngas can be used on its own as a transportation fuel, its energy content is only about half that of natural gas, so that normally, it is converted instead into a more energy-rich liquid fuel similar to gasoline, or else the hydrogen component is extracted and used to power fuel cells.

The conversion to a liquid akin to gasoline or diesel can be done through the Fischer–Tropsch (F–T) process, which involves a series of chemical reactions starting with syngas and resulting in the production of a variety of liquid hydrocarbons. The F–T process was invented by a pair of German scientists in the 1920s and was used by Germany in World War II, when it lacked access to gasoline supplies. In fact, by 1944, Germany was producing 124,000 barrels (bbl) of synthetic fuels per day. It was similarly used by the South Africans under apartheid when they were denied access to external supplies of gasoline. The process is still being used by South Africa today to make synthetic gasoline from coal—a process that accounts for 30% of their fuel needs. In most other nations, the production of synthetic fuels usually starts from natural gas rather than coal—this currently being the more economical alternative. Nevertheless, should a technological breakthrough make the coal process more viable, it could become an attractive backup possibility in the event of severe petroleum shortages or rise of prices. In fact, one Massachusetts Institute of Technology (MIT) study has projected that producing liquid fuels from coal could become economically viable in coal-rich nations in a few years (MIT, 2011a).

2.2.7 Coal Mining

Coal is mined by one of two basic methods: surface (or strip) mining and underground mining. Surface mining is utilized in places where the depth of the coal seam is such that removing the overburden to expose the coal seam is economically preferable to the more difficult and dangerous method of underground mining. In the United States, if a coal seam is less than 50 m deep, it will usually be surface mined, while depths below 100 m are usually underground mined. For depths between 50 and 100 m, the choice depends on the thickness of the coal seam. At one time, virtually all coal mining in the United States was done using underground mines, but today 60% are surface mined, sometimes using the highly controversial practice of mountaintop removal, which usually results in the complete disruption of natural ecosystems.

ARE BLACK LUNG DISEASE DEATHS REALLY 20 TIMES GREATER IN THE UNITED STATES THAN IN CHINA?

As already noted, Chinese miners tend to work in deep underground mines far more often than their US counterparts, which is where black lung disease is primarily contracted. Moreover, Chinese mines, especially smaller ones, tend to lack the kinds of high-tech safety measures that are common (required) in US mines. Additional circumstantial evidence that the Chinese statistic is bogus is the "hidden" nature of black lung disease compared to coal mine deaths from cave-ins and other disasters, which are much harder to keep secret and which are far less common in the United States than in China. Accurate statistics on black lung disease require a proactive monitoring system that is mandated by the national government and which results in penalties if it is not followed. The United States has just such a system in place. As a result of Federal Coal Mine Health and Safety Act of 1969 (amended in 1977), regular free chest X-rays are taken of all underground miners, with emphasis on early detection of the signs of black lung disease. There appears to be no such corresponding requirement in China. Even if there were, and even if it were complied with, workers, if given the option of coming forward to have free exams, might well feel not-so-subtle pressures that keep them from actually having the exams if they understandably fear that their jobs might be in jeopardy. Given all the aforementioned points, it would seem more plausible that Chinese black lung disease were far more common in China than in the United States rather than the reverse. If Chinese black lung deaths per miner occurred at the same rate as in the United States, there would be 50,000 such deaths there each year, but the true figure is probably far more.

Coal mining has historically been a dirty and dangerous occupation. The two largest coal-producing nations are China and the United States, which together produce 63% of the world's coal as of 2011, so it is worthwhile to look at some statistics for these two nations. It is estimated that 100,000 coal miners were killed in accidents over the last century in the United States, with the annual death rate dramatically declining with improved technology and safety measures. Most other developed nations have also seen great improvements in mine safety in recent decades, so that mine deaths have dramatically dropped—on average, there are now only about 30 mine-related deaths per year in the United States. In China, the number one coal producer, the story is quite different. Among China's estimated 5 million coal industry workers, an estimated 20,000 die each year in accidents—about 700 times the US number. Coal mining has been called the most dangerous occupation in China (*Shanghai Star Newspaper*, 2004). Moreover, these fatality figures are deaths in the mine and do not include deaths from pneumoconiosis (black lung disease), which is still quite common among coal miners, even in developed nations. In the United States, for example, over 10,000 coal miners have died from it in the past decade. According to Chinese self-reporting, black lung disease has claimed 140,000 coal miner lives in the last half century and close to half a million miners are now suffering from it (*People's Daily Online*, 2005). Given that China has around 50 times as many miners as the United States, these data surprisingly imply that the chances of a miner dying from black lung disease deaths are 20 times greater in the United States than in China. This comparison raises questions about the accuracy of self-reported data—particularly data that would prove embarrassing to a nation.

2.2.8 Environmental Impacts of Coal

Although coal miners may be the people having the greatest negative impact associated with coal, the health of the general population and the environment are both affected in a serious way when coal is mined, transported, stored, burned, even long afterward. The environmental effects associated with coal mining and coal burning include air, water, and land pollutions, resulting in very serious long-term consequences for both humans and ecosystems.

2.2.8.1 Atmospheric emissions from coal power plants Coal-fired power plants are prodigious emitters of pollution, although newer plants using *scrubbers* to filter the exhaust as it travels up the smoke stacks have significantly reduced some emissions. Nevertheless, as can be seen from Table 2.2, coal is still the dirtiest of the fossil fuels. For example, compared to gas-fired power plants, coal plants emit 1200 times more particulates and nearly double the CO_2.

Even apart from greenhouse gases contributing to climate change, air pollution due to coal-fired power plants has very serious consequences for humans. Just considering China alone, the world's largest coal producer,

Table 2.2 **Emissions from Power Plants Using Various Fuels in Kilograms/Gigajoule for Actual Power Plants, as Reported by the European Environmental Agency**

Pollutant	Hard Coal	Brown Coal	Fuel Oil	Gas
CO_2	94.6	101.0	77.4	56.1
SO_2	0.765	1.36	0.228	0.00068
NO_x	0.292	0.183	0.195	0.093
CO	0.0891	0.0891	0.0157	0.0145
Particulates	1.203	3.254	0.016	0.0001

Source: EEA, *Air Pollution from Electricity-Generating Large Combustion Plants*, EEA, Copenhagen, 2008.

a report from scientists at the University of California at Berkeley puts the annual Chinese coal-related death total at 420,000 (Zhang and Smith, 2007). The corresponding figure of annual coal-related deaths in the United States, the number two coal producer, is 24,000, many from air pollution (EPA, 2004). Moreover, an US Environmental Protection Agency (EPA)–funded study also concluded that 90% of those deaths are preventable with currently available technology. In fact, EPA had been given the power to regulate such emissions as far back as the Clean Air Act of 1990, but lobbying by the coal industry has delayed new regulations limiting emissions of mercury and other toxic substances until 2011. Under the new rules, affecting an estimated 40% of US coal plants, EPA estimates that 11,000 premature deaths will be avoided with a concurrent savings in healthcare costs of between $37 and $90 billion annually. The new rules are expected to cost the coal industry a one-time assessment of about $10 billion and will likely involve the closing of some older dirty coal plants that in many cases were slated for retirement anyway.

One of the authors is old enough to remember a time when coal was burned to heat the home that his family shared with his grandmother. Shoveling the coal to get it into the furnace was a dirty job that he occasionally had to do. Once burned, the smoke produced from the coal was quite appalling—even if most of the smoke was released up the chimney outside the home. Today, very few Americans use coal for home heating (outside of some rural areas of Pennsylvania), but this is not the case in China and much of the developing world. The Chinese situation is made even worse by the habit in many rural households of hanging vegetables to dry near the ceiling, which results in many cases of heavy metal poisoning when coal smoke is absorbed by the vegetables.

2.2.8.2 Other atmospheric emissions, including radioactivity Although one normally associates radioactivity with emissions and wastes from nuclear power plants, in one sense, the radioactive release is significantly worse in the case of coal power plants. According to a University of California at Berkeley study, even though nuclear wastes are much more radioactive than the fly ash wastes from coal burning, only the latter are routinely released to the environment in the many plants that lack the technology for capturing fly ash released in the flue gases after combustion. The study concluded that the fly ash from coal plants, in fact, releases 100 times more radiation to the environment than is the case for a normally operating nuclear plant per megawatt of power generated (Hvistendahl, 2011). Of course, nuclear plants do not always operate normally, and the matter of nuclear accidents will be considered in a subsequent chapter.

In addition to atmospheric emissions during coal burning, significant emissions also occur during the mining of coal, including methane gas, which often occurs in coal deposits. Apart from the direct hazard to miners, methane released into the atmosphere is a particularly potent greenhouse gas. While methane stays in the atmosphere a far shorter time than CO_2, it is considered the second leading contributor to the greenhouse effect, and has a global warming potential 21 times greater than CO_2.

2.2.8.3 Waterborne pollution and acid rain When coal is mined, water that comes into contact with coal surfaces leaches sulfuric acid, even after the mine has been shut down. The sulfuric acid pollutes streams, kills aquatic wildlife, and causes problems for the human water supply, which are especially serious for surface mining. Toxic trace elements dissolved in the water also contribute to the pollution, which, in addition to causing environmental damage, also cause serious economic losses by damaging both commercial and recreational agriculture and fishing. The geographical extent of the environmental damage can be greatly enhanced in the event of flooding.

Apart from water pollution resulting from mining, considerable amounts of water are needed during coal burning, but generally, this does not result in any significant pollution with the important exception of acid rain, which occurs when carbon dioxide and especially sulfur dioxide in the flue gases react with rainwater far from the plant to produce carbonic acid and sulfuric acid. We include these under waterborne pollution even though they leave the plant as gases, because they are deposited with rain. These corrosive substances (especially sulfuric acid) can kill trees and render lakes fish free. Since many power plant smokestacks are extremely tall, such pollution is locally reduced, but the net effect is the creation of acid rain often many hundreds of miles from the source—sometimes in another country. Through legislation, the acid rain problem has been considerably reduced in North America and the European Union at a fairly modest cost (about a quarter of what had been predicted), but it remains a significant problem in Russia, China, and elsewhere (Figure 2.9).

The extra CO_2 put into the atmosphere by coal burning also has a significant impact on the world's oceans, making them more acidic (lower pH), as they absorb extra CO_2 from the atmosphere. For example, since the industrial revolution, the average oceanic pH has decreased from 8.25 to 8.14. This change may sound insignificant, but it is actually a sizable change in acidity of nearly 30%. Moreover, by the year 2100, it is

Figure 2.9 Trees killed by acid rain in Germany.

expected that as a result of rising levels of atmospheric CO_2, the acidity is likely to rise to 227% of the preindustrial level, having profound effects on aquatic organisms.

2.2.8.4 Example 3: Connection between acidity and pH levels What pH change corresponds to a projected acidity increase of 127% by the year 2100?

Solution

The definition of pH is

$$pH = -\log_{10} H^+, \tag{2.4}$$

where H^+ is the hydrogen ion concentration in moles per liter. Based on the preindustrial pH level of the oceans of 8.25, using Equation 2.4, we find that the hydrogen ion concentration then was $H^+ = 10^{-pH} = 10^{-8.25} = 5.62 \times 10^{-9}$. If the acidity, i.e., the number of H^+ ions, were to increase by 127%, the acidity would rise to $2.27 \times (5.62 \times 10^{-9}) = 1.27 \times 10^{-8}$ and the pH would then be $pH = -\log_{10} 1.28 \times 10^{-8} = -7.89$, a change of -0.36 from its preindustrial value.

The greatest impact of increasing oceanic acidity is likely to be on the production rate of shells from calcium carbonate, since this calcification process is greatly inhibited by rising ocean acidity. Although the full impact of reduced shell production remains unclear, it is likely to be highly detrimental to the biology and survival of a wide range of marine organisms, as indicated by lab experiments on some specific species (Hardt and Safina, 2010).

2.2.8.5 Impacts on the land Surface (strip) coal mining has a severe impact on the landscape and usually destroys the preexisting ecosystems and habitat—a disruption that is generally permanent. It is not unusual for hundreds of surrounding acres to be affected and people living in the affected areas to be permanently displaced. Without rehabilitation efforts, the loss of topsoil and the toxic elements produced by mining may leave the land a vast infertile wasteland. With major rehabilitation efforts once a mine is shut down, some of the land may be reclaimed, but the result generally leaves the land unsuited to its original uses.

2.2.9 Carbon Sequestration and "Clean" Coal

Is "clean coal" just a slogan promoted by coal companies, or does it hold the promise of making coal an environmentally benign energy source? Certainly, the technology exists for significantly reducing pollutants other than CO_2. These technologies include scrubbers that remove gases (especially sulfur dioxide), toxic trace elements, and dust after combustion. They have been implemented for some coal-fired power plants, with the result that in many nations, many newer coal plants are significantly cleaner than they used to be. Scrubbers, for example, have been implemented in about half the world's coal-fired plants, and the simple expedient of burning low-sulfur coal can, when possible, also further cut down on pollution. However, the real question concerns the possibility of eliminating or significantly reducing CO_2 emissions. This is a much more difficult challenge whose feasibility and cost-effectiveness remains to be demonstrated. Moreover, once the CO_2 has been removed, there is the technically difficult matter of disposing of it (sequestering it) at a reasonable cost and in a manner that keeps it secure and does not allow it to enter the atmosphere at a later time.

Two of the disposal methods that have been extensively investigated include storing CO_2 in abandoned mines and injecting it into old oil or gas fields that are no longer producing or in repositories under the deep ocean bottom. If the gas is injected at depths in excess of 2700 m, its density when liquified would exceed that of seawater, so that it would presumably not rise to the surface. Nevertheless, despite the claims of industry spokespersons, there is no indication that carbon capture and sequestration (CCS) technology is anywhere close to fruition economically. Those technologies that have been developed could, if implemented, double the cost of generating electricity from coal, making their widespread implementation entirely dependent on subsidization or equivalently on setting a price on CO_2 emissions—either through a cap-and-trade system or, alternatively, a carbon tax. Given that many renewable energy technologies, especially wind power, are already approaching cost parity with coal and getting cheaper over time, the whole idea of promoting coal use (given all its other environmental problems besides climate change) is difficult to understand unless one's individual livelihood depends on the coal industry. One possibility, which might be a realistic, economically viable solution, would be carbon capture and utilization rather than CCS. The varieties of ways that captured CO_2 might be used include the following:

- As a feedstock to synthesizing various chemicals
- As a way to produce construction materials through mineral carbonation
- As a nutrient to promote algae growth for biofuel production

It has even been suggested to turn the CO_2 into an energy source itself through a variety of methods. One of these would involve injecting it at sufficient underground depth, where the temperature and pressure would convert the gas into a supercritical state giving the gas some liquid properties and then allowing it (when pulled up to the surface) to drive turbines. Of course, the question remains whether any of these schemes can be made cost effective, and whether the harmful environmental and health impacts of coal unrelated to climate change can be greatly reduced to the point where clean coal becomes more than an industry slogan (Figure 2.10).

US CCS PROGRAM

The CCS program has had a spotty record to date, although as of 2016, scientists at Columbia University have employed a new approach: dissolving the gas with water and pumping the resulting mixture into rocks and then essentially turning the mixture into stone through a chemical reaction. The key is to find the right kind of rocks, and the first signs are very encouraging with 95% of CO_2 turned into calcite.

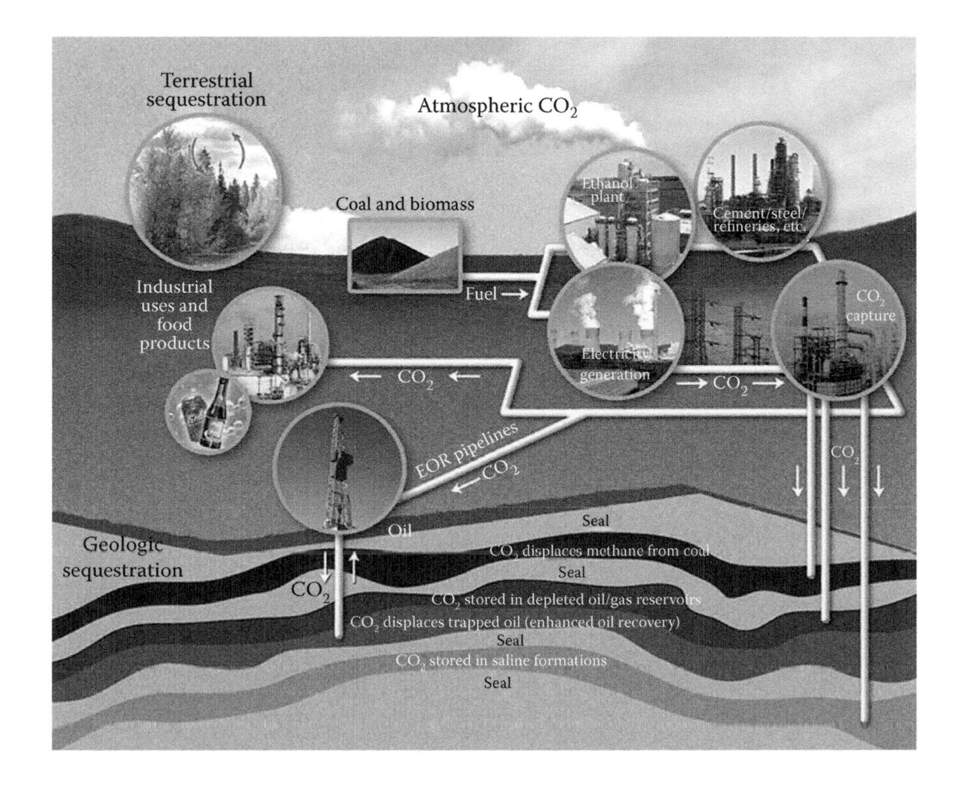

Figure 2.10 Various options for carbon sequestration.

2.3 PETROLEUM AND NATURAL GAS

Petroleum or crude oil is a liquid hydrocarbon consisting of many kinds of complex molecules. Its elemental composition includes 83–87% carbon, 10–14% hydrogen, 0–6% sulfur, and under 2% nitrogen and oxygen. Natural gas is a gaseous hydrocarbon, primarily methane, CH_4, with up to 20% higher hydrocarbons, primarily ethane. Recall that the complex molecules found in coal tend to have many strings of carbon atoms arranged in hexagons—see Figure 2.3. Oil, on the other hand, arose from the remnants of decaying microorganisms, mainly marine plants and animals, rather than long plant fibers. The microorganisms produce hydrocarbons with chains of various lengths. The shorter chain hydrocarbons exist as a gas (natural gas) and the longer chains, as a liquid (oil or petroleum). As with coal, petroleum or crude oil contains various amounts of impurities, such as sulfur.

2.3.1 History of Petroleum Use

In one form or another petroleum was known going back around 4000 years, but it was not until the 1850s when a process was invented to extract kerosene from it as an alternative to whale oil used for lighting lamps that the first commercial oil well was drilled in Poland. Initially, the natural gas that often accompanies petroleum was simply burned off and wasted, because there was no easy way to store and transport it, since gas pipelines came much later. In the United States, oil drilling got its start in 1859 when a black fluid was found oozing from the ground

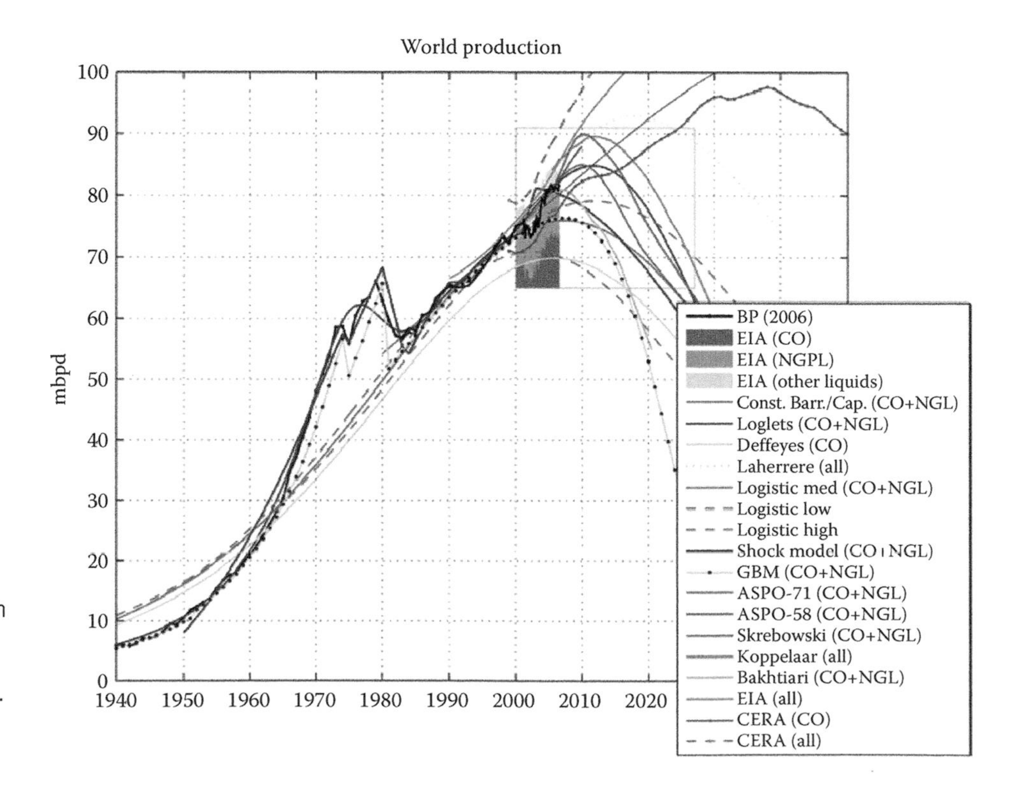

Figure 2.11 World oil production since 1940. Projections from various indicated sources mostly show the peak production year occurring between 2010 and 2020, this graph has been rendered obsolete in light of a signicant increase in worldwide oil production starting around 2010. (From Wikipedia, http://en.wikipedia.org/wiki/Peak_oil, July 1, 2017.)

in Titusville, Pennsylvania. Oil usage really began to take off when cars powered by an internal combustion engine began to be mass-produced during the early years of the twentieth century—see Figure 2.11.

On a worldwide basis, oil production has shown considerable growth since the early years of the twentieth century. One cannot know exactly how oil production will vary in the future, but projections can give estimates that take into account such factors as known proven reserves, the increasing costs of extraction (as less accessible locations are drilled), the costs of alternatives, environmental concerns and regulations, and market forces.

2.3.2 Resource Base of Oil and Gas

As in the case of coal, the world's petroleum reserves are quite unevenly distributed around the globe. While the nations of the Middle East have a majority of the proven reserves (56%), as seen in Figure 2.12, there are also very significant deposits in North America (16%), Africa (9%), South America (mainly Venezuela, 8%), and Eurasia (7%). The United States at one time did have a much larger fraction of the world's proven reserves, but the nation has been consuming them at a prodigious rate. The peak year for US oil production (1970) has long since passed, although due to the rapid rise since 2010, the United States could surpass that peak in only a few years. US oil imports as a share of consumption reached a high of 60% in 2005, and they have declined to 40% in 2014 (a 30-year low) largely due to increased domestic production. For any finite resource, a point in time is always reached when the extraction rate reaches a maximum, which is then followed by an inevitable decline. Considering oil on a worldwide basis, there is reason to believe that within a few years, one way or the other, the peak time is now—a topic discussed at greater length in Section 2.3.5.

Natural gas deposits can be found on their own, but they are often found in conjunction with oil. Thus, with some exceptions, some of the same nations that have large oil reserves also have large gas reserves. As of 2014, the world's proven natural gas reserves are about 188 trillion m^3, and the top six nations have the percentages as listed in Table 2.3.

Table 2.3 Percentage of the Proven Reserves of Natural Gas in the World, as of 2015

Country	World Gas Reserves (%)
Russia	24.2
Iran	17.2
Qatar	12.5
Turkmenistan	3.8
Saudi Arabia	4.2
United States	4.9

Source: US Energy Information Administration, 2015, https://www.eia.gov/naturalgas/.

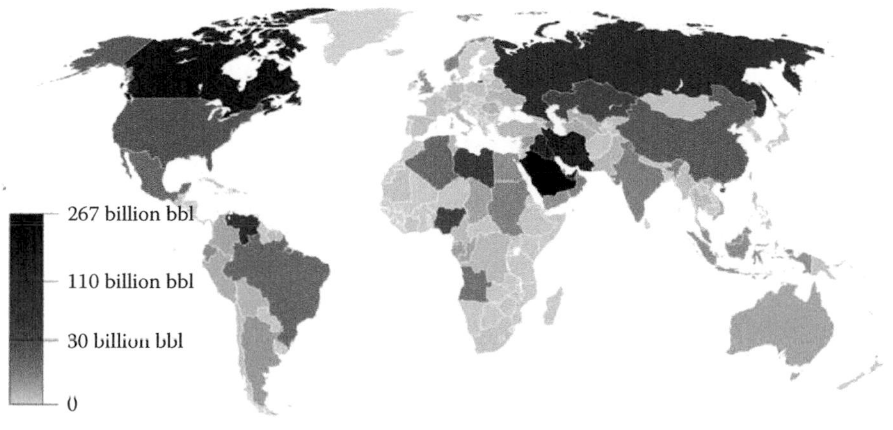

Figure 2.12 Proven oil reserves by country in billions of barrels, as reported in the *US Central Intelligence Agency Factbook*, 2010. (Courtesy of Wikimedia Foundation, San Francisco, California, http://en.wikipedia.org/wiki/File:Oil_Reserves.png.)

We see that Russia has greater gas reserves than any other nation, even though its oil reserves are dwarfed by some Middle East nations; likewise, the United States has virtually the same gas reserves as Saudi Arabia, even though its oil reserves are less than a tenth those of the Saudi's. One of the major energy developments for the United States in the last decade has been a near doubling in the nation's reserves of natural gas owing to new technologies that make it possible to extract gas from deposits (mainly in natural gas shale) that was previously considered unreachable. The technology known as horizontal drilling hydraulic fracturing is discussed in Section 2.3.6.1.

2.3.3 Formation and Location of Oil and Gas

Oil and gas have been primarily formed from decaying microscopic marine organic matter (plankton) including algae. Another difference from coal is that oil and gas can seep upward through porous rock layers. Thus, usually, oil and gas form within a source bed and then migrate up into a porous reservoir bed. For the deposits to survive, it is further necessary to have a nonporous cap rock above the reservoir formation to trap them, or the gas and oil would not be concentrated and the oil field would not exist. This requirement, of course, is not necessary for coal, which is a solid. Unlike the implication of Figure 2.13, the oil and gas may not be present in literal pools, but instead be present in the tiny pores of the porous rocks.

Oil shales or gas shales accumulate differently. In this case, the hydrocarbon accumulates in the nonporous shale and is trapped there, dispersed throughout the rock. Hydrofracturing creates porosity, and the fluids can be extracted.

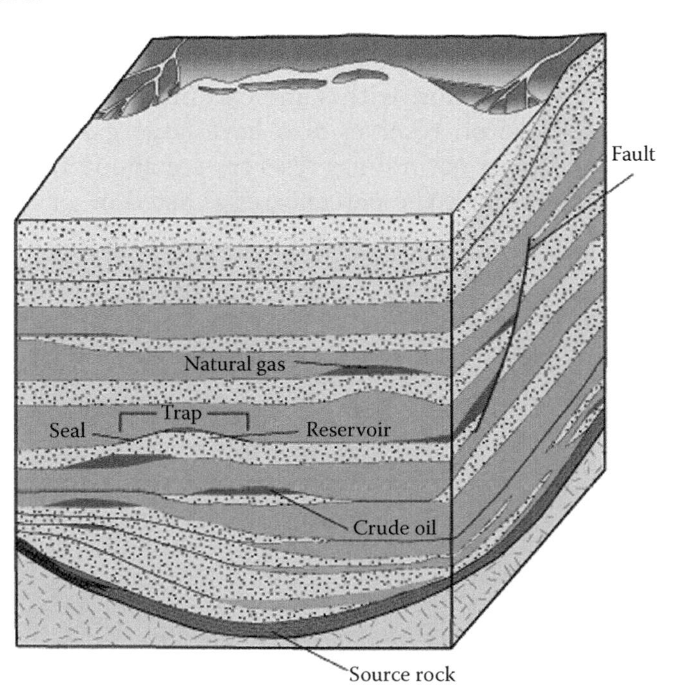

Figure 2.13 Depiction of oil and gas deposits formed under heat and pressure from source rocks and migrating upward through porous rocks and faults until they are trapped below a layer of nonporous rock. (Courtesy of US Geological Survey, Reston, VA, http://teeic.anl.gov/er/oilgas/restech/dist/index.cfm.)

Given the very specific arrangement of rock layers necessary for oil and gas deposits, geologists can explore for them by discerning the stratigraphy and structure of rock layers in different locations using seismic studies and then drilling test holes. Drilling can be an expensive process, depending on the well depth. Many promising locations can be undersea, which of course makes the oil drilling and extraction process even much more expensive and risky.

Undersea locations are also the place where large repositories of methane can be found in the form of clathrates, or methane hydrates, which are crystalline water-based solids similar to ice that can trap methane. In fact, estimates are that 6.4 trillion tons of methane is trapped in deposits of methane clathrate on the ocean floor, but extraction is far from being economically feasible. Petroleum formation requires certain temperature and pressure conditions. If the temperature is too low, gas may form, but not oil. At higher temperatures (the oil window), both oil and gas may form. Higher temperatures than this will destroy the oil, converting it back to methane. Many of the world's older oil and gas deposits, such as those in the United States and Russia, have experienced temperatures beyond the oil window, so that many of their fields are now gas fields.

Unlike oil, natural gas, in addition to being formed biogenically, can also have an abiogenic origin. According to theory, at very great depths in the Earth, the high temperature and pressure can convert buried organic matter into natural gas thermogenically. Without such an alternate non-biogenic formation process, it would be difficult to understand how abundant quantities of methane ever arose on the gas giant planets (including Jupiter) and their moons, where no one believes that life has ever existed.

2.3.4 Are Coal, Oil, and Gas Really Fossil Fuels?

The notion of being a fossil fuel implies that it was mainly formed from what originally was living material that, over the course of time, was transformed through the application of heat and pressure. We need, however, to remind ourselves that this commonly accepted belief is simply a theory of the formation of coal, oil, and gas and that the evidence for this theory needs to be weighed against possible competing theories. One alternative possibility disbelieved by most geologists is that the dominant process for producing hydrocarbons was abiogenic, with the organic material from which they were formed being part of the original composition of the planet. In this view championed by the late Cornell University physicist Thomas Gold and many others, hydrocarbons are formed deep in the Earth and seeped upward through the Earth's crust and either reached the surface or formed underground deposits—some of which later solidified to become coal (Gold, 1999).

As noted at the end of the last section, geologists readily admit that this scenario can and does occur for methane and even solid carbonaceous material (diamonds), but they are loath to extend it to liquid petroleum and most especially to coal. Nevertheless, the authors believe that the hypothesis, while unproven, continues to be plausible. Apparently, the abiogenic hypothesis was actually quite widely believed in the past among geologists, but it was abandoned at the end of the last century after it failed to be useful in predicting where oil deposits could be found (Glasby, 2006). These are some of the arguments in favor of the abiogenic theory:

- Lab experiments show that hydrocarbons can be synthesized under the temperature and pressures found in the Earth's upper mantle.
- Porosity in rocks provides migration pathways upward for liquids and gases, but not for coal, which is solid.
- If methane is known to have an abiogenic origin, why not petroleum as well, since the same basic mechanism is used (in the biogenic theory) to account for them.
- So-called biomarkers found in petroleum, supposedly giving evidence for its biogenic origin, have been found in some meteorites.
- Petroleum tends to be found in large patterns related to deep large-scale structural features of the crust rather than a patchwork of sedimentary deposits.
- There are indications that gas and oil deposits spontaneously refill over time—possibly from below.

There are, of course, many arguments against the theory. For example, geologists note that the spontaneous refilling of oil deposits need not imply upwelling from deep inside the Earth, but could indicate seepage upward from another layer below the deposit. Moreover, chemists note that when they analyze crude oil, they always find steroid molecules, and there is no known way to create such molecules other than from living creatures.

WHY SHOULD WE CARE IF COAL, OIL, AND GAS ARE REALLY FOSSIL FUELS?

Whether or not the abiogenic origin theory of petroleum can serve as a guide to locating new promising locations for finding oil, it still has important real-world implications if true. First, as the last point in the arguments given earlier indicates, future petroleum sources might simply hinge on leaving wells fallow for a while. Second, if the total amount of petroleum and other fossil fuels is vastly greater than we now imagine (as the abiogenic theory suggests), the potential impact on climate change will also be vastly greater without a shift away from them. Third, if the supply of fossil fuels is indeed near infinite (i.e., with as much as a 1000-year supply), the problem discussed in the next section becomes moot.

2.3.5 Peak Oil

The concept of peak oil was first raised by geologist M. King Hubbert in 1956. His basic idea was quite simple, namely, that as long as the quantity of oil is finite, then for any given location, or a nation, or even the planet as a whole, the rate of oil production will tend to follow a bell-shaped curve. Consider, for example, a single oil field—how would we expect the annual production from it to vary in time? Following its initial discovery, efforts are ramped up to exploit it, and as more and more wells are drilled, the production rate will rapidly rise. At some point, extraction becomes increasingly difficult, the rate of increase in production tapers off, and a maximum production rate is reached, to be followed by an inevitable decline, as the field begins to become exhausted. Although Hubbert initially did not suggest a specific mathematical function to describe the bell-shaped curve of production from a particular oil field versus time, he later suggested the derivative of the logistic function $Q = Q_0 (1 + ae^{-bt})$, which agrees better with observations than the more familiar Gaussian curve. Applying Hubbert's function to an entire nation rather than one oil field still gives good agreement with his theory, provided the nation's oil or gas supply tends to be not too geographically diverse—see Figure 2.14a for the nation of Norway. On the other hand, a large nation such as the United States, whose oil exploration efforts are in geographically different regions, might be expected to agree less well with his theory—see Figure 2.14b. It is noteworthy that Hubbert's theory of peak oil was made at a time when the climb up the bell curve in many nations was in its early stages, i.e., well before the peak. Moreover, Hubbert applied his theory to the world as a whole for which he predicted a maximum oil production around the year 2000,

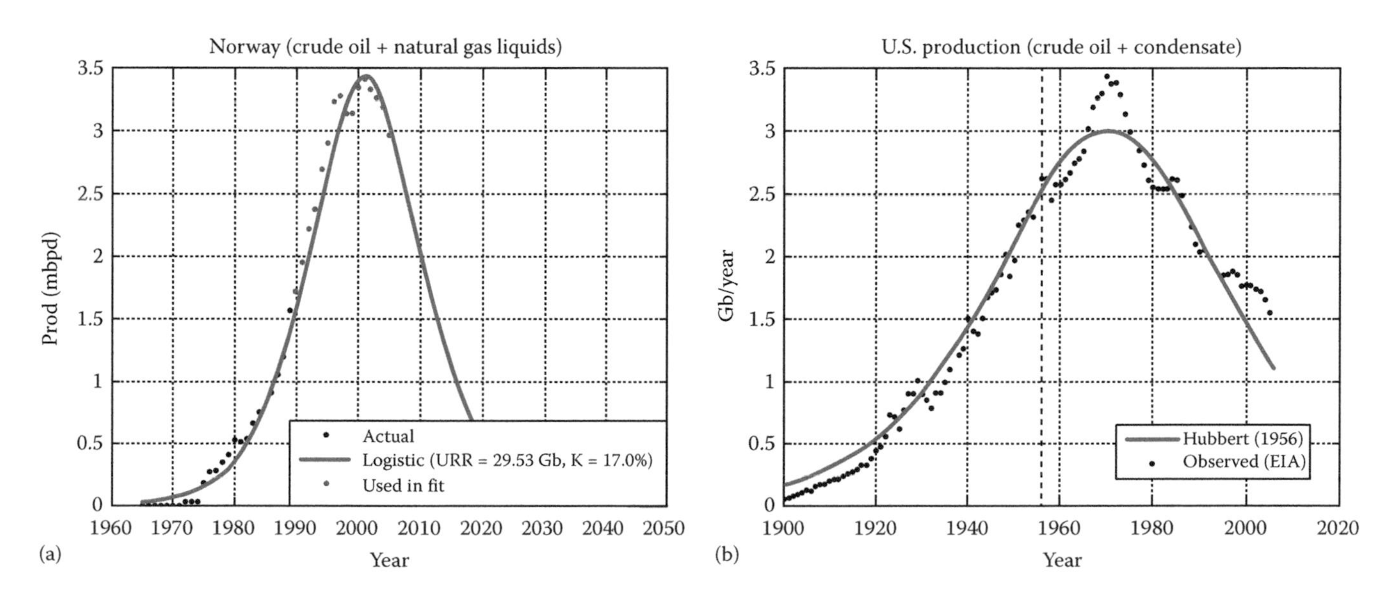

Figure 2.14 Oil production for (a) Norway and (b) the United States versus time with fitted Hubbert curves. Although it is not shown in the figure, the most recent available data for the United States shows a remarkable rise with domestic production in 2014 reaching a level of production about as high as that in 1987. (Courtesy of Wikimedia Foundation, San Francisco, California, http://en.wikipedia.org/wiki/File:Hubbert_Norway.svg, http://en.wikipedia.org/wiki/File:Hubbert_US_high.svg.)

although he did not foresee that a simple bell-shaped curve would apply in this case. If anything, one might expect a sum of bell-shaped curves displaced in time, as new more challenging sources are exploited, and new extraction technologies, developed.

Additionally, for the world as a whole, one needs to take into account many factors that do not apply to a single oil field, given the very long timescale involved. These factors include supply and demand issues as well as the cost of competing alternatives to oil—factors Hubbert could only guess about in 1956. In other words, as oil becomes increasingly scarce, its price will rise due to both market forces and the need to drill in more challenging deposits (such as under the sea floor). At some point, the rising oil price will make alternative fuels more desirable, which will further depress the rate of oil extraction. Not surprisingly, Hubbert's 1956 prediction for world peak oil was imperfect, since the peak of world oil production did not occur around the year 2000 as he suggested 44 years earlier, but rather at least a decade later. Nevertheless, the essentials of his prediction would seem to be correct. Today, many geologists believe that there is around a 40-year supply of oil at the present rate of consumption (production), which is quite consistent with a 2011 peak, given a bell-shaped curve having Hubbert's predicted width. There are, of course, many dissenters, and some analysts put the peak at around 2040 when taking into account oil from unconventional (and more expensive) sources (Darrell, 2007). Figure 2.11 shows a variety of predictions regarding when peak production will be reached. Figure 2.14 for the United States, however, offers a good lesson in the dangers of extrapolating from current trends. As of 2005 (the last year of data shown there), and as a result of hydrofracturing and horizontal drilling, the production of US crude oil has skyrocketed, making the Gaussian drawn in Figure 2.14 (right) a terrible fit.

The domestic and international socioeconomic implications of peak oil are extremely important, because it suggests that moving away from fossil fuels (petroleum in particular) is not just a matter of choice but, eventually, one of necessity. It also implies that if the world does not make the transition very soon, the result could mean profound societal disruption, as oil supplies become increasingly scarce relative to demand, with ever-increasing potential for international conflict arising from competition for access to oil. The pressures are likely to be aggravated by the rapid economic rise of China and India and by having so much of the world's petroleum reserves in politically unstable regions. It will be especially acute for nations such as Japan, which is entirely dependent on imported energy, and to a much lesser extent for the United States, which has extensive domestic oil production. The potential for conflict is further exacerbated by the impact of climate change on food availability and access to water in many developing countries, even if they have little need for oil themselves. In fact, the quartet of food, energy, water, and climate (to which might be added economic and political turmoil) has been referred to as "a perfect storm for global events" (Beddington, 2009).

2.3.5.1 Example 4: How many years are left? Show that if the present world consumption of a resource is now at its absolute peak and consumption follows a Gaussian curve, then T defined as the number of years left at the present rate of consumption R_0 is roughly equal to half of the full width of the Gaussian at half max (FWHM). Note that it can easily be shown that the FWHM and standard deviation are related by FWHM = 2.35σ.

Solution

If we are now at the peak (mean zero), the future annual consumption can be written as the following function of time t, where σ is the standard deviation:

$$R = R_0 \exp\left(-\frac{t^2}{2\sigma^2}\right). \tag{2.5}$$

Integrating Equation 2.5 over all future time gives us the remaining amount A of the resource:

$$A = \frac{1}{2}\sigma R_0\sqrt{2\pi} = \frac{1}{2.35}(\text{FWHM})R_0\sqrt{\pi/2} = R_0 T, \tag{2.6}$$

where the last equality follows from the definition of T. Solving Equation 2.6 yields $T = 0.53 \times$ FWHM. Thus, if the peak is occurring about now, the number of years left at the current rate of usage is (within 7%) equal to half the FWHM, or about 40 years. This, of course, is only a mathematical exercise, not a prediction, since various observers have very different estimates of when the peak of oil production will occur.

2.3.6 Petroleum and Natural Gas Processing

Starting from the original petroleum or natural gas deposit that has been located based on its geology and confirmed by test wells, there are many steps before the products of the crude oil or gas can be utilized, and we shall discuss here these three: extraction, transport, and refining.

2.3.6.1 Extraction of oil and gas In its initial stages after drilling an oil well, the pressure is usually sufficient to spontaneously force the oil to the surface. Typically, this primary recovery stage lasts up to 5–15% of the capacity of the reservoir. Once the pressure drops and the extraction reaches its second stage, pumps must be used or water must be injected into the well to bring oil to the surface. At some point, even these methods fail, and it is necessary to use enhanced recovery methods to make the oil easier to extract, which include injecting steam heat (to reduce the oil viscosity) or surfactants (detergents) to lower its surface tension. One enhanced recovery method that has been around since the 1940s is known as hydraulic fracturing, hydrofracking, or simply fracking.

In this method, fluids are injected into rock formations in order to induce fractures in them and to provide a pathway for oil trapped in the pores of the rock to reach the surface. Thus, the injected fluid (usually a mixture of water, chemicals, and suspended sand-like particles) has the dual purposes of opening and extending fractures and aiding in the transport of suspended particles in the fluid so as to keep the pathways open. Fracking is used with natural gas as well as oil, and it can greatly enhance the size of the recoverable reserves by making it possible to extract oil and gas from rock formations very deep in the Earth up to 20,000 ft and in formations not previously considered economically feasible, such as shale, which has very low natural permeability. The greater economy comes about when fracking is combined with the new technique of horizontal drilling. This combination has allowed extraction over an extended area using only one vertically drilled well instead of 10. As was noted earlier, owing to these new technologies, the extent of the proven reserves in the United States has approximately doubled within the span of a decade. Nevertheless, the practice of fracking remains a highly controversial one due to environmental considerations—see Section 2.3.8.

2.3.6.2 Refining of gas and oil Generally, crude oil and natural gas need to be processed or refined before they are useful. Refineries are sprawling extremely complex chemical plants with miles of piping connecting various processing units. Natural gas processing is designed to clean raw natural gas by separating impurities and various nonmethane hydrocarbons so as to produce pipeline-quality dry natural gas, but the complex processes will not be described here. Oil refineries can typically process several hundred thousand barrels of crude oil per day and usually operate on a continuous process, rather than in batches. The basic process in an oil refinery involves separating out the various useful components of crude oil according to their degree of volatility through a process known as fractional distillation. The different useful components of crude oil include many components including, among others, gasoline, kerosene, diesel oil, fuel oil, lubricating oil, wax, and asphalt. Each of the crude oil components consists of many different molecules whose structure give the substance its desirable properties, which may make it suitable as a fuel, lubricant, tar, or feedstock for producing petrochemicals, among other purposes. Unlike a pure substance consisting of a single molecule with a single boiling point, the distillates of crude oil are each defined in terms of a range of boiling points. Thus, kerosene is that distillate whose boiling point lies in the interval of 150–200°C. The fractional distillation process in cartoon version is shown in Figure 2.15.

The petroleum distillates having low boiling points tend to be lighter, and commonly, they are grouped into the three categories: light, middle, and heavy or residium (what is left after the lighter distillates are driven off). For example, the light petroleum distillates include liquefied petroleum gas, gasoline, and naphtha, and the middle distillates include kerosene (and related jet fuels) and diesel oil. Among those distillates intended as fuels, it is important to remove nonhydrocarbon components such as

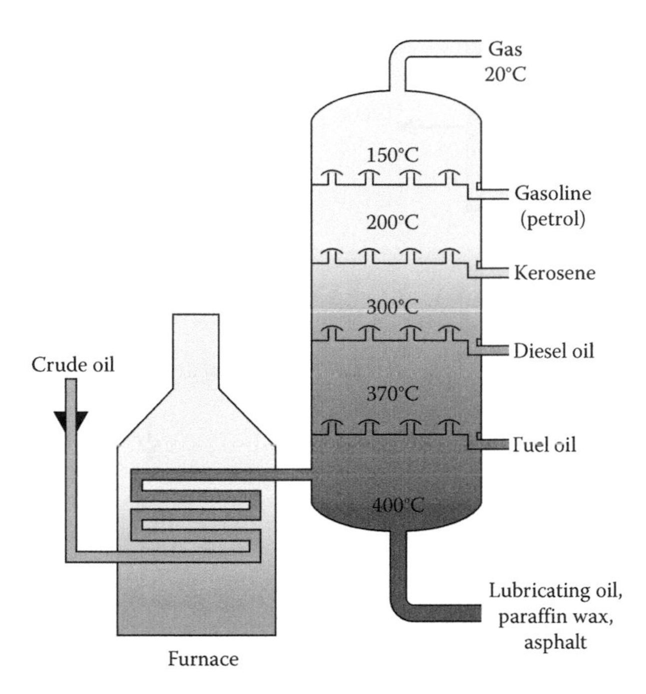

Figure 2.15 Fractional distillation process in an oil refinery in the case of straight run refining. Most gasoline today is made by catalytic cracking of large molecules to small molecules and then reforming them into gasoline.

sulfur, which can be a useful by-product for other purposes (such as making sulfuric acid). Fuels are also further processed in other ways, including blending them to achieve the desired octane rating, which measures how much compression the fuel can withstand before spontaneously igniting.

There were 140 oil refineries in the United States as of 2015, and nearly half of them are in three states (Louisiana, Texas, and California), all near coastal areas—making them highly vulnerable to hurricane damage in the event of a landfall. Owing largely to environmental concerns, very few new refinery have been built in the United States since 1976, although many existing refineries have been expanded. In terms of economic damage, disruption of the economy, and danger to surrounding communities, oil refineries may also represent a more desirable target for terrorists than a better protected nuclear plant—even if they lack the psychological dread factor associated with all things nuclear. Such vulnerability is even more acute for nations such as India that have their oil refineries even more highly concentrated—with India's Reliance Petroleum refinery handling half the capacity in the entire nation.

2.3.7 Gas and Oil Power Plants

Electric power from generating plants that use natural gas as the fuel is environmentally less damaging than coal, and these plants have been especially useful in supplying power during peak times. The ability to supply peaking power arises because unlike coal plants, their power output can be varied on a short timescale by adjusting the gas flow. Newer gas turbine power plants rely on a combined cycle using several turbines in series. In a combined cycle plant, some of the heat expelled from the first cycle is converted into work (driving a second generator). Since more

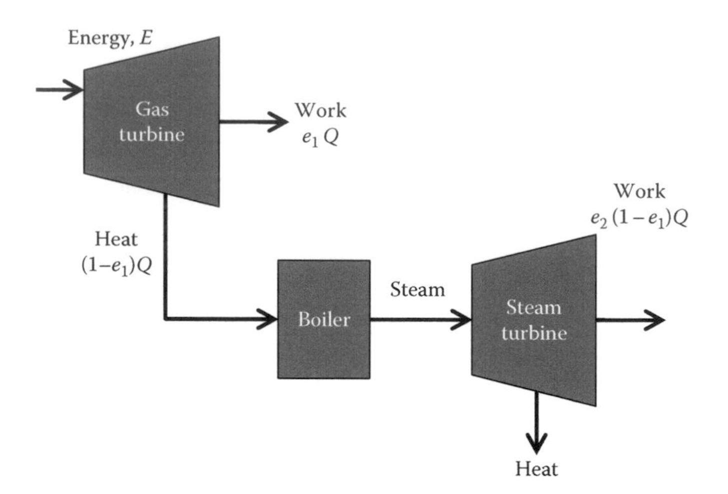

Figure 2.16 Schematic representation of a combined cycle (binary) power plant.

work is produced for the same input of heat, the combined cycle has significantly higher efficiency than a single cycle, and they are, therefore, much more economical. A combined cycle plant works best when the firing temperature in the first turbine (the gas turbine) is relatively high, so that the exhaust flue gas temperature is still quite high (450–650°C), and it is able to supply sufficient heat to provide the heat to the second stage; often, the second stage uses steam as the working fluid and operates on a Rankine cycle—see Section 2.2.5.2. A schematic representation of a binary cycle plant is shown in Figure 2.16. Triple cycle plants having a high-, medium-, and low-pressure turbine also exist, and they are still more efficient. From the input and output of each cycle in Figure 2.16, it is easy to see that the efficiency of a binary cycle plant can be expressed in terms of the efficiencies for each cycle as

$$e_{CC} = \frac{\text{Work}}{\text{Heat}_{\text{in}}} = \frac{e_1 Q + e_2(1 - e_1)Q}{Q} = e_1 + e_2 - e_1 e_2. \tag{2.7}$$

The much higher efficiency of combined cycle gas-fired power plants also translates into both greater economy and lower CO_2 emissions per megawatt generated. For example, while conventional coal-fired plants are less expensive than conventional (single cycle) gas-fired plants, this is not the case when conventional combined cycle gas-fired plants are considered. The cost comparison is even less favorable to coal if CCS were required. It should, therefore, not be surprising that a number of nations, including the United States, Germany, and the United Kingdom, plan to shift increasingly toward gas-fired combined cycle plants and much less heavily on coal plants in the future.

2.3.7.1 Example 5: A binary cycle plant Suppose that the first stage of a binary cycle power plant has an efficiency of 35%. What is the maximum possible overall efficiency if the second stage operates at a temperature of 227°C = 500 K and expels heat to the environment at 27°C = 300 K?

Solution

The maximum efficiency for the second stage is found from the Carnot efficiency, which gives $e_2 = 1 - (300/500) = 0.4$. Thus, using Equation 2.7, for the overall efficiency, we find $e_{CC} = e_1 + e_2 - e_1e_2 = 0.35 + 0.40 - (0.35)(0.4) = 0.61$ or 61%.

2.3.8 Environmental Impacts of Oil and Gas

The environmental impacts of oil and natural gas are considerable, and they occur at various points including their extraction, transport, refining, and eventual usage, mainly as transportation fuels or in electric power generation. In considering the environmental impacts from coal, we have seen earlier that its atmospheric emissions are significantly greater than those for oil and gas. For the transportation sector, it is possible to fuel cars and trucks using liquefied natural gas or compressed natural gas instead of gasoline with some engine modification. Studies of such vehicles show reductions in pollution levels of typically 49% for nitrogen oxides and 90% for particulates. Although natural gas CO_2 emissions are significantly less for natural gas than coal (see Table 2.2), the reduction for natural gas compared to gasoline is more modest, although still significant—up to 25% depending on the natural gas source. Thus, proposals to convert part of the transportation sector (such as all heavy trucks) to natural gas make sense in terms of some reduction in oil imports, as well as resulting in less air pollution—both CO_2 and, especially, non-CO_2. Although CO_2 emissions from the use of natural gas are certainly less than those of coal, a sometimes-overlooked source of emissions occurs due to gas leaks that can release prodigious amounts of methane into the atmosphere. Like CO_2, methane is a greenhouse gas and, in fact, has 25 times the global warming impact on a per kilogram basis. According to the EPA's estimate, the leakage rate for natural gas is about 2.4%. This is below the critical value of 3.2%, above which natural gas would actually be a more potent source of global warming than coal!

The chief environmental concern regarding natural gas probably involves the practice of fracking and the resultant contamination of groundwater by chemical additives to the fluid injected into the oil or gas wells. Quite apart from chemical additives, wastewater from fracking is also often laced with highly corrosive salts, carcinogens such as benzene, and other naturally occurring radioactive elements, such as radium, that can naturally occur deep underground. One 2016 study by the EPA found specific instances where one or more mechanisms led to impacts on drinking water resources, including contamination of drinking water wells. The number of identified cases, however, was small compared to the number of hydraulically fractured wells (EPA, 2016).

Another recent study by MIT came to a similarly mixed conclusion, namely, that "the environmental impacts of shale development (by fracking) are challenging but manageable." It further concluded that "here has

RECONCILING CONFLICTING CLAIMS ABOUT FRACKING

The natural gas industry claims that no case of groundwater contamination caused by fracking has ever been documented, and yet, some environmentalists maintain that thousands of cases of groundwater contamination due to oil and gas drilling have been documented. Surprisingly, both of these claims could be true. There are many ways that groundwater can become contaminated as a result of drilling, including surface spills followed by percolation down to an aquifer—a much more likely route of contamination than fracking for which the induced fractures occur at depths of thousands of feet. Since fracking results in less vertical wells drilled than previously, its net impact actually could be less groundwater contamination.

been concern that these fractures can also penetrate shallow freshwater zones and contaminate them with fracturing fluid, but there is no evidence that this is occurring" (MIT, 2011b). Other university studies of the problems associated with water contamination are more ominous. For example, according to one 2011 study, the methane concentration in water samples from 68 wells near shale gas drilling were found to be at a dangerous level, based on US Department of Interior standards. In some cases, the levels are high enough for homeowners to ignite the methane contained in the water coming out of their faucets by putting a match near the water stream. Another report on fracking by the US Department of Energy (DOE) (2011) supports using hydraulic fracturing (fracking) but with a variety of safeguards and continuous monitoring of wells for emissions. Other specific measures cited in that DOE report include disclosure of the composition of fracking fluids and the prohibition of specific fluids (diesel fuel). Drilling gas or oil wells is an expensive proposition,

NATURAL GAS: FRIEND OR FOE TO RENEWABLE ENERGY SOURCES?

There are two schools of thought concerning the relationship between natural gas and renewable energy sources such as wind and solar. Some observers see natural gas as an important bridge fuel that can replace coal and, to some extent, oil, as a much cleaner alternative during the decades necessary for a complete switch to renewable sources. Other observers take the contrary view that abundant cheap natural gas will impede the transition to renewable sources. However, given the rate at which the costs of renewable sources are decreasing and the rate at which they are penetrating the marketplace, this concern may be unwarranted. Moreover, natural gas can be a power source that fills in the gaps left by variable renewable sources in order to provide steady electrical output to the grid. Thus, it offsets the variability of renewable sources, which will become increasingly important as the clean energy portfolio expands. One argument that some environmentalists give against the notion of natural gas as a bridge fuel is that it will probably stand in the way of limiting the global temperature rise to at most 2°C by the year 2100, which they view as a maximum level that will not be catastrophic. The merits of this argument will be discussed after the matter of climate change is fully considered in later chapters.

and some drilling companies have resorted to shoddy practices in their haste to begin extracting gas in the shortest possible time.
Another area of significant environmental impact associated with oil and gas is that of oil spills.

Many small oil spills occur on a regular basis and fail to be reported in the media, but when a large spectacular one occurs, it can absorb the public attention for a considerable time, as was the case in the 2010 BP Deepwater Horizon disaster. Large oil spills can devastate the wildlife in the affected area. For example, even after cleaning, probably less than 1% of oil-soaked birds survive, so the effort would appear to be primarily done for public relations purposes. The restoration of the ecosystem by cleanup efforts can be difficult and lengthy and may depend as much on human remediation efforts as on nature, including the weather, the ocean currents, and the presence of oil-consuming bacteria in the water. While the economic damage of a large oil spill may be relatively easy to assess, the full long-term impact on the ecosystem and to human health (especially among cleanup workers) may be much less so. While oil spills will remain a fact of life as long as the world relies on petroleum, it is however worth noting that while small oil spills occur on a daily basis, large spills have become less frequent with each passing decade since the 1970s (ITOPF, 2011).

The health and fatality costs associated with oil and gas pipelines can also be considerable. In 2011, for example, a gasoline pipeline exploded in Nairobi, Kenya, killing nearly 100 people. All things considered, while oil and natural gas are very far from environmentally benign, their impact on the environment is probably not nearly as bad as that of coal, particularly in the case of natural gas. This judgment should not be interpreted as a reason to stick with fossil fuels, since they all do have serious negative impacts on the environment—it is only a judgment regarding which impacts are worse among the three fossil fuels.

2.4 SUMMARY

This chapter considers the three fossil fuels, coal, oil, and gas, and their formation, uses, and especially their environmental consequences, which can be very detrimental—especially in the case of coal. It also considers a variety of controversial questions including whether coal, oil, and gas are really fossil fuels at all, whether there can be such a thing as clean coal, and whether natural gas, which may be cleaner than coal or oil, can really serve as a bridge fuel while we move toward renewable energy sources.

PROBLEMS

1. Starting from the definition for a small change in entropy $dS = dQ/T$, show that $W = \oint P\,dV = \oint T\,dS$.
2. Explain why the Carnot cycle is described by a rectangle in a T–S plot. Hint: Based on the definition of entropy, why must adiabatic processes be represented by vertical lines?
3. Explain physically what is happening for step 3 → 4 in Figure 2.8.
4. Explain this sentence: "Based on the values of the constants in Equation 2.1, the maximum energy density requires H to be as high as possible and O to be as low as possible, and the minimum energy requires the opposite."
5. Suppose that a coal-fired power plant burns lignite coal. Based on the energy content of lignite (see Table 2.1), which is one type of brown coal? Estimate the CO_2 emissions in kilograms per gigajoule and compare your result with the emission data of Table 2.2. Assume that all the carbon in the coal goes into CO_2, with only a negligible amount creating CO.
6. In Section 2.2.8, it is noted that since the industrial revolution, the average oceanic pH has decreased from 8.25 to 8.14. Show that this corresponds to an increase in oceanic acidity of nearly 30%. Note that based on its definition, the pH is 7.0 for neutral distilled water and that the percentage increases in acidity are relative to this neutral point.
7. It has been suggested that one way to sequester CO_2 removed during coal burning would be to store it on the deep ocean floor at depths greater than 2700 m. (a) Find the pressure in atmospheres at that depth, assuming a density of seawater of 1020 kg/m^3; (b) determine the state of the CO_2 (solid, liquid, or gas) under such pressure from some searching on the web, assuming a temperature of 280 K; and (c) find a density–pressure phase diagram on the web that allows you to estimate the density at that depth, so you can verify whether the density exceeds that of seawater.
8. The world's proven natural gas reserves are estimated to be 1.9×10^{14} m^3. Given that a ton of natural gas occupies a volume of 48,700 ft^3 at atmospheric pressure, how does the amount trapped in hydrates (estimated at 6.4 trillion tons) compare to the proven reserves?
9. Explain these two sentences from the text: "Today many geologists also believe that there is around a 40-year supply of oil at the present rate of consumption (production). This expectation is actually quite consistent with a 2011 peak, given a Gaussian shape having Hubbert's predicted width."
10. In a combined cycle gas-fired plant, does it matter which cycle has the higher efficiency? Which cycle might you expect in fact has the higher efficiency? What would be the analogous formula for overall efficiency for a triple cycle plant, in terms of the three efficiencies e_1, e_2, and e_3?

11. Consider a power plant having a fixed electrical power output of 1000 MW. Show that if the efficiency of the plant were to increase from 33% to 50%, the amount of rejected heat per megawatt generated is halved.
12. For a combined cycle (binary) power plant, show that if it is assumed that each cycle were a Carnot cycle, the overall efficiency is identical to what it would be for a single cycle having the same combustion and ambient temperatures.
13. The United States consumes about 400 million gallons of gasoline per day. Suppose that the nation's entire fleet of two million 18-wheeler tractor trailer trucks were converted to natural gas, by what amount would the fraction of US oil imports decrease? Assume that the US imports are 50% of all the oil it consumes and that the average 18-wheeler gets 6 miles per gallon and drives 60,000 miles per year.
14. If your view of fracking (hydraulic fracturing) is that it is too risky to be pursued, look up some sources that support this view and, in a one-page description, see if you can find any flaws in the arguments. Do the same if your view happens to be that fracking should be pursued.
15. Write a one-page analysis on the relative importance of political and economic factors behind our addiction to fossil fuels versus strictly technical issues, such as their rich energy density.
16. A coal fired power plant burns at 825 K and uses a reservoir at 300 K. What is the maximum efficiency of this power plant?
17. A gasoline-fueled automobile is burning at 400 K. What is the maximum efficiency of the internal combustion engine if the ambient temperature is 290 K?
18. How much coal is required for a 1 GW coal-burning electrical power plant? Assume it takes three times the thermal energy to produce the electrical energy and express your answer in kilograms per year.
19. A 750 MW power plant burns 500 t of coal every hour. If the energy content of its coal is 7500 Btu/lb., what is the efficiency of the power plant?

REFERENCES

Beddington, J. (2009) Food energy water and the climate: A perfect storm of global events? *The Guardian*. March 18, 2009, http://www. guardian.co.uk/science/2009/mar/18/perfect-storm-john-beddington-energy-food-climate.

Darrell, B. (2007) *Why Confusion Exists over When the Oil Peak Will Occur*. Feasta, Tipperary. January 12, 2007, http://www.feasta.org/2007/01/12/why-confusion-exists-over-when-the-oil-peak-will-occur/.

DOE (US Department of Energy) (2011) http://www.shalegas.energy.gov/resources/111811_final_report.pdf.

EEA (European Environment Agency) (2008) European Environment Agency (EEA) gives fuel-dependent emission factors based on actual emissions from power plants in EU. In *Air Pollution from Electricity-Generating Large Combustion Plants*, EEA, Copenhagen.

EIA (US Energy Information Administration) (2015) https://www.eia.gov/naturalgas/.

EPA (United States Environmental Protection Agency) (2004) Deadly power plants? Study fuels debate. *NBCNews.com*. June 9, 2004, http://www.msnbc.msn.com/id/5174391/.

EPA (United States Environmental Protection Agency) (2016) http://www.epa.gov/hfstudy.

Glasby, G. P. (2006) Abiogenic origin of hydrocarbons: An historical overview, *Resour. Geol.*, 56(1), 83–96.

Gold, T. (1999) *The Deep, Hot Biosphere: The Myth of Fossil Fuels*, Copernicus Books, New York.

Hardt, M. J., and C. Safina (2010) How acidification threatens oceans from the inside out. *Scientific American*. http://www.scientificamerican.com/article.cfm?id=threatening-ocean-life.

Hvistendahl, M. (2011) Coal ash is more radioactive than nuclear waste. *Scientific American*. December 13, 2007, https://www.scientificamerican.com/article/coal-ash-is-more-radioactive-than-nuclear-waste (Accessed March 18, 2011).

ITOPF (International Tanker Owners Pollution Federation) (2011) *Oil Tanker Spill Statistics 2016*. ITOPF, London. http://www.itopf.com/information-services/data-and-statistics/statistics/

MIT (Massachusetts Institute of Technology) (2011a) MIT, Cambridge, MA. http://blog.cleantechies.com/2011/06/22/liquefied-coal-may-become-an-economically-viable-fuel-option/

MIT (2011b) *The Future of Natural Gas: An Interdisciplinary MIT Study. MIT Energy Initiative 7, 8*, MIT, Cambridge, MA. http://web.mit.edu/mitei/research/studies/documents/natural-gas-2011/NaturalGas_Chapter%201_Context.pdf (Accessed July 19, 2011).

People's Daily Online (2005) Black lung disease claims 140,000 lives in China. March 18, 2005, http://english.peopledaily.com.cn/200503/18/eng20050318_177365.html.

Shanghai Star Newspaper (2004) Mining, China's most dangerous job. November 18, 2004, http://app1.chinadaily.com.cn/star/2004/1118/bz9-3.html.

Zhang, J., and K. Smith (2007) Household air pollution from coal and biomass fuels in China: Measurements, health impacts, and interventions, *Environ. Health Perspect.*, 115(6), 848–855.

Nuclear Power

Basic Science

CONTENTS

3.1 INTRODUCTION

Some books on renewable energy do not go into the subject of nuclear energy, but it is important to include it based on the need to compare renewable energy technologies with the other available energy sources, including nuclear. Moreover, some would argue that nuclear is in fact a form of renewable energy or, at least, a useful supplement to it. In this first of two chapters on nuclear energy, we consider the basic science, an understanding of which is essential to the technological issues considered in the following chapter. The chapter begins with a historical overview and then proceeds with the development of the basic science needed to understand nuclear energy; it also delves into the consideration of nuclear radiation, including its effects on humans.

3.2 EARLY YEARS

As with any new science, the early years of nuclear science were a period of confusion and accidental discovery. Although there were important contributions by many pioneers, we highlight here those by three individuals: Henri Becquerel, Marie Curie, and, especially, Ernest Rutherford. Antoine Henri Becquerel (who, along with Marie and Pierre Curie, was awarded the Nobel Prize in Physics in 1903) is generally acknowledged to be the discoverer of radioactivity. Becquerel's discovery was entirely accidental and occurred one day in 1896, while investigating phosphorescence in uranium salts (Becquerel, 1896). He happened to have placed some uranium salt above some photographic plates that were wrapped in very thick black paper to prevent light exposure. Becquerel found that the plates became fogged nevertheless. He also noted that

> If one places between the phosphorescent substance and the paper a piece of money or a metal screen pierced with a cut-out design, one sees the image of these objects appear on the negative.... One must conclude from these experiments that the phosphorescent substance in question emits rays which pass through the opaque paper and reduces silver salts. (Becquerel, 1896)

At the time of Becquerel's discovery, the nature of these *radioactive* emissions was completely unknown, as was their connection to the nucleus of the atom, whose existence would not be discovered for another decade. Becquerel's name is now attached to one important SI unit used in nuclear science, the becquerel (Bq), which is defined as one nuclear disintegration or one decay per second.

Marie Curie's important contributions to early nuclear science were her creation of a theory of the nature of radioactivity (a term she coined) and her realization that the phenomenon was due to the presence of several elements that were hitherto unknown. The first of these she named *polonium*, in honor of her native country Poland, and the second she called *radium*. Marie Curie collaborated with her husband Pierre, with whom she shared the Nobel Prize in Physics awarded in 1903. Remarkably, her daughter Irène Joliot-Curie later also shared a Nobel Prize with her husband Frédéric Joliot-Curie. Marie Curie (born Maria Skłodowska), who also won a Nobel Prize in Chemistry in 1911, was a truly remarkable woman and the first scientist to ever be awarded two Nobel Prizes. Marie Curie made her discoveries at the University of Paris, where she was the first female professor. Marie and Pierre's work on separating the element radium from the raw pitchblende that contained it involved difficult physical work conducted under unbelievably primitive conditions in a windowless unheated leaky shed (Figure 3.1).

MARIE CURIE'S OTHER LEGACY?

In most nations, the representation of females in physics is among the lowest in the sciences. For example, in the United States, around 18% of PhDs in physics were granted to women in 2012, according to data compiled by the American Institute of Physics (AIP). One AIP report looked at comparable statistics in 19 nations (AIP et al., 2001). Interestingly, the two nations where Marie Curie was born and did her great work (Poland and France, respectively) topped the list at numbers 2 and 1, with 23% and 27% women physics PhDs, respectively. Is this merely a coincidence?

RADIATION-INDUCED CANCER?

Cancers caused by radioactivity are no different from those that are spontaneously caused, so an unambiguous claim that Marie Curie died by a radiation-induced cancer cannot be made. However, it is also true that her work undoubtedly exposed her to very high levels of radiation, which would certainly significantly increase the probability of cancer. Moreover, throughout her adult life, she was in a constant state of ill health (Coppes-Zatinga, 1998). It is ironic that while ionizing radiation can cause cancer, it is also used in its treatment—a field that Marie Curie pioneered.

One indication of the difficulty of the work is the fact that a ton of raw pitchblende was needed to extract a mere one-tenth of a gram of radium chloride. Of course, in those early years, the dangers of radioactivity were not realized, a fact that later probably cost Marie Curie her life to what was likely cancer. In Curie's honor, we have the radioactivity unit curie (Ci), which is 37 billion nuclear decays per second or becquerels, the number roughly corresponding to the activity of 1 g of pure radium.

(a)

(b)

Figure 3.1 (a) Extraction of radium in the old shed where Marie and Pierre Curie first obtained the element. Photo is taken from Marie Curie's autobiographical notes. (From Curie, M. and Curie, P., *Autobiographical Notes*, Macmillan, New York, 1923; Courtesy of AIP Emilio Segrè Visual Archives, College Park, MD.) (b) Marie Curie (born Maria Salomea Skłodowska), Nobel Prize awardee in chemistry and physics.

3.3 DISCOVERY OF THE ATOMIC NUCLEUS

In the early years of the twentieth century, the concept of matter consisting of atoms corresponding to the various elements was reasonably well established based on arguments from chemistry, even though some scientists doubted the actual physical existence of atoms, and none understood their structure. Nevertheless, some physicists, including J. J. Thomson, did postulate models of the atom, most notably his so-called plum or raisin pudding model (Thomson, 1904). Thomson (1897) had previously discovered the negatively charged electron in 1897. Knowing that normally atoms were electrically neutral, he surmised that the electrons could be thought of as raisins in a static mass (the "pudding") of an equal amount of continuously distributed positive charge. This model had a number of attractive features, but only an experimental test could reveal whether it had any basis in reality. This task fell to the physicist Ernest Rutherford, who led an experiment that is the prototype of much experimental work conducted today to reveal the properties of the fundamental particles of nature.

Having received the Nobel Prize in Chemistry in 1908, Rutherford conducted even more groundbreaking work the following year. Together with graduate students Hans Geiger (of later Geiger counter fame) and Ernest Marsden, Rutherford carried out the famous experiment that demonstrated the nuclear nature of atoms. The basic idea of the experiment is quite simple. Rutherford sought to probe the structure of the atom by using a collimated (directed) beam of particles fired at a thin sheet of material. Arranging to have a collimated beam was easy—by simply having a small hole in a thick lead container containing some radioactive radium. The so-called alpha particles that the radium emitted would then be reasonably well collimated, since only those alphas able to pass out of the narrow hole would escape the container. Rutherford chose gold as the atom to probe simply because a piece of gold foil could easily be made very thin (only a few atoms thick), which was essential so that the beam of alpha particles would usually encounter only one gold atom in close proximity in passing through the sheet (Figure 3.2).

Rutherford had established earlier that the electrical charge of the alpha particles is +2e (i.e., twice the charge of the electron in magnitude and opposite in sign), and its mass was roughly 4000 times greater. Alpha particles are now known to be the nuclei of helium atoms, which, of course, could not be known to Rutherford before he discovered the nucleus!

Rutherford wanted to observe how often a beam of alpha particles would be scattered through different angles when encountering gold atoms, and he planned to do this by simply counting the numbers of alphas deflected through different angles. In an age when no modern radiation detectors existed, measuring the angles along which deflected alpha particles traveled was challenging—certainly to the eyesight of his students Geiger and Marsden! Rutherford had earlier developed zinc sulfide scintillation screens, and he used them to detect the deflection angle when an alpha struck the screen placed at a given point and caused a brief flash of light there. What did Rutherford expect to find? Given the large mass of the alpha particles and their high speed, he expected that the vast majority of alphas would be deflected through very small angles by the electrical (Coulomb) force between an alpha and the nearest atom it encounters. In fact, to a first approximation on the basis of the Thomson raisin pudding

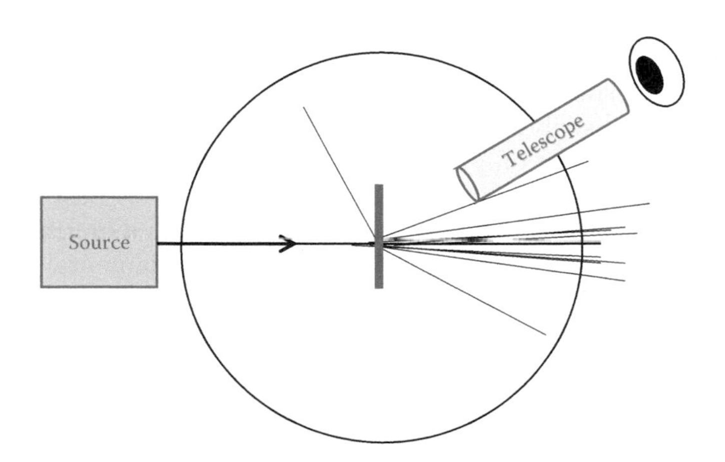

Figure 3.2 Drawing of the apparatus used in the Rutherford experiment. The telescope now making an angle of about 30° with the incident beam of alpha particles from a radioactive source is rotated about a vertical axis to count how many alphas are scattered through different angles after the beam strikes the thin gold foil target at the center of the apparatus.

model, the deflection force would be almost zero, since the atom as a whole is electrically neutral, and its positive charge is diffuse.

Day after day, Geiger and Marsden counted the numbers of flashes that they saw at various angles of deflection, and their observations confirmed Rutherford's expectation that the vast majority would be at very small angles. However, there was one strange anomaly in the data. Some alphas (albeit only 1 in 8000) were found to be deflected by very large angles (over 90°). In fact, a very tiny percentage of alphas were almost deflected through 180°, i.e., directly backward. Table 3.1 shows the number of counts found at various angles.

The fractional numbers for the numbers of counts for some angles appear in the original paper. This seeming impossibility reflects the fact that for angles greater than 90°, longer periods had to be observed in order to obtain statistically significant numbers, and fractional numbers of counts result when adjusting for different counting periods.

Rutherford, upon learning of Geiger and Marsden's observations that some counts were found at very large angles, was quoted as saying

> It was the most incredible event that ever happened to me in my life. It was almost as if you fired a 15-inch [cannon] shell at a piece of tissue paper and it came back at you. (Cassidy et al., 2002)

Rutherford realized that the explanation of this strange anomaly in the data was the existence in the atom of a small nucleus, which contained most of its mass. In that case, the tiny massive nucleus would be capable of occasionally deflecting alpha particles backward in the event that they were heading directly toward it. The rarity of these backward or near-backward deflections implied that the nucleus had an extremely small size compared to the atom itself. The usual description of Rutherford's discovery of the atomic nucleus ends here, but it does not do justice to Rutherford's magnificent achievement. Any scientist, if he or she is lucky, can observe an anomaly in the data and formulate a new revolutionary theory based on it, but only a great scientist will take the next step and rule out alternative theories by showing that the data fully support the new theory in all their quantitative detail.

Table 3.1 Data Recorded by Geiger and Marsden for Alpha Particle Scattering off a Gold Foil Showing the Number of Counts Recorded in 1° Intervals at Various Angles from 15° to 150°

Angle	150	135	120	105	75	60	45	37.5	30	22.5	15
Counts	33.1	43	51.9	69.5	211	477	1,435	3,300	7,800	27,300	132,000

Source: Geiger, H. and Marsden, E., The laws of deflexion (sic) of α particles through large angles, 25, 610, 1913, http://www.chemteam.info/Chem-History/GeigerMarsden-1913/GeigerMarsden-1913.html.

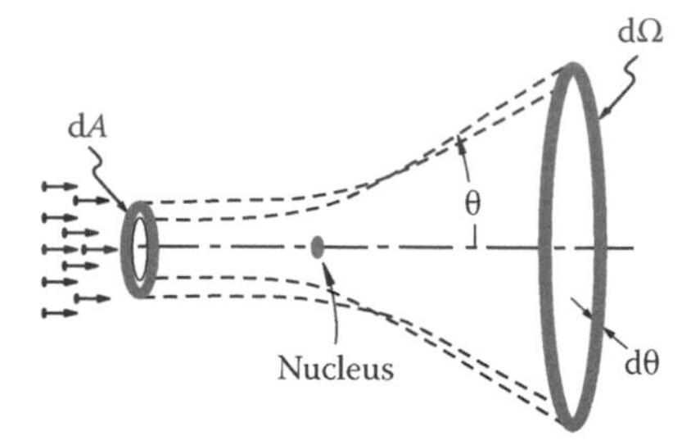

Figure 3.3 Rutherford experiment data and theory. The points are the number of counts observed in the experiment versus the scattering angle, and the smooth curve is a plot of the choice of $C \sin^{-4}(\theta/2)$, where C is a constant. The good agreement with the data for the appropriate choice of C is quite evident.

3.4 MATHEMATICAL DETAILS OF THE RUTHERFORD SCATTERING EXPERIMENT

Rutherford sought to explain the exact angular distribution of the alpha particles that his students Geiger and Marsden had recorded on the assumption that there exists a tiny massive nucleus at the center of the atom. The mathematics of this section is somewhat more challenging than most sections in this book, and some readers may wish to skip it on first reading, focusing only on the result of Rutherford's derivation for the number of particles scattered through different angles (Figure 3.3).

CONCEPT OF THE SOLID ANGLE

The analog of an angle in three dimensions is known as a solid angle, and it is measured in steradians, rather than radians. Recall that the basic definition of an angle in radians is the length of an arc along a unit circle surrounding a point. The corresponding definition of a solid angle, universally represented by the symbol Ω, is the amount of area on a unit sphere surrounding a point. Obviously, the largest possible solid angle would be $\Omega = 4\pi$.

Rutherford assumed that individual alpha particles are deflected through different angles strictly based on their impact parameter b, defined as the perpendicular distance from the x-axis to the incident alpha particle velocity vector when the particle is very far from the target. Thus, alphas that headed directly toward a nucleus (having $b = 0$) would be deflected through 180°, while those that had a large impact parameter would be deflected through a very small angle. Using classical mechanics and an inverse square Coulomb force, Rutherford was able to easily deduce the relationship between impact parameter b and scattering angle θ as (Goldstein et al., 2000)

$$b = \frac{kq_1q_2}{2E\tan(\theta/2)}, \tag{3.1}$$

where E is the kinetic energy of the alpha particle; $k = 9 \times 10^9$ N m^2/C^2 is the Coulomb force constant; and the respective charges of the alpha particle and nucleus are $q_1 = +2e$ and $q_2 = +79e$.

In deriving Equation 3.1, Rutherford assumed that the alpha particles in encountering an atom experience a force almost exclusively due to the positively charged nucleus that is of sufficiently small size, so that when the alphas are inside a spherical cloud of many electrons, they will exert no force on the alphas. Rutherford, by further assuming a random distribution of impact parameters, was able to deduce the fraction of particles scattered through each angle. In modern parlance, this is written in terms of the differential cross section:

$$dA = \sigma(\theta) = 2\pi b\, db, \tag{3.2}$$

which represents the area (of a ring) surrounding a target nucleus (the scattering center) that an incoming projectile needs to pass through to be deflected (scattered) by an angle θ or, more exactly, into the interval from θ to θ + dθ.

Figure 3.4 illustrates the concept of differential cross section for particles scattered into a small angular range dθ or in three dimensions into a solid angle range dΩ = (2π sin θ)dθ. The number of particles, scattered into a small interval of solid angle, is given by

$$dN = N_0\sigma(\theta)d\Omega = N_0\sigma(\theta)(2\pi\sin\theta)d\theta, \tag{3.3}$$

where N_0 is the incident intensity (number of particles per unit area per unit time).

Even when it is inappropriate to imagine particles traveling along trajectories (as in quantum mechanics), one can still define a measured cross section for scattering from Equation 3.3 by using

$$\sigma(\theta) = \frac{1}{N_0}\frac{dN}{d\Omega}. \tag{3.4}$$

However, recall that so far, we have assumed a single target nucleus. In general, for a foil having N_t nuclei within the area of the beam S, we have

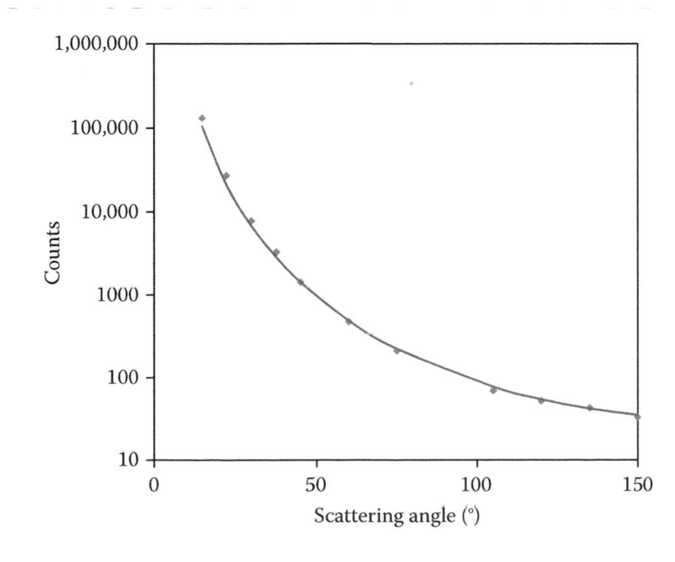

Figure 3.4 Collimated beam of particles incident on the ring area d*A* is scattered into a differential solid angle dΩ defined by this diagram.

for the actual number of particles dN scattered into a given solid angle range:

$$dN = N_t N_0 \sigma(\theta) d\Omega. \tag{3.5}$$

The only unfinished business is finding the number of target nuclei N_t in terms of known quantities. It can easily be shown using dimensional analysis that the number of target nuclei in the foil lying within the area of the beam can be expressed in terms of the density of the foil ρ, its thickness d, Avogadro's number N_A, the area S, and the atomic weight of the material A, i.e.,

$$N_t = \rho d N_A \frac{S}{A}. \tag{3.6}$$

Note that Equations 3.2 through 3.6 apply to any force, but Equation 3.1 applies only to the inverse square force. Using Equation 3.1, and the conservation laws of classical mechanics, it can be shown that (Goldstein et al., 2000)

$$\sigma(\theta) = \frac{dN}{d\Omega} = \left(\frac{kq_1q_2}{4E}\right)^2 \frac{1}{\sin^4(\theta/2)}. \tag{3.7}$$

The essential point of Equation 3.7 is that the number of alphas scattered per unit solid angle is proportional to the negative fourth power of the sine of half the scattering angle. Thus, for example, the number of particles through an angle of 60° should be 16 times greater than the number scattered through 180°. Thus, Rutherford had perfectly explained the alpha scattering data, which makes his claim of a nucleus to the atom not just an explanation that qualitatively fits an anomaly in the data (presence of some alphas deflected through large angles) but a detailed quantitative description.

3.4.1 Example 1: Setting an Upper Limit to the Nuclear Size

Using the data recorded by Geiger and Marsden and the dominant energy of alpha particles emitted by radium (4.75 MeV), determine the experimental upper limit that Rutherford was able to set for the radius of the gold nucleus. Compare this value with the actual radii of (1) the gold nucleus and (2) the gold atom. Note that 1 eV – 1.6×10^{-19} J.

Solution

In order for the Rutherford scattering formula to fit the data even for scattering angles approaching 180°, the size of the nucleus would need to be smaller than the distance of closest approach r. At its closest distance (for a 180° scattering), the initial kinetic energy E of the alpha particle would be entirely converted to electrostatic potential energy (since it is momentarily brought to rest), so that

$$E = \frac{kq_1q_2}{r}. \tag{3.8}$$

Using $E = 4.75$ MeV, $q_1 = 2e$, and $q_2 = 79e$, we find $r = 4.79 \times 10^{-14}\ m = 47.9$ fm. According to data on the web, the true nuclear radius for gold is now known to be 7.3 fm, while that of a gold atom is 0.144 nm, making Rutherford's upper limit to the nuclear radius about 1/3000 the size of the gold atom—a truly tiny object.

In addition to the differential cross section that we have considered at length, one can also define the total cross section for any process by merely integrating overall angles:

$$\sigma_{TOT} = \int_0^{4\pi} \frac{d\sigma}{d\Omega} d\Omega. \tag{3.9}$$

Note that the total cross section may be thought of as the effective size (cross-sectional area) of the target nucleus for any impact parameter. However, the total cross section, in general, can be different for different incident particles, and the meaning of the total cross section is not limited to the problem of elastic scattering, but it can be applied to any nuclear process induced by some projectile. The cross section is a measure of the probability that incident particles will cause that process to occur.

3.5 COMPOSITION AND STRUCTURE OF THE ATOM AND ITS NUCLEUS

Having established the existence of a tiny massive nucleus to the atom, Rutherford went on in 1911 to postulate his planetary model of the atom, whereby electrons orbit the nucleus, much like a miniature solar system (Rutherford, 1911). Two years later, Niels Bohr introduced his own model incorporating some new radical elements into the planetary model. These radical elements included quantum jumps between so-called stationary states (Bohr, 1913). Bohr's model, still taught in most introductory physics courses, was an important bridge on the road to a full understanding of atomic structure, based on quantum mechanics.

In the meantime, Rutherford and others continued their work on the structure and composition of the atomic nucleus. In 1919, Rutherford discovered that he could change (transmute) one element into another by bombarding it with alpha particles. In subsequent experiments, Rutherford and others found that often during these nuclear transmutations, hydrogen nuclei were emitted. Clearly, the hydrogen nucleus (now known as the proton) played a fundamental role in nuclear structure. By comparing nuclear masses to their charges, physicists realized that the nuclear positive charge could be accounted for by an integer number of these protons. Ernest Rutherford in 1920 then postulated that there were neutral particles in the nucleus of atoms (now known as neutrons), which he thought of as electrons bound to protons (Rutherford, 1921). The need for these neutral particles having about the same mass as protons was the observed disparity between the atomic number (the charge Z) and the atomic mass A, which was often twice the former for many elements. It was not until 1932 that James Chadwick was actually able to detect Rutherford's neutron and confirm its existence (Chadwick, 1932).

With the experimental discovery of the neutron, the constituents of the atom were now apparently complete: Z electrons outside a nucleus containing most of the atom's mass, with the nucleus consisting of Z protons plus $N = A - Z$ neutrons.* A given element is characterized by the atomic number Z, which determines its chemical properties, and various isotopes of that element have different numbers of neutrons, or different A-values. Neutrons and protons shared many characteristics, including having the same mass (to within 0.1%) and the same spin, and hence, they are collectively known as *nucleons*.

3.6 NUCLEAR RADII

Recall that Rutherford was able to set an upper limit to the radius of the gold nucleus, based on his scattering experiment. By using alpha particles of somewhat higher energies, which could approach the nucleus even closer, it is possible to actually measure the size of the nucleus and not merely set an upper limit. However, in practice, one usually uses electrons, rather than alpha particles in these experiments, since electrons have a strictly electromagnetic nuclear interaction, which is very well understood, and they do not feel the strong force. Such experiments have been conducted for many different target nuclei, and they show a striking regularity. For all the nuclei whose radii have been measured, the following simple dependence on mass number A holds:

$$r = r_0 A^{1/3}, \tag{3.10}$$

where r_0 = 1.25 fm, although admittedly, it is a bit of a simplification to regard the nucleus as having a sharp well-defined surface. To understand the significance of this basic formula, simply calculate the volume of a spherical nucleus from its radius (Equation 3.10), and you will see that the volume is proportional to A. What does this fact imply?

* We ignore here the fact that neutrons and protons are now known not to be fundamental particles (like electrons), but are themselves made of quarks. There are many good popular-level books about quark theory, and the still more current theory of everything known as string theory.

Note that this behavior (volume proportional to the number of particles) is quite different from the atom outside the nucleus, since the volume of an atom is certainly not proportional to the number of electrons it contains. Thus, unlike electrons, nucleons seem to behave like incompressible objects that are packed together in close proximity. Another way to express the situation is to note that all nuclei have precisely the same density, which, using Equation 3.10, is found to be the astonishing value of 2×10^{17} kg/m³ or 200 trillion times that of water. Does matter of such density exist anywhere in the universe, apart from the nucleus itself? The surprising answer is yes, inside of the strange astronomical objects known as neutron stars, which are the remnants of stars that have undergone supernova explosions toward the end of their lives.*

BASIC FACTS ABOUT NUCLEAR ISOTOPES

1. All isotopes of an element have the same number of protons Z, i.e., the same chemical identity. However, they have different numbers of neutrons and are distinguished by their atomic mass numbers A, e.g., $^{235}_{92}U$ or $^{238}_{92}U$, the two isotopes of uranium, the element with $Z = 92$ and $A = 235$ or 238.
2. Some isotopes are stable and some unstable (radioactive), but some stable isotopes just have half-lives too long to be observed—a half-life being the amount of time $\tau_{1/2}$ for half the original number of radioactive nuclei to decay.
3. Some elements have many stable isotopes; e.g., tin and xenon each have the most at 9, while other elements have none, e.g., uranium. As of 2010, the isotope with the longest half-life yet known is tellurium-128, with a half-life of $\tau_{1/2} = 8 \times 10^{24}$ years; the one with the shortest half-life is beryllium-13, whose half-life is $\tau_{1/2} = 2.7 \times 10^{-21}$ s.
4. Isotopes can be separated only by physical (not chemical) means that are sensitive to small nuclear mass differences.

3.7 NUCLEAR FORCES

One issue that Rutherford and other nuclear scientists of his day wrestled with is the question of what holds the nucleus together? If only forces of an electromagnetic nature were present, clearly the positively charged protons would repel each other, so that no assemblage of protons could stably exist, even with the presence of neutrons that do not "feel" the electromagnetic force. Clearly, some attractive force must be present that is strong enough to overcome that Coulomb repulsion. This additional force has been given the unimaginative name of *strong force*. Table 3.2

* A teaspoonful of nuclear matter would weigh 10 billion tons on Earth. Finding a material to make the spoon out of would be quite a challenge!

Table 3.2 **Comparison between the Four Fundamental Forces**

	Strong Force	Coulomb Force	Weak Force	Gravitational Force
Strength	1	1/137	10^{-13}	10^{-40}
Range	Around 1 fm	Infinite (~1/r2)	Around 0.01 fm	Infinite (~1/r2)
Sign	Always attractive	Attractive for opposite sign charges	Repulsive	Always attractive
Felt by	Nucleons	Any charged particle	Any particle	Any mass
Mediated by	Gluons	Photons	*W* and *Z* bosons	Gravitons

summarizes the nature of the strong force in comparison with the more familiar Coulomb and gravitational forces and the less familiar weak force.

The strength of 1 for the strong force is an arbitrary choice. Note the extreme weakness of the gravitational force compared to all the other three, which is why gravity is of no significance when considering nuclear reactions. There is one other fundamental force known as the weak force, which does come into play inside the nucleus but, like the strong force, has no role outside it given its short range. In particle physics, all forces are assumed to be mediated by exchanged particles. Thus, as indicated in Table 3.2, the force of an electron on another electron is due to photons exchanged between them. The exchanged particles, however, are not observed, and are referred to as "virtual" particles in contrast to "real" particles that are observed in detectors. In recent years, good arguments have been presented to show that at sufficiently high energies, all the four fundamental forces become unified.

3.8 IONIZING RADIATION AND NUCLEAR TRANSFORMATIONS

Radioactive nuclei by definition emit radiation when they decay. Often, this radiation can penetrate matter and leave a trail of ionization, hence the term *ionizing radiation*, which is preferred to the looser term *nuclear radiation*. In fact, not all ionizing radiation, e.g., X-rays, emanates from the nucleus, and not all radiation that emanates from the nucleus, e.g., neutrinos, is ionizing. Although many forms of radiation can be harmful to biological organisms if the dose is high enough, ionizing radiation can be especially harmful. It is not simply its penetrating power (radio waves are also quite penetrating), but rather the cell damage associated with the trail of ions left in the wake of the radiation.

Three common types of ionizing radiation are known as alpha, beta, and gamma. As already noted, an alpha particle is a helium nucleus, which has $A = 4$ and $Z = 2$. Therefore, after a parent nucleus (A, Z) decays by emitting an alpha particle, its daughter nucleus has atomic mass $A = 4$ and $Z = 2$, as illustrated in the case of the decay of the isotope $^{238}_{92}U$, which we would write as $^{238}_{92}U \rightarrow {}^{234}_{90}Th + {}^{4}_{2}He$. Beta rays are either electrons or their positively charged counterparts (known as positrons). During beta-plus nuclear emission, a proton transforms to a neutron and a positron and a third particle, the ghostly electron neutrino,* according to the reaction $p \rightarrow n + e^+ + \nu$, while in the beta-minus case, it is a neutron that gets transformed into a proton, electron, and antineutrino, according to the reaction $n \rightarrow p + e^- + \bar{\nu}$. Note that in both cases the identity of the nucleus containing the transformed n or p must change, since its atomic number changes by $\Delta Z = \pm 1$. Gamma rays also originate from nuclei when they undergo a change of energy level, but no change in their identity, i.e., their A or Z value.

You may be wondering under which conditions a proton is transformed into a neutron and when the reverse occurs in the cases of beta$^\pm$ decays. Nuclei spontaneously tend to transform themselves from less stable states to more stable states. In general, the most stable nuclei having a given mass number A tend to have a specific neutron–proton ratio, which is 50/50 ($N = Z$) for light nuclei, i.e., up to around $Z = 20$, and favoring neutrons ($N > Z$) to an increasing degree for heavier nuclei. These most stable nuclei lie along the "valley of stability"—the red (black) region in Figure 3.5. Beta-plus decay (changing a proton into a neutron) occurs when a particular isotope is above and to the left of the valley of stability, as they are too proton-rich, and conversely beta-minus decay (changing a neutron into a proton) occurs when a particular isotope is below or to the right of it.

The nuclei that are stable are represented by the dark dots in Figure 3.5, and they lie along a curve known as the valley of stability. As one moves away from the valley on either side, the nuclei get less and less stable (shorter half-lives). These gray regions are where the beta-plus and beta-minus emitters lie. Very massive nuclei at the upper end of the valley tend to be alpha emitters, as this process allows them to shed mass very efficiently. The reader may have wondered why it is that up to around $Z = 20$, the nuclei tend to have equal numbers of neutrons and protons, i.e., $N = Z$, while for $Z > 20$, neutrons are increasingly favored over protons. The reason is that as with electrons in an atom, neutrons and protons in a nucleus fill a set of energy levels starting with some lowest level, but each of these particles fills its own set of levels (two to each level just like electrons). Thus, suppose there were 10 protons in a nucleus but only six neutrons; in this case, it would be energetically favored for two of the protons to convert to neutrons and fill the lowest vacant neutron levels, giving rise to equal numbers of neutrons and protons. However, the situation changes above $Z = 20$, because the repulsion protons feel for one another becomes increasingly important as Z increases. The reason is because the number of proton–proton interactions varies as $Z(Z - 1)/2 < Z^2$, while the nearest-neighbor short-range strong attraction varies only as Z. The increasing importance of proton–proton repulsion over attraction as their number Z increases has the effect of raising the proton energy levels over those of neutrons.

* At this very moment, there are trillions of neutrinos passing through your body each second, but they cause no harm whatsoever, because neutrinos interact with matter so weakly. Detecting neutrinos requires a truly massive detector, because of their weak interaction.

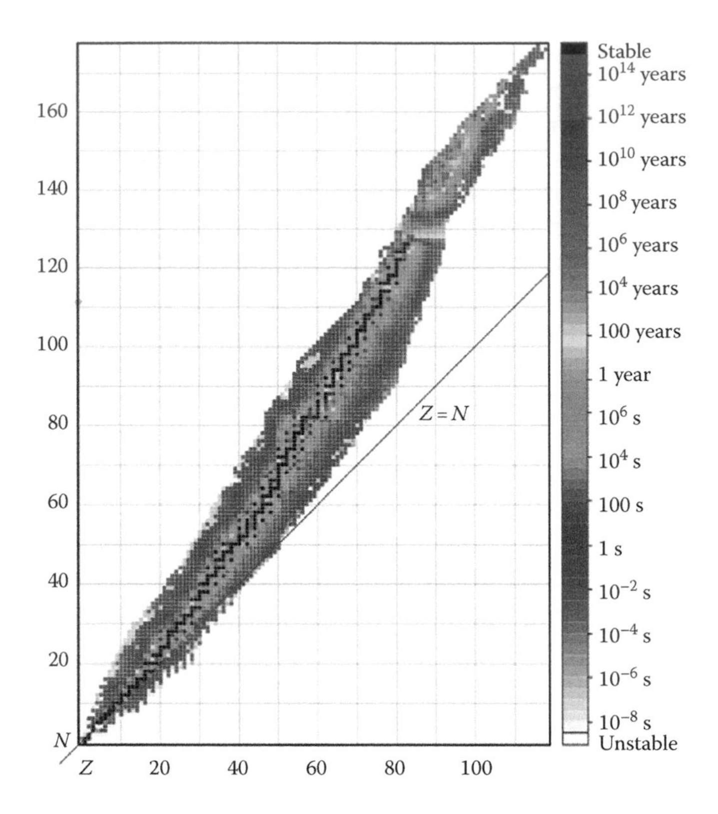

Figure 3.5 Plot of number of neutrons *N* versus number of protons *Z* for all known nuclei, which have been coded by their lifetime. (Courtesy of BenRG, http://commons.wikimedia.org/wiki/File:Isotopes_and_half-life.svg.)

3.9 NUCLEAR MASS AND ENERGY

In any of the three decay processes so far discussed, the amount of energy that is released when the parent nucleus decays is enormous—about a million times that of chemical processes on a per atom or per kilogram basis. Yet despite the enormous energy, Rutherford, the father of nuclear science, is famously reputed to have said about his nuclear studies, "… anyone who expects a source of power from the transformation of these atoms is talking moonshine" (Hendee et al., 2002). Remarkably, his comment was made in 1933, on the verge of the discovery of nuclear fission. However, Rutherford's failure to appreciate the practical impact of his work is understandable, because the types of processes known to him would indeed not be appropriate for generating large amounts of power. In order to understand the origin of the energy released in nuclear reactions, we need to consider Einstein's 1905 special theory of relativity and what is perhaps the most famous equation ever written: $E = mc^2$. One way to interpret this equation is to note that mass (m) is a form of energy (E), and the conversion factor between mass (in kilograms) and energy (in joules) is the quantity c^2, with c being the speed of light, 3×10^8 m/s.* An equivalent way to understand $E = mc^2$ is to note that since mass is just a form of energy, in any reaction in which energy E is released, the net mass of the reactants must decrease by an amount $m = E/c^2$. Given the enormous size of the quantity c^2, such mass changes will be noticeable only in cases where a truly prodigious amount of energy is released,

* Were it possible to somehow convert 1 kg of stuff entirely into energy, the amount available would be 9×10^{16} J or enough to supply all of New York City's electricity for nearly 2 years.

e.g., in nuclear processes. In such processes, the total number of nucleons always stays constant; however, despite this fact, the mass of the system does change in accordance with $E = mc^2$ because of differences between the binding energies of the initial and final states.

3.10 NUCLEAR BINDING ENERGY

The strong force holds the nucleus together against the much weaker Coulomb repulsion (between protons), so it should not be surprising that it would require a massive amount of energy to disassemble a nucleus into its constituent nucleons. The energy of total disassembly represents the binding energy of the nucleus, which simply equals the difference in mass between all the constituents (Z protons and N neutrons) and the original nucleus times c^2, or as in the following equation:

$$E_B = (Zm_P + Nm_n - M)c^2. \quad (3.11)$$

Consider a plot of E_B/A (binding energy per nucleon), shown in Figure 3.6.

Nuclei having the largest values of E_B/A tend to be the most stable, and hence, they are the most tightly bound. These nuclei also have the smallest nuclear mass in relation to that of the constituents by virtue of Equation 3.11. According to Figure 3.6, the most stable nucleus is iron (Fe^{56}). Many students are confused by the sign of the binding energy. Obviously, it must be positive based on its definition as the work needed to disassemble a bound system (the nucleus). However, the potential energy responsible for binding the system together is negative earlier—see Figure 3.7.

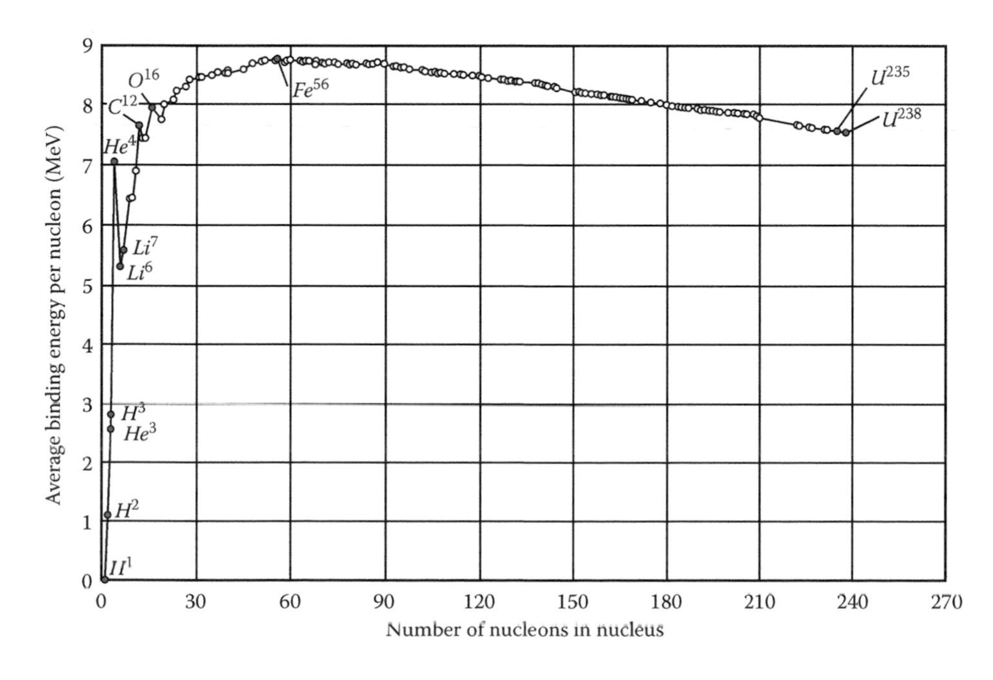

Figure 3.6 Curve of nuclear binding energy per nucleon, or E_B/A versus number of nucleons A. (Courtesy of Mononomic, http://en.wikipedia.org/wiki/File:Binding_energy_curve_-_common_isotopes_with_gridlines.svg.)

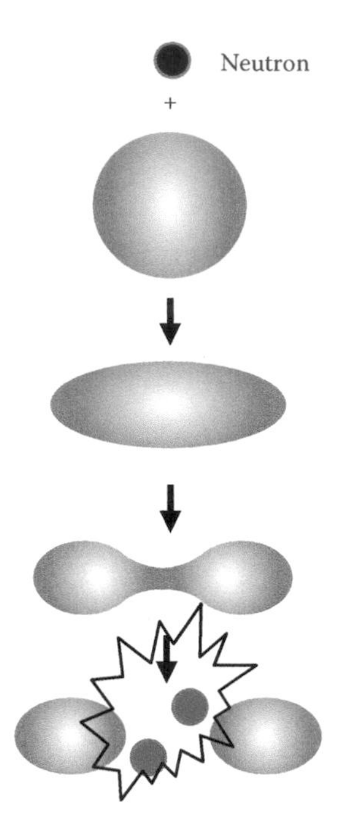

Figure 3.7 Potential energy $V(r)$ as a function of r for one deuteron approaching another shown with a thick curve. At distances greater than a few femtometers (10^{-15} m), $V(r)$ has the form of the Coulomb repulsive potential $1/r$, but at shorter distances (inside the potential well), the potential is dominated by the strong attractive potential—represented by the steep walls. If the approaching deuterons are described by classical mechanics, refer to right half of figure; if they obey quantum mechanics, refer to the left half.

3.11 ENERGY RELEASED IN NUCLEAR FUSION

The shape of the curve of binding energy suggests a way of extracting nuclear energy during two types of processes: fission and fusion. Very heavy nuclei have less binding energy per nucleon than those closer to iron, and therefore, were a heavy nucleus such as uranium to split (fission) into two lighter ones, the combined mass of the two lighter ones would be less than the original parent nucleus, with the mass loss converted into the released energy. In a similar manner, if two light nuclei were to combine (fuse), energy would also be released by exactly the same argument. To illustrate, consider the d–t fusion reaction, where d and t stand for the hydrogen isotopes known as deuterium and tritium, respectively, which are also often written as ^{2}H and ^{3}H. The d–t reaction can be written as ^{2}H + ^{3}He + ^{1}n, where ^{1}n is a neutron. Given the known respective binding energies of the initial nuclei, i.e., 2.2 and 8.5 MeV, and the final nuclei, i.e., 28.3 and 0 MeV, we find that the reduction in binding energy is 17.6 MeV, so that the mass lost in the reaction is 17.6 MeV/c^2, and hence, the energy released is 17.6 MeV. Note that it is convenient here to consider the c^2 as simply being part of the units of mass, i.e., MeV/c^2.

3.11.1 Example 2: Estimating the Energy Released in Fusion

How does the energy released in this hydrogen fusion reaction compare with the ordinary burning of hydrogen?

Solution

If 1 kg of hydrogen is burned, the energy released is 130 MJ, but the energy equivalent of the original 1 kg by $E = mc^2$ is $9 \times 10^{16} = 9 \times 10^{10}$ MJ. Thus, the fractional change is a mere 1.5×10^{-7}%. In contrast, consider the d–t fusion reaction where 17.6 MeV is liberated. The original two nuclei have a combined mass of $A = 5$, whose energy equivalent is approximately 5 × 938 MeV = 4690 MeV, for a percentage change of 0.38%. A mass decrease of 0.38% may not sound like a lot, but it is roughly 2.5 million times more energy released (per reaction) than for ordinary hydrogen combustion!

3.12 MECHANICS OF NUCLEAR FISSION

The fission of a heavy nucleus into two lighter ones can either take place spontaneously, as in the case of $^{238}_{92}$U, but with an extremely long half-life (4.5 billion years), or it can be induced, usually with the absorption of a neutron. Neutrons are especially effective at inducing a nucleus to fission, because unlike positively charged protons, they can easily penetrate the nucleus unhindered by any Coulomb repulsion, and unlike electrons, they feel the strong nuclear attractive force. An artist's conception of a neutron-induced fission of a heavy nucleus is illustrated in Figure 3.8.

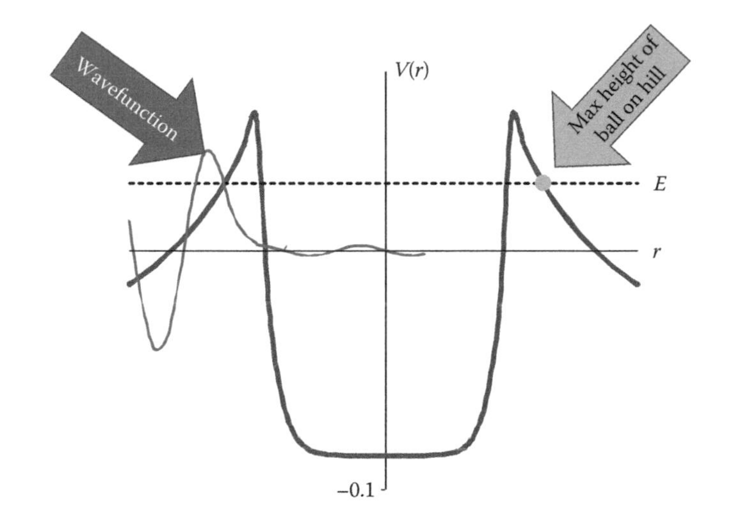

Figure 3.8 Time sequence of events leading to the fission of a large nucleus induced by the absorption of a neutron and ending with the formation of two daughter nuclei and two neutrons, and the release of energy. This drawing hints at how during an intermediate stage of the process, the parent nucleus undergoes oscillations, forming a dumbbell shape prior to the actual fission. In practice, the two fission fragments (or the daughters) tend to be of unequal sizes, unlike what is shown in the drawing.

The oscillations of the dumbbell shape—with the short dimension first oriented horizontally (not shown) and then vertically—are very much like the oscillations that can actually occur in a liquid drop in a weightless environment such as a space shuttle. One way to understand the source of the energy released is to consider that when the dumbbell is most elongated during its oscillation, the Coulomb repulsion between the two pieces drives them apart and wins out over the very short-range strong force, causing a complete rupture, followed by an acceleration of the pieces due to the Coulomb repulsion between them.

The emission of two or three neutrons following a nuclear fission is required by the fact that heavier nuclei tend to have a greater percentage of neutrons than lighter ones. Thus, when they fission into two fragments, these nuclei will tend to be too neutron rich and will in very short order emit neutrons to reach a more stable nucleus.

3.12.1 Example 3: Estimating the Energy Released in Fission

Consider the spontaneous fission of a $^{238}_{92}U$ nucleus. Imagine that it simply fissions into two equal mass fission fragments. Use Figure 3.6 to estimate the energy released in such an event and compare your estimate with the usually reported value. Hint: A straight-line approximation to the curve in Figure 3.6 between $A = 90$ and $A = 240$ would approximately go from an E_B/A value of 8.7–7.6.

Solution

Using the values given in the hint, we see that E_B/A changes by 1.1 MeV when A changes by 150 units. Given an assumed constant slope over this portion of the curve if the $^{238}_{92}U$ nucleus splits into equal-size pieces having $A = 119$, the change from the original A is also 119, so that the E_B/A rise during fission would be $(1.1 \times 119)/150 = 0.87$ MeV per nucleon. Given a total of 238 nucleons in the original nucleus, we find an increase in binding energy of 0.87 × 238 = 207 MeV when the $^{238}_{92}U$ splits; 207 MeV is therefore also the estimated energy released, which is very close to the usual estimate during $^{238}_{92}U$ fission.

3.13 MECHANICS OF NUCLEAR FUSION

This emission of neutrons accompanying fission is extremely important, because it makes the concept of a chain reaction possible and is the key to generating large amounts of nuclear energy. Claims of cold fusion aside, nuclear fusion unlike fission cannot be initiated without heating the atoms to be fused to an extremely high temperature, comparable to that at the center of the sun.

The need for very high temperatures to initiate fusion follows from the Coulomb repulsion between the positively charged nuclei that you seek to fuse. This repulsion can only be overcome if the nuclei collide with sufficiently high speed and energy. Let us consider the d–d fusion reaction $^2H + ^2H \rightarrow ^4He^*$, where $^4He^*$ represents an extremely short-lived nucleus that decays virtually instantly into one of three pathways. When one deuterium nucleus approaches the other head on from afar, it sees the potential energy function shown in Figure 3.7.

COLD FUSION?

In 1989, electrochemists Martin Fleischmann and Stanley Pons announced to the world that they had a tabletop method of producing nuclear fusion at close to room temperature (Fleischmann and Pons, 1989). The experiment involved electrolysis of heavy water on a palladium electrode, and their claim was based on (1) the anomalous heat production (excess heat) and (2) the observation of small amounts of nuclear by-products, including neutrons and tritium. There are many theoretical reasons for disbelieving this claim, and it later transpired that there was no convincing evidence for nuclear reaction by-products allegedly produced. Following a number of attempts by others to confirm these results—some positive, and some negative, the DOE (1989) convened a panel that same year to review the work. A majority of the panel found that the evidence for the discovery of a new nuclear process was not persuasive. Moreover, a second 2004 DOE review panel reached conclusions similar to the first (DOE, 2004). Cold fusion has now been renamed by the true believers as low-energy nuclear reactions or condensed matter nuclear science, to escape the disrepute the field is held by most physicists.

This potential energy graph (shown with a thick curve) is symmetrically plotted around $r = 0$ (the location of the target deuteron). For r greater than a few femtometers (at the cusps of the potential), $1/r$ Coulomb repulsion between the deuterons acts to repel them. The left and right halves of Figure 3.7 contrast what would happen if classical or quantum mechanics described the interaction. For the classical case (right half of the figure), the approaching deuteron, based on the value of its energy (height of the dotted line) stops at the point indicated by the small circle and then turns back—much like a ball rolling up a hill of this shape. Only if the energy were above the top of the hill would a classical deuteron come within the range of the strong (attractive) force and fuse with the other one.

Now consider the correct quantum mechanical description suggested in the left half of Figure 3.7. Here a deuteron approaching a second one from the left is described by a wave function. The amplitude of the wave function exponentially decays as it tunnels through the forbidden region (where $V > E$), but its nonzero amplitude inside the well indicates that fusion is possible, even though classical mechanics would not allow it for this energy E.

3.13.1 Example 4: Find the Temperature Needed to Initiate d–d Fusion

Solution

At very high temperatures, as in the core of the sun, matter is in a state known as plasma with the nuclei and electrons moving at random, somewhat like the molecules in a gas. In such a state, we may define the temperature using the relation from kinetic theory:

$$E = \frac{3}{2} k_B T, \tag{3.12}$$

where E is the average energy of the nuclei or electrons; T is the absolute temperature; and k_B is the Boltzmann constant.

The two deuterium nuclei have radii given by Equation 3.10. In the case of $A = 2$, the radius works out to be $r = 1.57$ fm. Thus, fusion is guaranteed if the two nuclei approach each other head on with a center-to-center separation equal to 3.14 fm. Let us stick with the head-on collision case, which makes the calculation simplest. If the two nuclei have the same energy E when very far apart, and the kinetic energy is entirely converted into electrostatic potential energy when they just make contact, we have

$$2E = \frac{kq^2}{r} = 2 \times \frac{3}{2} k_B T, \tag{3.13}$$

so that

$$T = \frac{kq^2}{3k_B r} = \frac{(9 \times 10^9)(1.6 \times 10^{-19})^2}{(3 \times 1.38 \times 10^{-23})(3.14 \times 10^{-15})} = 1.77 \times 10^9\,K. \tag{3.14}$$

The calculated temperature of 1.77 billion K for ignition to occur for the d–d reaction is much higher than the value reported in the literature, i.e., *only* 180 million K. As previously noted, the discrepancy arises because itis necessary to use quantum mechanics, not classical mechanics, here.

3.14 RADIOACTIVE DECAY LAW

Radioactive decay is a completely random process, implying that a radioactive nucleus has no memory of how long it has been waiting to decay. In fact, the number of decays per second for a given radioisotope depends only on the number of nuclei present at a given time. As a consequence, the number of nuclei N that survive to a time t can be expressed in terms of the initial number N_0 and either the decay constant λ or the so-called mean lifetime T according to

$$N = N_0 e^{-\lambda t} = N_0 e^{-t/T}. \tag{3.15}$$

Still another way to express this result is in terms of the number of half-lives $n = t/\tau_{1/2}$:

$$N = N_0 2^{-n}. \tag{3.16}$$

The differentiation of Equation 3.15 shows that the activity at any given time also satisfies the exponential decay law:

$$\frac{dN}{dt} = -\lambda N = -\frac{0.693N}{\tau_{1/2}}. \tag{3.17}$$

3.15 HEALTH PHYSICS

Health physics refers to the field of science concerned with radiation physics and radiation biology, with special emphasis on the protection of personnel from the harmful effects of ionizing radiation. The field of health physics is complicated by its use of various units—both current SI units and an older set of units that still appear in many books and articles (Table 3.3).

3.16 RADIATION DETECTORS

It has often been noted that we can neither see, feel, smell, nor sense by any other means the presence of ionizing radiation, which may be one

Table 3.3 **Some More Important Units for Various Radiation Quantities**

Name	SI Unit (Abbreviation)	Old Unit	Conversion
Activity (decays/s)	Becquerel (Bq) 1	Curie (Ci) 3.7×10^{10}	
Radiation dose (energy absorbed)	Gray (Gy) 1 J/kg	rad 100 erg/g	1 cGy = 1 rad
Biologically effective dose	Seivert (Sv)	rem	1 cSv = 1 rem
Dose rate	Gray/s	rad/s	

BIOLOGICALLY EFFECTIVE DOSE

The biologically effective dose is the dose adjusted for the type of radiation. For example, the ionization trail left by some particles tends to be more localized, which is sometimes more harmful than if it is not. Thus, for an equivalent dose, neutrons are roughly 10 times as harmful as gamma rays. The dose rate can also be important when considering biological effects. Since cell repair can spontaneously occur at low dose rates, a given total dose is likely to be more harmful if rapidly received.

reason that it is so greatly feared. It is therefore necessary to rely on various kinds of radiation detectors such as the well-known Geiger counter, which essentially counts the number of particles per second that pass through a detecting tube.

One can easily imagine Hans Geiger's motivation for inventing the Geiger counter: he and Marsden spent many days observing counts by looking at the scintillation screens in the experiment that they performed under Rutherford's direction (Figure 3.9).

Nowadays, radiation detectors come in a variety of forms—some small enough to fit on your keychain. There is even a simple app to convert your smartphone into a radiation detector! Apparently, all you need to do

Figure 3.9 Drawing of Hans Geiger by Kevin Milani, included with his permission, and modified by adding the "thought balloon."

is install the app and stick some opaque black tape such as electrician's tape over the camera lens. Since the sensors used in smartphone cameras pick up not just visible light but also gamma rays and X-rays from radioactive sources, then covering the lens allows only the latter to make it to the sensor. The application then counts the number of impacts the sensor receives and translates it into a value in microsieverts per hour.

3.17 RADIATION SOURCES

Ionizing radiation is continually present in the natural environment. The three primary natural sources of radiation are (1) space—in the form of cosmic rays (mostly shielded by atmosphere); (2) the Earth—both food* and water and building materials; and (3) the atmosphere—mostly in the form of radon gas that is released from the Earth's crust. Both the level of radon and the amount of cosmic rays to which you are exposed can greatly depend on your circumstances; thus, the latter significantly increases with your altitude, and the former depends on the local geology. Radon is usually the largest of the natural sources, amounting to around 200 mrem or 2000 μSv annually in a typical case, but with very large variations.

The most significant radiation you receive from human-made sources is received in medical tests or treatment. These medical exposures can enormously vary depending on their purpose, and responsible physicians will always weigh the diagnostic or treatment benefits against the risks of receiving the radiation exposure. For example, while a simple chest X-ray might expose you to the equivalent of 3 days of normal background radiation, a computerized tomography scan of your abdomen might give you the equivalent of 4.5 years' worth. These estimates assume that the technician administering the test adheres to accepted guidelines and that the machine is not defective, which regrettably is not always the case.

3.17.1 Example 5: Comparison of Two Radioactive Sources

A 1 Ci source with a half-life of 1 week has the same number of radioactive nuclei at time t as a second source whose half-life is 2 weeks. What is the activity of the second source at this time? What will be the activities of the two sources after 4 weeks have elapsed? Which source is more dangerous to be around?

Solution

Since the two sources have the same number of radioactive nuclei, by Equation 3.17, the second source must have half the activity of the first

* Sometimes food is exposed to extremely high levels of radiation that kill all the bacteria that they contain, but this exposure does not make the food itself radioactive. In fact, irradiated meat can even be left outside a refrigerator and stored on a shelf.

or 0.5 Ci. The activity of both sources exponentially decays based on their respective half-lives. Therefore, at the end of the 4-week period, the activity of the first source has declined to 1/16 Ci and that of the second source has reached 1/8 Ci. Initially, and up to 2 weeks, the first source had the higher activity (was more dangerous), but after 2 weeks, it was the second—see Figure 3.10. The total dose received integrated over time up to a period of many weeks would be the same for both sources if you were continually in their presence.

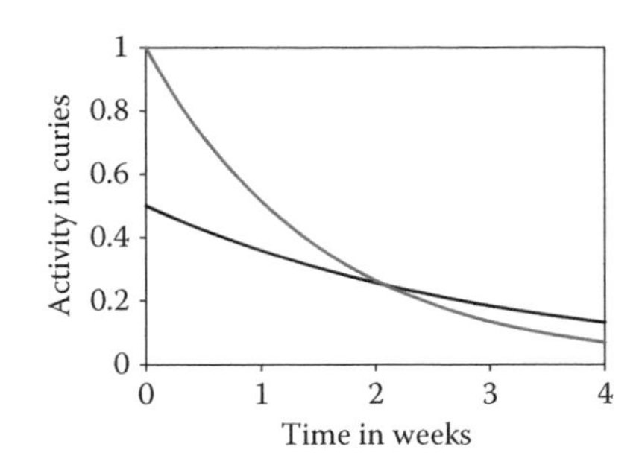

Figure 3.10 Activity in curies versus time in weeks for the two sources discussed in Section 3.17.1.

3.18 IMPACTS OF RADIATION ON HUMANS

In general, the danger of being in the presence of a radioactive source depends on many factors:

- The length of time you are in the presence of the source
- Whether the source is in a well-shielded container
- How far away you are from the source
- Type of radiation (alpha, beta, or gamma)
- Ventilation (in the event of a radioactive gas such as radon)
- Your own situation (whether you might be pregnant, for example)

It is also very important whether the radioactive source has become internalized to your body by either breathing a radioactive gas such as radon or eating food that was contaminated by radiation. Very large short-term exposures can cause radiation sickness and death, and longer-term exposures can cause both genetic mutations and cancer. In the case of cancer, it is important to note that the risk increases with increasing dose, and the occurrence of cancers does not appear for many years after radiation exposure.

3.18.1 Safe Radiation Level and Cancer Risks

It is often stated that there is no known safe level of radiation, which can be translated as there being no known level of radiation below which it is known that no harm whatsoever results. The reason that it is so difficult to establish whether a threshold for harm exists is that the harmful effects of radiation are so small at very low levels. An observation would need to have extraordinary statistics to reveal anything meaningful. For example, consider the question of whether having a chest X-ray (dose about 2 mrem) increases your risk of dying from cancer. Even though it is not morally possible to do experiments to see how humans are affected by doses of any given magnitude, extensive data have been compiled on the survivors of the Hiroshima–Nagasaki bombings during World War II (Preston et al., 2007). Based on these data, a dose of 1 Sv or 100 rem would increase your chances of dying from cancer by about 50%. If the harm done is assumed to be linearly proportional to the dose received, your chances of dying from cancer would increase by 0.001%.

Since roughly 25% of the population normally die from cancer anyway, such an increase would be impossible to establish.

Even for much larger exposures than 2 mrem, it is difficult to pin down whether a threshold for harm exists. For example, the largest source of natural radiation exposure is from radon gas seeping up from the ground. In fact, radon is second only to smoking as a cause of lung cancer. Radon levels considerably vary depending on the local geology. A study performed by physicist Bernard Cohen (1997) reported that in US counties having higher average radon levels, the rate of lung cancer tends to be lower, which some have interpreted as being evidence for the controversial hypothesis of radiation hormesis. This interpretation, however, can be challenged based on the possibility of confounding variables, which cannot be definitively ruled out in an epidemiological study of this type.

Hormesis, which is well established for agents or substances such as sunshine, iodine, and iron, is the notion that while very high levels are harmful to health, low doses are actually beneficial. Whether at very low levels the effects of radiation are in fact harmful, beneficial, or neutral, the risks of radiation need to be assessed in comparison to other risks. For example, taking a plane ride or moving to Colorado will very slightly increase your exposure to radiation, but few people would let this factor dictate their decision-making about these activities. In the former case, the risks of flying (or more importantly driving to the airport) likely far outweigh the extra risk of dying from cancer, and in the latter case, moving to Colorado, despite its slightly higher background radiation level, will probably improve your health in view of the climate, lack of smog, and healthy lifestyle of the populace.

RADIATION PARADOX

The highest level of natural background radiation recorded in the world is from areas around Ramsar, in Iran, where levels can reach 200 times greater than the worldwide average level. Most of the radiation in the area is due to dissolved radium-226 in hot springs. This high level of radiation has not had any observed ill effects on the residents, who live healthier and longer lives than average. This strange fact and similar reports from other high natural radiation areas, as well as studies like Bernard Cohen's, have been called the *radiation paradox*. They may not definitively prove that radiation hormesis is correct, but they certainly call into question the validity of linear no-threshold hypothesis, which is currently the basis of all radiation regulations.

3.18.2 Relative Risk

The possibility of hormesis aside, it is useful to have some sense of the relative risk to one's life resulting from various radiation exposures compared to other hazardous situations. Table 3.4 is highly instructive in this regard. It shows how much the average person's life tends to be shortened

Table 3.4 **Comparison of Shortened Life Expectancy for Various Hazards**

Health Risk	Shortened Life Span
Smoking a pack of cigarettes/day	6 years
Being 15% overweight	2 years
Consuming alcohol	1 year
Being a farmer	1 year
All accidents	207 days
All natural hazards	7 days
Receiving 300 mrem per year	15 days
Living next to a nuclear plant	12 h

as a result of various causes. Concerning the last table entry, it is assumed that the nuclear plant functions normally and that you live your whole life adjacent to it (Table 3.4).

3.19 SUMMARY

This chapter reviews the highlights of the science needed to understand how nuclear reactors operate. It begins with a brief overview of early nuclear history, including Rutherford's discovery of the nucleus, and then describes the properties of the atomic nucleus, including nuclear forces, and the role that they play in nuclear transformations, such as alpha, beta, and gamma decay. It considers the relation between mass and energy and how to estimate the energy released in the processes of fission and fusion, which is perhaps a million times greater than for chemical reactions. The chapter concludes with a discussion of the topic of health physics, namely, the biological effects of radiation.

PROBLEMS

1. Show by direct integration that the total cross section, according to the Rutherford formula, is infinite. Hint: Up to angles of say 15°, the small angle approximation holds very well, which allows the integration of Equation 3.7 to be easily performed over that interval.
2. Suppose that in Rutherford's experiment, he observed 1000 scattering events for a 1° interval centered on 30° in a given time interval. How many events would he have found for a 1° interval centered on 90° in the same time interval?
3. Find on the web the data contained in Geiger and Marsden's original paper for thin foils made of silver rather than gold and show how they do not fit the Rutherford scattering formula.
4. What energy alpha particles would have to be used in Rutherford's experiment for them to come within range of the strong force (about 1 fm) for a $b = 0$ scattering from a gold nucleus? Hint: First

find the approximate radii of an alpha particle and a gold nucleus and remember that $b = 0$ implies a head-on collision.

5. (a) Prove that the number of target nuclei of a sheet of material of thickness d lying within an area S is given by $N_t = \rho d N_A S/A$ (Equation 3.6). (b) Using this equation, show that the number of gold nuclei per square meter in a thin sheet of thickness dx is $5.9\ dx \times 10^{28}\ m^{-2}$.
6. A radioactive source consisting of a single radioisotope has an activity of 1000 Bq at a certain time and 900 Bq after 1 h. What is its half-life? What will be the activity after 10 h have elapsed?
7. Verify the figure 10^{-40} in Table 3.2 by considering the relative strengths of the gravitational and Coulomb forces between a pair of electrons any given distance apart.
8. The process of alpha emission from a nucleus has been explained on the basis of quantum tunneling. How might the explanation go? How did an alpha particle get to be in the nucleus? Why is alpha emission from nuclei observed, but proton emission is virtually never observed?
9. Another possibility instead of there being neutrons in the nucleus to account for the fact that for many nuclei, $A = 2Z$, is to have electrons inside the nucleus instead. Why according to the uncertainty principle of quantum mechanics is this not credible?
10. Do the calculation mentioned at the end of Section 3.6 to find the density of nuclear matter.
11. Why do alphas emitted by a particular radioisotope have a fixed energy (line spectrum), but betas have a continuous spectrum? Hint: How many particles are present after the decay in each case?
12. When a uranium-238 nucleus decays via alpha emission, use the known masses of the parent and daughter nuclei to determine the amount of energy liberated. How much of this energy is given to the alpha and how much to the daughter nucleus?
13. What form of radiation is a nucleus having $Z = N = 50$ likely to emit?
14. Look up on the web how much electricity New York City uses annually and do the calculation in the footnote in Section 3.9.
15. Show that if the activity of a source is merely proportional to the number of radioactive nuclei present, then the exponential decay law must follow.
16. Using the longest known half-life found to date (for beryllium-13), $\tau_{1/2} = 8 \times 10^{24}$ years, estimate what fraction of the original amount has decayed in a time equal to the age of the universe.
17. Very slow neutrons are especially effective in causing nuclei to fission. Can you think of a reason why the total cross section for nuclear absorption of neutrons might inversely vary with their velocity at low velocities? Hint: How does the likelihood of a neutron inducing a fission depend on the time it is close to the nucleus?

18. Show that the excess cancer deaths due to a chest X-ray is 0.00056%.
19. Suggest an alternative plausible hypothesis to hormesis that might explain the results Bernard Cohen found in his radon study in Section 3.18. Hint: The leading cause of lung cancer is smoking.
20. Why would you not expect the data in the Rutherford experiment to fit the formula for very small angle scattering—say 0.01°?
21. Consider two decays in sequence A → B → C, having two different half-lives. Derive an expression for the amount of B nuclei remaining after a time t, in terms of the initial number of A nuclei and the two half-lives.
22. Suppose you are inadvertently in the presence of a radioactive source whose half-life is 3 h for a 4 h interval. Calculate your total exposure over the 4 h if the initial dose rate was 5 rad/h.
23. Estimate the percentage increase in your long-term risk of cancer from a medical exposure equal to 3 months' worth of the average background radiation—assuming the linear no-threshold hypothesis is true.
24. Based on Figure 3.6, estimate the relative amount of energy released per kilogram in fission and fusion. Note that the vertical axis is in units of E_B/A, not E_B.
25. Calculate the energy density of ^{235}U given that one atom will produce about 200 MeV of energy and 1 mol is 0.24 kg.
26. What is the yield of a nuclear bomb whose shock wave generates a pressure of 10^3 N/m^2 across a volume of 4×10^9 m^3?

REFERENCES

AIP (American Institute of Physics), R. Ivie, R. Czujko, and K. Stowe (2001) *Women Physicists Speak: The 2001 International Study of Women in Physics*, American Institute of Physics, College Park, MA. http://www.aip.org/statistics/catalog.html (Accessed Fall 2011).

Becquerel, H. (1896) Sur les radiations émises par phosphorescence, *C. R. Acad. Sci.*, 122, 420–421, http://en.wikipedia.org/wiki/Henri_Becquerel (Accessed Fall 2011).

Bohr, N. (1913) On the constitution of atoms and molecules: Parts I, II, III, *Philos. Mag.*, 26, 1–24, 476–502, 857–875.

Cassidy, D. C., G. J. Holton, and F. J. Rutherford (2002) *Understanding Physics Harvard Project Physics*, Birkhäuser, Basel, p. 632.

Chadwick, J. (1932) The existence of a neutron, *Proc. R. Soc. Lond. A*, 136, 692.

Cohen, B. L. (1997) Problems in the radon vs lung cancer test of the linear-no threshold theory and a procedure for resolving them, *Health Phys.*, 72, 623–628.

Coppes-Zatinga, A. R. (1998) Radium and Curie, *Can. Med. Assoc. J.*, 159, 1389.

Curie, M., and P. Curie (1923) *Autobiographical Notes*, Macmillan, New York.

DOE (1989) *US Department of Energy, A Report of the Energy Research Advisory Board to the United States Department of Energy*, DOE, Washington, DC. http://files.ncas.org/erab/ (Accessed Fall 2011).

DOE (2004) *US Department of Energy, Report of the Review of Low Energy Nuclear Reactions*, DOE, Washington, DC.

Fleischmann, M., and S. Pons (1989) Electrochemically induced nuclear fusion of deuterium, *J. Electroanal. Chem.*, 261, 2A, 301–308.

Geiger, H., and E. Marsden (1913) The laws of deflexion (sic) of α particles through large angles, *Philos. Mag.*, 6, 25, 148, http://www.chemteam.info/Chem-History/GeigerMarsden-1913/GeigerMarsden-1913.html (Accessed Fall 2011).

Goldstein, H., C. P. Poole, and J. L. Safko (2000) *Classical Mechanics*, Addison-Wesley, Reading, MA, http://en.wikipedia.org/wiki/Rutherford_scattering.

Hendee, W., R. Ritenour, and E. Russell (2002) *Medical Imaging Physics*, John Wiley & Sons, New York, p. 21.

Preston, D. L. et al. (2007) Solid cancer incidence in atomic bomb survivors: 1958–1998, *Radiat. Res.*, 168, 1–64.

Rutherford, E. (1911) The scattering of α and β particles by matter and the structure of the atom, *Philos. Mag.*, 21, 669–688.

Rutherford, E. (1921) The mass of the long-range particles from thorium, *Philos. Mag.*, 41(244), 570–574, http://en.wikipedia.org/wiki/Ernest_Rutherford (Accessed Fall 2011).

Thomson, J. J. (1897) Cathode rays, *Philos. Mag.*, 44, 293, http://web.lemoyne.edu/~GIUNTA/thomson1897.html (Accessed Fall 2011).

Thomson, J. J. (1904) On the structure of the atom: An investigation of the stability and periods of oscillation of a number of corpuscles arranged at equal intervals around the circumference of a circle; with application of the results to the theory of atomic structure, *Philos. Mag.*, 7, 237–265, http://www.chemteam.info/Chem-History/Thomson-Structure-Atom.html (Accessed Fall 2011).

Nuclear Power

Technology

4.1 INTRODUCTION

The history of nuclear power dates back to the World War II era, and it was inextricably linked to the development of atomic (actually nuclear) weapons. Thus, this chapter will discuss nuclear weapons as well as nuclear power. Only after the war, simultaneous with a significant expansion of the US arsenal during the Cold War with the Soviet Union, was the "atoms for peace" slogan coined to broaden the focus to include commercial nuclear power. During that period, one commissioner of the old US Atomic Energy Commission, Admiral Lewis Strauss, optimistically proclaimed that nuclear power would prove to be "too cheap to meter," meaning that it soon would be supplied at no charge to consumers (Pfau, 1984). Although Strauss' prediction was very far from the truth, the fact remains that despite concerns some people have had about nuclear power, it currently generates 20 times the electricity produced by solar power and 20% of all electricity produced in the United States, a figure that is close to the world average. Moreover, as with wind and solar and other renewable forms of energy, nuclear power contributes virtually no greenhouse gases during operation of the plants, which is one reason it is now being looked on more favorably, including by some environmentalists especially in nations such as France (Figure 4.1).

Apart from contributing a negligible amount of greenhouse gases (neglecting the contribution associated with the construction of a nuclear plant), nuclear does share a number of other properties with renewable forms of energy, which arguably allows us to consider nuclear to be a form of renewable energy. Whether or not you believe this controversial assertion or the claim that new generations of nuclear reactors are expected to lack many of the problems with earlier ones, the inclusion of nuclear power in a book on renewable energy can be easily justified, because in making the case for renewable energy, we need to consider the relative merits of all energy sources. Moreover, nuclear has a property that it shares with no other energy source—namely, an extraordinary high energy density. Specifically, the energy liberated in nuclear reactions is roughly a million times greater per unit mass of fuel than that liberated in any chemical process. It is this extraordinary energy density that makes nuclear potentially simultaneously attractive as an energy source and dangerous if not carefully controlled. In this chapter, the primary focus will be on nuclear fission—the splitting of the atomic nucleus—but some attention will be given to the other prospective way of extracting nuclear energy, namely, nuclear fusion. Nuclear fusion, the combining of light nuclei, has been an active field of research for many years, but at the time of this writing, no commercial fusion reactors exist—nor are they likely to exist for at least several decades according to most estimates.

CONTENTS

Figure 4.1 Nuclear power plant and its reactor cooling towers in France—a nation that leads the world in the percentage of its electricity generated by nuclear power at 78%. The large plumes emitted from the reactor cooling towers consist of water vapor.

4.2 EARLY HISTORY

Nuclear fission was discovered just before the outbreak of World War II in Nazi Germany by Otto Hahn, Lise Meitner, and Fritz Strassmann (Hahn and Strassmann, 1939; Meitner and Frisch, 1939). Meitner, who was Jewish, barely escaped Germany with her life after foolishly remaining there until 1938. After taking up residence in Stockholm, Meitner continued a secret collaboration with her former colleague Hahn who performed difficult experiments to find chemical evidence for fission. Hahn was baffled by his results, and he relied on Meitner to explain them. Politically, however, it was by then impossible for them to coauthor a publication on their results, and Hahn therefore published with Strassmann, with Hahn receiving the lion's share of the credit. As a result, Meitner was unjustifiably overlooked by the Nobel Committee when they later awarded the Nobel Prize to Hahn for discovering fission. Although Meitner soon thereafter realized the potential for using fission to build an enormously destructive weapon, she was not the first to do so. That idea had come to the remarkable Hungarian refugee Leo Szilard in 1933, a full 5 years before the discovery of fission. Szilard conceived the concept of a nuclear chain reaction in a bolt out of the blue that struck him one day while waiting for a London traffic light (Figure 4.2). He was granted a patent on the idea and later received a patent with Enrico Fermi on the idea of a nuclear reactor to release nuclear energy in a controlled manner.

After the actual confirmation by German scientists on the eve of World War II that fission could occur, Szilard wanted to alert the US government to the possibility of building a nuclear weapon, lest Germany do so first. He enlisted Albert Einstein in the effort on the grounds that a US president would be more likely to pay attention to the world's most

Figure 4.2 Physicist Leo Szilard conceived the idea of a nuclear chain reaction while crossing a London street in front of the Imperial Hotel. (Image courtesy of Brian Page.)

famous physicist than an unknown Hungarian refugee. Szilard, however, actually drafted the letter to President Roosevelt for Einstein to sign, which he did in August 1939, and the top-secret US project to build the bomb (the innocuous sounding Manhattan Project) was the eventual outcome (Figure 4.3).

Figure 4.3 Albert Einstein, official 1921 Nobel Prize in Physics photograph. (Public domain image.)

EINSTEIN AND THE BOMB

Although Einstein's relativity, specifically $E = mc^2$, was the theoretical underpinning behind nuclear energy, and his famous letter to Roosevelt may have started the U.S. project to develop the bomb, he played no role in its actual development. In fact, Einstein had long regarded himself as a pacifist—a position he no longer held to absolutely once Hitler assumed power in Germany in 1933. Nevertheless, Einstein later deplored the bomb's use against Japan, and toward the end of his life he noted: "I made one great mistake in my life… when I signed the letter to President Roosevelt recommending that atom bombs be made; but there was some justification—the danger that the Germans would make them" (Clark, 1953). Given that understandable fear it is ironic that Germany's progress in developing a nuclear weapon during the war was negligible, in part perhaps due to its well-documented disdain for Einstein's "Jewish physics."*

* The term *Jewish physics* was coined by physics Nobel Prize laureate Philipp Lenard, who was also a dedicated Nazi in the 1930s, when he derided Einstein's theories.

4.3 CRITICAL MASS

Szilard's idea of a chain reaction is quite easy to understand, given (1) the existence of nuclear fission and (2) the emission during fission of two to three neutrons—neither of which had been empirically demonstrated when he conceived the idea! Suppose we imagine that two of the emitted neutrons in a fission are absorbed by other fissionable nuclei, and as a result, they undergo fission and each emits two to three neutrons, two of which are again absorbed creating further fissions. Clearly, as this process continues from one generation to the next, the number of nuclei undergoing fission would exponentially grow, reaching 2^n after n generations have elapsed. If the time between generations (the time between neutron emission and subsequent absorption) is extremely short, the result would be a gigantic explosion.

You might wonder what is to prevent a mass of fissionable material from exploding all the time. It all depends on whether or not a mass of fissionable material exceeds a value known as the critical mass. In general, we may define f as the fraction of emitted neutrons that are absorbed by other nuclei causing them to fission, and N_0 as the average number of neutrons emitted per fission—typically, a number from 2 to 3 depending on the isotope. Those neutrons that fail to cause fissions escape the mass of material before they are absorbed by a fissionable nucleus. We, thus, have for the number of fissions in the nth generation

$$N_n = (fN_0)^n. \tag{4.1}$$

Clearly, in order to have exponential growth over time, it is necessary that $f > 1/N_0$—which is one way to define the condition of criticality. What will influence the actual value of f for a mass of fissionable material? The main variable would be the mass of material present, since the larger the mass, the less likely it is that emitted neutrons will escape before being absorbed, and the larger f will be. However, there are many variables besides mass itself that determine whether criticality is reached, and an explosion will occur. These include the density, geometric shape, and the level of *enrichment* in the fissionable isotope. The first of these factors should be obvious if we consider a fixed mass in the shape of either a thin pancake versus a sphere. In the case of a pancake, a much larger fraction of emitted neutrons would leave the surface of the material and not cause subsequent fissions. The importance of the enrichment level is also easy to understand, because the greater the percentage of a fissionable isotope, the less distance neutrons have to travel before causing a fission—and the less likely they will leave the surface of the material before doing so.

4.3.1 Neutron Absorption by Uranium Nuclei

In order to understand the manner in which neutrons are absorbed in passing through some thickness of uranium, it is easiest to start with a very simple geometry: a parallel beam of neutrons incident on a very thin slab of uranium—see Figure 4.4. Define the intensity of the beam to be I (which is simply the number of neutrons per second per unit area). Let the slab have a unit area and a thickness dx that is so small that the chances of a neutron being absorbed in traveling the distance dx through it are negligible.

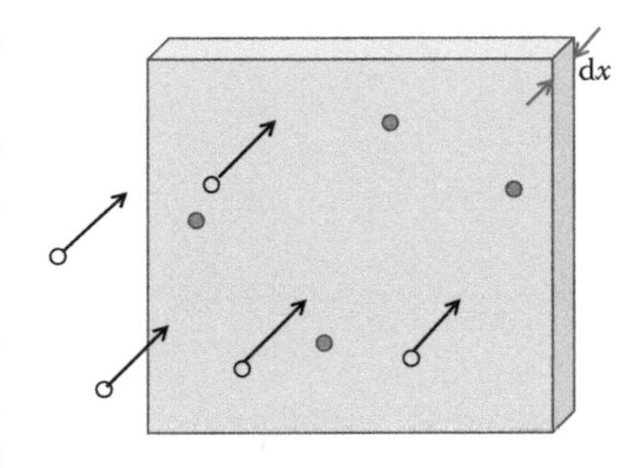

Figure 4.4 Thin slab of uranium of thickness dx on which a parallel beam of neutrons is incident. Open circles are the approaching neutrons, and closed circles are a few of the uranium nuclei in the slab.

By the definition of the total cross section for neutron absorption σ, only neutrons incident on an area this size will be absorbed by the nucleus. Suppose that the uranium slab has n nuclei per cubic meter so that the number of nuclei inside the slab will be ndx. Thus, the chance that one incident neutron is absorbed in the total cross section of all the nuclei in the slab is $\sigma n\, dx$. Remember, however, that we are not dealing with just one neutron but I neutrons per second incident on the unit area slab, making the absorbed intensity also proportional to I, and hence, the intensity loss is $dI = -\sigma nI\, dx$, from which we find

$$\frac{dI}{I} = -\sigma n\, dx. \tag{4.2}$$

Finally, if we have a thick slab of width x (which we imagine consisting of many thin slabs), we can easily find the total absorbed intensity by integrating Equation 4.2 and then expressing the result as

$$I = I_0 e^{-\sigma n x}, \tag{4.3}$$

where I_0 is the initial intensity before striking the slab. We can rewrite Equation 4.3 as

$$I - I_0 e^{-x/d}, \tag{4.4}$$

where the quantity d, known as the mean free path, satisfies

$$d = \frac{1}{\sigma n}. \tag{4.5}$$

It can be shown that physically the mean free path represents the mean distance neutrons will travel before being absorbed.

4.3.2 Why Does Density Matter in Determining Critical Mass?

The importance of the density of the material in determining the critical mass of a piece of uranium requires a bit more explanation. The distance d that a neutron will travel before having a 50% chance of being absorbed by a fissionable nucleus is given by Equation 4.5, and notice that it inversely varies with the density of fissionable nuclei per unit volume. Recall that whether a sphere of this fissionable material of radius R is critical or not depends on the ratio d/R being sufficiently small. Let us suppose that a sphere has a radius R so that the d/R ratio is just above the critical value. If the sphere were compressed to a fraction f of its original radius, the density of the sphere n would increase to nf^{-3} so that the mean free path for neutron absorption would decrease to df^3 and the ratio of the mean free path to the radius would then become $f^2(d/R)$.

Thus, if d/R was initially just above the critical value, a relatively small compression factor f would be needed to cause the sphere to become critical. Detonating a nuclear bomb by compressing a subcritical sphere can be achieved by surrounding the core of the bomb with shaped explosive charges that when detonated, cause the sphere to implode and increase its density. However, the detonations must occur virtually simultaneously and the charges must be precisely shaped; otherwise the implosion will not be symmetrical and no detonation will occur if the compressed material becomes significantly nonspherical.

4.3.2.1 Example 1: Estimation of critical mass The cross section for the absorption of a neutron having megaelectron volt energies in ^{235}U is on the order of a few barns, where 1 barn = 10^{-28} m^2. Estimate the mean free path for neutron absorption and the critical mass for a spherical shape.

Solution

As you can easily verify by considering the units of each quantity, the number of nuclei per cubic meter in uranium or any other element can be expressed in terms of its density ρ, atomic weight A, and Avogadro's number N_A:

$$n = \frac{\rho N_A}{A}, \tag{4.6}$$

which here yields $n = 4.9 \times 10^{28}$ nuclei per cubic meter. Using Equation 4.5, we find a mean free path of around 0.1 m = 10 cm. If two neutrons were emitted during a fission occurring at the center of a sphere having a 10 cm radius, the chances of zero, one, or two being absorbed before leaving the sphere would be 0.25, 0.5, and 0.25, respectively. Thus, the average number of neutrons absorbed in this case is one. This implies that 10 cm is roughly the radius of a sphere having the critical mass—in this case around 80 kg. In contrast, the critical mass listed in the open literature for this isotope is listed as 52 kg, which is significantly less because we have made a number of simplifying assumptions in arriving at the 80 kg estimate.

As noted earlier, the critical mass strongly depends on the level of enrichment, so that with only 20% ^{235}U, it would be over 400 kg, which is generally considered the minimum enrichment level needed for a crude nuclear weapon. Our preceding list of ways to achieve criticality through changes in the mass, shape, density, and enrichment level are not exhaustive. Two other methods applying to nuclear reactors but not bombs would include (1) the introduction of a medium known as a moderator to slow neutrons down or alternatively (2) the presence or absence of so-called control rods made of a material that absorbs neutrons.

THE ESSENTIAL DIFFERENCE BETWEEN NUCLEAR BOMBS AND REACTORS

The essential difference between bombs and reactors concerns the critical mass. In a bomb you wish to be able to achieve the critical mass as quickly as possible, so as to have a rapidly rising exponential growth in energy released. If the critical mass is not achieved quickly, the bomb would detonate prematurely and the result would be a "dud," i.e., it would blow itself apart from the heat released before a large fraction of the nuclei fission. In a reactor, the goal is to never exceed the critical mass. However, for maximum power generation it would be desirable to approach the critical mass as closely as possible.

4.4 NUCLEAR WEAPONS AND NUCLEAR PROLIFERATION

The link between nuclear power and nuclear weapons established in World War II continues to this day, because exactly the same technology to produce enriched uranium is needed in both cases, although the level of enrichment required in the case of fuel for a nuclear reactor is only around 4% (up from the 0.7% found in nature for ^{235}U)—far less than the 90% required for a military-grade weapon. Another fissionable isotope that can be used in both reactors and bombs is ^{239}Pu, although unlike ^{235}U, it is not found in nature.

Given the common enrichment technology for creating fuel for reactors and bombs, it is not surprising that among the eight declared nuclear weapons states, several have developed nuclear weapons under the pretense of developing a nuclear power or a nuclear research program.* One nation (Israel) not included in the eight declared weapons states has chosen not to confirm that it has a nuclear arsenal, but no informed observer doubts that fact. Of the eight declared nuclear weapons states only two have developed them since the mid-1970s: Pakistan (in 1998) and North Korea (in 2006)—giving some hope to the notion that the spread of nuclear proliferation can be slowed or halted. However, this relatively slow pace of nuclear proliferation could abruptly change if Iran were to develop a nuclear arsenal in the volatile Middle East, since at least four other nations in that region almost certainly could build a nuclear arsenal if they so chose. The general rule of thumb is that any nation that has an engineering school could build the bomb. In late 2015, Iran signed an international agreement to constrain its nuclear activities to peaceful uses. The final chapter on this diplomatic effort remains to be seen.

HOW SECURE IS THE WORLD'S NUCLEAR WEAPONS MATERIAL?

There is much speculation about the highly enriched nuclear material in certain nations being diverted and either stolen or deliberately sold. A 2012 report by a nonprofit advocacy group (the Nuclear Threat Initiative) has ranked 32 countries based on their levels of nuclear security (NTI, 2012). A nation needs to have at least a kilogram of highly enriched uranium or plutonium to be included on the list. Nuclear security is evaluated based on the degree to which procedures and policies exist to prevent theft, as well as societal factors, e.g., those affecting the government's degree of corruption and stability, which could undermine security. Not surprisingly, this last factor places North Korea, Pakistan, and Iran at the bottom of the list of 32 nations. According to the NTI report, the top countries in terms of nuclear security are Australia, Hungary, Czech Republic, Switzerland, and Austria, with the United States not showing up until 13th place. The United States would have ranked in second place were it not for its large quantity of highly enriched material and the number of locations where it is stored—both of which contribute to vulnerability to theft.

Following World War II, the United States began to amass a very large nuclear arsenal as a means of deterring the Soviet Union, in a possible conventional conflict in Europe. At the peak in the mid-1960s, the US arsenal numbered over 30,000 nuclear weapons. The Soviets, who started their nuclear arsenal later (with the help of some spies in the US and UK programs), eventually amassed an even larger number of weapons. Although both arsenals have been scaled considerably back, the numbers of nuclear weapons in the US and now Russian arsenals are still believed to dwarf those of any other country. A question that continues to divide

* The eight declared weapons states in order of which they first conducted a nuclear test are United States, Russia, United Kingdom, France, China, India, Pakistan, and North Korea.

many analysts concerned with national security is the optimum (and the minimum) number of weapons a nation needs to protect itself and deter threats against it.

There are many nations who are quite capable of building a large nuclear arsenal if they so choose, but who have concluded that this optimum number is exactly zero. Whether that choice will ever be realistic for the world as a whole is a matter that is tied to such controversial questions as world government and/or the prevention of war as an instrument of national policy. Clearly, in our present-day world, where nations need to deter not only so much threats from conventional nation states, but also terrorist groups, the relevance of a large nuclear arsenal becomes less certain, apart from discouraging collaborations between rogue nations led by rational leaders and terrorists.

WHAT IF THE WORST HAPPENS?

The "worst" used to be defined in terms of an all-out nuclear exchange between the United States and its superpower rival the U.S.S.R. Nowadays, it is considered much more likely that if nuclear weapons are used the threat would involve either a rogue state or terrorist group, so it is instructive to consider what would happen in such a case. In 2004, the Rand Corporation did a study for the U.S. Department of Homeland Security involving the detonation of a 10 kton bomb brought in a shipping container to the port of Long Beach, California. Such a bomb would be about two thirds the size of the Hiroshima bomb, and would be within the capabilities of a rogue state to produce. According to the study, the result would be around 60,000 short-term deaths—a horrific number. However, the study also found that an additional 150,000 people would be at risk from fallout carried by the wind. The "good news" of the study, if one can call it that, is that many deaths in that latter group could be avoided with some simple precautions, such as taking shelter for a few days in an ordinary basement if one were available.

INTERNATIONAL NUCLEAR AGREEMENTS

The most important international agreement for controlling the spread of nuclear weapons is the Nuclear Nonproliferation Treaty (NPT). The NPT with 189 nations participating in it is essentially a bargain between most of the nuclear weapons states and those nations not possessing them. In accordance with the NPT, the weapons states agree to help the nonweapons states with the peaceful applications of nuclear technology, and in return the nonweapons states promise not to pursue their own weapons program. In addition, the weapons states promise to work toward eventual nuclear disarmament. Unfortunately, however, three states with nuclear weapons (Pakistan, India, and Israel) never signed the treaty, one state that had signed chose to withdraw and develop a weapon (North Korea), and several other states that had signed were found to be in noncompliance with the treaty (Iran and Libya). Thus, it is clear that regardless of treaties controlling the spread of nuclear weapons, nations will pursue what they regard to be in their national interest. Only one nation (South Africa) has at one time developed nuclear weapons on its own, and later chosen to dismantle them. Presumably, however, should it ever feel the need to reconstitute an arsenal and withdraw from the NPT, this option would remain open. Another important international agreement, the Atmospheric Test Ban Treaty in force since 1963, bans testing of nuclear weapons aboveground, where the amount of radioactivity released to the atmosphere is significantly greater than in underground tests.

Many technologies exist for enriching uranium, but fortunately, they tend to be expensive and time-consuming, since they all rely on the very small mass differences between isotopes. A common one used by many nations involves ultrahigh-speed centrifuges filled with uranium hexafluoride, a gaseous compound of uranium (UF^6). If the gaseous centrifuge is spun at extremely high speed, the slightly lighter ^{235}U isotope tends to concentrate closer to the spin axis on average than ^{238}U. The operation of the centrifuge is illustrated in Figure 4.5a. UF^6 enters from the left; slightly enriched gas and slightly depleted gas exit through separate pipes as the centrifuge spins. The spin rate is so high that the walls of the rotor are moving at almost the speed of sound, which requires it to be made from an extremely strong type of material, and the casing containing the rotor must be evacuated of air to avoid frictional losses. Since the degree of concentration from one pass through the centrifuge is extremely small, either the gas must be run through the centrifuge many times, or else, many of them must be used in series with the very slightly enriched gas from one being piped to the next. A single centrifuge might produce only 30 g of highly enriched uranium per year, so the usual practice is to use many in series. A cascade of 1000 centrifuges of them operating

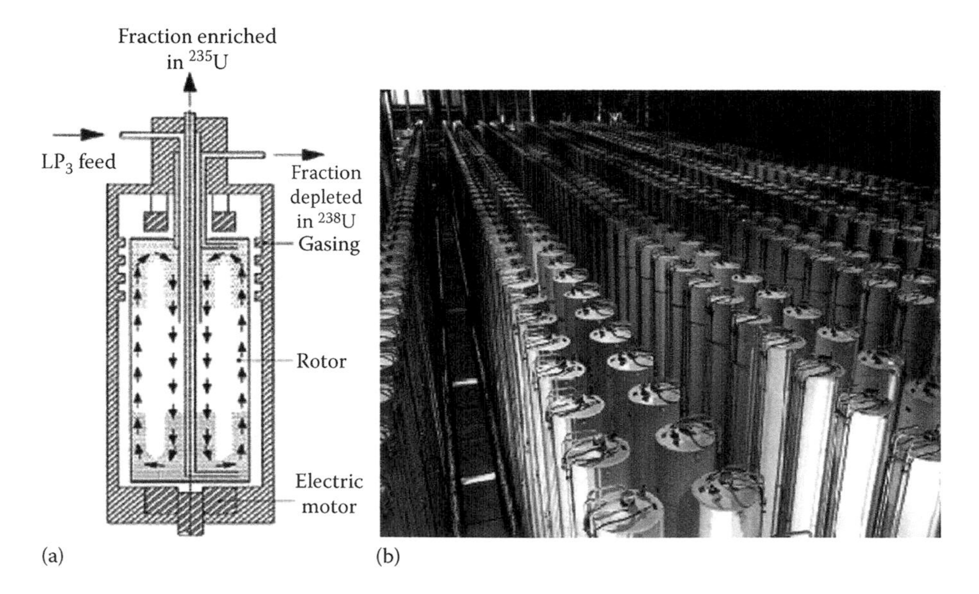

Figure 4.5 (a) Cross-sectional drawing of a gaseous centrifuge for uranium enrichment. (Courtesy of NRC, Delta, Pennsylvania.) (b) Cascade of many gas centrifuges. (Courtesy of DOE, Washington, DC.)

continuously might yield 30 kg per year, enough for one weapon. During the World War II Manhattan Project, the race to amass enough fissionable material for several bombs was pursued by all available means including gaseous centrifuges, but these were abandoned during the project in favor of using a reactor (the world's first) to "breed" plutonium from uranium.

4.5 WORLD'S FIRST NUCLEAR REACTOR

In 1938, after receiving the Nobel Prize for work on induced radioactivity, Enrico Fermi fled his native Italy to escape the dictatorship of fascist Italy that was then allied with Nazi Germany and took a position at the University of Chicago, where he led the effort to design and build an atomic pile—essentially the world's first nuclear reactor (Figure 4.6). The purpose of the first nuclear reactor was to breed plutonium (^{239}Pu). This fissionable isotope has a much smaller critical mass than ^{235}U, which is a considerable advantage in creating a bomb that is easily deliverable. Fermi's atomic pile was secretly constructed under the stands at Stagg Field at his university. However, unlike almost all reactors built since then, Fermi's design had neither radiation shielding to protect the researchers, nor a cooling system to prevent a runaway chain reaction. As the neutron-absorbing control rods were withdrawn and the power level increased, the reactor came ever closer to the point of criticality. Fermi, however, was sufficiently confident in his calculations that he was given the go ahead to conduct what could have been a potentially disastrous experiment in the midst of one of the nation's largest cities! After the

Figure 4.6 Drawing of the first nuclear reactor, which was erected in 1942 in the West Stands section of Stagg Field at the University of Chicago. On December 2, 1942, a group of scientists achieved the first self-sustaining chain reaction and thereby initiated the controlled release of nuclear energy. (Courtesy of DOE, Washington, DC.)

Figure 4.7 Sailors aboard USS *Enterprise* spell out "E = mc^2 × 40" on the flight deck of the carrier to mark 40 years of US Naval nuclear power. (Courtesy of US Navy, Washington, DC.)

happily successful result, he reported the outcome using the previously agreed upon coded phrase that the "Italian navigator has landed in the New World." In terms of power produced, in its first run, Fermi's reactor produced a meager 50 W—although power production was not its intended purpose, of course.

There are several other reasons for building nuclear reactors aside from the obvious ones of producing electric power or breeding fuel for bombs. These include conducting nuclear research and as propulsion systems (Figure 4.7). The first nuclear powered submarine was, for example, built in 1954—the USS *Nautilus*. The twin advantages of nuclear power for propulsion in subs is their ability to stay submerged for much longer times—given the long period before the reactors need to be refueled and their much quieter operation than diesel-powered subs. The United States even at one time considered building a nuclear powered aircraft, and perhaps the strangest of all was the project considered by the Ford Motor Company to build a nuclear-powered car, the Nucleon. Readers will need to find pictures of this vehicle on the web because the Ford Motor Company, perhaps understandably, declined to respond to my request for permission to use a picture of this vehicle in this book.

4.6 NUCLEAR REACTORS OF GENERATIONS I AND II

The first nuclear reactor ever built for the purpose of generating electricity (which produced a meager 100 kW) was not constructed until 1951—6 years after the end of World War II. Nuclear reactors built over the next two decades were the early generation I prototypes. These later led to generation II reactors currently still in use today in the United States and many other nations. Although the current generation II reactors are more sophisticated and safer than the early prototypes, they also have had their problems over the years, including some very serious ones.

A nuclear reactor used for producing electricity begins with creating heat—a result of the enormous energy release during fission. Once the heat is generated, the rest of the process for creating electricity is very similar to what occurs in many fossil fuel power plants: the heat boils water to produce steam, which drives a turbine that runs a generator producing the electricity. Thus, the components of the most common type of existing reactor that are exclusively nuclear are in the reactor vessel that is normally placed within a containment structure with thick concrete walls—the last line of defense in case of a severe reactor accident.

Inside the reactor vessel itself is the core of the reactor consisting of fuel (usually in the form of rods filled with pellets [Figure 4.8]) and the neutron-absorbing control rods that can be partially withdrawn to bring the reactor closer to criticality and increase the power level. The water that flows through the reactor core serves three purposes: (1) it prevents the reactor from overheating, (2) its heat creates the steam used to power the turbine, and (3) it acts as a moderator whose function is to slow down the neutrons emitted in fission and thereby increase their cross section for absorption. Notice that the water that actually flows through the reactor vessel (and becomes radioactive) never comes into contact with the steam turbine, because there are two separate closed water loops that are connected only through a heat exchanger (Figure 4.9).

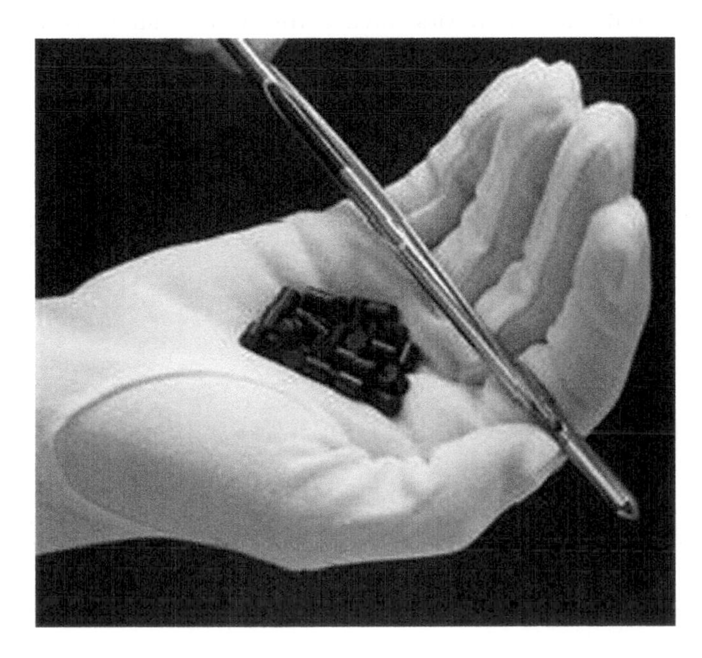

Figure 4.8 Around 20 fuel pellets for a nuclear reactor shown together with a section of a fuel rod into which they are inserted. One of those tiny pellets contains the energy equivalent of nearly a ton of coal. (Courtesy of DOE, Washington, DC.)

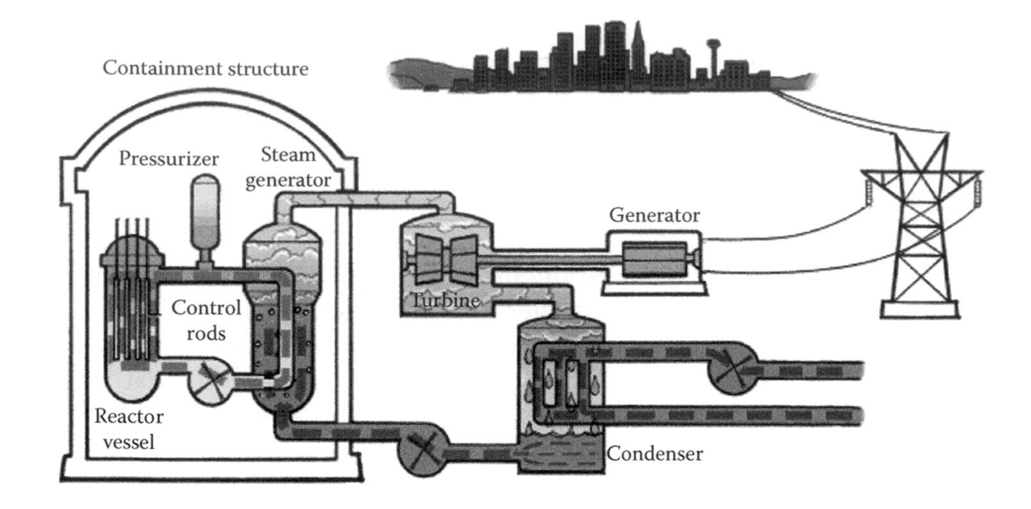

Figure 4.9 Main components of a pressurized water reactor. (Image courtesy of the U.S. Nuclear Regulatory Commission, is in the public domain.)

4.7 EXISTING REACTOR TYPES

Although upward of 85% of today's reactors are of the light water variety—some of which (5%) use a graphite moderator—a number of other types are also in use around the world, including 9% that use heavy water. Light water, of course, is not something dieters should drink! Light and heavy water are distinguished according to whether the hydrogen nucleus in the water molecules is a single $A = 1$ proton (light) or an $A = 2$ deuteron (heavy). Water found in nature consists of 99.97% of the light variety and 0.03% heavy. In addition to reactors cooled by light or heavy water, there are 5% that are cooled by gas rather than water, and 1% that are so-called fast breeders. These various reactor types will be discussed in the following sections based on the choice of moderator, fuel, and coolant.

4.7.1 Choice of Moderator

The reason that most reactors have a moderator such as water has to do with the dramatic variation of neutron absorption cross section with energy (Figure 4.10). Neutrons emitted in fission have energies on the

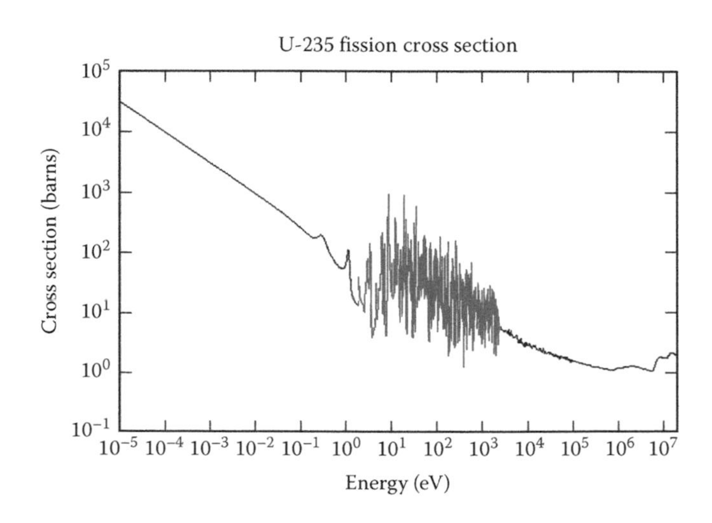

Figure 4.10 Neutron absorption cross section in barns for ^{235}U versus neutron energy in electron volts. A barn is an area of 10^{-28} m^2. (Courtesy of DOE, Washington, DC.)

order of megaelectron volts, where their cross section is around a barn. As neutrons make elastic collisions with the nuclei in the moderator, they transfer a fraction of their energy to those nuclei and gradually slow down. This has the effect of increasing their absorption cross section and making it much more likely that they will be absorbed by a fissionable ^{235}U nucleus they encounter. The section of the plot in Figure 4.10 between around 1 eV and 1 keV where the cross section wildly fluctuates is the resonance region. In this region, there are large variations in the absorption cross section depending on whether the neutron energy matches thye spacing between the energy levels in the nucleus. At energies below around 1 eV, the cross section resumes its steady rise, reaching around 1000 barns at the energy of 0.025 eV, which is in thermal equilibrium with the environment. Neutrons having energies near 0.025 eV are therefore known as thermal neutrons, and reactors that have moderators that slow neutrons to these energies are known as thermal reactors. The importance of a large cross section cannot be overstated, because it means the mean free path for fission-inducing absorption is correspondingly less, and the amount of fuel needed to achieve a critical mass is therefore less.

One reason that water is often chosen as a moderator is because it has hydrogen, and therefore, the proton nuclei being of almost the same mass as the neutrons are particularly effective in slowing them in elastic collisions. In contrast, if we imagine a neutron elastically colliding with a heavy nucleus, that nucleus would recoil with only a small fraction of the neutron's energy. Water also has other advantages, namely, that it is an effective coolant and that it is nonflammable—unlike graphite, for example, which greatly increased the environmental consequences at the Chernobyl disaster.

4.7.1.1 Example 2: How much energy does a neutron lose on average during elastic collisions? Suppose we had an elastic collision between a neutron of energy E and a stationary atomic nucleus of a moderator having atomic mass A. Consider the two extreme types of collisions (1) where the scattering angle of the neutron is very close to zero and (2) where it is 180°. In the first case, the energy lost by the neutron is essentially zero. In the second case, if $A > 1$, the neutron recoils backward. Let the original momentum of the neutron be p and assume that the nucleus it hits is initially at rest. After impact, let p' be the recoil momentum of the neutron so that $q = p + p'$ is that of the recoiling nucleus. Using the equations of conservation of momentum and energy, it can easily be shown that the ratio R of the kinetic energy of the recoiling nucleus to that of the original neutron is given by

$$R = \frac{4A}{(A+1)^2}. \tag{4.7}$$

Recall that this result applies in the case of a 180° scattering, where the neutron loses the maximum amount of energy. Thus, since all scattering angles are possible, to obtain an approximate estimate, we shall assume that all angles are equally likely and, furthermore, that for the average collision angle, a neutron would lose half this much energy. Clearly, the smaller the A value of the nucleus, the greater the R will be, meaning that the neutron loses more energy in the collision. Note that when $A = 1$, R also is 1—why is that?

FOUR TYPES OF NEUTRON INTERACTIONS WITH MATTER

So far we have been considering three sorts of neutron interactions: (1) where a neutron is absorbed by a fissionable nucleus, such as ^{235}U and causes it to undergo fission; (2) where a neutron elastically scatters off nuclei of the moderator, which slows them down and makes fission more likely, due to the dependence of fission cross section on energy; and (3) where neutrons are absorbed by certain materials that "poison" the chain reaction by removing them. A fourth type of neutron reaction that occurs in breeder reactors is where a neutron is absorbed by a "fertile" nucleus that it converts into a fissionable one. One example of this process is the three-step reaction of neutron absorption followed by two beta decays: $n + {}^{238}U_{92} \rightarrow {}^{239}U_{92} \rightarrow \beta^- + \bar{\upsilon} + {}^{239}Np_{93} \rightarrow \beta^- + \bar{\upsilon} + {}^{239}Pu_{94}$. It is important to realize that the likelihood of each of the four neutron processes occurring depends on the nuclear cross section for the process, which in general will vary with the neutron's energy, and the particular nuclear isotope it encounters.

4.7.2 Choice of Fuel

Nuclear reactor designers also need to decide what fuel to use. The most common choice is uranium, which has been enriched to around 3–5% of the fissionable isotope ^{235}U. A noteworthy exception is the Canadian Canada deuterium uranium (CANDU) reactor, which was originally designed to use natural (unenriched) uranium, since Canada at the time lacked enrichment facilities. CANDU is a trademarked abbreviation standing for "CANada Deuterium Uranium," and unlike most reactors, it uses heavy water as its moderator and coolant. The reason for the heavy water is that in a normal light water reactor, while the water may be a very effective moderator in slowing neutrons to energies where they cause fission, it also has the unfortunate side effect of sometimes absorbing neutrons and decreasing the probability that they will reach ^{235}U nuclei and cause fissions. Heavy water—a much poorer neutron absorber, but still an excellent moderator—avoids this problem and allows a reactor to operate at the enrichment level of 0.7% found in nature. Despite its capability, however, CANDU reactors now do operate using enriched uranium, which allows them to operate at higher power levels.

Another choice of fuel in some reactors is plutonium, especially in breeder reactors. However, unlike uranium, plutonium is not found in nature.

Rather it needs to be created in nuclear reactions when, for example, a reactor core is surrounded by a blanket or a layer of a so-called fertile isotope such as ^{238}U or ^{232}Th. By definition, fertile isotopes can be converted to fissionable ones by neutron absorption. During the operation of a breeder reactor, more fissionable material is created than is consumed. In addition, one of the fissionable isotopes that is bred (^{239}Pu) itself fissions afterward and contributes to the reactor power generated. Moreover, the portion of the ^{239}Pu that is not consumed can be reprocessed afterward (removed from the reactor waste) and then mixed with natural uranium to refuel the reactor.

The great advantage of breeders is that by breeding new fuel, they allow reactors to use a far greater fraction of the original uranium, whereas other reactors make use of only the 0.7% that is ^{235}U. After a number of breeders were built in the United States and other nations several decades ago, they fell out of favor. One reason was that breeders imply waste fuel reprocessing, which was considered to have a significant risk of diversion of plutonium that could lead some nations to stockpile it for bomb making. In addition, during that period, uranium was cheap and abundant, and there seemed to be no need for breeders. Currently, however, there is renewed interest in them—and a number of nations (India, Japan, China, Korea, and Russia) are all committing substantial research funds to further develop fast breeders. Fast breeders use neutrons having megaelectron volt energies to cause fission rather than thermal neutrons and, hence, have no need for a moderator.

One final fuel choice briefly discussed here is thorium. Unlike uranium, which is a fissionable nucleus, thorium is a fertile one that needs to be bred. The chief advantages of thorium as a reactor fuel are twofold. First, all thorium is a single isotope which can be readily bred into fuel. Second, thorium is three times as naturally abundant as uranium. Taking both of these factors into account, a ton of thorium can produce as much energy as 200 tons of uranium. But the advantages of thorium do not stop there. Thorium also has better physical and nuclear properties when made into a reactor fuel; it has greater proliferation resistance (since the waste is poisoned for bomb making); and it has a reduced volume of nuclear waste. At the moment, thorium reactors are being researched in a number of nations including the United States, and one commentator has suggested that thorium-fueled reactors would "reinvent the global energy landscape … and an end to our dependence on fossil fuels within three to five years," and he has called for a Manhattan Project scale effort to implement this vision (Evans-Pritchard, 2010).

4.7.3 Choice of Coolant

Whether a water-cooled reactor uses heavy or light water as a coolant would make almost no difference in terms of its cooling properties, since they both have the same specific heat. On the other hand, the pressure at which the reactor core operates does make a difference. Two common types of light water reactors are the boiling water reactor (BWR) and the pressurized water reactor (PWR). In the BWR, the cooling water has a pressure of about 75 atm and a boiling point of about 285°C, and it is allowed to boil so as to produce steam, which then drives a turbine. In contrast, in the PWR, the higher pressure of about 158 atm increases the boiling point enough so that boiling does not occur in the primary water loop.

An alternative liquid coolant to water in some advanced reactors is liquid metal, which has the advantage of higher power density for a given reactor size and greater safety owing to the lack of need to operate the reactor under high pressure. It also has some significant disadvantages, including having a coolant that may be corrosive to steel—depending on the choice of metals. Given the higher power densities of liquid metal-cooled reactors, an early application was in submarine propulsion systems, and both the US and Russian fleets have used them. Liquid metal coolant reactors tend to be of the fast neutron variety, because they need to have a lower neutron absorption cross section in view of their high energy density. In other words, if thermal neutrons were used, their power level would skyrocket, and the reactor could not operate safely. The earliest liquid metal used in a reactor was mercury; however, mercury has the disadvantage of being highly toxic and emitting highly poisonous vapor at high temperatures and even at room temperature. Two other choices that have suitable low melting points and suitably high boiling points are lead (327°C and 1749°C, respectively) and sodium (98°C and 883°C, respectively). However, these choices also have their problems. Sodium, for example, undergoes violent reactions with both air and water, and it emits explosive hydrogen gas. Nevertheless, despite their problems, liquid metal-cooled reactors have enough advantages to be planned for many advanced generation IV reactor designs.

Another advanced design built in Britain in 1983 is the advanced gas-cooled reactor (AGR), which used high pressure CO_2 (40 atm) as its coolant. Gas cooling results in higher temperatures of operation and, hence, higher thermal efficiency. In addition, since a significant fraction of the cost of water-cooled reactors is in the cooling system, gas-cooled reactors should be much more economical. The graphite-moderated AGR, while not using water as a coolant, still relies on it to generate the steam needed to drive the turbines. Unfortunately, the British AGRs took far longer than expected to build due to their complexity, and the cost overruns led them to prove uneconomical, although seven of them continue to operate. They are also being planned for some generation IV reactor designs.

It should be obvious that "very bad things" can happen to a reactor should it lose its coolant. However, unlike your car, where a loss of coolant at the worst will result in fatal damage to the engine, for a reactor, a loss of cooling accident is potentially devastating for the environment. While it may be true that reactors are incapable of a nuclear explosion, they can have (and have had) meltdowns and release a very large amount of radioactivity to the environment.

WHAT IS A MELTDOWN?

A meltdown has a large number of meanings in English, but as applied to a nuclear reactor it refers to the core of the reactor partially melting due to the extreme heat generated (particularly if there should be a loss of coolant), or the reactor momentarily becoming critical. Even in the extreme case, however, in which an explosion occurs due to the rapid buildup of energy, the reactor core will blow itself apart before a very significant fraction of the core undergoes nuclear reactions, and hence the explosion would be nonnuclear. In the Chernobyl disaster, it has been estimated that vastly more radiation was released during the meltdown than was released from the Hiroshima bomb dropped on Japan.

4.8 REACTOR ACCIDENTS

The seriousness of nuclear accidents is rated on a scale of 1–7, by the International Atomic Energy Agency (IAEA), with the two most serious categories being 6 = serious and 7 = major, based on the impacts on people and the environment. There have been two major accidents in the nuclear age: Fukushima in 2011 and Chernobyl in 1986, as well as one serious one at Mayak in the former Soviet Union in 1957, which many people in the West may never have heard about. On the same rating scale, the accident at Three Mile Island (TMI) in the United States in 1979 was rated as a 5 based on the amount of released radiation qualifying as being in the "limited" category. In the following sections, we discuss the three accidents, TMI, Chernobyl, and Fukushima, starting with the last one, which is probably the most familiar to the majority of readers.

4.8.1 Fukushima

The accident at Fukushima was a direct result of the earthquake and tsunami that hit Japan on March 11, 2011. These led to a series of nuclear meltdowns among some of the reactors at that six-reactor complex. Although the reactors operating at the time automatically shut down when the earthquake occurred, there was a loss of power both from the grid as well as from backup generators (due to flooding). These power failures caused a loss of coolant to the shutdown reactors, which triggered meltdowns in three of the six reactors. In fact, according to a 2015 report, the Fukushima disaster was entirely preventable, in that the critical backup generators were built in low-lying areas at risk from tsunami, despite warnings from scientists. The meltdowns were followed by hydrogen gas explosions and fires with releases of radioactivity to the environment both locally and, eventually, over a much wider area. The released radiation led to an evacuation of Japanese living in a 20 km radius around the plant. Fukushima will certainly cause long-term health, environmental, and economic problems for the Japanese for years to come. Initially, Japan resolved to phase out their reliance on nuclear power, but the government has (as of 2015) reversed its position and started up some of its reactors not involved in the accident. The following facts about Fukushima need to be considered to put the accident in perspective:

- There were zero known deaths or serious injuries from direct radiation exposures, even though 300 plant workers are judged to have experienced *significant* radiation exposures. The few plant workers that did die were killed as a result of the earthquake.
- Most of those residents in the mandatory evacuation zones are estimated to have received *annualized* doses of perhaps 20 mSv, which is the dose that would have been received had there been no evacuation. However, since most residents left this zone after perhaps no more than 2–3 days, their actual dose received was perhaps 0.2 mSv, which is equivalent to 6% of what is received from background radiation each year.
- Some residents were not evacuated for up to a month and not evacuated very far; thus, definitive assessments of the doses received cannot yet be made. Despite this word of caution, for most people living in Fukushima, their total radiation exposure was relatively small and unlikely to result in any observable increase in the long-term cancer death rate.
- The total amount of radiation released from the Fukushima accident has been estimated as being about 1/10 that which was released by the Chernobyl accident.
- The total number of Japanese killed by the tsunami and earthquake was 28,000.

4.8.2 Chernobyl

The Swedes are reputed to be a very safety conscious society, and yet, they heavily depend on nuclear power for 45% of their electricity—a fact that speaks to either the safety of nuclear power when carefully controlled or the hubris of the Swedes, depending on your point of view. On April 27, 1986, the alarms triggered by high levels of radiation went off at the Swedish Forsmark nuclear plant prompting concerns of a leak. However, the source of the radiation was found not to be at the Forsmark plant, but rather, it was wind-borne fallout originating 1100 km to the southeast—from one of the Chernobyl reactors in the town of Pripyat in Ukraine.

Initially, the Soviet Union (to which Ukraine then belonged) tried to cover up what had happened, but after the Swedes reported their detection, the Soviets finally had to acknowledge to the world that a nuclear catastrophe had taken place, and only then they belatedly ordered an evacuation of Pripyat a full 36 h after the April 26 disaster—and only after the town's citizens had received the early (most intense) radiation exposures. The Swedish radiation detectors were triggered a day after the disaster, because it took that long for the radioactive dust cloud to reach them.

The total radioactivity released by Chernobyl into the atmosphere has been estimated to be 50–250 million Curies, which is about the equivalent of that released by between 100 and 400 Hiroshima bombs. Apparently, 70% of this radioactivity was deposited in the neighboring country of Belarus, whose border is 7 km from the Chernobyl site. People living in the immediate area would have received extremely high doses before they were evacuated, and for people at a considerable distance, the dose received would depend on their distance but, even more strongly, on whether they happened to be in the path of the wind-borne radioactive dust cloud and whether any precipitation occurred over them. As of December 2000, over 350,000 people had been evacuated from the most severely contaminated areas and resettled.

The number of fatalities in the immediate aftermath of Chernobyl includes 57 workers who met an agonizing (often slow) death from radiation sickness and an estimated 9 children who died from thyroid cancer—the one cancer where the increase due to the radiation exposure is most evident—see Figure 4.11. Thyroid cancer, however, is rarely fatal, with a 5-year survival rate of 96%. The eventual death toll from other cancers over time has been estimated to be 4,000 among the 600,000 persons receiving more significant exposures, plus perhaps another equal amount among 5 million people in the less contaminated areas (UNSCEAR, 2010). Both of these estimates are based on the linear no-threshold model. However, since cancer has a long latency period, and since the number of spontaneously occurring cancers will eventually number in the tens or hundreds of millions, the actual percentage of rise in the cancer death rate will be very modest and almost certainly not detectable.

4.8.2.1 Causes of chernobyl It may be inaccurate to call Chernobyl an *accident*, because it was probably bound to occur sooner or later for this reactor design, so if it was an accident, it was one that was "waiting to

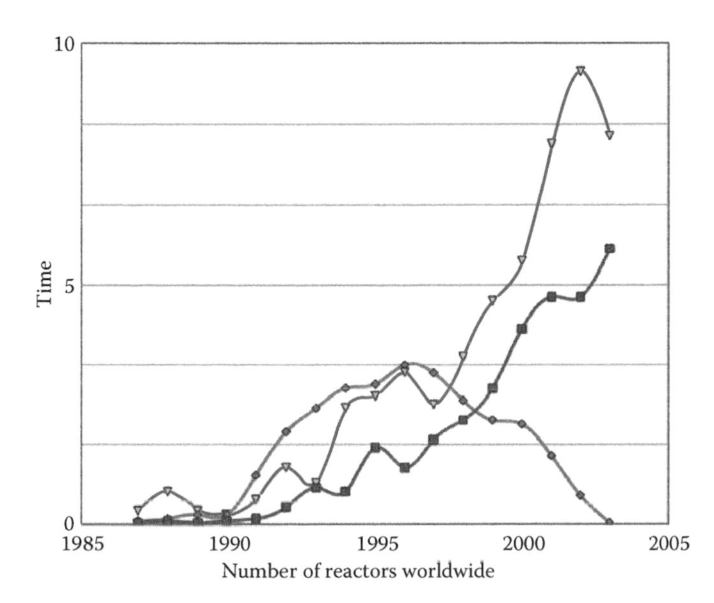

Figure 4.11 Thyroid cancer incidence in children and adolescents from Belarus after the Chernobyl accident per 100,000 persons. *Triangle*, adults (19–34); *square*, adolescents (15–18); *circle*, children (0–14). (Reprinted from *Int. Congr. Ser.*, 1299, Demidchik, Y.E. et al., Childhood thyroid cancer in Belarus, 32–38, Copyright (2007), with permission from **Elsevier**; Cardis, E. et al., *J. Radiol. Prot.*, 26, 127, 2006, **Institute Of Physics**.)

happen." The catastrophe occurred during a systems test that had not been properly authorized at a time when the chief engineer of the reactor was home asleep and the man in charge was Anatoly Diatlov. The purpose of the test was to see if the reactor could safely be shut down if there was a loss of power from the grid to the pumps that supplied the reactor with cooling water. In principle, emergency diesel generators should automatically come on in such an event to supply the needed 5.5 MW of power to run the pumps. However, there was an unacceptable 60 s time delay between the signal that grid power had been lost and the emergency generators coming on and reaching their full power. The engineers thought that the residual rotational momentum of the massive turbines might be enough to bridge the bulk of that 60 s gap, and the purpose of the test was to check this idea—even though three previous tests had given a negative result.

During the start of the test, owing to an error (inserting the reaction quenching control rods), the reactor power level had precipitously dropped to 30 MW—a near-total shutdown and only 5% of the minimum safe power level to conduct the test. Below the authorized level of 700 MW, owing to a known design defect, reactors of the Chernobyl design were unstable and prone to a runaway chain reaction, whereby a small increase in power leads to a still larger increase. Anatoly Diatlov, however, was unaware of this fatal design flaw, and against the advice of others in the control room, he ordered that the test proceed anyway. In an attempt to raise the power level back up to the mandated 700 MW, nearly all the neutron-absorbing control rods were raised out of the reactor also in violation of standard operating procedure. This action essentially disconnected the reactor's "brakes."

After some minutes had elapsed, the power in the reactor was steadily rising and the cooling water began boiling away, leading to a number of low water-level alarms going off. These were foolishly ignored by a crew all too used to false alarms, and the power level rose still further as less and less water was cooling the core. When the crew finally realized what was happening, and they tried to slam on the brakes, the control rods descended far too slowly taking a full 20 s to reach the core after being activated—another design flaw. Moreover, those same control rods, which should have never been fully removed in the first place, had a further flaw. Their graphite tips (the first part to enter the core) actually caused an increase in the reaction rate, not a decrease. With their insertion, the power level in the reactor increased at one point to over 100 times its normal level, and the result was an immense pressure buildup followed by a series of massive explosions and the destruction of the reactor.

To compound matters even further, the use of a graphite core, a reactor roof made of combustible material, and the absence of a containment dome (standard in all US reactors) led to a fire that burned for days and the release of much of the reactors radioactive core into the environment. The firefighters who were called in to put out the fire and many of the other emergency workers had no idea what they were dealing with, and many of them died from radiation sickness. Here is how one French observer sums up the accident in terms of an analogy with a bus careening down a mountain road:

> To sum up, we had a bus without a body careening down a mountain road with a steering wheel that doesn't work and with a brake system that speeds up the vehicle for a few seconds and then takes 20 seconds to apply the brakes, that is, well after the bus has slammed into the wall or gone off into the ravine. (Frot, 2001)

Remarkably, after Chernobyl, Ukraine continued to run the other reactors at Chernobyl for many years, and the last one was not shut down until the year 2000.

HOW MUCH RADIOACTIVITY WAS RELEASED BY CHERNOBYL?

The amount of radioactivity released has been a subject on which a wide range of opinions has been offered. One estimate is 3 billion Curies, which corresponds to a third of the total radioactivity in the reactor core. Despite such an incredibly large release, however, it is noteworthy that the more "slow motion" release that occurred over the years during the period of atmospheric nuclear testing of the 1950s and 1960s was actually about a thousand times greater! However, even those atmospheric tests during their most intense period of 1963 increased the worldwide background radiation level by only 5% (Thorne, 2003).

4.8.3 Reactor Accidents: Three Mile Island

Although the accident that occurred at TMI paled besides Chernobyl and Fukushima in terms of its seriousness (Table 4.1), and its impact on the growth of nuclear power worldwide (Figure 4.12), it was the most serious on American soil. In the minds of many Americans, it is probably considered on a par with Chernobyl given its location in the United States near Harrisburg, Pennsylvania. The accident began on March 28, 1979, as a result of a stuck-open pilot-operated relief valve. The open valve allowed a significant loss of coolant, which went unrecognized by the operators for some time. Eventually, the reactor was brought under

Table 4.1 Comparison of Effects of Chernobyl and TMI

Consequence	Chernobyl	TMI
Radioactivity released	Up to 3 billion curies	Up to 13 million curies
Impact on immediate area	Immediate area uninhabitable	0.3% rise in background
Global fallout	Much of Europe and Asia contaminated	Zero
Health effects (short term)	56 deaths	Psychological distress
Health effects (long term)	Estimated 4000 excess cancer deaths	<1 excess cancer death

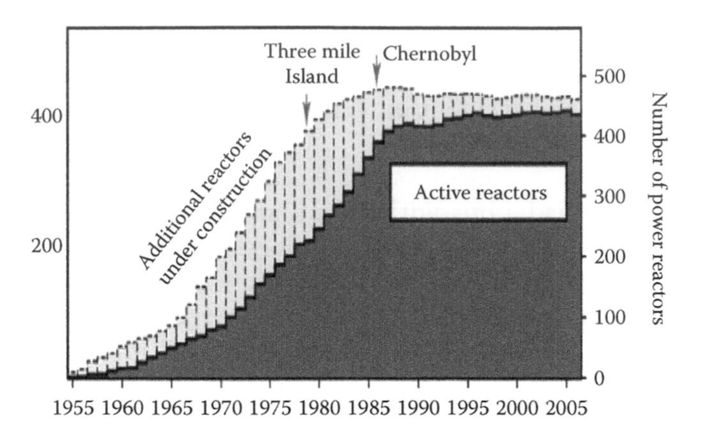

Figure 4.12 Number of reactors worldwide versus time. The impact of the TMI and Chernobyl accidents on the growth of nuclear power is evident from this graph. Since 2007 (the last year of data in the graph), there has been a modest drop of a few percentage with the total in operation being 438 in 2015. (Courtesy of Robert A. Rohde, http://en.wikipedia.org/wiki/File:Nuclear_Power_History.png, image created for the Global Warming Art project and licensed under the Creative Commons Attribution-Share Alike 3.0 Unported license.)

control, but not before a small (nonnuclear) explosion occurred and up to 13 million curies of radioactive gases was released to the atmosphere. The main reason for the absence of global fallout is that the release from TMI was in the form of radioactive gases, not dust, and hence, it did not return to ground level.

4.9 FRONT END OF THE FUEL CYCLE: OBTAINING THE RAW MATERIAL

Deposits of uranium ore can be found in many nations, and the world supply has been estimated to be quite abundant. At the current mining costs and ore grades presently mined, there is about 100 years' worth, but this estimate is misleading because there is much more uranium available if we go to less economical lower grade ores. Even though they may be more expensive (on a per unit energy basis), lower grade ores would have a negligible impact on the cost of nuclear energy, given that the cost of fuel is a very small contributor to the total cost (mainly personnel, construction, and maintenance). At present, the three countries that supply the largest share of uranium are Australia, Canada, and Kazakhstan, which between them, account for 63% of uranium production in 2010. During the twentieth century, the United States was the leading producer of uranium in the world, but given that the best high-grade ores in known deposits have been depleted in the United States, it is cheaper to import from other nations. However, it is worth emphasizing that relying on imports of uranium is quite a different proposition than relying on imported oil, since if the need arose, the United States could use its domestic reserves to satisfy its needs with only a negligible impact on cost, and having Australia and Canada as our main suppliers is less worrisome than relying on oil imports from the Middle East.

Uranium mining tends to be similar to many other hard rock mining operations, and it is extracted either in open pits or underground mines—the latter being more hazardous to miners' health, given the concentrations of radon gas and radioactive dust. In the past, especially during the early period of the 1950s, there was an increase in the cancer death rate among uranium miners due to their exposure to radon. Of course, underground mining of any sort is a dangerous occupation, which continues to get safer over the years. For example, in 1907, there were over 900 deaths among US coal miners due to mine disasters plus many more deaths due to long-term exposure—which is a far cry from the experience of recent years even though accidents continue to happen. Of the various types of mining and drilling to provide energy—both fossil fuel and nuclear—nuclear is certainly the safest on a per kilowatt-hour generated basis. The reason is that due to the extremely high energy density, much less uranium is needed than any fossil fuel—recall the equivalence of one tiny uranium pellet and a ton of coal.

A particularly intriguing method of obtaining uranium involves mining the oceans. The world's oceans have a staggering amount of uranium—about 1000 times what is found on land, but in exceedingly low concentrations, of about three parts per billion. There has been considerable work demonstrating the feasibility of such ocean extraction since the mid-1990s, particularly by Japanese and US scientists. The method is more costly to implement by a factor of 5–10 than mining on land, given the much lower uranium concentration in the oceans. Such a high cost would essentially make it out of the question to extract uranium in this manner, were it not for the fact that the cost of the uranium fuel, as previously noted, is a small fraction of the cost of the energy being generated (see Problem 15). Thus, the existence of this uranium supply in the oceans reminds us that the world has a supply that will last 1000 times longer than can be provided by land-based reserves, even though it is cheaper not to extract it from the oceans for now. Interestingly, even that factor of 1000 greater abundance severely understates the amount of uranium available! It is believed that the crust of the Earth and the sea tend to be in equilibrium chemically, so that as uranium is extracted from the sea, it would tend to be replenished by uranium in the Earth's crust (that is not accessible to mining)—40 trillion tons worth. On the basis of such continued ocean replenishment, it has been suggested that a source of uranium might be available for billions of years.

4.10 BACK END OF THE FUEL CYCLE: NUCLEAR WASTE

A major concern of many people who worry about nuclear power is the waste that reactors generate. This high-level (intensely radioactive) waste consists of many different radioisotopes having many different half-lives, and the radioactivity, therefore, does not decay according to the simple radioactive decay law. Most nuclear waste is classified as low level, in terms of its radioactivity, but the waste generated by nuclear reactors is much more radioactive and considered high level. Although it is sometimes noted that given the very long half-lives of some fission products, high-level wastes will remain hazardous in some cases for hundreds of thousands of years, such statements ignore the fact that after such time spans, the magnitude of the danger is negligible, since the wastes will have decayed to well below the level of radioactivity of the original ore and be equivalent to a small rise in background radiation. Decay of high-level waste to the level of the radioactivity in the original ore requires about 7000 years, but when transuranic elements (elements with $A > 92$) are first removed, the remaining waste decays to the level of the original ore in around 500 years. As a consequence, the prior extraction of the transuranic elements that can provide fuel for breeder reactors could simplify the waste disposal problem, given the much shorter decay times of the remaining wastes.

The three main approaches for dealing with high-level nuclear wastes depending on the length of time after being removed from a reactor are (1) isolation often in water-filled pools for some years when they are initially most radioactive, (2) later storage in casks on-site (often in the open) at the power plant, and (3) final eventual geological disposal underground in a nuclear waste repository. The third and final stage remains hypothetical (at least in the case of the United States), because the Yucca Mountain repository, which has been constructed and approved by Congress in 2002, has been blocked for some time based on various concerns, including geologic stability of the site and the hazards associated with shipping high-level wastes across country by truck or rail. Moreover, all work on Yucca has been halted since 2009, after $15 billion has been spent on the facility. Meanwhile, the imperative to do something with the high-level waste that continues to accumulate (at 2000 tons per year) on-site at the nation's reactors has most recently been recognized by a 2012 Presidential Commission, but the identification of an alternative site remains elusive. Of course, it is also possible that Yucca (the officially designated repository) could be reactivated under a Republican senate and president.

The actual means of storage envisioned would be in a vitrified form in which the wastes become bonded into a glass matrix, which should be highly resistant to water—thus eliminating the possibility of the wastes finding their way into groundwater—a method in use in several countries. For this reason, some observers consider the nuclear waste disposal problem essentially a political one rather than a technical one. Whether it is essentially a political problem or not, the failure of the US government to resolve the impasse surrounding the Yucca Mountain waste repository also causes both the general public and many potential investors in nuclear power to question its viability.

While there continues to be NIMBYism ("not in my backyard") on the part of some citizens toward all things nuclear, opposition toward a long-term nuclear waste repository site anywhere nearby tends to be especially strong compared to public opposition toward a new nuclear power plant. The reasons are understandable, since nuclear power plants bring many more jobs than would a waste repository, and if there is only one in the nation why put it in *my* backyard? Given this reality, the lesson of Sweden—a nation much more dependent on nuclear power than the United States—is instructive. In the past, the Swedes had been rather opposed to nuclear power. However, in 2009, the government, despite some continuing opposition, lifted a 30-year ban on new nuclear plants. In addition, a number of small towns changed their attitude toward a nuclear waste repository in their area from NIMBY to PIMBY ("please in my backyard!") when they became convinced that the facility planned was not a hazard to their health, or at least that the financial incentive the government offered was more than worth the risk. As of 2014, Sweden generates 40% of its electricity from nuclear power—among the world's highest percentage.

4.10.1 Shipping Nuclear Waste

Quite apart from the issue of storing nuclear waste in a permanent repository, many people are much more concerned with the matter of getting it there, especially since that would often involve shipping it cross-country and possibly right through their own town. If or when the nuclear waste repository in the United States should start being used, there might be perhaps 600 cross-country shipments by rail or 3000 by truck annually—up from the present 100. On the other hand, on a worldwide basis, there have been more than 20,000 such shipments annually involving high-level wastes (amounting to over 80,000 tons) and over millions of kilometers by rail, road, or ship. In none of these shipments was there an accident in which a container (always a very sturdy and fire resistant) filled with highly radioactive material has been breached or has leaked.

The steel and lead containers carrying the high-level wastes are protected by armed guards and are designed to withstand serious crashes and fires, but an accident that breached them cannot be ruled out. According to the DOE, in such a worst-case scenario occurring in a major city, there might be as many as 80 deaths from a year's exposure to the radiation—although why someone would remain in a high-radiation area for a full year is unclear. It is instructive to compare the nuclear situation with that of certain other dangerous cargoes—especially chlorine, which is a highly poisonous gas (used in combat in World War I) and for which 100,000 shipments are made annually. The Naval Research Laboratory has done a study that concluded that

> the scenario of a major chlorine leak caused by a terrorist attack on a rail car passing through Washington, D.C., could produce a chlorine cloud covering a 14 mile (23 km) radius that would encompass the White House, the Capitol, and the Supreme Court, endangering nearly 2.5 million people and killing 100 people per second. (NRL, 2004)

At that rate, within the first half hour before any evacuation got underway, there could be 50,000 deaths. Moreover, unlike the case of nuclear waste shipments, aside from predictions of models, there have been real fatalities

Figure 4.13 Shipment of a 90 ton pressurized chlorine gas rail tank within four blocks of the Capitol building in Washington, DC. Helpfully (to terrorists), the nature of the cargo is clearly labeled with a sign on the side, and there are no obvious armed guards onboard. (Courtesy of Jim Dougherty.)

in some very serious accidents involving chlorine shipments. Note that chlorine being heavier than air tends to stay near ground level where most people are, rather than drift away. Surely, no city, especially Washington, DC, would be so foolish as to permit such cargo to transit the city!

After taking the photo in Figure 4.13, Dougherty, a lawyer, wrote and lobbied for the passage of a law that the Washington, DC, council later adopted restricting the movement of ultrahazardous rail cargo through the center of Washington, DC. This law was subsequently challenged by the federal government on the grounds that states and localities cannot interfere with interstate commerce. Dougherty helped defend the DC law in court, and the DC law was upheld. However, since there is no federal law on the books (as of 2007), such ultrahazardous cargo is not prohibited from moving through cities generally.

The main point of this apples and oranges comparison (nuclear waste versus chlorine shipments) is that as a society, we seem to insist on a level of safety in the former case that goes far beyond what we insist on in other categories that actually involve far greater risks and consequences.

4.11 ECONOMICS OF LARGE-SCALE NUCLEAR POWER

Many opponents of nuclear power believe that apart from any environmental or safety issues, it simply makes no sense economically. The opponents could be right, since electricity from nuclear power is currently more expensive than either coal or gas. However, one can also find studies making the opposite claim made depending on whether one is speaking about existing reactors or new reactors. The obvious question is then why is there so much uncertainty regarding the economics of nuclear power relative to other ways to generate electricity?

There are four categories of cost for electricity generation from nuclear reactors: (1) construction, (2) operating and maintenance (including fuel cost), (3) decommissioning, and (4) waste disposal. In the United States, eventual nuclear waste disposal costs are funded by a charge to the utility of a 0.1 cent/kWh. The cost of the fuel alone is a small fraction of the total—amounting to about 0.5 cent/kWh, and the cost to decommission a reactor tends to be around 15% of the construction—also a relatively small contribution of the total. The main costs of nuclear power are for construction of the plant, which typically range from 70% to 80% of the total. In part, the higher relative fraction of costs for construction for nuclear plants is a reflection of a much lower cost for the fuel owing to the enormous energy density of nuclear, but it is also a consequence of the higher construction costs for nuclear power. Capital construction costs for nuclear tend to be higher than for other energy sources for many reasons, including the higher skill level needed for construction workers, and the need for more stringent safety precautions, but the two main reasons have to do with the length of the construction time and the discount or interest rate paid to borrow the money used to construct a plant.

In the United States, in particular, there have been unexpected changes in licensing, inspection, and certification of nuclear plants that have lengthened construction time (in some cases by many years) and increased costs, due to the interest paid on borrowed money. Additionally, the interest rate for these projects tends to be higher than for other capital projects owing to either the perceived greater risks or the greater uncertainties, which in some cases, have been created by unwise policies. For example, the US Nuclear Regulatory Commission (NRC) used to approve new nuclear plants in a two-step process: first, granting approval to begin construction, and then only after completion, granting approval to begin operation of the plant. Investors who loan money for the capital construction phase had no guarantee that the completed plant would ever be allowed to operate. Understandably, this uncertainty might lead investors to demand a higher interest rate particularly after the Shoreham reactor met a fate of exactly this kind! As one incentive to promote nuclear power in the United States, the Congress passed in 2005 a program of loan guarantees for new clean energy plants, making nuclear power potentially more attractive to lenders. Regrettably, the funds for such loan guarantees tends to dry up in bad economic times, which impacts not only nuclear, but large-scale initial investments in various renewable energy technologies, such as solar–thermal plants.

THE FATE OF THE SHOREHAM REACTOR

Shoreham was built on highly populated Long Island, NY, between 1973 and 1984, a construction period of 11 years. Such a long construction time obviously led to skyrocketing costs. The delay, in part, was due to the intense public opposition to the reactor (located only 60 miles from Manhattan), particularly after the 1979 TMI accident in nearby Pennsylvania. The public opposition eventually led to the Governor of the state refusing to sign the mandated evacuation plan for the surrounding area. This action then led the NRC to deny the utility a permit to operate the reactor, which was then taken over by the state, and later decommissioned in 1994.

Subsequent to Shoreham, the NRC wisely changed its policy on the two-step process, and it now grants permission to both start construction and operate the completed reactor in a single step. It has also put in place other rule changes that permit for a more logical and streamlined approval process without compromising safety. Most importantly, each reactor that had been built in the United States was done so as a one-of-a-kind design, and it had to be approved individually. Following the long-term practice in France, as of 1997, the NRC finally approved applications for standardized designs, with four different approved designs (by two companies) as of 2010.

Regardless of national policies, however, the construction of nuclear plants (assuming they are funded by private investors not a government) is always likely to be higher than nonnuclear plants because of their greater complexity, longer construction times, and likely higher interest rates paid for loans. In addition, nuclear plant construction costs have dramatically escalated in recent years, as a result of such factors as a lack of experience in building plants with the recently approved designs and a strong worldwide competition for the resources and manufacturing capacity to build such plants. Thus, whether nuclear is economically favorable entirely depends on what one assumes about the (1) construction cost, (2) interest rate for loans, and (3) construction time.

Based on Table 4.2, the range in electricity costs for new nuclear plants spans an enormous range—a range that is large enough for optimists to say it quite favorably compares with other alternatives and pessimists to make the contrary claim.

No one can predict the future, but unless memories of Fukushima fade fairly quickly, the nuclear pessimists seem more likely to be correct. In fact, the actual costs of new reactors have tended to be far greater than the cost projections of the enthusiasts, as seen in Figure 4.14. Moreover, it is worth noting that virtually all the 31 new plants that had been proposed in the United States by 2009 have been shelved due to the confluence of low gas prices, high costs of nuclear power, and weak demand for new electricity capacity. The 33-year-long stoppage has been apparently ended by two new reactors being planned in Georgia—a project that will undoubtedly

Table 4.2 Cost of Electricity in Cents per Kilowatt-Hour Based on Construction Costs of $5 Billion and $2.5 Billion, Interest Rates from 5% to 10%, and Construction Times from 3 to 7 Years

	Construction Cost							
	$5 Billion				$2.5 Billion			
Interest Rate (%)	3 Years	4 Years	5 Years	7 Years	3 Years	4 Years	5 Years	7 Years
5	5.8	6.1	6.5	7.6	3.7	3.9	4.2	4.9
6	6.6	7.1	7.6	9.2	4.1	4.4	4.8	5.9
7	7.5	8.1	8.9	11.3	4.6	4.9	5.5	7.2
8	8.5	9.4	10.5	14.2	5.0	5.5	6.3	9.0
9	9.5	10.7	12.4	18.2	5.6	6.3	7.4	11.7
10	10.7	12.3	14.7	24.1	6.2	7.1	8.6	16.1

Source: Nuclearinfo.net, *Cost of Nuclear Power*, http://nuclearinfo.net/Nuclearpower/WebHomeCostOfNuclearPower.

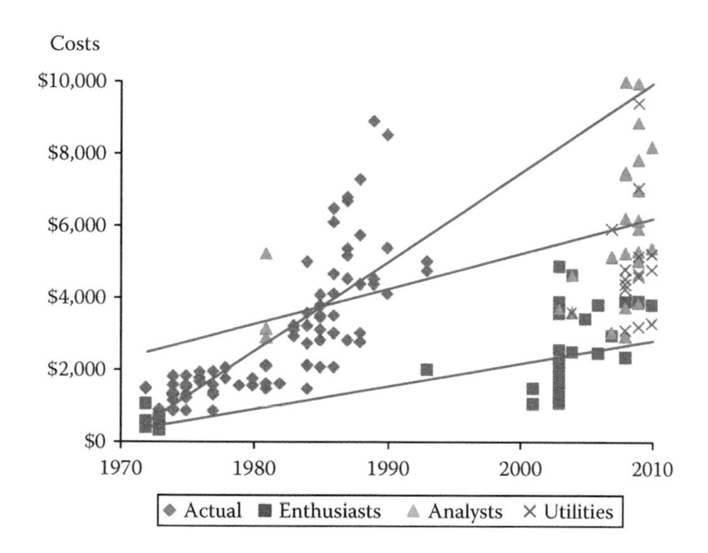

Figure 4.14 Actual costs and cost projections in 2009 dollars per kilowatt for new US nuclear reactors. The cost projections are identified by their source. Note that estimates for other nations particularly China and France tend to be lower than for the United States. (From Cooper, M., *Nuclear Monitor*, 692–693, August 28, 2009. With permission.)

be closely watched by utilities around the country. Nevertheless, there are also grounds for some optimism if you are a proponent of nuclear power. According to the Nuclear Energy Institute, worldwide, 150 nuclear energy projects are in either the licensing or the advanced planning stage as of 2012, and 63 reactors are under construction (NEI, 2012).

4.12 SMALL MODULAR REACTORS

While the economics of traditional nuclear power plants remains uncertain and highly subject to one's choice of assumptions, that for a new type of reactor—the small modular reactor (SMR)—appears to be much more favorable, because it is not subject to escalating on-site construction costs. Although nuclear power is a politically contentious issue, it is interesting that the polarization on the nuclear issue tends to be more correlated with gender than politics (Bisconti, 2010). In fact, there was one nuclear energy-related proposal by President Obama in 2011 that met a very different fate in Congress from almost all his other proposals—energy-related or not. The president, calling for a "new generation" of nuclear power plants, proposed money for research and loan guarantees to help build SMRs. That request for funding was approved by the US Senate by a vote of 99 to 1 with not a single Democrat or Republican opposed.

As of 2010, there are at least eight nations developing SMRs according to 16 different designs. One example is the 25 MW reactor being built by Hyperion Power Generation in the United States, based on designs developed at Los Alamos National lab (Figure 4.15). This small liquid metal-cooled reactor would produce enough power for around 20,000 homes. The reactor has no moving parts and has a sufficiently small amount of fuel that a meltdown is said to be impossible. In fact, Hyperion envisions that the reactors would be made at the factory, shipped in one piece to the site where they will be used, and then buried underground, where they would run with little or no human intervention required during the 10 years for the fuel to burn up.

After that period, a reactor would be shipped back to the factory perhaps on a flatbed truck to have the spent fuel replaced. The goal of the company is to produce power for under 15 cents per kWh, which while not competitive with grid parity in most areas of the United States would be highly competitive for remote off-grid communities and government installations.

Another somewhat higher power reactor (165 MW) of a very novel design is the pebble bed modular reactor (PBMR), first developed in Germany and

Figure 4.15 Conceptual drawing of the Hyperion nuclear battery (the central component of its power generating plant), which stands about the height of a human. (Courtesy of US Los Alamos National Lab, Los Alamos, New Mexico.)

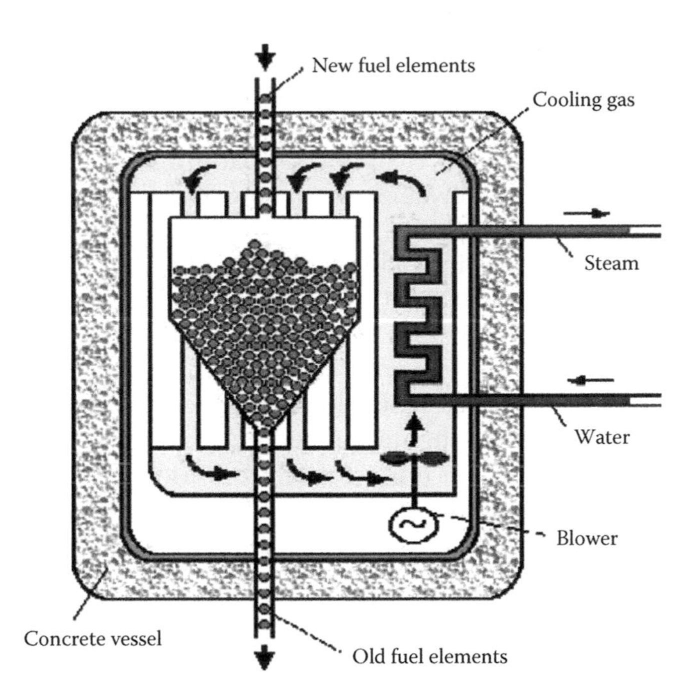

Figure 4.16 Pebble bed modular reactor. (Image created by Picoterawatt and released to the public domain; http://en.wikipedia.org/wiki/Pebble_bed_reactor.)

now being mainly pursued by the United States and China (Figure 4.16). The PBMR is cooled by helium gas, and its fuel is in the form of spherical pellets about the size of tennis balls. Each pellet consists of the nuclear fuel, surrounded by a fission product barrier, and graphite moderator. Simply piling in enough pebbles will allow the reactor to approach criticality. The pellets, because of their size and composition, never get hot enough to melt, so that a meltdown is said to be impossible. In fact, should there be a coolant failure, the effect would be to slow the reaction rate and cause the reactor to shut down. This passive safety feature is diametrically opposite the unfortunate design feature of Chernobyl-type reactors, which become more reactive when they heat up. In the PBMR, at any one time, the reactor vessel contains around 450,000 of the pellets, with new ones continually entering from above and spent ones leaving from the bottom of the reactor vessel. Thus, the reactor is continually being refueled online, and costly shutdowns for refueling are never necessary. Defects in the production of pebbles can, however, cause problems, and in fact, an accident at a German PBMR in 1986 resulting from a jammed pebble did cause a shutdown and resulted in a small release of a small amount of radioactivity.

Other types of SMRs are of a more conventional light water reactor design (such as those made by NuScale Power and by Babcock and Wilcox) that better lend themselves to being scalable so as to provide large amounts of power when a number of them are added as the demand grows. Virtually all the SMR designs rely on passive safety features (no operator intervention required) to maintain safe operation and prevent a catastrophic meltdown. Passive safety features, in general, rely on either the laws of physics, the properties of particular materials, or the reactor design to prevent accidents. In addition to passive safety systems, simplicity of operation, and (perhaps) lower cost, modular nuclear reactors have many other advantages. They can also be placed very close to the need—thus requiring smaller transmission costs and fewer new power lines. Effectively, modular reactors can be thought of as "nuclear batteries," having an energy density millions of times greater than normal batteries that can also provide backup to intermittent renewable energy sources.

The future of SMRs does indeed look bright, but they will probably not come on line for another decade (2020) due to the lengthy NRC approval process required. This process is necessary for usage either in the United States or in other nations in order to comply with IAEA safeguards. Although the US Navy has had a long experience with small reactors for propulsion, the fuel configuration and enrichment levels required for civilian commercial use are quite different, which is the reason that a lengthy process of evaluating new designs is necessary. Moreover, when considering the economics of SMRs, we should remember that many new technologies fail to live up to initial expectations. Recall that large-scale nuclear power, initially thought to be too cheap to meter, may turn out to be too expensive to compete, so one cannot be certain about either the economics or the public acceptance of modular reactors until they meet the realities of the marketplace.

4.13 NUCLEAR FUSION REACTORS

Nuclear fusion, should it prove technically and economically feasible, would be an ideal energy source for many reasons, including an inexhaustible supply of energy in the hydrogen in the world's oceans and the lack of any long-lived fission decay products. The main technical difficulty is (1) achieving the high temperatures needed for controlled nuclear fusion and (2) confining the fuel for a long enough time for self-sustaining ignition to occur. In the core of the sun, gravity is able to provide the confinement, but on Earth, the only two known means of achieving confinement are either a magnetic field or inertial confinement. In the latter case, pellets of fuel are bombarded by powerful lasers from many directions, and the pellet heats up so fast that the inertia of its parts prevents it from blowing itself apart before its temperature is raised to the ignition point. In the other technique pioneered by the Russians in their Tokamak reactor, a diffuse plasma is confined using a magnetic field of a toroidal geometry, which keeps it away from the walls of the vessel while energy is added to heat it. Although considerable progress has been made since the first Tokamak, we are still at least a decade away from a commercially viable fusion reactor. The usual way to measure progress in this field is based on the ratio

$$Q = \frac{\text{power produced}}{\text{power input}},$$

where Q = 1 represents the break-even point; Q = 5 is the point for a self-sustaining reaction, i.e., where power input equals power produced plus power lost; and Q = 22 is what is considered necessary for reactor conditions.

A simplified way for measuring how close a given design is to these Q values is based on the Lawson criterion, i.e., the triple product of the plasma density, temperature, and confinement time. For ignition, i.e., for Q = 1 to occur with the deuterium–tritium (d–t) fusion reaction, the Lawson criterion is

$$nT\tau > 10^{21} \text{ keV s/m}^3, \qquad (4.8)$$

where the temperature T is chosen to have its optimum value. Currently, a seven-nation $15 billion effort is taking place involving an international collaboration known as International Thermonuclear Experimental Reactor (ITER) based in France. It is hoped that ITER will come quite close to the self-sustaining point for a time of 10 s by the year 2018, although as of 2015, its progress is said to be less than promising. There are competing (or complementary) efforts—one known as Ignitor using a "riskier" design originating at MIT that might beat ITER at its own game at only 2% of the cost. Although research continues on fusion using inertial confinement as well, the inertial approach currently appears further from reaching a commercially feasible reactor.

It will be seen from Figure 4.17 that the Lawson criterion (Equation 4.8) is not the whole story in deciding how close we are to achieving ignition, since if it were, curves for constant Q would be shown as horizontal lines rather than the parabolas they appear to be. In other words, the temperature variable T is unlike the other two in Equation 4.8, in that we cannot say that the higher, the better, since above some optimum value, there becomes a decreased chance of ignition—can you think of a reason for this strange fact?

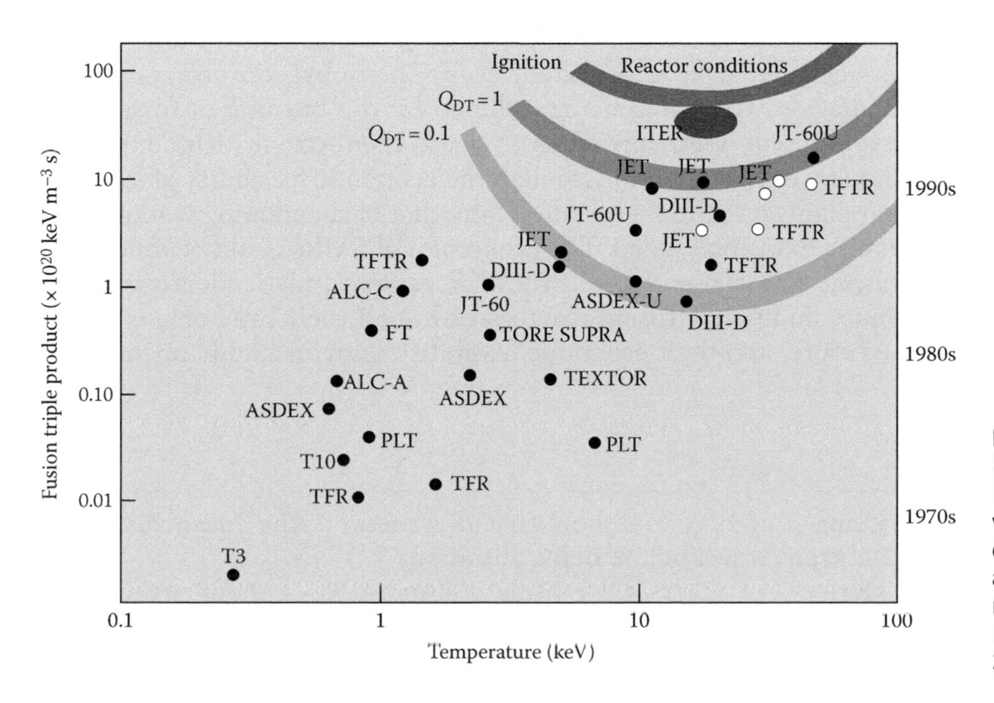

Figure 4.17 Progress toward achieving a fusion reactor: plots of the Lawson triple product (Equation 4.8) versus temperature. The thick curves correspond to particular *Q*-values, and the points show the performance of various designs. (From *Physics World*, Institute of Physics, 2006. With permission.)

UNDERSTANDING THE LAWSON CRITERION: BUILDING A CAMPFIRE

Imagine that you are camping outdoors on a cold evening. There are at least four things you need to build a camp fire: (a) a suitable source of kindling (with a low enough ignition temperature), (b) enough fuel gathered in one place, (c) an initial spark or source of heat allowing the fuel to reach the ignition temperature, and (d) a long enough interval of time without a wind strong enough to snuff the fire out. Each of these four conditions corresponds closely to what is needed to create a self-sustaining nuclear fusion reaction, namely, a choice of light nuclei that can be made to fuse without too high an ignition temperature; a high enough density of the fuel, n; a high enough temperature to which the fuel is heated, T; and a long enough time, t, the fuel is confined before ignition is reached. In fact, it should not surprise us that the product of these three variables needs to exceed some critical value for a self-sustaining ignition to occur—hence the basis of the Lawson criterion.

4.14 SUMMARY

This chapter traced the history of nuclear technology beginning with the World War II era, when the main goal was developing nuclear weapons, not power. Only after the war was the goal broadened to include nuclear power, which has become an important contributor to the generation of electricity in many nations—roughly 20% of all electricity generated. Following several high-profile nuclear accidents, including TMI and especially the much more serious Chernobyl and Fukushima accidents, public opinion has moved sharply away from new nuclear reactors. Many citizens now consider them more of a hazard than coal plants, which, by objective measures, is demonstrably false. According to a May 2, 2014, *New York Times* editorial, "The dangers of nuclear power are real, but the accidents that have occurred, even Chernobyl, do not compare to the damage to the Earth being inflicted by the burning of fossil fuels." Still, so far, only Germany plans to phase them out, which could prove quite costly to that nation. Although the economic feasibility of large new nuclear plants remains highly uncertain, that of abandoning working reactors is also extremely steep. The economics of SMRs could be much more promising than those of the 1000 MW variety. Independently, research continues on nuclear fusion reactors, although their timetable is further in the future, and their economic feasibility remains highly uncertain.

PROBLEMS

1. Using Equation 4.4, show that d represents the mean distance neutrons travel before being absorbed.
2. Derive Equation 4.7 by using conservation of momentum and energy.

3. Discuss the specific simplifications made in Example 1 that led to a too high a value of the critical mass and indicate whether each simplification leads to a value that is too high or too low.
4. Show that if the smallest critical mass for ^{235}U is 52 kg and that if the enrichment level is only 20%, the critical mass rises to over 200 kg.
5. Using Equation 4.7, quantify how effective hydrogen versus carbon would be as a moderator.
6. Approximately, how many elastic collisions would a 1 MeV neutron need to make with nuclei of a moderator in order to have its energy reduced to thermal energies? Do the calculation for light water ($A = 1$), heavy water ($A = 2$), and graphite ($A = 12$) as moderators.
7. The most abundant form of uranium, ^{238}U, spontaneously fissions, so why is it not suitable as a reactor fuel?
8. The specific heat of carbon dioxide at constant pressure is about 1200 J/kg K. An AGRuses CO_2 at 40 atm pressure as its coolant. How much does its temperature rise if the reactor is putting out 2000 MW of heat when operating and the flow rate of the CO_2 through the reactor is 1000 kg/min?
9. Suppose a beam of neutrons containing 10^{15} particles per second is incident on a gold foil. The cross-sectional area of the beam is 1 cm^2. Assume a total cross section for neutron absorption of 10^{-24} m^2 and a foil thickness of 0.2 mm. What percentage of the neutrons in the beam will be stopped by the foil?
10. The walls of the rotor in a gaseous centrifuge spin at almost the speed of sound. If the rotor has a diameter of 0.5 m, how fast does it spin?
11. Explain clearly how you would empirically determine the total cross section for some process.
12. The section on the causes of Chernobyl compares the accident with a bus careening down a mountain road. Explain each of the features of the analogy.
13. Given that Chernobyl is estimated to cause 4000 cancer deaths long term, and the radioactivity released at TMI was 250 times less, one might have expected the long-term cancer deaths caused by TMI to be around 4000/250, i.e., around 16, not less than one as stated in Table 4.1. Explain.
14. Suppose that the intensity of a neutron beam is reduced by 10% when passing through a 10 cm thickness of lead. (a) Find the mean free path for neutrons in lead. (b) If lead atoms are 500 pm apart, find the total cross section for absorption of neutrons by a lead nucleus.
15. Suppose that nuclear plants produce electricity at 7 cents/kWh and that the cost of nuclear fuel is 0.7 cents/kWh, of which 35% is the cost of the ore. Now suppose that the ocean extraction of uranium proves to be 10 times as costly as mining the ore. How much would the cost of electricity rise as a result?

16. Suppose the mean free path for neutron absorption in concrete is 0.5 m. How large a thickness slab of concrete would you need between you and the reactor so that no more than one neutron out of a hundred coming out of a nuclear reactor reaches you?
17. Do some searching on the web to ascertain the following data: the number of nuclear plants that Germany now has and their average power rating and age. Assume that nuclear plants have an average lifetime of 30 years, make a model to calculate the cost to Germany to terminate their reliance on nuclear power assuming they choose to build new renewable power plants based on solar or offshore wind and that they do it (a) on an immediate basis or (b) on a phased basis as existing nuclear plants reach the end of their assumed 30-year lifespan.
18. If a 1000 MW nuclear plant is 35% efficient, how many gallons of water would need to flow through the reactor per minute if the water temperature is raised by 10°F?
19. Explain why there is an optimum temperature to achieve fusion (see Figure 4.17), i.e., why you need to have a larger triple product (Equation 4.8) if the actual temperature is either lower or higher than the optimum. What does the optimum appear to be for the d–t reaction from Figure 4.17?
20. Find the loss in mass of the nuclear fuel in a 35% efficient 1000 MW reactor running for 1 year.
21. Assume that a 35% efficient 1000 MW reactor has its fuel rods replaced with new ones after about 12 years. Uranium-235 has an energy density of 80×10^{12} J/kg, and typical nuclear fuel has 4% enrichment in this isotope. Find the fraction of the original energy that has been removed from the fuel rod during its 12 years in the reactor, assuming only 10% reactor downtime during that period.
22. Prove the correctness of Equation 4.6 from the units on each quantity.
23. If your view of nuclear power is that it is too risky to be pursued, look up some sources that support this view, and in a one-page description, see if you can find any flaws in the arguments. Do the same if your view happens to be that nuclear power should be pursued.
24. How much fuel will a 1 GW nuclear power plant require per year? Assume that the uranium fuel is only 5% of the uranium and the energy density of ^{235}U is that as calculated in the problem section of Chapter 3. Also, how much uranium is this expressed in tons?
25. If you were located 100 km from a 1 GW nuclear power plant, what would the neutrino flux be at your location? Assume that a 1 GW nuclear power plant releases 10^{21} neutrinos per second and you present a 1 m^2 surface to the neutrino flux.

REFERENCES

Cardis, E. et al. (2006) Cancer consequences of the Chernobyl accident: 20 years on, *J. Radiol. Prot.*, 26, 127–140.

Cooper, M. (2009) The economics of nuclear reactors: Renaissance or relapse? *Nuclear Monitor*, August 28, 2009, 692–693.

Demidchik, Y. E. et al. (2007) Childhood thyroid cancer in Belarus, *Int. Congr. Ser.*, 1299, 32–38.

Evans-Pritchard, A. (2010) (Article about thorium reactors and about Nobel laureate Carlo Rubbia's work on thorium reactors), *The Telegraph*, December 21, 2011.

Frot, J. (2001) *The Causes of the Chernobyl Event.*

Hahn, O., and F. Strassmann (1939) Über den Nachweis und das Verhalten der bei der Bestrahlung des Urans mittels Neutronen entstehenden, Erdalkalimetalle (On the detection and characteristics of the alkaline earth metals formed by irradiation of uranium with neutrons), *Naturwissenschaften*, 27(1), 11–15.

Meitner, L., and O. R. Frisch (1939) Disintegration of uranium by neutrons: A new type of nuclear reaction, *Nature*, 143(3615), 239.

NEI (Nuclear Energy Institute) (2012) Nuclear Energy Institute advertisement, *Newsweek*, May 21, 2012.

NRL (Naval Research Laboratory) (2004) Scenario drawn from a Naval Research Laboratory study, cited by "Hazardous Proposals," *Traffic World Magazine*, February 23, 2004.

Nuclearinfo.net (___) *Cost of Nuclear Power*, Nuclearinfo.net. http://nuclearinfo.net/Nuclearpower/WebHomeCostOfNuclearPower.

Pfau, R. (1984) Quoted in R. Pfau (Ed.), *No Sacrifice Too Great: The Life of Lewis L. Strauss*, University Press of Virginia, Charlottesville, VA, p. 187.

_____ (2006) *Physics World.*

UNSCEAR (United Nations Scientific Committee on the Effects of Atomic Radiation) (2010) *Chernobyl's Legacy: Health, Environmental and Socio-Economic Impacts and Recommendations to the Governments of Belarus, Russian Federation and Ukraine.* International Atomic Energy Agency—The Chernobyl Forum, 2003–2005, Vienna.

Biofuels

5.1 INTRODUCTION

The surface of the Earth is bathed by solar energy totaling 3.8×10^{24} J during the course of a year, which is equivalent to a power of about 120,000 TW. Of that amount, less than 0.1% is converted via photosynthesis to plant matter, yet that tiny percentage is over six times greater than all the energy used by humans in a year. The term *biomass* is used to describe plant and animal matter, living or dead, the wastes from such organisms, and the wastes of society that has been generated from these organisms. The stored chemical energy that biomass contains is referred to as bioenergy. Fossil fuels could be said to contain bioenergy from ancient plant material, but unlike recently created biomass, fossil fuels are, of course, nonrenewable. Thus, the three terms *bioenergy, biomass*, and *biofuels* (fuels made from biomass) generally exclude fossil fuels. Biofuels include liquids, such as ethanol alcohol, biodiesel, and various vegetable oils; gases, such as methane; and solids such as wood chips and charcoal.

An important advantage of biofuels over fossil fuels, apart from their being renewable, is that biofuels in principle can be used without adding any net CO_2 to the atmosphere provided that during the growth of the biomass, CO_2 was removed from the atmosphere by no more than the amount it is later added when the fuel is consumed. In this case, we would describe the biofuel as being carbon neutral (or perhaps even carbon negative) over their life cycle. Of course, this assertion assumes that the CO_2 emitted during the planting, cultivating, and harvesting the biomass, along with that released when it is converted to a biofuel and finally transported and used, are not great enough to alter the balance, which need not be the case. Nevertheless, biofuels are generally considered to come under the heading of renewable energy sources and, in fact, comprise about 63% of them.

On the other hand, that amount mainly includes traditional fuels such as wood and dung used by about half the world's population for heating and cooking. In fact, in some developing countries, firewood accounts for 96% of their total energy consumption. When used in this manner, biofuels are hardly renewable and can be extremely destructive to the environment, since trees are usually harvested as needed for firewood, with no replanting. Regrettably, there are excellent low-tech alternatives to firewood readily available for cooking in developing nations that need to be better known—see the solar cookers described in Chapter 10.

Apart from wood and dung for heating, most biofuels in use today throughout the world are still of the "first-generation" variety, meaning that they are made from the sugars, starches, and vegetable oils found in food crops, from which they can be readily extracted using well-established technology. The two biofuels in greatest usage worldwide are bioethanol and biodiesel, which are almost always produced

CONTENTS

using edible crops such as corn, sugarcane, or sugar beets. Starting from near zero in 1975, bioethanol production has dramatically risen in recent years, reaching over 30 billion gallons annually. These biofuels are also surprisingly concentrated in particular nations or regions; thus, 90% of all bioethanol is produced in just two nations: Brazil and the United States; while a majority of the world's biodiesel is produced by the European Union (EU). Most biofuels are of course used in the transportation sector, where they can offset a certain percentage of a nation's petroleum consumption, either as an additive to gasoline or diesel fuel or, in some cases, as a one-for-one replacement, provided the engines have been modified to allow this substitution. Worldwide, biodiesel and bioethanol now account for just under 3% of the fuel used for road transportation, although the International Energy Agency has estimated that they could supply more than 25% of the world demand by 2050 (EIA, 2011). Thus, the use of biofuels, especially biodiesel and bioethanol, has been rapidly growing and is expected to continue its increase in the future (Figure 5.1).

In general, biofuels come in either liquid, gaseous, or solid form. However, since the large majority of biofuels are used in the transportation sector, liquids or gases are far preferable to solids, which tend to be largely used for either heating or electric power generation. When gaseous biofuels are produced, it is useful if they can be easily liquefied at room temperature, owing to the greater energy density of liquids and the difficulties of storing gases under very high pressures. Liquids also have many other advantages as transportation fuels, including being cleaner burning, easier to transport and store (because they can be pumped and sent through pipelines), and, of course, usable in an internal combustion engine.

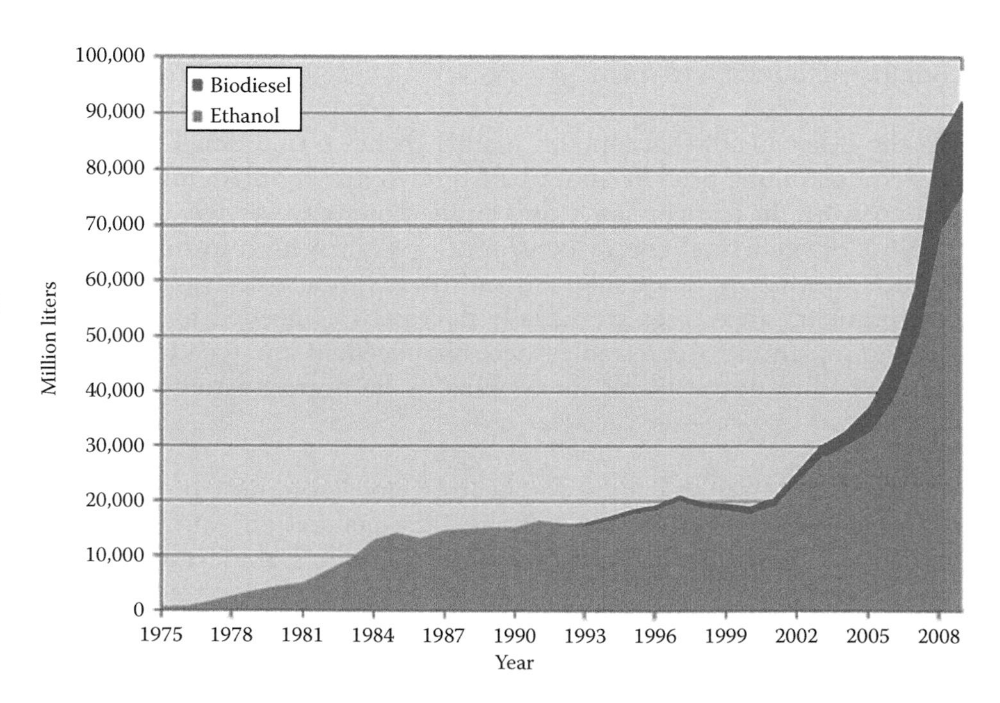

Figure 5.1 Worldwide, bioethanol and biodiesel production. In the years since 2009 (the last year in the chart), biofuel production has continued to grow, but at a slower pace due in part to a slow economy. Still one 2014 projection expects production to reach 135 billion liters by 2919—a 2.7% annual growth rate. (Courtesy of Worldwatch Institute, Washington, DC, http://arizonaenergy.org/News_10/News_Nov10/GrowthofBiofuelProductionSlows.html.)

5.2 PHOTOSYNTHESIS

Photosynthesis is the process whereby energy-rich organic compounds such as sugars and starches are created from carbon dioxide and water by sunlight. When sugar and oxygen are the end products of the reaction, the overall chemical equation may be written as

$$6CO_2 + 6H_2O \rightarrow C_6H_{12}O_6 + 6O_2. \tag{5.1}$$

Thus, photosynthesis plays an important part of the carbon cycle, whereby the element carbon is exchanged among the biosphere, atmosphere, oceans, and land. The process of photosynthesis occurs in plants; algae; and some species of bacteria, such as cyanobacteria (also known as blue-green algae), and it is the ultimate source of energy for most life on Earth, including us.

CYANOBACTERIA

Cyanobacteria often caused by runoff of sewage or fertilizers sometimes causes algae blooms in lakes and rivers. In fact, according to the US Department of Agriculture, such nonpoint sources as sewage and fertilizers constitute the primary pollution hazard in the nation. The bacteria causing these algae blooms can be harmful to humans and other living creatures and cause significant environmental and economic damage, e.g., through its impact on seafood harvests. In humans, studies show that these bacteria are harmful to the liver and nerves and can be responsible for several serious diseases.

The only organisms not ultimately dependent on photosynthesis are some bacteria and other single-celled organisms known as archaea, living deep underground or undersea. Many deep undersea organisms live near thermal vents and use heat as their energy source, although recently, researchers have discovered that some of them, in fact, also live off the dim light from thermal vents as well (Blankenship, 2005). Of course, most photosynthesis in the oceans occurs at the surface or at moderate depths (the euphotic zone) that typically extend between 10 and 200 m depending on the murkiness of the water. The extent of oceanic photosynthesis also depends on the amount of incident sunlight reaching the ocean surface, which, in turn, depends on the latitude and time of year. On land, the amount of photosynthesis also depends on the richness of the soil, which is a function of specific geographic features, such as forests, deserts, and mountain ranges—Figure 5.2 shows the amounts of photosynthesis occurring on both land and sea. Sea-based photosynthesis mainly occurs as a result of single-celled plants known as plankton. It has been estimated that these marine plankton produce perhaps half of the Earth's oxygen, even though their total biomass is orders of magnitude below that of terrestrial plants.

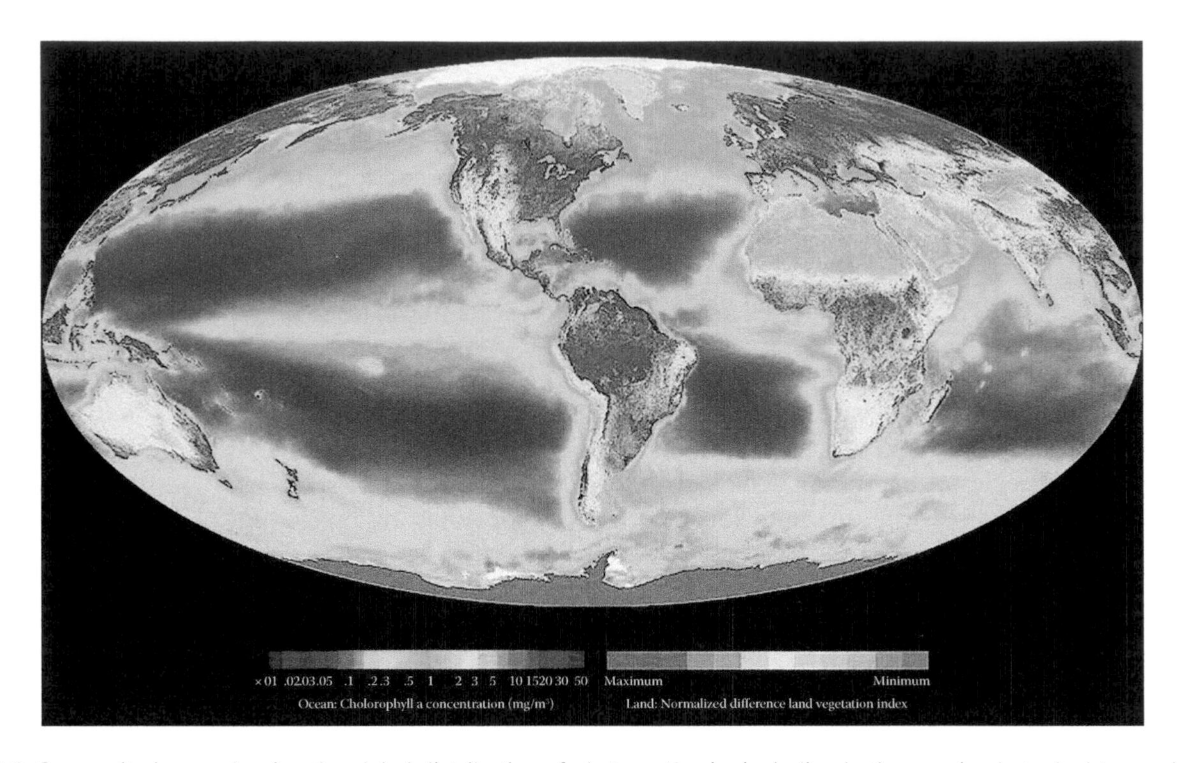

Figure 5.2 Composite image showing the global distribution of photosynthesis, including both oceanic phytoplankton and vegetation. The composite color-coded image gives an indication of the magnitude and distribution of global primary production of chlorophyll in units of milligrams per cubic meter chlorophyll. It was compiled from satellite imagery taken over the period of September 1997–August 1998. (Courtesy of National Aeronautics and Space Administration/Goddard Space Flight Center, Washington, DC, and Orbimage, Herdon, Virginia; Courtesy of Wikimedia Foundation, San Francisco, California, http://en.wikipedia.org/wiki/Photosynthesis.)

Photosynthesis absorbs about 10^{14} kg of CO_2 per year from the atmosphere converting it to biomass, which, of course, is returned to the atmosphere when that biomass decays. Thus, there is no net removal or addition to the atmospheric CO_2, unless the total biomass should shrink or grow, e.g., through deforestation or reforestation.

The process of photosynthesis is known to occur in two stages. In the first stage, light is captured primarily using the green pigment chlorophyll, and the energy is stored in energy-rich molecules such as adenosine triphosphate and nicotinamide adenine dinucleotide phosphate (Figure 5.3). In the second stage, light-independent reactions occur and CO_2 is captured from the atmosphere. It is in this stage that the carbon is fixed or converted to plant matter such as sugar or starch in a series of reactions known as the Calvin cycle. Certain wavelengths of light are especially important in the first stage process, and while most photosynthetic organisms rely on visible light, there are some that use infrared or heat radiation—hence, the existence of those organisms capable of living deep underground or near undersea thermal vents. The details of the reactions in both stages of photosynthesis need not concern us here, as they involve some complex biochemistry.

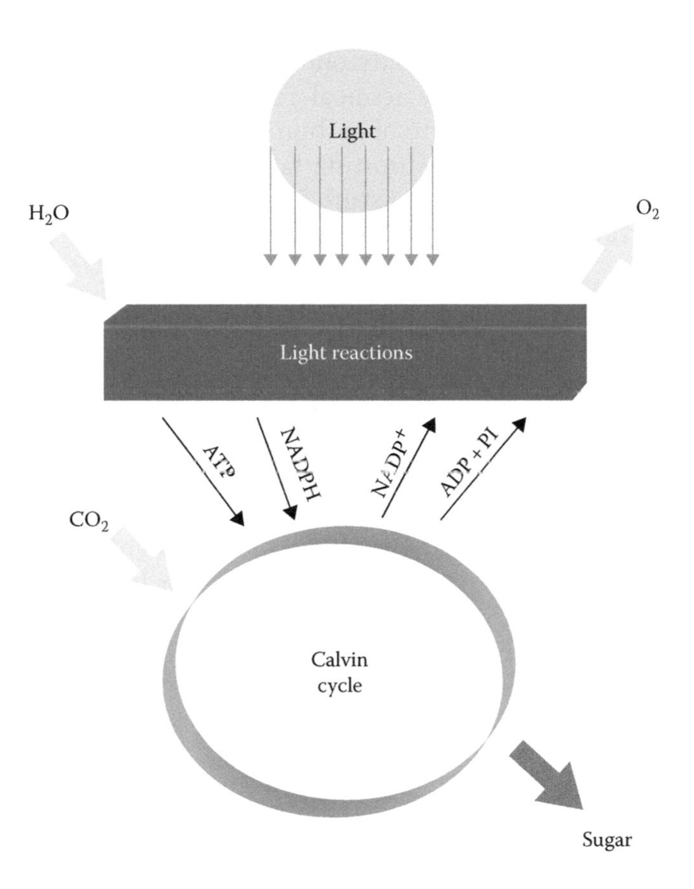

Figure 5.3 Simplified version of the two phases of the chemical reactions taking place during photosynthesis.

5.2.1 Example 1: Efficiency of Photosynthesis

The total energy stored worldwide in the sugar produced by photosynthesis is about 8.4×10^{21} J per year. Assuming that the process is 3–6% efficient in collecting sunlight, find the fraction of sunlight that is stored in sugars and approximately what area of the Earth's surface is covered by photosynthetic organisms.

Solution

As previously noted, the annual incoming solar energy totals 3.8×10^{24} J in a year, so the fraction of incoming solar energy creating sugars via photosynthesis is therefore about $8.4 \times 10^{21}/(3.8 \times 10^{24}) = 0.002$ (0.2%). If the actual efficiency of the overall reaction is 3–6%, this implies that the fraction of the surface covered by photosynthetic organisms is between $(1/3)(0.2) = 0.067$ and $(1/6)(0.02) = 0.033$ or between 3.3% and 6.7%.

What does the actual efficiency of photosynthesis depend on in a given location? The most important variables are as follows:

- Light intensity
- Light spectrum, i.e., wavelengths of light present
- Atmospheric carbon dioxide concentration
- Ambient temperature

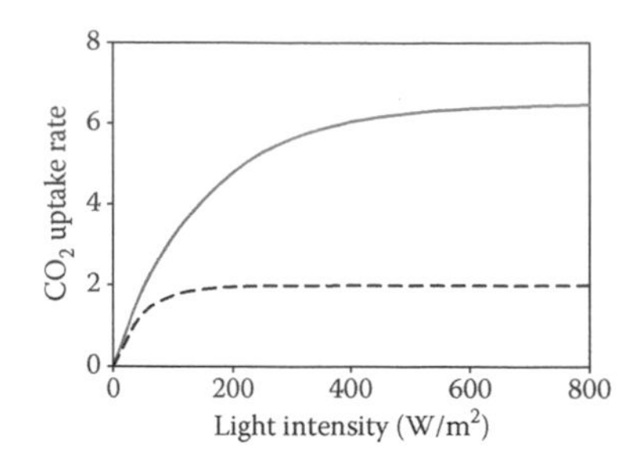

Figure 5.4 CO_2 uptake rate (a measure of photosynthetic activity level) versus the light intensity in watts per square meter for a typical sun plant (*solid curve*) and shade plant. Full sun might correspond to around 750 W/m^2. Note how typical shade plants have a photosynthetic activity level that saturates at levels much less than full sun.

Some of these variables affect only one part of the overall process; thus, for example, although photochemical reactions of stage one are unaffected by temperature, the rate of carbon fixation during stage two does. There is also the matter of a limiting factor in regard to plant growth, whereby a variable such as moisture or light intensity assumes much greater importance when it is below some critical value.

How well does nature do in gathering solar energy compared to humans' best efforts? Although photosynthetic efficiency lies in the 3–6% range, commercial solar panels convert between 6% and 20% of sunlight into electricity, and their efficiency can reach over 40% in the laboratory. Of course, while less efficient in terms of energy conversion, plants are also capable of assembling themselves out of the absorbed energy and soil nutrients—something that humans have not yet designed solar panels to achieve! One of the other reasons that photosynthesis has a rather low efficiency is that the reaction saturates at a low light intensity—much less than full sunlight (Figure 5.4). It is not known why evolution would favor this fact, but of course, evolution favors reproductive success, which need not be the same as the efficiency of energy conversion by a plant.

5.2.2 Example 2: Best Wavelengths of Light for Photosynthesis

What parts of the visible spectrum would you imagine are most effective in promoting photosynthesis in green plants?

Solution

The color of any opaque surface viewed under white light is directly linked to light reflected from that surface. Thus, a surface that appears green, such as nearly all plant leaves, is reflecting green wavelengths more so than those at either larger and smaller wavelengths, i.e., toward the red and blue ends of the spectrum—see Figure 5.5. If plants have evolved to maximize the rate of photosynthesis under white light illumination, this must mean that the wavelengths away from the green middle of the spectrum are more effective in promoting photosynthesis. In fact, studies have shown that chlorophylls *a* and *b* (the two primary compounds responsible for photosynthesis) absorb wavelengths of light with highest efficiency in wavelength regions λ_1 = 439–469 nm (in the blue) and λ_2 = 642–667 nm (in the red). For this reason, plants grown under artificial lighting (using light-emitting diodes with the appropriately chosen wavelengths can be made to grow with greater efficiency than if one attempts to recreate natural lighting indoors. Note that for plants grown under natural light, the fraction of the incident solar energy that they utilize is the ratio of the areas under the black absorption curve to that of the full spectrum in Figure 5.5.

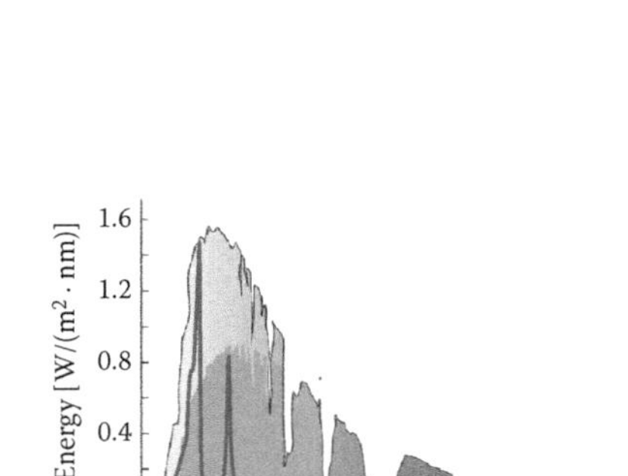

Figure 5.5 Spectrum of sunlight at sea level and the absorption curve for chlorophyll *b*, one of two important pigments found in almost all plants, algae, and cyanobacteria.

IS CO_2 "GREEN"?

As already noted, atmospheric CO_2 levels are a factor in the rate at which photosynthesis occurs, with higher CO_2 levels having a fertilizing effect, especially for some types of plants. Climate change skeptics therefore sometimes argue that atmospheric CO_2 is "green" in promoting plant growth, so the more atmospheric CO_2, the better. In fact, one nonprofit organization known as CO_2 Is Green, with ties to the oil and gas industry, lobbies against limiting CO_2 on precisely these grounds. The claim that CO_2 is green like any effective propaganda effort does have some element of truth. Thus, experiments with crops grown in plots of land subject to normal and elevated CO_2 levels do show that at least for some types of crops, growth is enhanced by about 13% when CO_2 levels are elevated (Chandler and Le Page, 2007). However, the effects appear to level off after a few years, and for most crops, the other variables—especially moisture and temperature—assume greater importance. For example, a 20-year study of rainforest plots in the tropics has shown that local temperature rises of more than 1°C reduce tree growth in half (Fox, 2007). Thus, in a world with higher CO_2 levels, which, by the greenhouse effect, will also be a warmer world, the higher temperatures will very likely have more of an inhibitory effect on plant growth than any benefit that higher CO_2 levels might have. This negative effect is likely to be compounded by drier conditions in continental interiors, which would also have a negative impact on 95% of plants that fix carbon according to the C3 metabolic pathway, since such plants, including rice and barley, do very poorly in hot dry climates.

5.3 BIOFUEL CLASSIFICATIONS

Three ways to characterize biofuels are in terms of (1) their feedstocks, i.e., the inputs; (2) the process used to produce the fuels from their feedstocks; and (3) the outputs of these processes. The importance of considering the feedstocks and processes as well as the end product is nicely illustrated by a comparison of two nations' approaches to the production of the same product, bioethanol.

5.3.1 Choice of Feedstock for Biofuels

The United States and Brazil between them produce 88% of all the bioethanol in the world, but the experiences of the two nations have been quite different. Brazil has been in the ethanol production business longer than the United States—ever since the 1973 Arab oil embargo—and its program is far larger in scope considering the relative sizes of the two nations. Thus, while the United States produces about a third more ethanol than Brazil, that amount offsets only a meager 4% of the US demand for gasoline, while in Brazil, it offsets fully half the demand. In the United States, ethanol is primarily used as a gasoline additive—up to 15%, while in Brazil, many cars can now run on any blend up to 100% ethanol (E100). In fact, such cars now constitute over 90% of all new cars and light trucks sold in Brazil.

Ethanol can be produced from a variety of crops, including sugarcane, cassava, sorghum, sweet potato, corn, and wood. The primary difference between ethanol production in the United States and Brazil involves their choice of feedstock—corn for the United States and sugarcane for Brazil (Table 5.1).

The Brazil–United States ethanol experiences also differ in many other respects, with the Brazilians generating nearly twice as much ethanol per acre of crop—in part, the result of the different sugar content of the two crops, but also the result of a 40-year Brazilian research and development program for agriculture. Brazil has now achieved the most efficient technology for sugarcane cultivation in the world, and it has tripled its production per acre over a recent 30-year period. This high production efficiency also directly translates into energy efficiency, which is often defined in terms of the net energy ratio (NER) or the ratio of the energy a biofuel supplies to that required to produce it. For Brazilian ethanol, the NER is as high as 10, while for the United States, it is a meager 1.3. Effectively, this means that US corn ethanol yields 30% more energy than was used to create it, while for Brazilian ethanol, it is 900% more. Part of the high-energy efficiency for Brazilian ethanol results from the practice of harvesting the sugarcane residue (bagasse) and burning it to produce electricity that powers the operation.

An equally impressive comparison (favoring the Brazil model over the US model) lies in the relative greenhouse gas (GHG) reductions, which is far greater for Brazil—even when the loss of Amazon rainforest to create agricultural land is taken into account. In fact, studies have shown that the extra GHG emissions resulting from the loss of rainforest when the land is used for sugarcane to produce ethanol can be recouped in

Table 5.1 Brazil and the US Ethanol Production

Criterion Feedstock	Brazil Sugarcane	US Corn
Fuel production	6472 Mgal	9000 Mgal
Share of gas market	50%	4%
Arable land used	1.5%	3.7%
Fuel per hectare	1798 gal	900 gal
NER	8.3–10.2	1.3
GHG reduction	61%	19%
Blend for existing vehicles	E25	E10
Blend for new vehicles	E 25–E 100	E10
Subsidy	None	Substantial
Use of waste	Energy generation	Livestock feed

Note: GHG reductions include land use changes.

about 4 years if the ethanol replaces gasoline, while in the case of US forest land replaced by corn ethanol, the corresponding figure is 167 years (Searchinger et al., 2008).

Finally, all the Brazilian sugarcane used for ethanol production has been accomplished without government subsidies, while the United States has taken the opposite course, largely for political (not economic or environmental) reasons. Interestingly, these subsidies totaling about $5 billion annually are paid not to farmers (who would plant the corn anyway) but to the oil industry to induce them to include the ethanol into their product. The $6 billion ethanol fuel subsidy was allowed to expire in 2011. However, one could say that the subsidy is just hidden a bit better, since corn grown for ethanol is now covered by the equally unfortunate renewable fuel standard.

Let us now broaden the possible feedstock choices beyond just sugarcane and corn and consider the six feedstocks shown in Table 5.2, three of which could be used in the production of ethanol and three in biodiesel. These feedstocks have been rated here on their greenness based on three important criteria: (1) their contribution to CO_2 emissions per unit energy the fuels contain, (2) their usage of various resources (water, fertilizer, pesticide, and energy), and (3) their availability. Availability has been expressed in terms of the percentage of existing US cropland they would consume to produce enough fuel to displace half the gasoline needed for road transportation. The entries in the GHG emissions column of Table 5.2 are those for the complete life cycle of the fuel and should be compared to those of gasoline, i.e., 94 kg CO_2/MJ, but note that some are actually negative, which requires that

Table 5.2 Comparison of Six Biofuel Feedstocks Used to Make Ethanol and Biodiesel

Crop	NER	GHG (kg CO_2/MJ)	Resources Used (W, F, P, E[a])[b]	Yield (L/ha)	Percentage of US Cropland (%)
For Ethanol					
Corn	1.1–1.25	81–85	H, H, H, H	1,135–1,900	157–262
Sugarcane	8–10.2	4–12	H, H, M, M	5,300–6,500	46–57
Switchgrass	1.8–4.4	−24	L, L, L, L	2,750–5,000	60–108
For Biodiesel					
Soybeans	1.9–6	49	H, L, M, M	225–350	180–240
Rapeseed	1.8–4.4	37	H, M, M, M	2,700	30
Algae	–	−183	M, L, L, H	49,700–109,000	1.1–1.7

Source: Groom, M. et al.: Biofuels and biodiversity: Principles for creating better policies for biofuels production. *Conserv. Biol.*. 2007. 22. 602–609. Copyright Wiley-VCH Verlag GmbH & Co. KGaA. Reproduced with permission.

Note: The percentage of US cropland is that needed to supply half the nation's fuel for its road transportation.

[a] Resources include water (W), fertilizer (F), pesticides (P), and energy (E).

[b] H: high; M: medium; L: low.

more carbon be removed from the atmosphere during their growth than is later returned to it. This carbon sequestration occurs because some grasses have been found to store carbon in the soil through their roots during their growth.

The vast differences between the numbers in Table 5.2 for the various feedstocks are quite striking, especially in regard to the percentage of cropland needed to satisfy half the US transportation needs. The values here range from 1% to 2% for biodiesel made from algae to an impossible 262% of US arable land for ethanol made from corn. It is clear from this table that if one wanted to select the worst possible choice of crop from these seven to use to generate a transport biofuel, corn ethanol probably would be at the top of the list. The best current candidate for an advanced biofuel (miscanthus) does not appear in Table 5.2 (Figure 5.6). It, like the best ones listed there (switchgrass and algae), involves converting the nonsugar components of plants (cellulose and lignin) into biofuels, and this step involves technology not yet ready for economically viable widespread use.

Figure 5.6 Field of *Miscanthus giganteus*. (Courtesy of Pat Schmitz, http://en.wikipedia.org/wiki/Miscanthus_giganteus#cite_note-5, image licensed under the Creative Commons Attribution-Share Alike 3.0 Unported license.)

MISCANTHUS

Miscanthus is a tall grass that can grow to heights of more than 3.5 m in one growing season. It has been long hailed as a superb candidate for biofuel production, because of its rapid growth, high yield per acre (about 25 tons), and low mineral content. Most importantly, it is not used for food and can grow in some places not well suited to many food crops. *Miscanthus* has very low nutritional requirements, which enables it to grow well on barren land without the aid of heavy fertilization.

5.3.2 Biofuel Production Processes

Having seen the importance of the choice of feedstock for a given biofuel, here we examine the differences in the various biofuel production processes. These processes can be classified into three broad categories, thermochemical, biochemical, and agrochemical, each of which has several subcategories (Figure 5.7).

Thermochemical processes based on their name obviously use heat to induce chemical reactions. The most well known of these would be the direct combustion of biomass—either for heating, cooking, generating electric power, or supplying energy to drive some industrial process. Direct combustion in its simplest form is thus not a process for creating a biofuel, but rather a usage of the original biomass as the fuel to produce energy. In this case, it is important that the biomass be completely dried and preferably homogeneous in composition. Pyrolysis (*pyr* for "fire" and *lysis* for "separating") is the process of anaerobic decomposition of organic material using heat. Pyrolysis differs from direct combustion in three important ways:

- Being anaerobic, the process occurs in the absence or near absence of oxygen.
- Moisture may be present and is sometimes essential.
- The decomposed material retains its stored energy afterward.

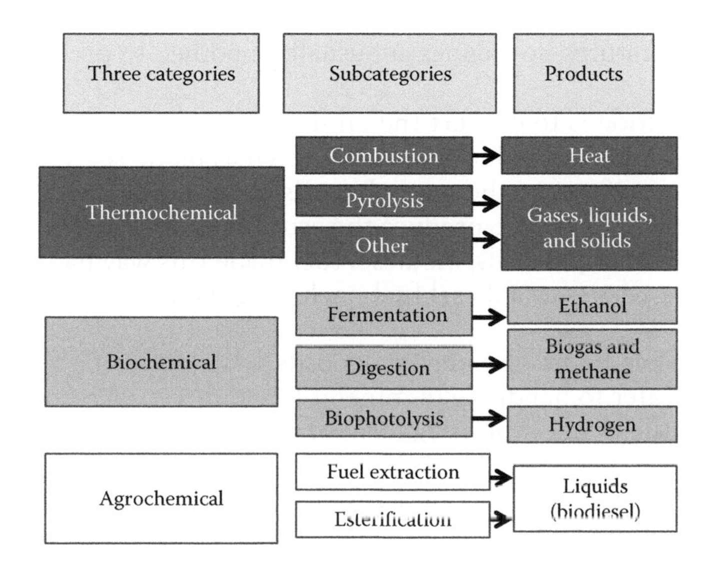

Figure 5.7 Summary of the categories of biofuel production processes.

The end product of pyrolysis may be a combustible solid, liquid, or gas, in which case, the process is referred to as *gasification*. A variety of other industrial thermochemical processes exists besides combustion and pyrolysis that involve sophisticated chemical control.

Biochemical processes obviously use biological organisms including bacteria, yeasts, or other microorganisms to induce chemical reactions in the original biomass. One biochemical subcategory involves the process of digestion. Our bodies use this process to convert food into substances that can be absorbed and assimilated. In its general meaning, *digestion* refers to the decomposition of organic matter by bacteria, either in the presence of oxygen (aerobically) or its absence (anaerobically). Anaerobic digestion, which also occurs in the stomachs of cattle and other ruminant animals, where one end product is biogas, a mixture of methane and CO_2 that is also known as sewage or landfill gas. Another biochemical process, fermentation, usually refers to the conversion of carbohydrates such as sugars into ethyl alcohol. The third subcategory of a biochemical process for producing biofuels is biophotolysis. Photolysis as its name suggests involves making use of the energy in light to cause a chemical decomposition. Thus, biophotolysis involves having microorganisms help drive the process. In the present context, this involves splitting the water molecule into its constituent hydrogen and oxygen gas—the former being an energy-rich store of energy.

Agrochemical processes are the third way that biofuels can be created. One subcategory involves the direct extraction of useful products from living plants by tapping into their trunks or stems or by crushing them—molasses and latex rubber production being two examples of such extracts or exudates. In many cases, these plant exudates are fuels, which may serve as petroleum substitutes. Oils, for example, directly extracted from plant (or animal) matter may be used to power a diesel engine. In fact, some owners of diesel-powered cars fill up their tanks for free using the used vegetable oil from restaurants after it has been filtered. On the other hand, the high viscosity of these oils can cause engine problems, especially at low temperatures, so engines are usually modified to preheat the oils.

A better solution is to convert the oil into a chemical compound known as esters in order to produce the fuel known as biodiesel. In this process known as *esterification*, the vegetable oils or animal fats are chemically reacted with an alcohol to produce the ester. In addition to having a lower viscosity than the pure oils, biodiesel fuel made this way has many highly desirable properties as a diesel fuel, including the following:

- Being able to dissolve engine deposits
- Being safer to handle than mineral-based diesel
- Being the cleanest burning form of diesel

Although normal (mineral) diesel fuel is known for having high emissions, according to the EPA, biodiesel has between 57% and 86% less

GHGs compared to mineral diesel, depending on the feedstock (EPA, 2010). Particulate emissions, a significant health hazard, are also about half those of mineral diesel. In the United States, although bioethanol is the main biofuel for transport, the use of biodiesel is rapidly growing, and there remains considerable room for further growth in the United States, given that 80% of trucks and buses run on diesel.

5.3.3 Example 3: Loss of Energy When Combustible Material Is Moist

The energy density of dry wood is about 15 MJ/kg, while that of undried green wood is about 8 MJ/kg—the difference being due to some of the mass being simply water and the energy needed to drive off the water as vapor during combustion. Given the two energy densities, estimate the fraction f of moisture in green wood, assuming that it is essentially water.

Solution

A mass m kilograms of green wood will contain fm kg of moisture that needs to be driven off as vapor. Assume that the wood is initially at an ambient temperature of T = 20°C and that it needs to be raised from 80°C to its boiling point 100°C and then vaporized. This will require 80 + 539 = 619 cal/g of water or 619,000fm cal = 2.60 fm MJ total. Thus, the energy content in megajoules of a mass m of green wood is $8m$ and can be expressed as

$$8m = E_{dry} - 2.60fm = 15m_{dry} - 2.60fm. \tag{5.2}$$

The mass of the dry wood is $(1 - f)m$, so that Equation 5.2 yields

$$8m = 15(1 - f)m - 2.60fm, \tag{5.3}$$

which when solved for f yields f = 0.40 (40%).

HAD ENOUGH FIBER TODAY?

Lignin and cellulose, notoriously difficult to decompose, are complex chemical compounds that comprise an integral part of the cell walls of plants and many algae. They form the structural component of plants and trees and are often derived from wood. Cellulose is the most common organic compound on Earth, and together, lignin and cellulose comprise a majority of all plant matter by weight, so that finding an economic way to harvest the stored energy they contain is essential to the future of second- and higher-generation biofuels. Although the digestive systems of some animals can decompose lignin and cellulose with the aid of helpful bacteria, the human ability to do so is much more limited. Nevertheless, these compounds do play a useful role in the human digestive process, because they are the fiber that plays such an important part of our diets—especially as we age!

5.3.4 Generation of Biofuels

The choice of feedstock for biofuels is closely related to the generation to which they belong (Table 5.3). At least four generations have been defined, although their definitions vary according to the source. For example, by some definitions, biofuels in the second generation are said to come from a sustainable feedstock, which is more broadly defined than merely being a nonfood crop. Part of the reason for the confusion is that the vast majority of biofuels in commercial use still belong to the first generation, so the definitions of generations two, three, and four (often referred to as *advanced* biofuels) are a bit of a hypothetical exercise. The reason that generations two and higher are not yet widely used commercially has to do with the much greater difficulty of converting cellulose and lignin into a biofuel as compared to sugar, since there is the extra step of first converting these compounds into a sugar. This process is technically more difficult and remains to be perfected, although much research on the subject is underway. Scientists doing the work might learn much about this process if they could find a way to emulate nature, because livestock such as cattle, through a slow digestive process, turn the grass they eat into sugar.

Research is also underway in seeking to accomplish the goals of the third- and fourth-generation biofuels, which are likely to be still further in the future, with the science of genetic engineering playing a vital role in both cases. Algae are believed to be an especially promising feedstock for third-generation biofuels, and claims have been made that they might yield up to 100 times more energy per unit area than second-generation crops (Greenwell et al., 2010). In fact, the DOE estimates that if algae-based fuel replaced all the petroleum-based fuel in the United States, the land area required would total 39,000 km^2, which is a mere 0.42% of total area—a far cry from the figure for corn-based ethanol (Hartman, 2008). Not surprisingly, the technology is not yet mature enough to economically produce algae-based biofuels, and more advanced genetics is probably needed to successfully engineer synthetic microorganisms. Optimists, however, predict that algae-based biofuels may reach cost parity with conventional fuels within this decade. On the other hand, even if the economic optimists are right, there are also the environmental effects to consider, since algae-based fuels so far require substantial amounts of water and their production emits more GHGs than fuels generated than many second-generation feedstocks. These negative impacts are mainly

Table 5.3 Possible Definitions of the Four Generations of Biofuels

Generation	Characteristics
First	Made from edible feedstocks (sugar, starch, and vegetable oil); unfavorable NER; and CO_2 balance
Second	Use nonfood components of biomass, such as trees and grasses
Third	Specially engineered energy crops relying on genomics, e.g., algae based
Fourth	Crops that are very efficient in capturing CO_2—carbon-negative

the result of the heavy use of fertilizers needed to boost the algae production rate and the fossil fuels consumed in making those fertilizers (Clarens and Colosi, 2011).

Biofuels of the so-called fourth generation would be carbon negative, meaning that they would remove more carbon from the atmosphere during their growth than they would later return to it when they are consumed. In effect, they would need to somehow sequester captured carbon and not release it all when the fuel is consumed. Various methods for doing this have been proposed. In one scheme, small pieces of wood would be pyrolyzed to produce charcoal and a gas. The gas would then be condensed into oil, which after processing is blended into biodiesel. The charcoal residue could be used as a fertilizer and put back into the ground where it would remain. Scientists have demonstrated by experiment using selected grassland plants that carbon sequestration in the soil works quite well for sandy soil that is agriculturally degraded and nitrogen poor (Tilman et al., 2006).

5.4 OTHER USES OF BIOFUELS AND SOCIAL–ENVIRONMENTAL IMPACTS

Although the primary use of biofuels is in the transportation sector, they can also play a role in electric power generation. For example, *Miscanthus*, which promises to be one of the best plants for producing biofuels, is now grown in Europe mainly for mixing 50/50 with coal in electric power generation. Estimates are that it could supply 12% of the EU's electrical energy need by 2050 (Dondini et al., 2009). Direct combustion of solid biomass fuel does of course result in airborne pollutants, but their environmental impact is considerably less than fossil fuels. Apart from direct combustion of biofuels, there is also research underway to explore how the plant and other biomass sources can be used to make plastics and other products usually made from petroleum.

There are many social, economic, and environmental issues with biofuels including their impacts on such matters as oil prices, availability (and price) of food, CO_2 emissions, deforestation and biodiversity, water resources, and energy usage. As we have seen, some biofuels are far preferable than others in terms of minimizing the negative environmental impacts.

5.4.1 Biofuels from Wastes and Residues

One category of biofuels that still needs to be discussed is that produced from agricultural, residential, and industrial wastes. Various processes are used here depending on the input, with digestion (producing methane or landfill gas) followed by direct combustion to produce electric power being especially common. In order for the process to prove to be economically viable, widely distributed wastes need to be aggregated as part of

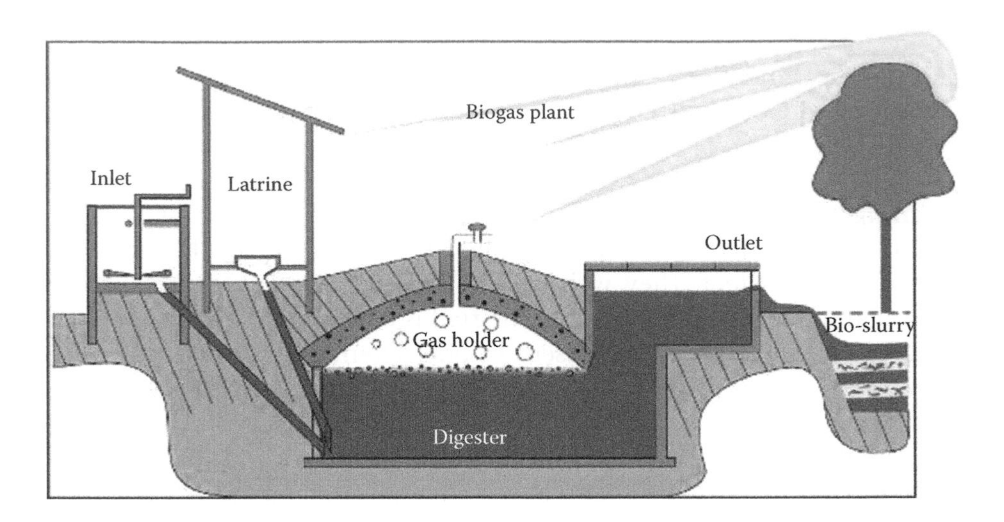

Figure 5.8 Design of a simple small-scale plant for generating biogas. (Courtesy of SNV, http://en.wikipedia.org/wiki/Biogas.)

some other goal, such as collection of garbage in a landfill. In the United States, for example, there are hundreds of landfills where methane is captured from decomposing trash and used to generate electricity totaling 12 billion kWh/year. Burning landfill gas does create airborne emissions, which can widely vary, depending on the nature of the waste and the state of the technology. However, the CO_2 released from burning landfill gas is considered to be a part of the natural carbon cycle, and it is less harmful as a GHG than if the methane were released to the atmosphere. The generation of power from the wastes of society also is quite suitable in a rural setting in developing nations even in a community whose livestock produce 50 kg manure per day, which is an equivalent of about six pigs or three cows. In such a setting, a typical biogas plant that supplied gas for cooking could be built by a rural household with an investment of as little as $300, depending on region. Several countries, especially China and India, have embarked on large-scale programs for producing biogas for domestic use in rural areas (Figure 5.8).

5.4.2 Agricultural Wastes

The same process of digestion of wastes followed by combustion of methane to produce electricity can also be done in an agricultural setting, which, as noted earlier, is routinely done in Brazil as a by-product of their ethanol production from sugarcane. Such usage of agricultural wastes is less common in the United States, however. One notable exception is the Mason–Dixon farm in Gettysburg, Pennsylvania. The Mason–Dixon farm, where the motto is "Change is inevitable; success is optional," stands as a model to the world for innovation in agricultural efficiency. Their 2000 cowherd of dairy cattle are housed in an extremely large barn where they are milked by robots that the cows seek out every few hours when their udders become uncomfortably full. Housing the cows in a barn also makes it easy to automatically gather their manure—not by robots but by slowly moving scraper bars that run the full length of the barn. The cow manure is then piped to a digester where methane is produced—enough to supply the whole farm with electricity and sell

Figure 5.9 Sculpture made from the residue of cow manure by the Mason-Dixon farm.

some back to the power company. The residue of the cow manure is also sold as fertilizer, and what is left over from that is used to make sculptures sold to tourists who visit the farm from all over the world to learn about its practices! Visitors to the farm on noticing that the sculptures look strikingly like a recent US Democratic president are likely to believe that they reveal the Republican leanings of the farm owner, particularly if they notice the inscription on the bottom (Figure 5.9). However, they might be of a different opinion if they knew that an equal number of sculptures of a recent Republican president had also been made, but they sold out very quickly.

5.4.3 Central Role of Agriculture in a Sustainable Future

The field of agriculture is of vital importance for the future sustainability of the human race. In fact, it plays a central role in many of the problems humanity faces and will face in the future (Figure 5.10). Advances in agriculture are what have made possible the past growth in the number of humans on the planet this past century—for example, during 1950–1984, a period known as the *Green Revolution*, agriculture was transformed,

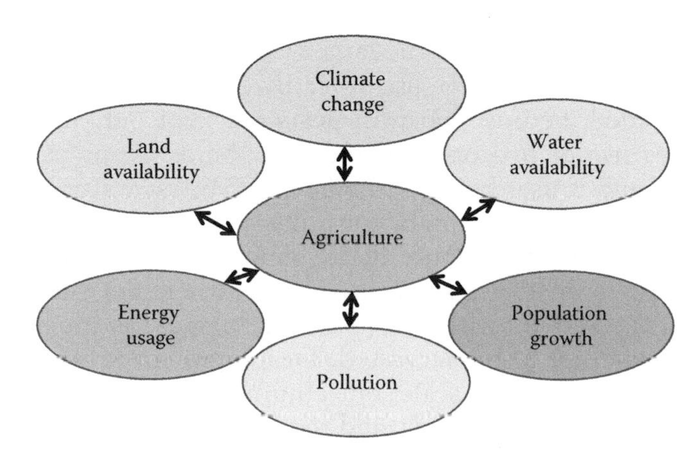

Figure 5.10 Central role of agriculture in relation to six major global problems.

and world grain production increased by 250%. Demographers now predict a continued growth in world population from the present 7 billion to as much as 10 billion by 2050. Whether these projections of population growth (or the less pleasant alternative massive famines) are realized will hinge in large measure over whether further comparable advances in agriculture are possible.

As it is, however, about 70% of the world's land area is either already used for agriculture or unsuitable to it, so simply planting crops on unused land does not offer a prospect for great expansion. Apart from the scarcity of land, that of freshwater is also becoming an increasingly serious problem in many heavily populated parts of the world plagued by drought. These problems may be exacerbated by climate change—which will likely decrease water availability and suitable cropland. Moreover, agriculture is a major factor in our usage of energy, another factor in promoting climate change. The Green Revolution was only made possible by a vast increase in the amount of energy used in agriculture, which increased by about 50 times what is used in traditional agriculture. This high energy usage (about 17% of all fossil fuel use in the United States) is primarily used for making fertilizer and operating farm machinery. Finally, the heavy usage of fertilizers relates to the sixth global problem—pollution, which, as previously noted, is caused by a major degree by agricultural runoff. All these global problems, in which agriculture plays a pivotal role, call out for a new way of doing agriculture if the human race is going to be able to sustain itself in the future.

5.4.4 Vertical Farming

The idea of *vertical farming* pioneered by Columbia University Professor Dickson Despommier (2010) has the potential to revolutionize agriculture, and help solve all the six problems linked to it identified in Figure 5.10. The idea is to bring agriculture to the cities and house it indoors in multistory buildings specially constructed for this purpose, using a combination of natural and artificial lighting. As Despommier explains, not a single drop of water, bit of light, or joule of energy is wasted in the operation, and in fact, there is no waste, with everything continually being recycled. The vertical farm essentially brings the farm to the grocery store and avoids both the large transportation costs (and energy expenditures) and the massive use of fertilizers and pesticides, which are no longer needed, because harmful pests are kept out. In this scheme which makes use of hydroponics and aeroponics, crops can be grown without soil, and it uses between 70% and 95% less water than conventional farming—one of the main consumers of freshwater in the world. The enclosed temperature-controlled system allows for year-round crops and eliminates crop failures due to bad weather. Between such crop failures and those due to disease (also eliminated), as much as 70% of crops worldwide fail to be harvested. Despommier's scheme has yet to be implemented on a large scale, but a number of projects have already begun in Japan, the Netherlands, and the United States (Figure 5.11).

Figure 5.11 Three proposed designs for a large-scale vertical farm designed by Chris Jacobs, Gordon Graff, SOA Architectes. (Courtesy of Wikimedia Foundation, San Francisco, California, http://en.wikipedia.org/wiki/Vertical_farming, image is made available under the Creative Commons CCO 1.0 Universal Public Domain Dedication.)

5.5 ARTIFICIAL PHOTOSYNTHESIS

The idea of artificial photosynthesis goes back to 1912 in the form of a challenge by Giacomo Ciamician to other scientists to search for a series of photochemical reactions that would mimic the process that plants use in storing energy. Recently, Daniel Nocera (2011), an MIT chemist, has met the challenge by developing an artificial leaf that uses photochemical reactions initiated by sunlight to produce hydrogen—an important energy-rich fuel. Essentially, the process is a form of water splitting, i.e., separating the hydrogen and oxygen in water using sunlight and catalysts that facilitate the reaction. Of course, real leaves do not generate hydrogen, but store the energy in other chemicals, such as carbohydrates, but the artificial leaf conforms to the basic functions taking place in nature, and it relies on earth-abundant materials and requires no wires. Quoting from an MIT press release:

> Simply placed in a container of water and exposed to sunlight, it quickly begins to generate streams of bubbles: oxygen bubbles from one side and hydrogen bubbles from the other. If placed in a container that has a barrier to separate the two sides, the two streams of bubbles can be collected and stored, and used later to deliver power: for example, by feeding them into a fuel cell that combines them once again into water while delivering an electric current. (MIT, 2011)

5.6 SUMMARY

Following an overview of biofuels, and the process of photosynthesis from which nearly all biomass is created, the chapter considers various categories of biofuels, including their feedstocks, the processes used to make them, and their end products and uses. It is seen that all biofuels are far from equal, whether the measure be energy supplied, GHG emissions, or other impacts on society and the environment. Although much research on biofuels is ongoing, and some possibilities appear particularly promising (especially algae-based biofuels), most of the biofuels used worldwide continue to be either bioethanol or biodiesel produced from first-generation feedstocks.

PROBLEMS

1. Suppose you wanted to supply half the US demand for gasoline using ethanol from corn. It was stated in the chapter that this might require as much as 262% of US cropland. Using data available on the web for the yield per acre of corn, its energy content, and the amount of energy needed to meet the needs of US road transport, see if this estimate is about right.
2. How is it possible that marine plankton produce perhaps half of the Earth's oxygen from photosynthesis, even though their total biomass is orders of magnitude below that of terrestrial plants?
3. Consider this statement made in the chapter: "The amount of energy trapped by photosynthesis is approximately 100 TW, which is about six times larger than the power consumption of human civilization." Assume that the basic photosynthesis reaction can be written as $CO_2 + H_2O + \text{energy} \rightarrow CH_2O + O_2$. (a) Given the original statement, how many tons of CO_2 does photosynthesis remove from the atmosphere each second, assuming that the absorbed energy consists of visible photons whose energy is about 2 eV each and that absorption of one photon is sufficient to induce the aforementioned reaction? (b) What is the net effect on atmospheric CO_2 levels from photosynthesis over the course of a year? (c) Explain why the net effect on CO_2 levels during the year due to plants is actually zero. (d) If 100 TW is in fact six times larger than the power consumed by all humans, what does that imply about the average power consumed per person? How does that compare with the average power consumed by an American?
4. A farmer has a small herd of 100 pigs and wishes to use their wastes to produce methane to generate part of the electricity used by the farm. Assume that each pig generates 1 kg of solid waste per day, which yields 0.8 m^3 of methane at standard temperature and pressure. Methane contains about 38 MJ/m^3. Assume that you can convert 25% of this energy into electricity; find how many kilowatts would the farmer be able to generate in this manner?
5. During anaerobic digestion, glucose is converted to methane according to the overall formula $(CH_2O)_6 \rightarrow 3CH_4 + 3CO_2$. Calculate the percentage of the gas produced that is methane by mass.

6. Over some range of sunlight power density and atmospheric CO_2 concentration, assume that the rate at which plant leaves take up CO_2 is a linear function of both variables. Suppose that the rate of CO_2 uptake by a leaf is 0.05 μmol/min/cm^2 when the solar power density is 50 W/m^2 and the atmospheric concentration is 330 ppm and that the rate of CO_2 uptake by a leaf is 0.15 μmol/min/cm^2 when the solar power density is 100 W/m^2 and the atmospheric concentration is 400 ppm. Find the rate of CO_2 uptake when the solar power density is 200 W/m^2 and the atmospheric concentration is 450 ppm.
7. Using advanced fermentation technology for converting cellulose to ethanol can yield about 900 gal/acre. Assume that on average, vehicles get 20 miles per gallon from gasoline. Estimate how many acres would need to be planted to replace 10% of the miles driven in the United States with ethanol generated in this way. The number of vehicles in the United States is about 250 million, and each vehicle is driven on average of 12,000 miles. Note that ethanol provides one-third less miles per gallon than gasoline.
8. Estimate the number of tons of CO_2 added to the atmosphere for each square kilometer of Amazon rainforest that is cleared and burned to provide room for agriculture. You will need to make some assumptions on the average size and spacing of trees in the forest.
9. Normally, it would not make any sense to produce a biofuel that had an energy content that was lower than the energy required to produce it or an NER less than one. Are there exceptions and what are they?
10. The state of Massachusetts at one time had considered generating electric power by harvesting energy crops and burning them. Assume that the state requires 4000 MW of electricity and that planted crops yield between 10,000 and 20,000 lb of dry biomass per acre per year that could be burned to produce electricity at 35% efficiency. How many acres would the state need to plant to supply all its electricity in this way?
11. Do some searching on the web to find estimates on the amount of carbon sequestered per acre of Amazon rainforest and the amount of CO_2 emissions saved by using the ethanol produced each year by one acre's worth of sugarcane instead of gasoline. Based on the aforementioned two numbers, see if the time (4 years) given in the chapter is correct for the time it takes for the extra GHG emissions resulting from the loss of rainforest to be made up for by using sugarcane-based ethanol instead of gasoline.
12. Ethanol has a 38% lower energy density by volume than gasoline. Partially offsetting this disadvantage is the higher octane rating of ethanol, which allows it to be used in engines having a higher compression ratio. In fact, a standard gasoline-powered engine typically runs at a compression ratio $r = 10$, while an ethanol-powered one can run at $r = 16$. Internal combustion engines can be approximated by the ideal Otto cycle, for which the efficiency

is given by $e_o = 1 - r^{-0.4}$. Assume that a real engine has one-third the ideal Otto efficiency, and calculate how much improvement this would make to the efficiency of the ethanol-fueled engine over a standard gasoline engine. Is it enough to offset the lower fuel energy density of ethanol?

13. Assume that the rate of photosynthesis in some plants depends on the CO_2 concentration C (in parts per million) according to $R = 50\ (1 - e^{-C/200})$. How does R change per unit change in C when $C \ll 200$ ppm and when $C \gg 200$ ppm? The units of R are milligrams CO_2 per square meter per hour (mg $CO_2/m^2/h$).
14. Estimate the fraction of all sunlight harvested by the chlorophyll *b* pigment of plants—see Figure 5.5.
15. Find a pair of equations that describe the two curves in Figure 5.4. If *saturation* is defined as the point where each curve is at 80% of its asymptotic value, what is the ratio of the saturation light intensities for sun and shade plants based on this figure?
16. How much BTU could be generated in a year in the United States if all biomass waste were incinerated for heat energy. Assume that there is 1000 lb. of biomass waste per person per year and the energy content is 4300 Btu/lb. of biomass waste. Approximately what percentage of the total energy consumption of the United States does this biomass waste heat generated represent?
17. How much cornfield area would be required if you were to replace all the oil consumed in the United States with ethanol from corn? Use the following assumptions: a cornfield is 1.5% efficient at converting radiant energy into stored chemical potential energy; the conversion from corn to ethanol is 17% efficient; assume a 1.2:1 ratio for farm equipment to energy production; a 50% growing season; and, 200 W/ m^2 solar insolation. What would the length of each side of a square be to have such an area?

REFERENCES

Blankenship, R. (2005) *Researchers Find Photosynthesis Deep within Ocean*, Arizona State University, Tempe, AZ. http://www.asu.edu/feature/includes/summer05/readmore/photosyn.html.

Chandler, D. L. (2011) "Artificial leaf" makes fuel from sunlight, *MIT News*. http://web.mit.edu/newsoffice/2011/artificial-leaf-0930.html (Accessed Fall 2011).

Chandler, D., and M. Le Page (2007) Climate myths: Higher CO_2 levels will boost plant growth and food production, *New Scientist*. http://www.newscientist.com/article/dn11655-climate-myths-higher-co2-levels-will-boost-plant-growth-and-food-production.html.

Clarens, A. F., and L. M. Colosi (2011) Environmental impacts of algae-derived biodiesel and bioelectricity for transportation, *Environ. Sci. Technol.*, 45(17), 7554–7560.

Despommier, D. (2010) *The Vertical Farm: Feeding the World in the 21st Century*, St. Martin's Press, New York.

Dondini, M., A. Hastings, G. Saiz, M. B. Jones, and P. Smith (2009) The potential of *Miscanthus* to sequester carbon in soils: Comparing field measurements in Carlow, Ireland to model predictions, *Glob. Change Biol. Bioenergy*, 1–6, 413–425.

EIA (Energy Information Agency) *Annual Energy Outlook 2011*, EIA, Washington, DC.

EPA (US Environmental Protection Agency) (2010) *U.S. Environmental Protection Agency, Renewable Fuel Standards Program Regulatory Impact Analysis*, EPA-420-R-10–006, EPA, Washington, DC.

Fox, D. (2007) CO_2: Don't count on the trees, *New Scientist*. http://www.newscientist.com/article/mg19626271.900-co2-dont-count-on-the-trees.html.

Greenwell, H. C. et al. (2010) Placing microalgae on the biofuels priority list: Are view of the technological challenges, *J. R. Soc. Interface*, 7(46), 703–726.

Groom, M., E. Gray, and P. Townsend (2007) Biofuels and biodiversity: Principles for creating better policies for biofuels production, *Conserv. Biol.*, 22(3), 602–609.

Hartman, E. (2008) A promising oil alternative: Algae energy, *The Washington Post*. http://www.washingtonpost.com/wp-dyn/content/article/2008/01/03/AR2008010303907.html (Accessed June 10, 2008).

Nocera, D. et al. (2011) Wireless solar water splitting using silicon-based semiconductors and earth-abundant catalysts, *Science*, 334, 645.

Searchinger, T. et al. (2008) Use of U.S. croplands for biofuels increases greenhouse gases through emissions from land use change, *Science*, 319, 1238–1240.

Tilman, D., J. Hill, and C. Lehman (2006) Carbon-negative biofuels from low-input high-diversity grassland biomass, *Science*, 314(5805), 1598–1600.

Geothermal Energy

CONTENTS

6.1 INTRODUCTION

6.1.1 History and Growth of Usage

Geothermal power has been used since ancient times—at least in those places on Earth where geysers and hot springs spontaneously bubble up from the Earth. In fact, the town of Bath, in England, is named for the thermal baths developed there by the ancient Romans. It was, however, not until the twentieth century that the vastness of the reserves of underground heat throughout the planet's interior was appreciated and the first usage of geothermal power for electricity generation was demonstrated in 1904 and put to use, generating significant amounts in 1911 when a power plant was built in Larderello, Italy. Subsequent progress of installing geothermal electrical-generating capacity has continued since that pioneering development at an exponentially expanding rate. In fact, the capacity during the 80 years preceding 2000 has been growing approximately exponentially with an 8.5% annual growth rate. Since 2000, the global usage of geothermal power has accelerated just as rapidly because of both the push for energy alternatives and recent technological advances.

6.1.2 Geographic Distribution

As far as the direct heating usage of geothermal power is concerned, the leading nation is China in 2010, with the United States a close second. Interestingly, China is nowhere to be found among the top 15 nations in geothermal electricity production for which 10.7 GW was produced in 2010 worldwide—a 20% increase over the last 5 years. Despite the past rapid growth, geothermal power now accounts for a meager 0.5% of the world's electricity, which is about the same as solar cells. The United States produces the most geothermal electricity (3.1 GW), with the Philippines in second place at 1.9 GW. Strictly in terms of percentages, Iceland is the world leader, where approximately 53.4% of the total national consumption of primary energy is from geothermal power (Figure 6.1).

The global distribution of the most productive geothermal sources is largely dictated by geography, since this energy source is most abundant and accessible at places on Earth near tectonic plate boundaries or in major volcanic regions. In most other places, geothermal power has not yet proven to be economically competitive to exploit—at least for electricity generation. Given that many of the best sites for geothermal electricity generation have already been exploited, its future rapid growth is contingent on making technological advances that will allow lower-grade resources to be exploited at costs competitive with other sources.

Figure 6.1 Geothermal borehole outside the Reykjavik Power Station. (Courtesy of Yomangani, http://en.wikipedia.org/wiki/Geothermal_power_in_Iceland.)

6.1.3 Sources of the Earth's Thermal Energy

Much of the Earth's stored heat never makes it to the surface spontaneously. In fact, on average, only around 0.06 ± 0.02 W/m^2 geothermal power reaches the Earth's surface on its own, which is a tiny fraction of the energy reaching the surface from the sun. However, the available stored thermal energy in the Earth's interior is enormous. According to an MIT study, the US total geothermal energy that could feasibly be extracted with improved technology in the upper 10 km of the Earth's crust is over 2000 ZJ or 2×10^{24} J or 4000 times the energy humans use per year (Tester, 2006).

Heat is created underground by at least six different mechanisms, but around 80% of it is generated due to radioactive decay mostly from very long-lived isotopes of uranium and thorium—although that estimated 80% could be as little as 45% or as much as 90%. Given that decay of radioisotopes having multibillion-year half-life is the primary source of the Earth's heat, even if it were extracted in sizable quantities, there need be no concern of it running out, since it is being continually replenished.

6.1.4 Comparison with Other Energy Sources

Geothermal energy as a means of generating electricity has the great advantage of not being intermittent like most other renewable sources, such as wind and solar; in fact, its average capacity factor is around 73%. This means that a plant produces full power 73% of the time—far higher than wind turbines, for example. As a result, geothermal electrical plants are capable of providing base load electricity, which is not the case for intermittent renewable sources such as wind. Moreover, in places where conditions are favorable, geothermal electric power can be produced at a very cost-competitive basis compared to other methods, either renewable or nonrenewable. As with other forms of renewable energy, the cost of electricity from an existing plant does not fluctuate like the price of gas or oil does, since the fuel is free. However, the cost of new geothermal plants is strongly dependent on the price of oil and gas because those costs influence the competition for drilling equipment—and drilling is the main contributor to capital costs (Figure 6.2).

The International Geothermal Association expects that in the coming 5 years, geothermal electric-generating capacity might be expected to expand by as much as 80%, with much of the expansion taking place in areas previously not considered favorable—a development made possible by recent technological improvements. However, even in areas where conditions are not favorable for generating electricity because drilling to reach high enough temperatures would be cost prohibitive, geothermal power can always be used for residential heating, which does not require very high temperatures. Geothermal heat has also proven to be useful for a wide range of other nonresidential uses, including district heating; hot water heating; horticulture; industrial processes; and even tourism, i.e., hot thermal baths.

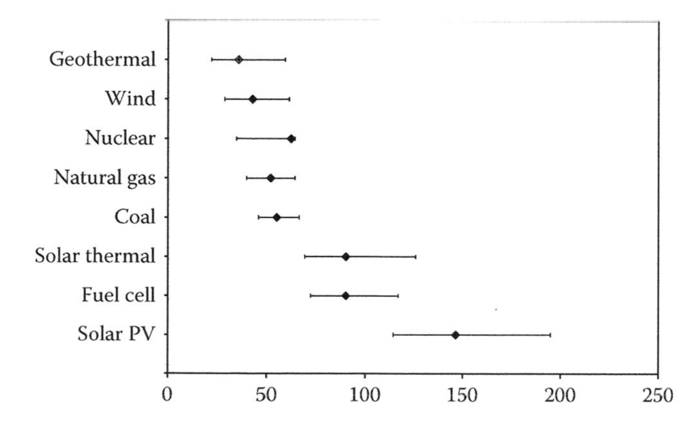

Figure 6.2 Comparison of electricity generation costs in dollars per megawatt-hour for eight fuel sources. The bars show high and low estimates for the ranges. The costs are for plants in the United States, and they include a $19/MWh tax incentive for renewable sources. Moreover, these are levelized costs, which assume that the same interest rates can be obtained for highly capital-intensive sources compared to others. These data are based on high and low estimates for a plant entering service in 2017, and they are based on US Energy Information estimates.

6.2 GEOPHYSICS OF THE EARTH'S INTERIOR

The underground composition, temperature, and pressure throughout the Earth's interior is a challenging problem that geophysicists have solved only indirectly, since direct underground exploration is limited by the depths to which boreholes can be drilled. Few oil and gas wells go deeper than 6 km, although the current depth record is the Kola research borehole at 12.262 km in Russia. Despite having directly explored only partway through continental crust, geophysicists are confident that they understand the entire interior of the Earth, based on results from the science of seismology. The chief method is to create seismic waves at one point on the Earth by detonating an explosive charge and then recording the arrival time at many other locations around the globe. These waves will be reflected at discontinuous interior boundaries and refracted in media whose properties continuously change. Moreover, the two types of seismic waves known as *s-waves* and *p-waves* differ in their nature, since while the former is longitudinal, the latter is transverse. This distinction is important since while both s-waves and p-waves can pass through solids, only the former can pass through liquids, allowing seismologists to deduce which interior layers of the Earth are liquid and which are solid. As a result of seismology studies, geophysicists now believe that the Earth's interior consists of the following three main regions:

- *Core*: The core extends out to half the Earth's radius (6400 km) and is made mostly of iron (80%) and nickel (20%), whose inner half (by radius) is solid and whose outer half is liquid. This iron and nickel core is the source of the Earth's magnetic field, which is believed to be created by electric currents in the core.
- *Mantle*: The mantle makes up most of the rest (83%) of the Earth's volume and made mostly of rocky material, whose inner part is semirigid and whose outer and cooler part is plastic and, therefore, can flow (think lava).
- *Crust*: The crust is the outermost thin layer (1% of the Earth's volume), whose average thickness is 15 km. The crustal thickness ranges from a high of 90 km under continental mountains to as little as 5 km under some parts of the oceans. On a scale where the Earth is the size of a soccer ball, the crust would be a mere 0.25 mm thick.

6.3 THERMAL GRADIENT

The thermal gradient is the rate of change of temperature with depth. The Earth has a radius of 6400 km, and at its center, the temperature is believed to be 7000 K, giving the convenient value of about 1 K/km or 1°C/km for the average gradient. The gradient, however, does vary enormously both as a function of depth and as a function of the particular location on Earth. Figure 6.3 illustrates the former variation, which is strongly correlated with the composition of each interior region. The largest gradient (the topmost section of the graph) is on the Earth's crust, where the gradient averages 25–30 K/km. Since the crust is solid and heat cannot be transferred by convection, we may apply the heat conduction equation for the flow across a layer (slab) of thickness Δz to find the thermal gradient. The heat flow per unit area across the slab is given by $\dot{q} = k\Delta T/\Delta z$, where k is the thermal conductivity and ΔT is the temperature difference across the slab. Hence,

$$\frac{\Delta T}{\Delta z} = \frac{\dot{q}}{k}. \tag{6.1}$$

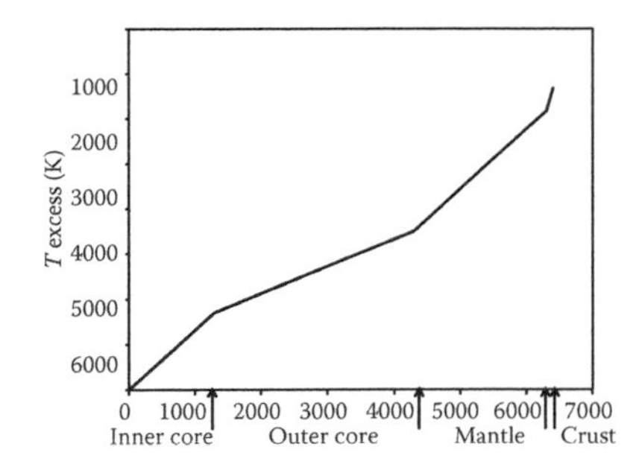

Figure 6.3 Excess temperature in kelvins or degrees Celsius above that on the surface as a function of distance from the center in kilometers showing the boundaries between the interior regions of the Earth.

For the inner core of the Earth, we see in Figure 6.3 that there is a rise in temperature of about 1200 K in the first 1000 km out from the center for a gradient of 1.2 K/km—a value that is 20–25 times smaller than that for the crust. This difference can be explained using Equation 6.1, since the thermal conductivities of iron and rocks are 55 W/m K and around 4 W/m K, respectively—making k for rock around 13 times smaller than that for iron. If we were to assume that the heat flow out of the core is the same as that which eventually passes through the crust, then based on the preceding ratio of k values, we would predict the thermal gradient in the crust to be around 13 times larger than that in the inner core, or around 15 K/km, which is within a factor of 2 with what is actually found. Expecting any better agreement than this is unrealistic given the large variation in conductivities for different types of rock.

How can we explain the sudden changes in thermal gradient (slope) that occur at the two boundaries of the outer core? Recall that the outer core is liquid not solid and that convective and conductive heat flows occur in parallel there. As a result, the thermal gradient will be smaller for the outer core than for the inner core. So far, we have been considering how the thermal gradients vary with depth on a very large scale—from the surface to the center of the Earth—most of which is inaccessible for energy extraction. For geothermal energy to be accessible, we are primarily concerned about the Earth's crust and how gradients there vary from place to place. As we can see from Figure 6.4, it can vary quite a bit—both from place to place and as a function of depth. Not surprisingly, the known location that has the highest thermal gradient, Lardorello, Italy, was the site for the first geothermal electricity source because the high thermal gradient found there means that one can reach very high temperatures (close to 200°C) in a mere 0.25 km.

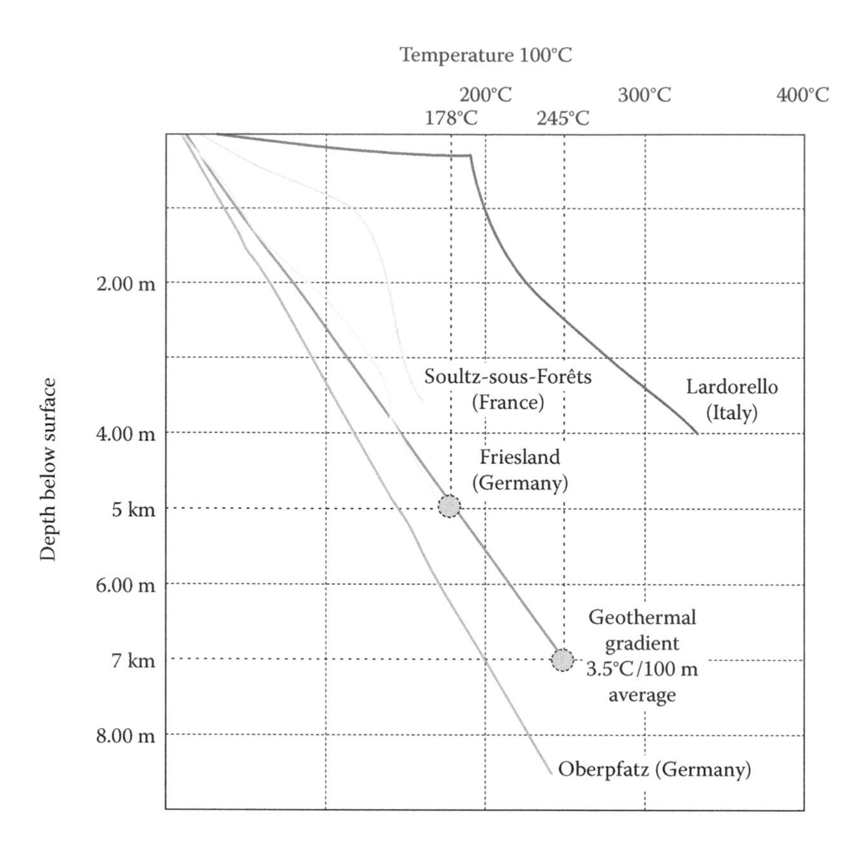

Figure 6.4 Earth's crust temperature in degrees Celsius versus depth in kilometers in selected places. (Courtesy of Geohil AG, U.K., http://www.mpoweruk.com/geothermal_energy.htm [modified by Electropaedia].)

The abrupt changes seen in thermal gradients for Lardorello and Oberpfalz in Figure 6.4 are a consequence of radical changes in rock composition at certain depths. For example, the sharp discontinuity seen for the Lardorello curve is due to a magma intrusion into a region where the rocks (such as granite) have high values of specific heat and density (so they hold a lot of thermal energy per unit volume). In addition, there is a layer of sedimentary rock above the granite having low thermal conductivity that tends to trap the stored heat below. It is natural to ask whether the abrupt changes in gradient (slope discontinuity) such as those occurring at Lardorello (initial gradient of an astounding 680°C/km) and Soultz-sous are rare or common. It must be the case anywhere on Earth where the thermal gradient is initially very high that its value will radically change at some deeper depth. Were this not the case, then given the initial gradient of 680°C/km found at Lardorello, the temperature would reach nearly that at the center of the Earth in a mere 10 km, which is clearly impossible.

Obviously, the most promising places to build geothermal plants are where the gradient is highest and the depth of wells to access high temperatures is the least. Figure 6.5 shows how the gradient varies across the continental United States, based on data from drilling numerous boreholes—sometimes in connection with gas and oil exploration.

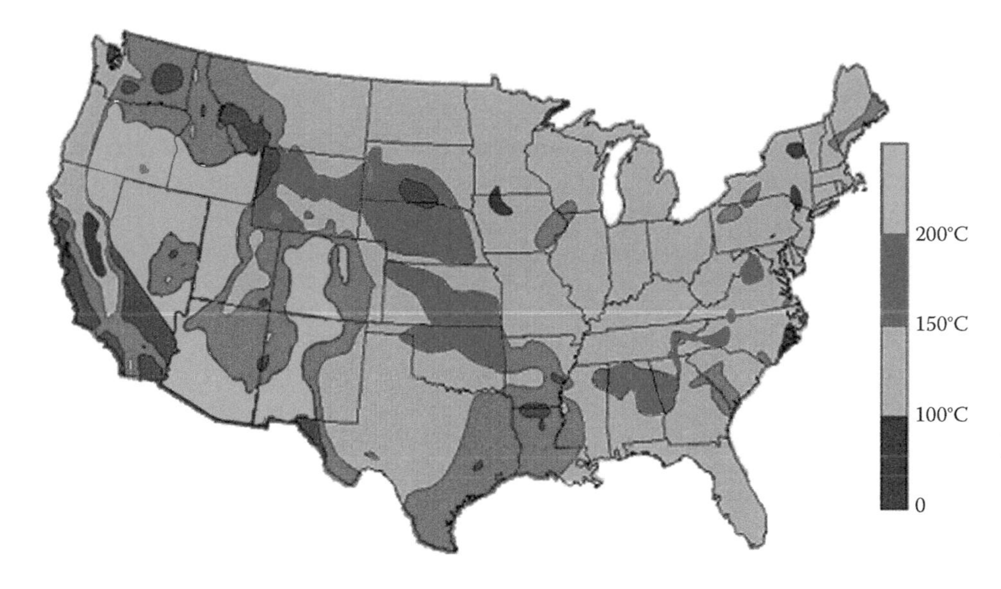

Figure 6.5 Temperatures at 6 km depth beneath the surface of the Earth. The thermal gradient in degrees Celsius per kilometer over that distance therefore equals the indicated grayscale-coded temperatures divided by 6. (Courtesy of DOE, Washington, DC.)

6.4 CHARACTERIZATION AND RELATIVE ABUNDANCE OF THE RESOURCE

6.4.1 Impact of the Thermal Gradient

As we have noted, to generate electricity, the most important characteristic is the thermal gradient because this quantity will determine the well depth that needs to be reached to access temperatures above some minimum needed for a power plant—typically 150°C, even though some types of power plants can operate at lower temperatures. For this reason, geothermal resources are often put into three grades: high, medium, and low, based on the gradient. High-grade resources have gradients in excess of 250°C/km, medium grade have gradients 150–250°C/km, and low grade have gradients below 150°C/km. These grades are quite arbitrary and depend on the intended usage of the resource—in this case, electricity generation. Thus, it is not surprising that some experts prefer other categories, such as "hyperthermal" (above 80°C/km), "semithermal" (40–80°C/km), and "normal" (below 40°C/km). It is clear from Figure 6.5 that in the case of the United States, it is mainly in some of the Western states that geothermal energy can most easily be exploited for producing electricity.

Let us now evaluate the amount of the thermal resource available under the assumption that the thermal gradient G at some location is constant with depth z, although as already noted, this assumption is questionable—especially for the case of a very high initial gradient. Suppose that z_1 is the minimum depth needed to reach temperatures of $T_1 = 150°C$ and z_2 is the maximum depth to which current technology allows wells to be drilled. Recall that by the definition of the specific heat of a substance c, the stored thermal energy in a mass m whose temperature excess ΔT

above some reference temperature can be expressed as $E = mc\Delta T = mc(T - T_1)$. Thus, the amount of stored thermal energy below a surface area A between a depth z and $z + dz$ can be expressed as $dE = \rho Ac\Delta T\,dz = \rho Ac(T - T_1)\,dz$, where ρ is the rock density. Finally, given the definition of the thermal gradient $G = dT/dz = (T - T_1)/(z - z_1)$, we can integrate dE to find the total energy stored between depths z_1 and z_2:

$$E = \int_{z_1}^{z_2} \rho Ac(T - T_1)dz = \int_{z_1}^{z_2} \rho AcG(z - z_1)dz = \frac{1}{2}\rho AcG(z_2 - z_1)^2 = \frac{1}{2}\rho AcG\left(z_2 - \frac{T_1}{G}\right)^2. \quad (6.2)$$

6.4.2 Example 1: Relative Energy Content for Two Gradients

Suppose we have two locations A and B for which the gradients are $G_A = 100°C/km$ and $G_B = 50°C/km$. What are the ratios of the energy content per unit area of surface down to a depth of 6 km at the two places, assuming that a minimum temperature of 150°C is needed? How can this be graphically illustrated?

Solution

$$E_A = \frac{1}{2}\rho AcG\left(z_2 - \frac{T_{min}}{G_A}\right)^2 = \frac{1}{2}\rho Ac(100)\left(6 - \frac{150}{100}\right)^2 = 703\rho Ac,$$

$$E_B = \frac{1}{2}\rho AcG\left(z_2 - \frac{T_{min}}{G_B}\right)^2 = \frac{1}{2}\rho Ac(50)\left(6 - \frac{150}{50}\right)^2 = 225\rho Ac.$$

Thus, the energy per unit area at A is 3.12 times that at B. The respective energies at A and B correspond to the areas of the white and shaded triangles (Figure 6.6).

6.4.3 Questioning Our Assumptions

Recall that the preceding analysis made the assumption that it is the maximum drillable depth that is the main limitation in exploiting a geothermal resource. This assumption may be incorrect. For example, the Kola borehole (the world's deepest) reached a far greater depth (12.3 km) than would be possible in most places on Earth. This was possible in Kola only due to the exceptionally low gradient there (13°C/km), so that even at 12.3 km, the temperature did not yet exceed 200°C. It therefore seems

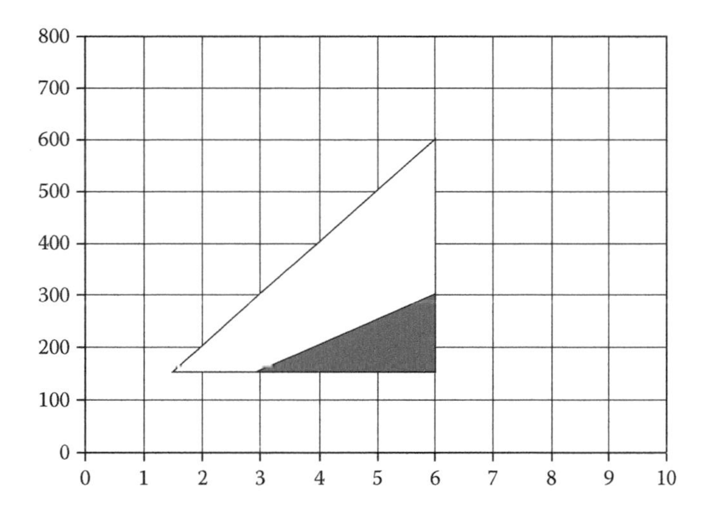

Figure 6.6 Temperature in degrees Celsius versus depth for two different gradients: 100°C/km (slope of hypotenuse of *white triangle*) and 50°C/km (slope of hypotenuse of *shaded triangle*). It is assumed that for both cases, the wells have the same maximum depth (6 km), but different minimum depths since z_1 = 150°C/G. We have also assumed for simplicity that the surface temperature is 0°C.

reasonable to believe that the state of drilling technology in reality does not limit the maximum drillable depth at all, but rather the maximum temperature, which currently seems to be around 300°C. This fact has a surprising impact on our earlier assessment of how the amount of available energy depends on thermal gradient.

It can be easily shown by integration that when T is the limiting factor, instead of Equation 6.2, we have

$$E = \frac{1}{2G}\rho Ac(T_2 - T_1)^2. \tag{6.3}$$

We can again use a graphical representation to understand this result. As illustrated in Figure 6.7, the respective areas for the high- and low-gradient cases now favor the low-gradient case by a 2-to-1 margin! Moreover, this very surprising result holds irrespective of the specific choice of maximum temperature (an unrealistic 600°C in the figure). Of course, the economic feasibility of extraction may make the exploitation of the

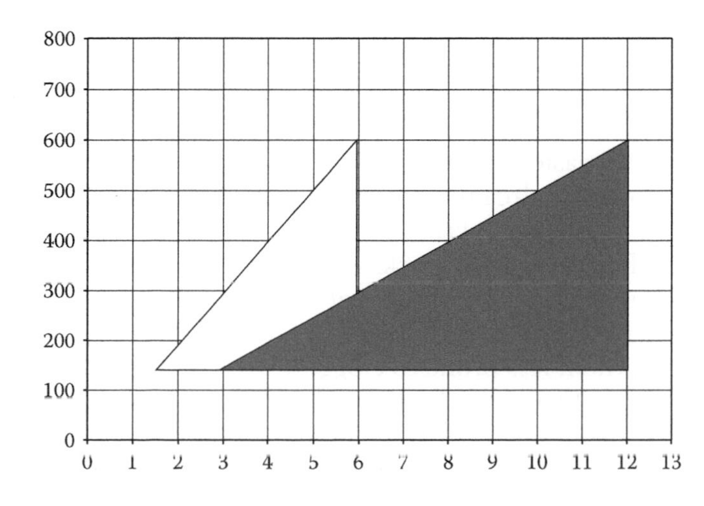

Figure 6.7 Temperature in degrees Celsius versus depth for two different gradients. 100°C/km (slope of hypotenuse of *white triangle*) and 50°C/km (slope of hypotenuse of *shaded triangle*). It is assumed that for both cases, the wells have the same maximum temperature (600°C) and the surface temperature (for z = 0) is 0°C for simplicity.

low-gradient location (with the need for deeper wells) out of the question—but much more on this topic later. To summarize the main point of the preceding discussion, the conventional wisdom that high-thermal gradient resources are more worthwhile to exploit in terms of extracting energy at reasonable cost crucially depends on whether the limitation on drilling technology is a matter of (1) maximum depth or (2) maximum temperature and the precise manner in which drilling costs depend on depth.

6.4.4 Other Geologic Factors Affecting the Amount of the Resource

In addition to the thermal gradient at a given location, a geothermal prospector will also want to know many other geological characteristics, including these five properties of the underlying rock formations: hardness; thermal conductivity; specific heat; density; and porosity, the latter being the fraction of the rock volume that is empty space, which is often filled with fluids, usually water with dissolved salts. A particularly desirable choice would be rocks with high values of the density, specific heat, and thermal conductivity (such as granite). The best locations are also important: that the rock be overlain by a layer of sedimentary rocks having low thermal conductivity, which acts to trap the heat. Rocks that are permeable, meaning that fluids can flow through them, are also more desirable. If there is in fact fluid present in the rocks, we have by definition an aquifer or hydrothermal system. A confined aquifer is one that is overlain by a nonporous rock layer.

In order to evaluate the thermal energy content of an aquifer by using Equation 6.2, it would be necessary to use only the average value of $c\rho$ for the rock and fluid, i.e., to make the substitution

$$c\rho = p'\rho_W c_W + (1 - p')\rho_r c_r, \tag{6.4}$$

where the subscripts W and r refer to water and rock, respectively, and p' is the porosity of the rock or the fraction of the aquifer volume taken up by fluid assumed to be water.

6.4.5 Hot Dry Rock Formations

Aquifers are the easiest geothermal resource to exploit for extracting energy, because they already contain fluid, but most of the stored thermal energy in the Earth's crust is in dry rock formations, which may lack porosity. The porosity of rocks can occur in one of two ways: either because of the spaces between grains of the rocks or because of large-scale fractures, which are far more favorable in terms of yielding greater permeability, making them less prone to clogging up over time when fluid flows through the rock. Starting in the 1970s, hydrofraction (or fracking) of rock was pioneered (using injected water under pressure to create rock fractures). This technique allowed engineers to extract energy from hot dry rock (HDR) formations, using what has also more recently been termed *enhanced geothermal systems* (EGSs). Although the fracking technique has been controversial in connection with oil and gas exploration, its use with geothermal power is not nearly as problematic since the chemical additives used to free up oil and gas in rock pores are not needed.

As explained earlier, in an EGS, water needs to be pumped down a well under pressure to induce thermal stresses in HDRs, causing them to fracture and create porosity. In addition to this injection well, additional production wells must be dug some distance away—but close enough for the injected water to reach them by flowing through the rock fissures—see Figure 6.8.

EGS geothermal power tends to be much more expensive in terms of its initial investment (about five times more than hydrothermal systems) because wells must usually go deeper to reach higher temperatures. Extra production and injection wells may be required if the induced pores in the rocks should clog. Such extra wells may also be useful in order to increase the extracted power. In the case of multiple injection and production wells, their spacing is quite important—if they are too close, they draw on the same thermal energy store, while if they are too far apart, they fail to adequately draw on a thermal energy store between them.

There are, of course, many other nongeological factors that a geothermal "prospector" need to consider that will determine whether it makes

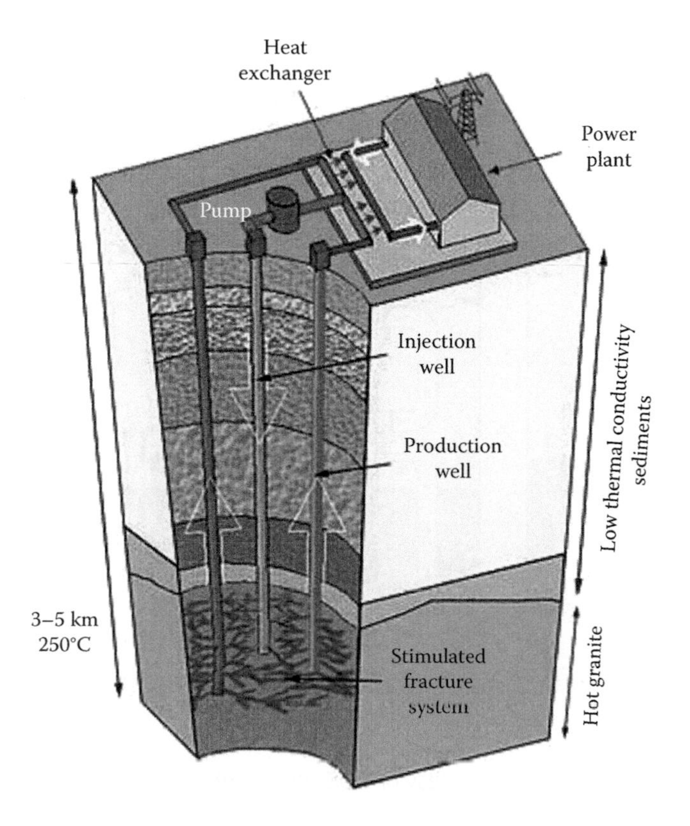

Figure 6.8 EGS with one injection well and two production wells. (Courtesy of Australian National University, Canberra [modified by Geothermal Resources].)

sense to exploit a geothermal resource. These include the current state of drilling technology, the state of the economy (and whether funds will be available for a large initial investment), the cost of natural gas (which affects the cost and availability of drilling equipment), the adequacy of existing transmission lines, the price of land or drilling rights, and the closeness to population centers.

6.5 GEOTHERMAL ELECTRICITY POWER PLANTS

There are three main types of geothermal power plants: dry steam, flash, and binary cycle, with the flash type being most common. The basics of the last two of these three types are depicted in Figure 6.9a and b. In the flash type of power plant, high-pressure water comes up the production well and vaporizes (flashes) when its pressure is reduced to produce a flow of steam that drives a turbine, which then generates electricity. The dry steam type (not depicted) is similar to the flash type, but without the first step, since the dry steam directly coming up from the production well directly drives the turbine. This type of plant is rare because it is generally used in very high-gradient locations where steam spontaneously rises out of the production well. Binary cycle power plants involve one additional step in the process. For these plants, high-temperature fluid coming up from a production well passes through a heat exchanger in which the secondary loop contains a low-boiling point liquid such as butane or pentane, which can vaporize at a lower temperature than water. This added step allows such plants to generate electricity at much lower temperatures than the other types. The current low temperature record for a binary plant is 57°C!—but, of course, the thermodynamic efficiency is very low

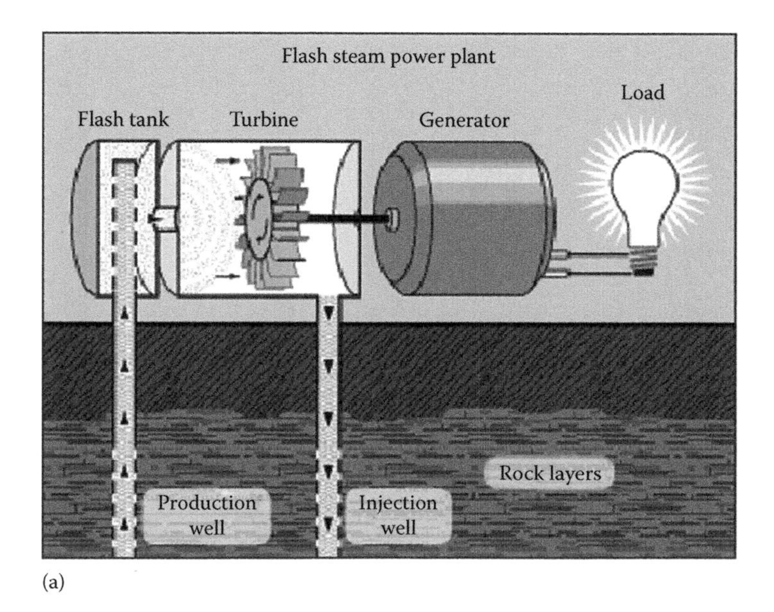

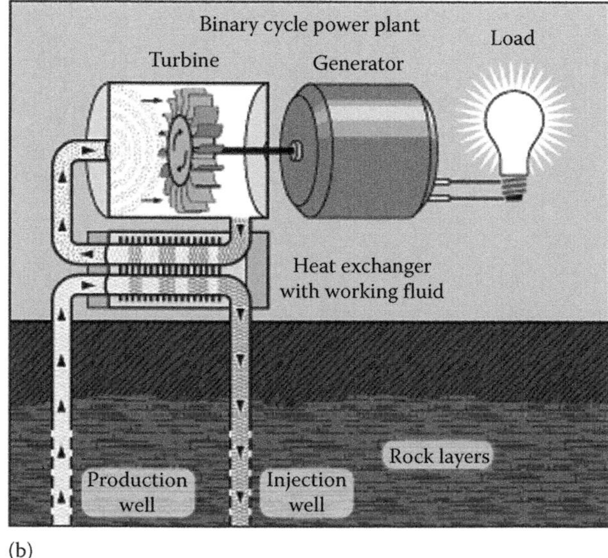

Figure 6.9 (a) Flash steam power plant; (b) binary cycle power plant. (Courtesy of DOE, Washington, DC.)

due to Carnot's theorem. For example, a plant planned for Alaska using a Rankine cycle uses a 75°C hot spring that ejects heat to a 3°C river water, and it has an expected efficiency of only 8%. Some binary plants have several flash loops in series, each using a lower-boiling point liquid, which allows thermal energy to be extracted several times and results in higher efficiency. Low efficiency is a significant detriment in geothermal plants because even if the fuel is free, this raises the cost per megawatt per hour of electricity to the point where it may be uneconomical. Thus, binary cycle plants tend to be significantly more expensive on a per megawatt-hour basis than the other types, even though they are useful in expanding the region over which geothermal electricity can be generated.

6.5.1 Example 2: Efficiency of a Geothermal Power Plant

Show that the efficiency of the proposed Alaskan power plant is a bit worse than half the maximum possible value and explain why that fraction would be even worse were the hot spring slightly colder or a nearby cold river unavailable.

Solution

The Carnot efficiency is given by

$$e_C = 1 - \frac{T_L}{T_H} = 1 - \frac{3+273}{75+273} = 0.207\,(20.7\%), \tag{6.5}$$

which is a bit more than twice the actual efficiency. Note that if either T_H were lower or T_C were higher, the efficiency would be less.

6.6 RESIDENTIAL AND COMMERCIAL GEOTHERMAL HEATING

The direct use of geothermal energy especially for home heating is probably the fastest growing application, primarily because it can be implemented virtually anywhere and requires neither high thermal gradients nor deeply drilled holes. In fact, it is not necessary to dig deeply enough to access temperatures even as warm as the desired temperature of your home; merely below the point where the ground temperature year-round is approximately the same. Given average soil conditions, this requirement means a depth of about 3 m where the annual variation is perhaps ±3°C from summer to winter.

A conventional heat pump works by extracting heat from the outside air (which in winter is probably colder than your house) and expelling the extracted heat to your home. In order to make heat "flow the wrong way," i.e., from cold to hot, it is of course necessary to have an input of energy

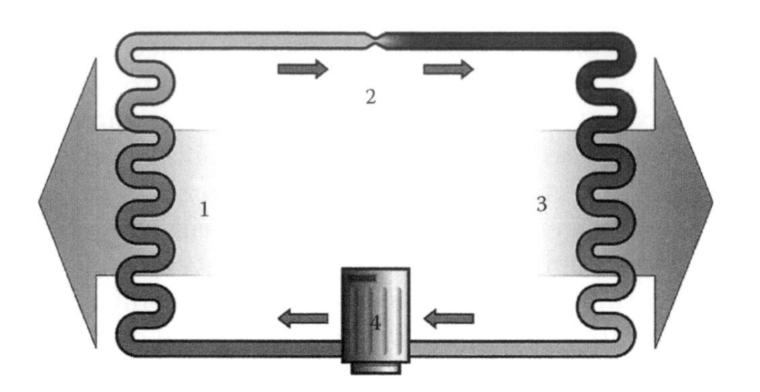

Figure 6.10 Simple stylized diagram of the vapor-compression refrigeration cycle of a heat pump: *1*, condenser; *2*, expansion valve; *3*, evaporator; and *4*, compressor expelling heat on the left (hot) side and absorbing heat or expelling cold on the right (cold) side.

in the form of electricity that drives a compressor. How exactly does this work? In a conventional heat pump, a volatile fluid (basically, a refrigerant) in its vapor state is compressed by a compressor (4 in Figure 6.10) so that it releases heat to its surroundings in the process of liquefying—the left coils in the figure. The high-pressure liquid then passes through a valve where the pressure drop allows it to vaporize and cool below the temperature of the ground which acts to heat it—the section of the coil on the right. The cyclic process continues as long as electrical energy is supplied to the compressor.

The performance of a heat pump is measured in terms of its coefficient of performance (COP), which is the ratio of the heat released to the home divided by the electrical energy supplied, a figure that is usually greater than 4, although much depends on how warm you want your home to be. In general, the coefficient of performance becomes greater, the smaller the difference between the ground and home temperatures, with the maximum possible (Carnot) value being given by

$$\mathrm{COP}_{\max} = \frac{T_{\mathrm{H}}}{T_{\mathrm{H}} - T_{\mathrm{G}}}. \tag{6.6}$$

In Equation 6.6, both the high (*H*) and the ground (G) temperatures must be in kelvins. Heat pumps are not specific to geothermal power, and in fact, many homes use electric heat pumps that use heat extracted from the cold outside air to heat the home.

Geothermal heat pumps are more efficient than conventional ones since they extract thermal energy from the ground rather than the outside air, and the ground below a few meters of depth is warmer than the outside air in winter and because the circulating fluid is a liquid, not air which has a lower specific heat. In addition, unlike conventional heat pumps, the circulating fluid can be simply water or water plus antifreeze rather than a refrigerant. An unfortunate aspect of standard heat pumps is that the colder it is outside, i.e., the more you need them, the lower their ideal coefficient of performance by Equation 6.6. This flaw, however, does not apply to geothermal heat pumps—do you see why? Heat pumps can also operate as air conditioners as well as suppliers of heat. In fact, the exact same schematic diagram would describe how this works. The only difference is that now, the hot (left) coil is understood to be inside your home and the cold (right) one is in the ground. Essentially, the volatile fluid needs to circulate in the opposite direction in Figure 6.10.

6.6.1 Economics of Residential Geothermal Power

There are several ways to characterize geothermal heat pumps, one being the layout of the pipes—either horizontal or vertical. Horizontal systems generally cost less than half the cost of vertical systems, given the much greater depth needed in the latter case. However, horizontal systems, such as the one shown in Figure 6.11 before it was buried, are not as suitable on many lots because of the greater land area they require. Here are some approximate cost figures in 2010 US dollars for these two types of systems.

- *Horizontal system*: The up-front costs of a residential system is about \$2500 per ton (1 T = 3.517 kW) or, roughly, \$7500 for a 3 ton unit on an average residential size system. This figure is about double the cost of a gas furnace, but since the fuel is free, typical annual energy savings are perhaps \$450/year, although obviously, it depends on many factors: the price of gas, the size of your house, and the temperature you choose to keep it at. Given the preceding cost figures, the breakeven time would be 7 years—longer if gas is very cheap, shorter if a sizable tax credit is allowed for the geothermal system.
- *Vertical system*: In this case, drilling can run anywhere from \$10,000 to \$30,000, so the cost of a vertical system might be \$25,000, and so, to make up the difference between this initial cost and that of a gas furnace would take 20,000/450 or around 45 years.

Figure 6.11 A horizontal closed-loop field is composed of pipes that run horizontally in the ground below the frost line—photo shows a slinky (overlapping coils) arrangement before it has been covered by dirt. The slinky style layout is a more space-efficient version of a horizontal ground loop than straight piping. (Courtesy of Marktj, http://en.wikipedia.org/wiki/Geothermal_heat_pump.)

The other way to characterize geothermal heat pumps is whether they are open- or closed-loop systems, with the former used only when the system is of the vertical type. In this case, water is pumped down a vertical injection well and passes through HDR to reach the production well in much the same way as in EGSs used for electric power generation.

6.7 SUSTAINABILITY OF GEOTHERMAL POWER

Given the way heat pumps work, and the relatively small amount of energy extracted from the ground, there is no question that a residential system can indefinitely operate without exhausting the heat stored there. What about the sustainability of large-scale geothermal extraction of heat from the Earth for producing electricity? Would the Earth begin to cool down in that case? Currently, geothermal power accounts for only around 10 GW electricity around the world. Since radioactive decay is continually replenishing most of the Earth's heat at about 30 TW—or 3000 times as much—there is no need to worry about running out on a timescale of billions of years even if we extracted a far greater percentage than at present.

Nevertheless, the depletion of an individual geothermal field is another story, which we shall now consider with the aid of a simplified model.

6.7.1 Depletion of a Geothermal Field

Here we shall assume that some number N production wells are sunk over a surface area A from which geothermal energy is extracted down to a depth z. The spacing between the wells is assumed to be such that the thermal energy in a given region is extracted by only one well. Let us further artificially assume that the entire resource from which energy is extracted is at a common temperature T, whose initial value at time $t = 0$ is taken to be T_0. The geothermal field is stimulated when water is injected, and the thermal power extracted from the HDRs of density ρ_r and specific heat c_r can be expressed as

$$\dot{q} = \frac{\mathrm{d}q}{\mathrm{d}t} = -m_r c_r \frac{\mathrm{d}T}{\mathrm{d}t} = -Az\rho_r c_r \frac{\mathrm{d}T}{\mathrm{d}t}. \tag{6.7}$$

This thermal power equals that in the water rising out of the production well, ignoring any losses or replenishment of the thermal energy from surrounding rock regions. Thus,

$$\dot{q} = \dot{m}c_w (T - T_S) = N\rho_w a v c_w (T - T_S), \tag{6.8}$$

where v is the speed of the water rising out of the N pipes, each of cross-sectional area a and T_S is the surface temperature.

Combining Equations 6.7 and 6.8 and rearranging terms yield

$$\frac{d(T - T_S)}{(T - T_S)} = -\frac{dt}{\tau}, \tag{6.9}$$

where the lifetime τ of the geothermal field is given by

$$\tau = \frac{\rho_r A z c_r}{N \rho_w a v c_w}. \tag{6.10}$$

Upon integrating both sides of Equation 6.9, we find that

$$T - T_S = (T_0 - T_S)e^{-t/\tau}. \tag{6.11}$$

Note that the lifetime of the resource can also be characterized in terms of a half-life given by $T_{1/2} = \tau \ln 2$. Furthermore, note that the thermal power extracted declines with the same exponential time dependence, assuming v is constant. Thus, if E_0 is the initial energy content at $t = 0$, we have for the power extraction at any given time t

$$\frac{dE}{dt} = -\frac{E_0 e^{-t/\tau}}{\tau}. \tag{6.12}$$

In practice, however, the actual lifetime will likely be greater than τ since we have ignored the replenishment from surrounding regions as energy is extracted. Typical values of the replenishment time (once energy extraction ceases) range from 1 to 10 times the lifetime. One can, of course, choose to extract energy at a lower rate to prolong the drawdown time, but then, the power output becomes proportionately smaller. Note that geothermal power plants do tend to have much smaller power output in any case than either fossil fuel or nuclear plants—50–100 MW being typical. Another approach to prolonging the drawdown time is to extract the energy over a larger surface area, but that would mean many more extraction wells, which tend to increase costs, although not necessarily on a per megawatt-hour basis.

6.7.2 Example 3: Lengthening the Lifetime

Suppose a given geothermal resource has a lifetime of 20 years and a replenishment time of 60 years. What are two ways to extend the lifetime of the resource to over 1000 years?

Solution

Given that the replenishment time is three times longer than the lifetime, thermal energy is being restored to the geothermal field at a third the rate it is being extracted, so if the power drawn were cut by two-thirds, the two would be in balance, and the field would essentially supply energy for many millions of years. An even better choice might be to supply electricity only at peak times each day, when the demand is highest and when it is in short supply. Still another option would be to sink more extraction wells, but then run the water through them at a lower rate so the same amount of power extracted is now extracted over an area three times as large, resulting in a drawdown of the thermal energy at one-third the original rate, and a near-infinite lifetime would be the result.

6.7.3 Example 4: A 100 MW Power Plant

(1) Find the useful heat content per square kilometer to a depth of 7 km. Assume a thermal gradient of 40°C/km, a minimum useful T = 140°C above that on the surface, a rock density of 2700 kg/m^3, and rock specific heat of 820 J/kg K. (2) What volume flow rate of injected water is needed for this power plant if it extracts heat over a surface area of 0.5 km^2? (3) After how many years will the power produced be half its initial value, assuming a constant water flow rate?

Solution

(1) Using Equation 6.2 yields 5.4×10^{17} J/km^2. (2) Using Equation 6.12, we first find the lifetime in terms of the initial power extraction (dE/dt = 100 MW) and the initial total energy stored $E = 0.5$ km^2 $\times 5.4 \times 10^{17}$ J/km^2 to be $\tau = 5.4 \times 10^9$ s. Finally, using Equation 6.10, we may solve for the mass flow rate of the water (the product va) to be 3500 kg/s. (c) Using $T_{1/2} = \tau \ln 2$, we find a half-life of 118 years.

6.8 ENVIRONMENTAL IMPACTS

6.8.1 Released Gases

A number of noxious gases are emitted during operation of a geothermal plant, but in relatively low concentrations. Plants where this is a problem may be required to install emission controls. The emitted gases may include a small amount of radon, which is a by-product of the decay of uranium—one of the main isotopes accounting for much of the Earth's stored heat. Radon (the second leading cause of lung cancer) is a well-recognized problem in some homes where it can seep up through cracks and become concentrated in the home, especially in the basement, and even more especially in tight, well-insulated homes. However, it is relatively harmless when it comes up from the Earth and is released to the outdoors. Moreover, areas having good geothermal potential are not associated with higher than average uranium concentrations. Radon gas can become dissolved in the drilling fluid used in geothermal wells, but in EGSs, this fluid is continually recycled back down the well, and hence, it is not released to the environment. GHGs (especially carbon dioxide) are also an issue of possible concern. However, the production of geothermal electricity usually results in far less CO_2 emitted than fossil fuel sources, but there is some variation depending on the type of power plant and the characteristics of the geothermal field. Typically, geothermal power plants have less than a tenth that of coal-fired plants (Bloomfield, 2003). The situation with respect to CO_2 emissions when geothermal power is used for residential heating is more complex and is discussed in Section 6.8.3.

6.8.2 Impact on Land and Freshwater

Small amounts of some harmful substances are found in the fluid after water is injected into a well during the hydrofraction process. However, these can be injected and recycled back from the production hole to reduce risk. Normally with HDR stimulation, the wells are deep enough so that groundwater should not be affected, unlike the case of using hydrofraction to extract natural gas. Geothermal power plants tend to occupy a relatively small land area (a small footprint), especially in comparison to other energy sources. For example, the comparable figures in units of square kilometers per gigawatt (km^2/GW) are 3.5 (geothermal plant), 12 (wind farms), 32 (coal), 20 (nuclear), and 20–50 (solar).

Geothermal plants do require a source of freshwater; however, unlike nuclear, coal, gas, or oil, the water is continually recycled, so that the amount used per megawatt-hour generated is negligible. Subsistence of land has occurred in some places due to the operation of a geothermal plant in New Zealand and several locations in Germany. Even worse, one geothermal plant built in Basel, Switzerland, was shut down after many small earthquakes were observed—10,000 of them up to a magnitude of 3.4 during its first week of operation. On the other hand, it must also be remembered that seismologists have been inducing artificial earthquakes since the 1960s, with the largest having a magnitude 4.5, but no large earthquakes have occurred as a result. During the hydrofraction process, earthquakes can be induced when the size of the fractures created is large, but can be controlled by adjusting water flow rates and pressures in fracturing the rocks. Most importantly, one must avoid intersecting a large natural fault that could trigger a large earthquake.

BASEL EXPERIENCE

After a series of small earthquakes occurred, the Swiss government did a study that concluded that if the project had been allowed to continue, there was a 15% chance of it triggering a quake powerful enough to cause damage of up to $500 million. As a result, the government brought criminal charges against the head of the company Geopower that did the drilling. However, the trial found that the company had not deliberately damaged property or carelessly acted. Nevertheless, the Swiss government's fears may not have been unfounded because there is an earthquake fault in Basel, and in the year 1356, the town experienced what may have been the most significant seismological event in Central Europe in recorded history. That earthquake led to the destruction of the town and all major churches and castles within a 30 km radius.

6.8.3 Do Heat Pumps Cut Down on CO_2 Emissions?

We have seen that geothermal plants emit negligible amounts of CO_2, but for geothermal residential heating, the issue is less clear. Geothermal heating systems usually rely on heat pumps that extract heat from the ground and deliver it to a higher temperature, i.e., the interior of your house. Thus, they make heat flow the wrong way through the input of work. The COP of ground source heat pumps is usually above 4. The actual average COP tends to be somewhat lower than 4 when we include the energy needed to power the water pumps. Let us assume a COP of 3—meaning that the heat supplied is three times the electrical energy used to power the compressor.

Let us further assume that electricity consumed was generated by a gas-fueled power station. Given that the electrical energy generated at the power plant and transmitted to your home is typically 31% of the thermal energy at the plant from burning the gas or 0.31Q and given that this electricity powers a geothermal heat pump with COP = 3, the thermal power it delivers to your home is three times as much or 0.93Q. But, now suppose that instead of a geothermal heating system, you chose to install a high-efficiency gas furnace with an efficiency of 95%. For the same amount of gas burned Q, the furnace delivers more heat, 0.95Q, so it will require less gas to deliver the same heat and would emit less CO_2. This comparison is, of course, invalid if the power plant uses coal (where the situation is much worse in terms of emissions) or if it uses nuclear or renewable energy, where it is much better. The comparison is also invalid if you chose not a high-efficiency gas furnace, but instead an electric heat pump or an oil-burning furnace.

6.9 ECONOMICS OF GEOTHERMAL ELECTRICITY

A major factor determining the cost of electricity from geothermal resources is the high initial costs associated with drilling wells, which can be very substantial—particularly with deeper wells where they may amount to over 60% of the initial investment, with the rest being mostly construction of the power plant.

6.9.1 Drilling Costs

The cost and technology for drilling geothermal wells have many similarities with drilling gas and oil wells, although they tend to be much higher—at least for shallow wells. Currently, there are not enough geothermal wells (especially at greater depths) to draw reliable conclusions on how their cost varies with depth over a wide range of depths. Therefore, it is common to examine the situation for gas and oil wells and then make models that take into account the differences for the geothermal case. The costs of wells having any given depth can enormously vary, depending on many factors. For example, for wells over 6 km depth, there is a difference of a factor of 10 between the least expensive and the most expensive wells. Another complication in estimating well drilling costs is that the cost of individual oil and gas is propriety information. Fortunately, however, the industry association does provide average data for a given range of depths for a given year. As can be seen in Figure 6.12, the average well drilling costs C in dollars for 2004 are nicely fit by an exponential function of depth z of the form (Augustine, 2009)

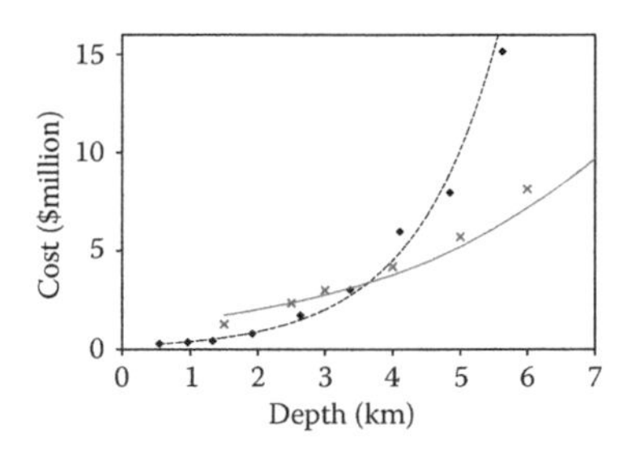

Figure 6.12 Black diamond points are the costs of drilling in millions of US 2004 dollars versus depth of hole in kilometers for oil and gas wells, and the dashed line is an exponential fit to them. The points shown by Xs correspond to a much more limited number of geothermal wells, and the solid curve is a fit to them based on the WellCost Lite model that takes into account differences between the two types of wells.

$$C = Ae^{Bz}, \tag{6.13}$$

where A = \$200,000 and B = 0.75 million/km. The values of the constants A and B can vary somewhat from year to year based on the availability of drilling equipment and labor, which strongly correlate with the price of oil and gas. Nevertheless, while the coefficients in the exponential function may change with time, the shape of the curve for gas and oil wells tends to remain exponential—at least over the range of depths drilled.

6.9.2 Beating the Exponential?

Beating the exponential increase in drilling cost with depth is of vital importance for tapping reserves that have a modest thermal gradient and are therefore found at deeper depths. For wells 5–6 km deep, drilling costs can be perhaps half the initial investment (the other half mostly being the plant itself). It may be true that reducing drilling costs will still leave all those other costs intact and can therefore have only a limited impact. However, if the exponential increase in cost with depth can really be avoided (made linear), vast reserves in the upper 10 km of the Earth might be accessed at only a modest increase in cost and not just in places having a high gradient. Moreover, if it can be empirically shown that improved technology allows the drilling cost to be a linear function of depth, the gradient of the resource would have no effect on drilling costs per megawatt-hour! This assertion is based on the discussion associated with Figure 6.7, where it was shown that if the drilling depth is limited only by some maximum temperature (not depth), the size of the accessible geothermal resource is proportional to the depth, regardless of gradient. Under these assumptions, it should be possible to economically exploit geothermal power for electric power generation anywhere on land.

Unfortunately, there is not enough direct empirical evidence for the drilling costs of geothermal wells (especially for very deep ones) to directly check if they are also exponential functions of depth. Instead, what has been done is to develop a model (known as WellCost Lite) based on a detailed analysis of the time and costs of each step in the drilling process and then correct for the variations in costs from year to year. Note that while geothermal wells tend to be more expensive than oil and gas wells for lower depths, as the depth increases, they are less expensive with the crossover point being around 4 km. Is the WellCost Lite model predictions (checked against actual drilling costs only up to 5 km) consistent with an exponential function? At first sight, they appear not to be—see Figure 6.12. On the other hand, as can be seen, the model results do fit an exponential—provided one subtracts a constant dollar amount, meaning that the costs of drilling geothermal wells can be expressed as

$$C = D + Ae^{Bz}. \tag{6.14}$$

6.9.3 Why the Exponential Dependence of Cost on Well Depth?

We have seen that drilling costs with depth empirically tend to be an exponential function of depth for oil and gas wells (upper dashed curve in Figure 6.12). Are there reasons to suspect such a dependence of cost with depth? As long as the cost to dig each additional increment of depth dz is a percentage of p more than the previous increment, it is easy to show that the overall cost of drilling a well to any depth will be an exponential function of depth. This would seem to be true for any factors that become more time consuming the deeper you drill. Some examples might include the following:

- Difficulty of the actual drilling wherein as you go deeper as it gets hotter and there is more wear on the bit and more likelihood of it getting stuck
- Flushing out debris—which has further to travel to surface
- Greater likelihood of losing circulation of drilling fluid at deeper depths
- Time needed to replace worn drilling bits that need to be hauled out

For such factors as these, the time (and cost) to drill an increment of depth dz is quite likely to be proportional to depth, so just like with compound interest, there will be an exponential growth in cost with depth z.

6.9.4 Is Spallation Drilling the Answer?

Spallation is the process by which fragments of a piece of material are ejected due to either impact or stress. Spallation drilling involves no drill bit that can wear out due to contact with the rock. Instead, a flame jet makes contact with a small area of the rock at the bottom of the borehole, and the induced thermal stresses in the rock cause small fragments of it (spalls) to be ejected.* The fragments are small enough that the injection of high-pressure water carries them up the water-filled drilling pipe.

* There is also a nice animation on spallation drilling available online at http://www.youtube.com/watch?v=mF1WeM4vZC8.

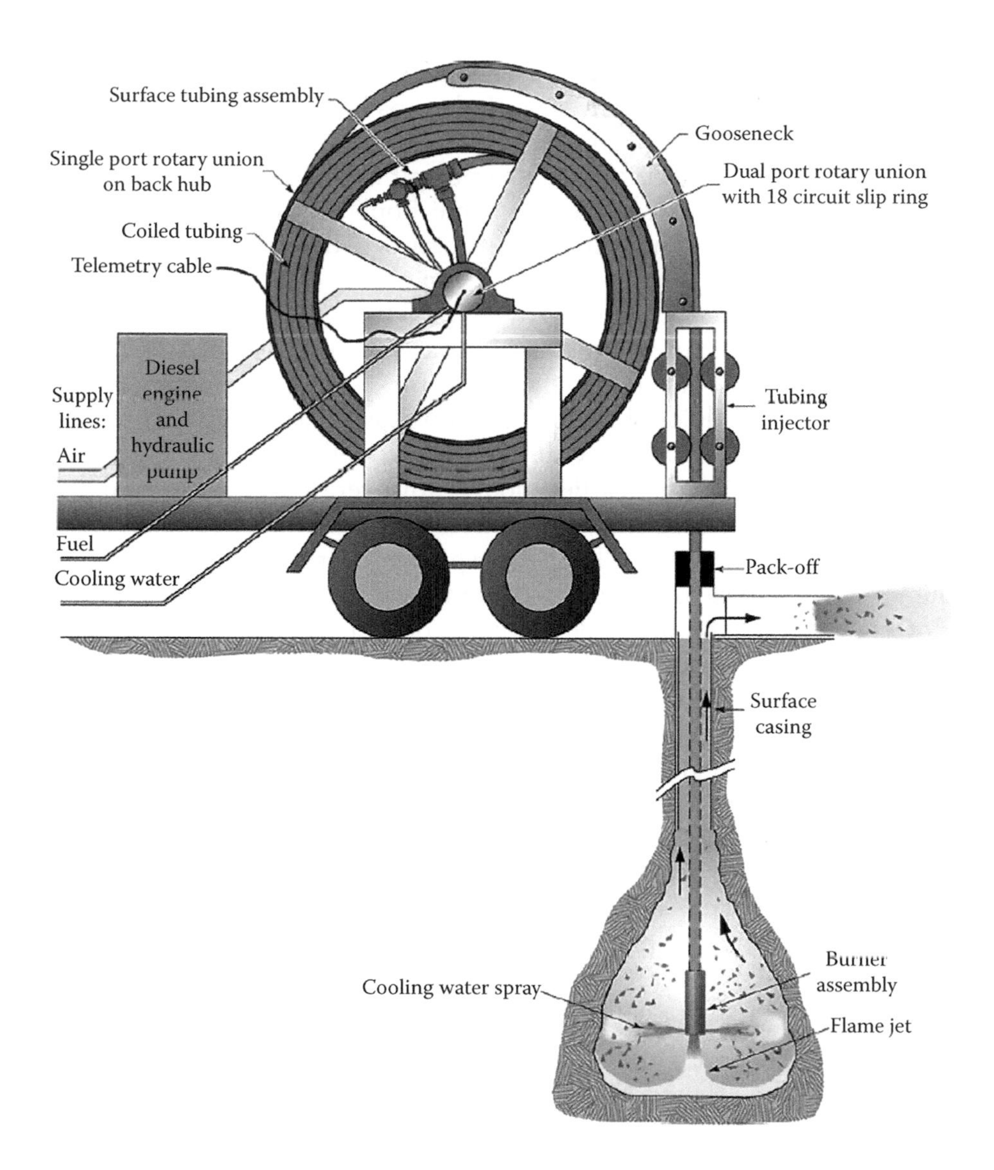

Figure 6.13 Diagram of spallation drilling of a borehole. (Courtesy of Los Alamos National Laboratory, Los Alamos, New Mexico.)

Oxygen must be supplied to allow combustion underwater in a similar manner as done in underwater welding. The spallation technique has been demonstrated and found to work well, and it is likely to be a significant improvement over conventional drilling methods, in terms of both speed and cost—especially if it can make the cost dependence on depth linear rather than exponential (Figure 6.13).

6.9.5 Why Spallation Drilling Cost Might Be a Linear Function of Depth

For a variety of reasons, it might be reasonable to expect that the drilling cost associated with spallation drilling might prove to be a linear function of depth, rather than an exponential one. Recall that a linear cost function means that the time and cost to drill a segment of length dz does not depend on how deep that segment is. Unlike conventional drilling, which requires a rotary bit, the spallation process would be a continuous one with no need to haul out worn out drill bits. The continuous nature of the process and the small size of the spallated particles coming off the rock mean that this debris is continuously removed with the water flushed down the hole. It would also permit a single-diameter hole even at great depths. These factors give some hope that drilling deeper segments should take no more time than shallower ones of the same length.

Chad Augustine and his thesis advisor, MIT chemical engineering professor Jefferson Tester (who shares a patent on the spallation drilling process), produced the linear model, using a similar approach as was used with the WellCost Lite model shown as a dashed curve in Figure 6.12, that fit actual drilling cost data fairly well—at least up to about 5 km depths. Unlike that earlier model, however, there is no actual drilling data to compare against the spallation linear model to check if the model conforms to reality. Thus, the spallation drilling process is still in the prototype stage and remains to be tested in real-world applications. However, should the linear dependence of drilling cost on depth prove to hold at great depths, the impact on the future of geothermal energy would be enormous, since it would mean that the drilling costs of geothermal electric power per megawatt-hour would be the same for low-thermal gradient regions as for high, and vast quantities of geothermal energy now prohibitively expensive to extract because of the exponential dependence on depth would become available. Geothermal power-produced electricity would then move from a source that is economical only at some special places having a very high geothermal gradient to one that is economically exploitable everywhere. Even more controversially, as Problem 11 illustrates, the cost of power could actually be less for areas having low thermal gradients than high, under the assumptions stated in the problem.

AS LONG AS THE HOLE IS ALREADY THERE?

It has been estimated that the United States has 2.5 million abandoned oil and gas wells, some of which are miles deep. Given that drilling costs are a not-insignificant fraction of total costs, Chinese scientists have come up with a way to retrofit these abandoned holes with new shafts containing pipes within pipes that would allow them to function as geothermal wells (Xianbiao et al., 2011). They estimate that the average abandoned well could generate 54 kW of electricity—not much compared to any central power station, but enough to collectively make them a source of a considerable amount of clean energy. They further estimate that the economic returns of electricity are about $40,000 per year for each retrofitted well having a thermal gradient of 45°C/km.

6.10 SUMMARY

Geothermal power has many advantages as a renewable energy source: it is economical, environmentally fairly benign, and sustainable. It can also be used for either residential or commercial heating virtually anywhere, and in places where conditions are right, it can generate electric power more cheaply than most other sources. Its usage over the last century has exponentially increased, even though it still accounts for a very small fraction of electric power production—about the same as solar power. Geothermal electricity production has the potential for even more widespread usage into areas with low thermal gradient, but this development entirely depends on whether drilling costs can be made linear with depth for some novel drilling method such as spallation drilling.

PROBLEMS

1. Assume that the thermal gradient in a given location is not constant but rather a linear function of depth of the form $G = G_0(1 + \alpha z)$. (a) Find by integration the thermal energy per unit area between two depths z_1 and z_2. (b) Find by integration the thermal energy per unit area between two temperatures T_1 and T_2.
2. In Section 6.1, it was noted that according to an MIT study, the total geothermal energy that could be feasibly extracted in the United States with improved technology in the upper 10 km of the Earth's crust is over 2000 ZJ or 2×10^{24} J. Show that this is roughly correct by calculating the energy stored in the upper 10 km per unit area of surface above a temperature of 150°C using the average thermal gradient of 30 K/km, values for the specific heat and density of average rocks, and the area of the United States found on the web.
3. Based on Equation 6.1, we explained the difference between the average thermal gradients in the Earth's inner core and its crust. However, that comparison assumed the same q for both cases (see discussion after Equation 6.1). How would this comparison change if we take into account that roughly 80% of the heat is generated from radioactive decay in the core and mantle?
4. Use the thermal conductivity of iron and the radius of the Earth's inner core and its thermal gradient (seen in Figure 6.3) to find the total heat flow out of the core integrated over the area of the core. Compare that value to the total heat flow that reaches the Earth's surface about 0.06 W/m^2 integrated over the surface of the Earth. Explain the discrepancy. Hint: See previous problem.
5. Given the reserves of geothermal heat in Section 6.1 according to the MIT study, how long would they last at the present annual world energy usage, taking into account the fact that 80% of the current heat supply is being continually replenished by radioactive decay of elements having multibillion-year half-lives?
6. (a) Justify Equation 6.3 by integration. (b) Consider the thermal energy in an aquifer for which the rocks have a porosity of 0.2, a density of 3000 kg/m^3, and a specific heat of 1000 J/kg K and for which the fluid is water. Find the average value of $c\rho$ for

this aquifer and the total thermal energy to a depth of 6 km that exceeds 150°C, assuming a gradient of 75°C/km.

7. A geothermal plant initially operates at 100 MW electric in its first year of operation. In the second year, the power is reduced by 2% due to drawdown from the geothermal reservoir. After 2 years, to prolong the life of the plant, the circulating water flow is reduced to the point that the electrical power produced is only 50 MW, and it is then found that in the third year of operation, the power reduction is only 0.7%. (a) Find the drawdown time for the plant power to be reduced to 25 MW. (b) Find the replenishment time if this geothermal field is left unused.
8. Using relevant data found on the web, determine the CO_2 emitted by a typical home-heating system under these three choices: electric heat pump, geothermal heat pump, and high-efficiency gas furnace and these three assumptions as to the source of your home's electricity: gas-fired, coal-fired, and nuclear power plants—a total of $3 \times 3 = 9$ combinations.
9. Show that as long as the cost to dig each additional increment of depth dz is a percentage of p more than the previous increment that the overall drilling cost to reach a depth z will be an exponential function of depth $C = Ae^{pz}$. What would p be if it costs $10 million to drill a 5 km deep well and it costs $100 million to drill a 10 km deep well? Why is it plausible that the actual cost of drilling a well to some depth z is fit better by including an additive constant—the D in Equation 6.14—rather than a pure exponential?
10. Explicitly show that if it can be empirically shown that improved technology allows the drilling cost to be a linear function of depth, the gradient of the resource does not matter in evaluating drilling costs per megawatt-hour.
11. Consider the primary cost of components of geothermal power, under the assumption that drilling costs can be made linear with depth, and use the parameters for the slope 0.433 $M/km and intercept 0.789 $M. Assume that owing to the economies of scale, the costs of constructing a geothermal plant (on a per megawatt basis) decline somewhat as the power of the plant increases. Thus, assume the plant cost is a linear function of P, given by $A + BP$, with A = $15 million and B = $2 million/MW. Let us further assume the plant operates for 25 years (over which, half the energy in the reservoir has been extracted), with an annual operations and maintenance (O&M) cost of $100,000/MW. Now consider two places to site the plant, one having a gradient of 50°C/km, and the other, 100°C/km, and assume that the minimum and maximum well depths are both governed by the minimum and maximum temperatures of 150°C and 300°C. For each location, calculate the total thermal energy available and the total cost of the plant (drilling, plant construction, and O&M) on a per megawatt basis. Which location is the better choice—the one with the high or low gradient? This illustration presupposes

a constant gradient for both locations, which may be less likely to be true in locations having a very high gradient.

12. Consider the comparison between a geothermal heat pump and a high-efficiency gas furnace in Section 6.8.3. Find the COP of a geothermal heat pump for which the amount of natural gas burned at the power plant to power the heat pump exactly equals that burned by the high-efficiency gas furnace.
13. Show that the right-hand side of Equation 6.10 has units of seconds and explain why each term on the right-hand side appears in either the numerator or the denominator.
14. Calculate the amount of thermal energy available by cooling 1 m^3 of rock from 240°C to 100°C. Assume that the rock specific heat is 2.4 J/cm^3 °C.
15. Imagine a steam turbine electric generator power plant that utilizes geothermal renewable energy. With a steam temperature of 120°C and a heat reservoir at 25°C, what is the maximum efficiency possible? Also, what percentage of the steam's energy will be dissipated as waste heat?

REFERENCES

Augustine, C. R. (2009) *Hydrothermal Spallation Drilling and Advanced Energy Conversion*. PhD thesis, Massachusetts Institute of Technology, Cambridge, MA.

Bloomfield, K. K. (2003) Geothermal energy reduces greenhouse gases, In K. K. Bloomfield (INEEL), J. N. Moore, and R. M. Neilson, Jr. (eds), *Climate Change Research*.

Tester, J. (2006) *The Future of Geothermal Energy—Impact of Enhanced Geothermal Systems (EGS) on the United States in the 21st Century*, MIT, Cambridge, MA. http://mitei.mit.edu/publications/reports-studies/future-geothermal-energy.

Xianbiao, B., M. Weibin, and L. Huashan (2011) Geothermal energy production utilizing abandoned oil and gas wells, *Renew. Energy*, 41, 80–85.

Wind Power

7.1 INTRODUCTION AND HISTORICAL USES

The global wind power resource is truly enormous, but much of the wind power is relatively inaccessible, being far out at sea or at high altitudes, where high winds continuously blow. However, the technically accessible amount has been estimated to be about 300 million GWh per year—around 20 times the current electricity demand. Wind is one of the oldest forms of energy harnessed by humans. The earliest applications include sailing ships, windmills to grind grain, and pumps for irrigation purposes or to prevent flooding. Wind used for propulsion goes back at least 5500 years, and its agricultural uses can be traced back to the seventh century in the Middle East and Asia.

The advent of wind power for producing electricity is much more recent and dates back to the late nineteenth century. Early pioneers include James Blyth of Scotland and Charles Brush of the United States, both of whom used wind power for this purpose since 1887. Brush's wind turbine looked nothing like today's three-bladed versions and might not even be recognized for what it was if a twenty-first-century time traveler stumbled across it. The scale of his 40 ton, 17 m diameter turbine can be gleaned from the tiny figure of a man standing on the right in Figure 7.1. The device built in Brush's backyard supplied 12 kW and was used to charge batteries so as to supply power to his home and laboratory. However, the earliest attempts to supply power to a nation's electric grid is much more recent (1931 in the former Soviet Union)—more than 40 years after its first electrical usage. Several formidable challenges had to be overcome before Blyth's and Brush's early electrical applications could extend beyond charging batteries.

As one indicator of the magnitude of improvements, a modern-day successor to Brush's device having the same diameter would be able to generate well over 100 times as much power, which is indicative of the great improvements made in this technology.

Those technological improvements, and the economies of scale as the technology becomes more widely used, have fueled a significant expansion in wind power in recent decades. As of 2010, wind capacity reached 175 GW, and wind now generates about 2% of the world's electricity. In Denmark that figure has now reached 20%—the highest in the world. Denmark is not a large nation and accounts only for a small absolute amount of the world's installed wind capacity. The six leading nations for installed wind capacity are China, the United States, Germany, Spain, India, and the United Kingdom, which collectively account for 76% of the world total. In fact, China's growth has been phenomenal, as its total installed wind power capacity has doubled every year since 2005 and is outpacing the growth of other competing energy sources.

CONTENTS

Figure 7.1 First automatically operated wind turbine, built in Cleveland, in 1887, by Charles F. Brush. The large rectangular shape to the left of the rotor is the vane, used to move the blades into the wind. (From Robert, W. R., *Wind Energy in America: A History*, University of Oklahoma Press, Norman, OK, 1996.)

Although wind power continues to be used for agricultural purposes, including pumping water, its application to electricity generation is by far the dominant one (Figure 7.2). In the United States and many other industrialized nations, wind power is one of the largest components of new electrical generating capacity in recent years, which is in part

Figure 7.2 Wind farm in Royd Moor in England. (Courtesy of Charles Cook.)

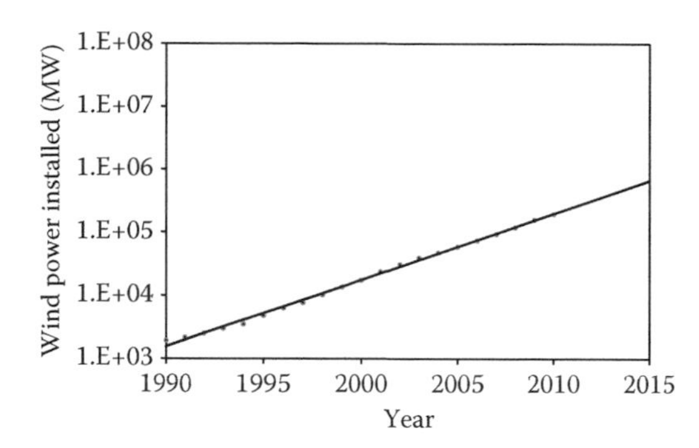

Figure 7.3 Cumulative installed wind power in megawatts versus year. Projected line based on exponential growth since 1990 reaches 2010 world total electric capacity by the year 2024.

attributable to a combination of cost and environmental considerations. In fact, the DOE has estimated the cost of new generating capacity using wind to be on a par with coal or natural gas—see Figure 6.2. An independent 2011 assessment by Bloomberg New Energy Finance agrees—at least for some regions where wind is particularly favorable in the United States and other nations (Bloomberg, 2011). By contrast, for example, solar power on a per kilowatt-hour basis is roughly 3.5 times as expensive. The growth in installed wind over the last three decades is in fact fairly well described by an exponential function, with a doubling time in wind capacity every 2.86 years—see Figure 7.3. It would probably be unrealistic to expect that exponential growth to continue unchecked for the indefinite future—especially as the use of wind begins to account for a significant fraction of the world's electric generating capacity. However, if we were to extrapolate the observed exponential growth forward for just five more doublings, we find an installed wind capacity equal to the world's total present electricity generation by 2024.

7.2 WIND CHARACTERISTICS AND RESOURCES

The energy of the wind is an indirect form of solar energy, since differences in heating of regions together with the rotation of the planet are what drive the winds. The average wind speed considerably varies from place to place, and the power of the winds varies by an even greater amount both spatially and temporally. The world wind resource is of course greatest at sea, where the ocean surface offers less viscous resistance to wind flow than the land, as can be seen in Figure 7.4. The high winds below the 40th parallel south are especially prominent due to the lack of land there to hinder them. Although placing wind turbines in the middle of the ocean may not be practical, onshore coastal areas and near offshore areas with their high average wind speeds are.

A similar wind resource map is shown in Figure 7.5 for the United States, where the grayscale is indicative of the wind resource potential, defined as the value of the wind power per unit area perpendicular to the wind velocity. Generally, it is not economical to install a wind turbine if the

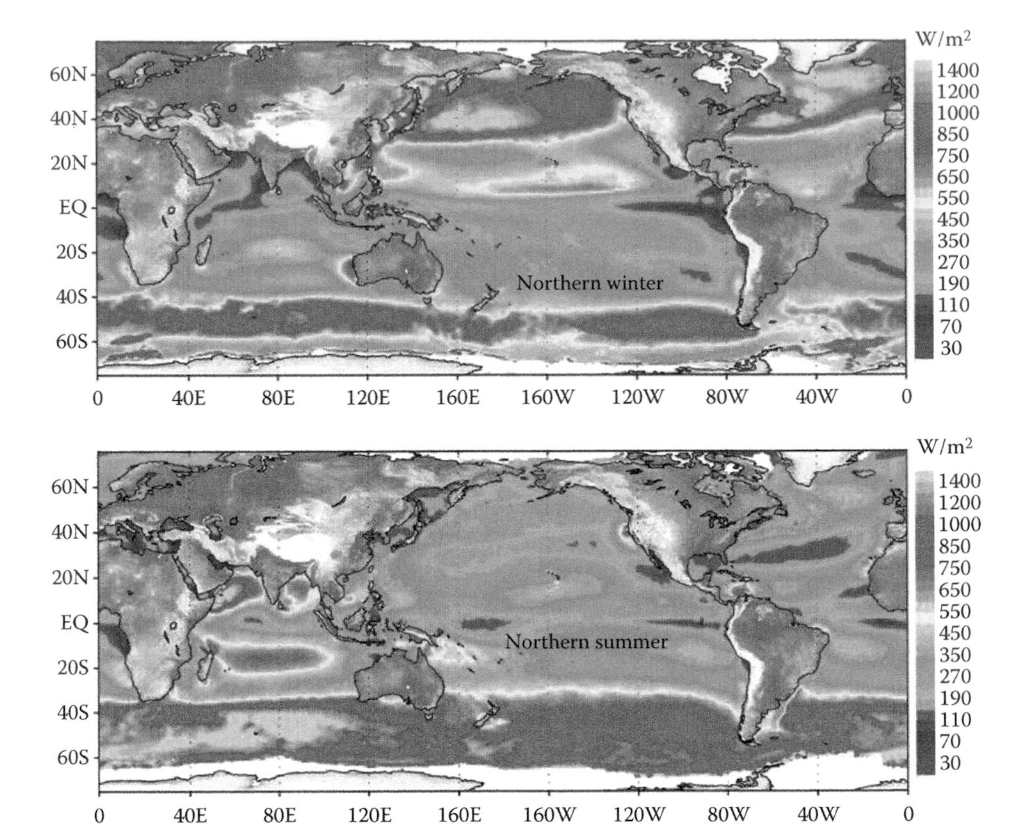

Figure 7.4 QuikScat data image shows wind power density for winter and summer in the Northern Hemisphere. (Courtesy of National Aeronautics and Space Administration/Jet Propulsion Laboratory, Pasadena, California, http://www.jpl.nasa.gov/news/news.cfm?release=2008-128.)

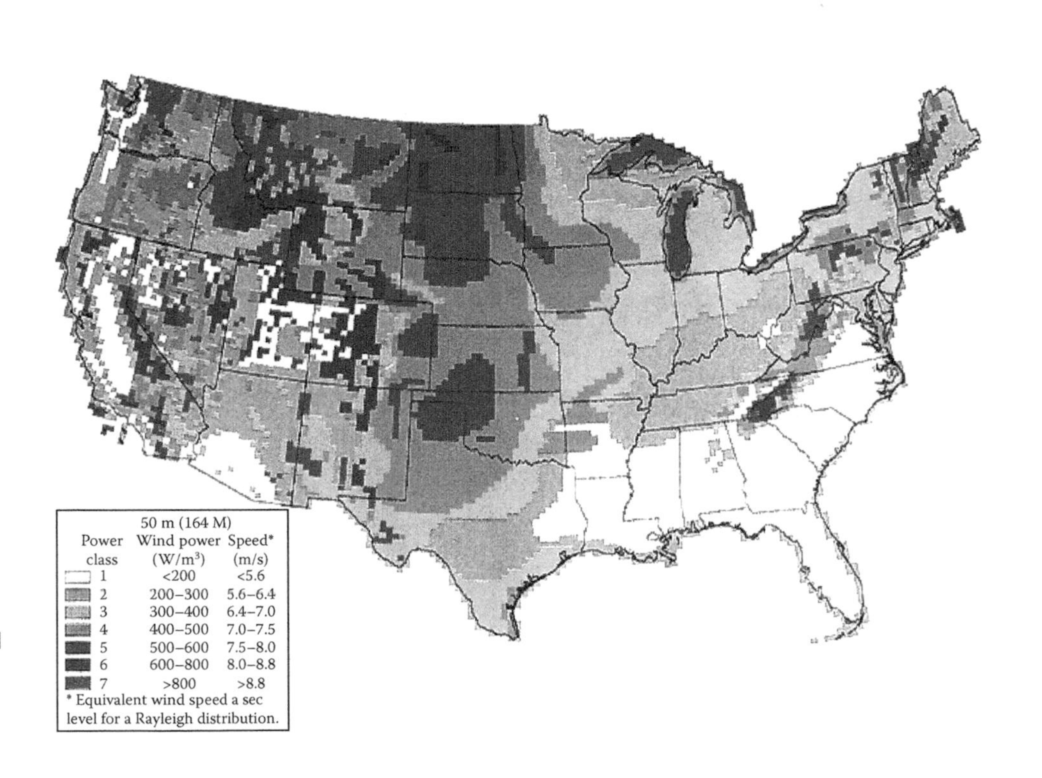

Figure 7.5 Wind resource map for the United States. (Image provided by the U.S. Department of Energy, Washington, DC, is in the public domain.)

resource is below class 3 ("fair"), which corresponds to an average power density in the wind from 300 to 400 W/m^2. Class 7, the highest one shown, corresponds to a power density 800–1600 W/m^2 and is considered "superb." As seen in the world wind map, the best wind potential is for coastal areas and offshore, but a swath of land running from the Dakotas to Texas also has good potential.

WIND POTENTIAL VERSUS SOLAR POTENTIAL

Other forms of renewable energy such as solar also have significant spatial variation in terms of their potential, but the variation tends to be much greater for wind. Thus, in the case of solar even a nation such as Germany that is not noted for abundant sunshine can still economically harvest solar energy, although the political commitment of the nation and the financial support in terms of subsidies are of course important as well. Nevertheless, it is fair to say that when the solar potential is not so good it is still not so bad either, but for wind when it is not so good it is usually terrible. Thus, in places where the wind potential was poor, it would be foolish and wasteful to have massive subsidies sufficient to entice individuals or utility companies to install wind farms. Another important difference in the wind–solar comparison is that while the potential is expressed in the same units in both cases, W/m^2, we must remember that the relevant area for solar is a square meter of the Earth's surface, while for wind it is a (vertical) square meter perpendicular to the wind flow.

7.2.1 *v*-Cubed Dependence of Power on Wind Speed

The very large variations in wind potential from place to place can be understood as a consequence of the v-cubed dependence of wind power on wind speed. Thus, in place A where the wind speed is 80% as great as in place B having a "good" wind potential, the power of the wind is 0.8^3 or only around half as great—putting it into the "poor" category. The basis of the v-cubed relation is very easy to understand. Consider a cylindrical mass of air of length Δx and end cap area A moving horizontally with a velocity v (the wind speed). In Figure 7.6, the air mass is incident onto circular area A that might represent the swept out disk of a wind turbine. A wind turbine is any device for harnessing wind power for useful purposes. The old term *windmill* is rarely used because of its connotation of a device used to mill grain.

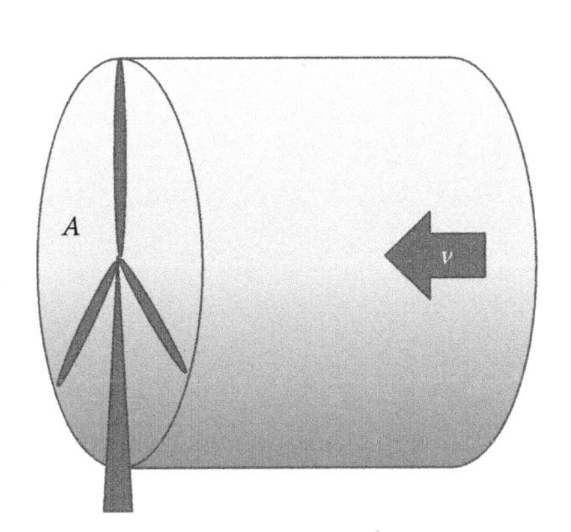

Figure 7.6 Moving cylindrical volume of air incident onto a circular area *A*.

The kinetic energy of the moving air mass is

$$\Delta E = \frac{1}{2}mv^2 = \frac{1}{2}\rho A \Delta x v^2. \tag{7.1}$$

Since $\Delta x = v\Delta t$ and the power may be written as $p = \Delta E/\Delta t$, we find for the power incident on area A as

$$p_{\text{wind}} = \frac{1}{2}\rho A v^3. \tag{7.2}$$

When the wind is incident on a turbine, only a fraction of this power C_P, known as the power coefficient, is extracted by the turbine. Thus, we may write with complete generality for the extracted power:

$$p = C_P p_{\text{wind}} \frac{1}{2} C_P \rho A v^3. \tag{7.3}$$

In addition to showing dramatic variations from place to place, the wind potential also, of course, varies in time at any given location, and those variations occur on all different timescales from the very short (sudden gusts and calms) to the very long (seasonal variations). These variations are important to understand, since the power in the wind does depend not only on the power for the average wind speed but also on the nature of its variation in time. Fortunately, while the wind speed may vary in an unpredictable manner from moment to moment, the distribution in wind speeds over a long period does follow a regular pattern.

7.2.2 Wind Speed Distributions

The Weibul distribution is a continuous probability distribution of a variable v, which can be written in the form

$$W(s, v_0, v) = \frac{s}{v_0}\left(\frac{v}{v_0}\right)^{s-1} e^{-(v/v_0)s}. \tag{7.4}$$

Equation 7.4 represents the probability density function, and it is normalized so as to give unity when integrated over all v for any choice of the shape parameter s and the scale parameter v_0. The distribution is named after Waloddi Weibul who studied its properties in detail. The effect on the shape of the distribution when the shape and scale parameters are varied can easily be seen in Figure 7.7.

The Weibul distribution has been found to give a good fit to a number of applications of practical interest including time to failure for manufactured parts and the distribution of wind speeds at a given location—our interest here. It should not be too surprising that wind speeds and many other phenomena are well described by a Weibul distribution, since with a suitable choice of the shape and scale parameters, it can nicely approximate any distribution having a single maximum that also goes to zero at both $v = 0$ and large v.

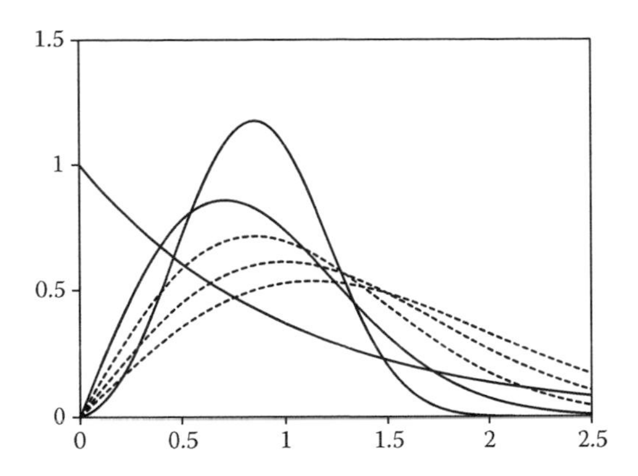

Figure 7.7 Six examples of the Weibul distribution with three choices of values for the shape parameter ($s = 1, 2, 3$) (*solid curves*) and the scale parameter ($v_0 = 1.2, 1.4, 1.6$) (*dotted curves*). In the latter three cases, the shape parameter $s - 2$, and in the former three cases, the scale parameter $v_0 = 1$.

As one might expect, the two parameters s and v_0 do vary with location. For example, for 10 German locations, the dimensionless shape parameter was reportedly found to vary between 1.32 and 2.13, while the scale parameter varies between 2.6 and 8.0 m/s. By comparing Figures 7.7 and 7.8, what would you guess the shape parameter was at the Lee Ranch? Since many locations have a shape parameter reasonably close to 2.0, it is sometimes convenient to set s to 2, in which case, Equation 7.4 reduces to the Rayleigh distribution. One advantage in using the Rayleigh distribution is that the scale parameter and the average wind speed then satisfy the simple relation

$$v_0 = \frac{2\bar{v}}{\sqrt{\pi}}. \tag{7.5}$$

For other s-values, only an approximate relationship exists between these quantities:

$$\bar{v} \approx v_0 \left(0.568 + \frac{0.434}{s}\right)^{1/s}. \tag{7.6}$$

It is important to realize that the distribution in power—essentially proportional to $v^3W(v)$—is entirely different from the distribution in wind

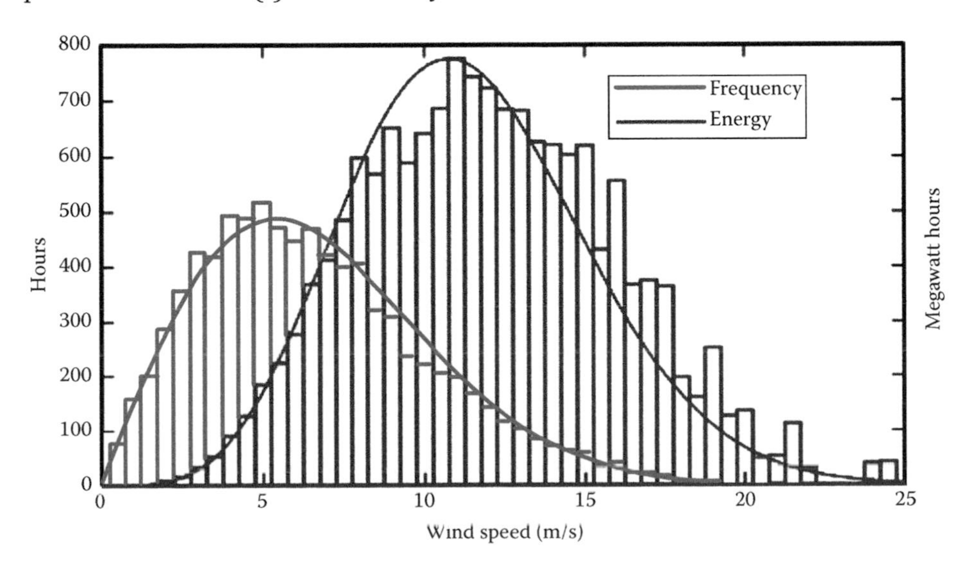

Figure 7.8 Wind speed histogram for a typical year (2002) at Lee Ranch, in Colorado, and fitted Weibul distribution (*left curve*). The right histogram and the associated fitted curve shows the distribution for the amount of power the wind contains at each speed interval. (Courtesy of Wikimedia Foundation, San Francisco, California, http://en.wikipedia.org/wiki/Wind_power, image is licensed under the Creative Commons Attribution-Share Alike 3.0 Unported license.)

speed itself. In fact, because of the v-cubed dependence, much of the power occurs in very short bursts of high wind speed, and these are not accurately represented if one only looks at hourly averaged speeds, but they can be captured if one averages $v^3W(s, v_0, v)$ each hour. It is interesting that in Figure 7.8, the fraction of power present at wind speeds greater than the most likely wind speed (maximum of the Weibul distribution) here constitutes perhaps 95% of the total power based on the area under the curve.

Since it is more common to encounter the average wind speed than the average of its cube, the following approximation is useful:

$$\bar{p} = \frac{1}{2}\rho A(v^3)_{\text{avg}} \approx \frac{1}{2}\rho A v_{\text{avg}}^3. \tag{7.7}$$

However, the extent to which this approximation is valid depends strongly on the shape parameter that best fits the wind speed distribution. It is most legitimate to use the approximation in Equation 7.7 when comparing the average power potential at two locations having wind speed distributions described by the same shape parameter so as to merely find the relative average wind potential of the two.

Wind turbines tend to produce their full (rated) power only for wind speeds above their so-called rated wind speed, typically around 11–14 m/s. What fraction of the time does this occur? If we assume that the distribution in wind speeds follows a Rayleigh distribution (Weibul distribution with $s = 2$), we find that the cumulative probability of having a wind speed greater than v when the average wind speed is $\bar{v}$ is given by

$$P(\geq v)\int_{v}^{\infty} W(2, v_0, v)\,dv = \exp\left(-\frac{\pi}{4}\left(\frac{v}{\bar{v}}\right)^2\right). \tag{7.8}$$

This function is plotted in Figure 7.9 for two values of v, 11 and 14 m/s. Based on this graph, we see, for example, that in a location where the

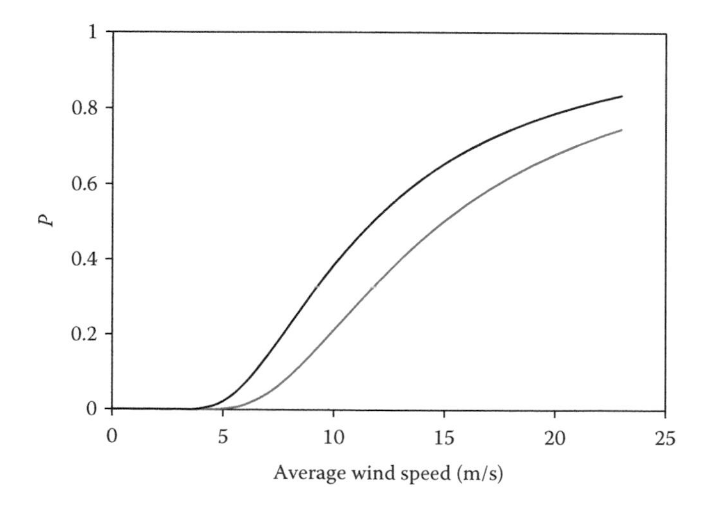

Figure 7.9 Cumulative probability of having a wind speed v in excess of 11 m/s (*upper curve*) and 14 m/s (*lower curve*) for various average wind speeds. Curves are based on a Rayleigh distribution of wind speeds.

average wind speed was 5 m/s, the percentage of time the wind speed is greater than the rated wind speed is negligible, while if the average were 10 m/s, the percentage of time it is greater than the typical rated values is still only 0.2–0.35, i.e., 20–35%.

7.2.3 Wind Speed as a Function of Height

Generally, average wind speeds increase with height above the ground, but the nature of the variation depends on many factors including the roughness of the terrain and the presence of human-made structures or natural ones such as trees. If such ground obstacles are present, it is recommended that the bottom of the wind turbine blades be no less than three times higher than any obstacles. This clearance is as much to avoid turbulence as it is to have a higher wind speed incident on the turbine blades. The average wind speed is particularly good along the crests of mountains, and on the downwind side, it can become twice as great as the average value.

Wind measurements are often made at a standard height of 10 m above the ground. Since many large wind turbines are erected at significantly greater heights, it is useful to have an approximate way of estimating the average wind speed at height h given its value at 10 m. One such relation is the empirical power law due to Hellmann that fits the observations for heights up to around 100 m, above which the average wind speed is roughly constant up to at least a kilometer:

$$v_h = v_{10}\left(\frac{h}{10}\right)^{\alpha}. \tag{7.9}$$

The value of the Hellmann exponent α in Equation 7.8 is, however, a function of the extent of ground roughness and the presence of obstacles. In the case of an open flat terrain, where large wind turbines are often erected, the value $\alpha = 0.14$ is used, but larger values, typically in the range $\alpha = 0.3$–0.6, are needed for populated areas. The large variation in possible turbine power with increasing hub height is illustrated in Figure 7.10.

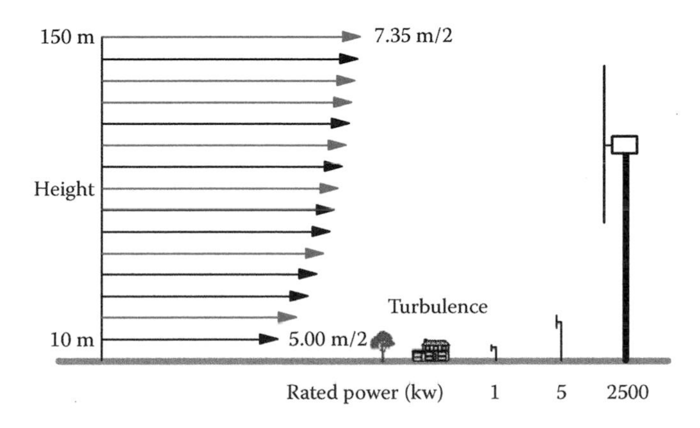

Figure 7.10 Example of the variation of wind speed and turbine power as a function of height of the turbine. (Image courtesy of David Mortimer.)

7.2.4 Example 1: Turbine Power versus Height

Suppose two wind turbines are erected: *A* at a hub height of 10 m and *B* at a hub height of 100 m at a location having an open flat terrain with no obstacles such as trees. What is the power ratio for turbines *A* and *B* assuming both have the same power coefficient C_P?

Solution

There are at least two reasons that turbine *B* will be able to generate much more power than *A*: the much larger diameter of its blades and the greater average wind speeds at its higher elevation. The average wind speed at 100 m height will be $10^{0.14} = 1.38$ times that at 10 m, based on Equation 7.9. However, since the power scales as the cube of the wind speed, the power increases by a factor of $1.38^3 = 2.63$ at the 100 m height. The increase in power due to the larger blade diameter is, however, far more significant since the power of a turbine scales as the swept out area. Normally, turbine blade diameters are about equal to the turbine hub heights, so this would imply a ratio of 10 for the two blade diameters and a ratio of 100 for the respective circular areas they sweep out. Combined with the previous factor of 2.63 based on wind speed, we therefore find that a turbine erected at a 100 m hub height could generate about 263 times as much power as one erected with a 10 m hub height.

Figure 7.11 Although wind turbines rated at 10 MW are under development by several companies, one of the largest ones built to date by the German company Enercon have a rotor diameter of 126 m (413 ft) and a power rating of 7.58 MW.

The actual power ratio in the previous example could be even greater than 263 if the Helmann exponent were larger than 0.14, or if the average wind speed at the lower height were marginal, since a minimum wind speed is necessary for turbines to even turn on. In light of this example, it is not surprising that the size and power of wind turbines have dramatically risen in recent decades—with a 10 MW behemoth being built in Norway as the current record holder. Largely because of this trend toward turbines of increasing size, the new turbines being sold in 2011 are about 300 times more powerful than those sold 15 years ago (Shahan, 2011).

SUPER WIND TURBINES

The manufacturing upper limit for conventional wind turbines is believed to be in the vicinity of 5–6 MW. Those having 8–10 MW power often make use of superefficient generators that rely on superconducting wire. This greatly reduces the size and weight of the turbine, which is a major advantage, especially for offshore turbines (Figure 7.11).

7.3 POWER TRANSFER TO A TURBINE

Until now it has been assumed here that the power coefficient C_P can have any value up to 1.0, but in fact, there is a theoretical maximum value, the Betz limit, which has the surprising value of 0.593 (59.3%).

WHO DISCOVERED THE BETZ LIMIT?

This Betz limit was independently discovered by three scientists in different nations: Frederick Lanchester (Britain), Nikolay Zhukowsky (Russia), and Albert Betz (Germany) between 1915 and 1920. Although Lanchester was actually the first to document his discovery, the limit is usually called the Betz limit, although some (non-Russians!) have proposed referring to it as the Lanchester–Betz limit.

In order to derive the limit, we consider the airflow through the plane of a wind turbine rotor and use a control volume analysis, meaning that all airflow goes in one end and out the other without passing through the side walls, i.e., that the flow is axial or along the turbine axis. As indicated in Figure 7.12, the cross section of that control volume continuously changes as the air approaches and then passes the rotor plane.

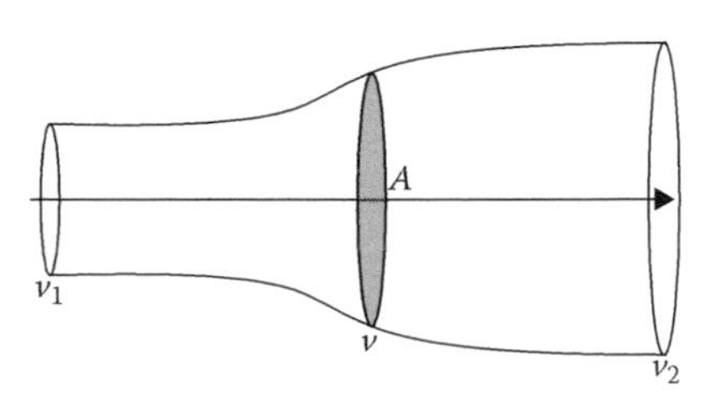

Figure 7.12 Airflow through the plane containing the rotor of a wind turbine.

We can see from Figure 7.12 that the power transferred to the turbine must be less than 100% in the steady state since we cannot have $v_2 = 0$ or the air would pile up there. The first part of the derivation of the limit involves proving the intuitively reasonable result that the air velocity right at the plane of the turbine v is simply the average of v_1 and v_2. Assuming that the air is incompressible, i.e., constant density, that the equation of continuity is satisfied, and that no air leaves the control volume, we have for the mass flow rate through the turbine

$$\dot{m} = \rho A_1 v_1 = \rho A_2 v_2 = \rho A v. \tag{7.10}$$

From Newton's second law, it then follows that

$$F = \dot{m}(v_1 - v_2) = \rho A v (v_1 - v_2), \tag{7.11}$$

so that the power transferred to the turbine is

$$p = Fv = \rho A v^2 (v_1 - v_2). \tag{7.12}$$

In the ideal case of no viscosity or turbine friction, mechanical energy is conserved, so that the power transferred to the turbine can also be expressed as the difference between the power of the air at points 1 and 2:

$$p = \frac{1}{2}\rho\left(A_1 v_1^3 - A_2 v_2^3\right) = \frac{1}{2}\rho A v\left(v_1^2 - v_2^2\right). \tag{7.13}$$

Equating the right-hand sides of Equations 7.12 and 7.13 gives the sought result:

$$v = \frac{(v_1 + v_2)}{2}. \tag{7.14}$$

If we now define $x \equiv v_2/v_1$, then substitution of Equation 7.13 into Equation 7.12 gives

$$p = \frac{1}{4}\rho A v_1^3 (1 - x^2 + x - x^3). \tag{7.15}$$

Since the power transferred to the turbine can also be expressed in terms of the power coefficient as $p = 1/2 C_P \rho A v_1^3$, we therefore find using Equation 7.15 that C_P is the following function of x:

$$C_P(x) = \frac{1}{2}(1 - x^2 + x - x^3). \tag{7.16}$$

Finally, the maximum for $C_P(x)$ is found by setting $d/dx\ C_P(x) = 0$. We find that $x = 1/3$ and $C_{P\_max} = 16/27 = 0.593$. Keeping this theoretical maximum value in mind helps put into perspective the power extraction levels attained by today's modern wind turbines, which commonly achieve C_P values in the range 0.4–0.5 or up to 86% of the maximum possible value. It needs to be stressed, however, that the C_P value is not a fixed number for a given turbine, but it depends on the wind speed and other variables discussed in subsequent sections. Although the derivation for the maximum possible C_P implicitly assumed in Figure 7.12 a turbine having a horizontal axis of rotation for its blades (the most common type), the result is completely general and applies to turbines of any type. One caveat is that area A refers to the actual area occupied by the entrance to the rotor. Thus, for turbines that have a cowl, which funnels the airstream onto a rotor, one must use the area of the cowl and not the rotor itself, or else, an apparent violation of the Betz Limit can result.

7.4 TURBINE TYPES AND TERMS

There are numerous ways to describe a wind turbine, including, among others, the following:

- Rotation axis (horizontal or vertical)
- Rotation speed (constant or variable)
- Number of blades
- Solidity (lack of empty space between the blades)
- Size or power rating
- Nature of rotor mounting (upwind or downwind of supporting tower)
- Maximum survivable wind speed
- Minimum wind speed at which turbine rotor begins to turn
- Purpose (e.g., electricity generation, pumping water, or wind measurement)
- Dominant driving force that turns it (lift or drag)

Some of these distinctions will now be discussed.

7.4.1 Lift and Drag Forces and the Tip-Speed Ratio

The aerodynamics of wind turbines has much in common with airplane wings, which is where one normally associates the concepts *lift* and *drag*—terms that are illustrated in Figure 7.13 for a wind turbine blade.

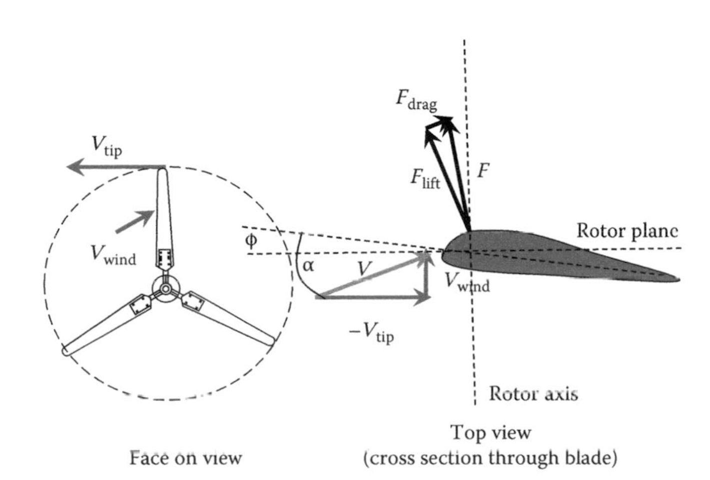

Figure 7.13 Relevant velocities and aerodynamic forces on a wind turbine blade. It is assumed that the wind is normally incident on the plane of rotation of the blade or along its rotation axis, assumed to be horizontal.

The air approaching the blade at speed v (in a reference frame in which the blade is at rest) has one component due to the tangential speed of the tip of the blade v_{tip} as well as the speed of the wind v_{wind} itself at right angles to v_{tip}. The ratio of these two components is known as the dimensionless tip–speed ratio:

$$\lambda = \frac{v_{tip}}{v_{wind}} = \frac{\omega R}{v_{wind}}, \tag{7.17}$$

where R is the blade radius and ω is its angular velocity in radians per second.

If we draw a chord through a cross section of the blade, the angle the chord makes with the resultant air velocity defines the angle of attack α—another phrase usually associated with airplane wings. A second angle is also defined in Figure 7.13, i.e., the pitch of the blade, which is the angle φ between the chord through the blade and the plane of the rotor. Note that while φ is purely a function of the construction of the turbine blade, α depends on the wind speed as well.

Now let us consider the aerodynamic forces acting on the blade. The net force of the moving air on the blade can be divided into two components—a drag force that lies along the direction of the net air velocity vector v and a lift force that lies along a perpendicular direction. Note that lift forces here are not necessarily upward, despite the connotation of the term. The components of F_{lift} and F_{drag} that point perpendicular to the rotor axis create the torque that drives the turbine. Turbines can be classified as to whether the dominant force that makes them turn is lift or drag, with the former being the more common situation. It can be seen in Figure 7.13, for example, based on the length of the vectors, that the component of F_{lift} perpendicular to the rotation axis is far greater than that of F_{drag}. One example of a turbine that turns in response to a drag force is the cup anemometer that is often used to measure wind speed based on its rotation rate (Figure 7.14).

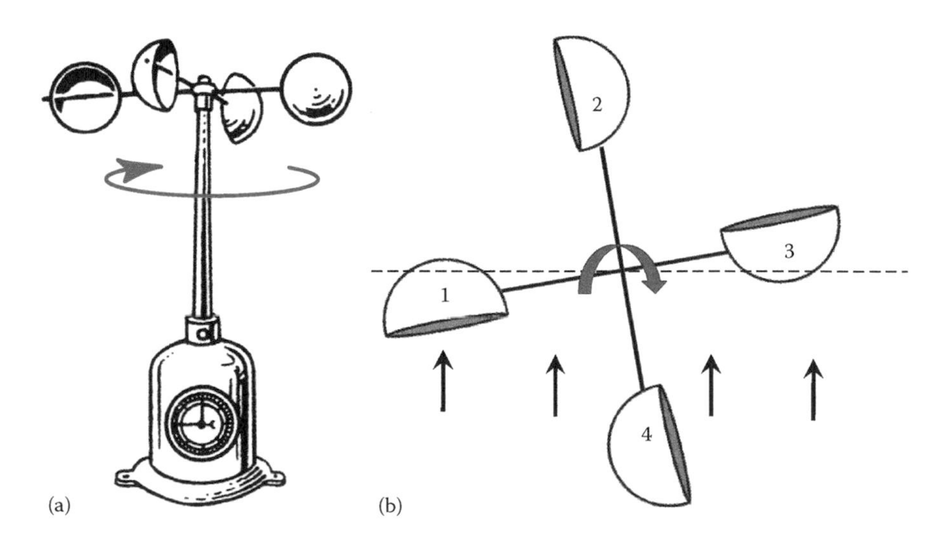

Figure 7.14 (a) A cup anemometer. Image provided by Pearson Scott Foresman is released to the public domain. (b) Top view of the anemometer with wind approaching.

7.4.2 Example 2: The Cup Anemometer

How does the rotation rate relate to the wind speed for a cup anemometer?

Solution

Clearly, two of the cups are always within no more than 45° of the dotted line in Figure 7.14. Let us, therefore, consider the drag forces on cups 1 and 3 for the simplified case where the angle is zero. This simplification means that our calculation is only an approximation. Drag forces can be expressed as

$$F = \frac{1}{2} C_d \rho A v^2, \tag{7.18}$$

where the drag coefficient C_d depends on how streamlined the object is. A hemispherical cup with its open side facing the wind offers much more air resistance than one whose open side faces away and the respective drag coefficients of cups 1 and 3 based on wind tunnel measurements 1.3 and 0.34, respectively. Remembering that the net velocity in Equation 7.18 needs to take into account both the tangential or tip velocity of each cup $v_{tip} = \omega r$ and the wind speed v, we find that the respective forces on those two cups can be expressed as $F_1 = 1/2\ (1.3)\rho A(v_{tip} - \omega r)^2$ and $F_3 = 1/2\ (0.34)\rho A(v_{tip} + \omega r)^2$. Although the two forces are in the same direction, the torques that they create (clockwise or counterclockwise) are oppositely directed. Thus, the equilibrium rotation speed is found when these forces are set equal to one another. On solving for the rotational velocity, we find a direct proportionality between wind speed v and the rotation rate in radians per second, i.e., $\omega = 0.586v/r$. The direct proportionality is what makes the anemometer a useful way to measure wind speeds. In practice, when the device is used for this purpose, a more accurate relationship between ω and v would need to be empirically found in order to calibrate the device. In addition to being a turbine driven by drag rather than lift forces, the cup anemometer is also an example of a vertical rather than horizontal axis device.

7.4.3 Horizontal versus Vertical Axis Turbines

The large majority of commercially available wind turbines rotate about a horizontal axis, but vertical axis wind turbines (VAWTs) do offer some advantages; most importantly, they need not be oriented to face the wind. This advantage is not as great as it might seem, however, since horizontal axis wind turbines (HAWTs) can accomplish this orientation either passively by means of a vane that is deflected by the wind or actively using a sensor and control system. Rotating a turbine to face the wind is known as *yawing*, and the misalignment of the normal to the rotor plane with the wind direction is known as the yaw angle ϑ. The need for good alignment is clear because the dependence of power on the cube of the wind velocity means that if the approaching wind velocity is $v \cos \vartheta$ rather than v, the power will be reduced by the factor $\cos^3 \vartheta$. Thus, a misalignment of 30° would cause the power to drop by 35%.

RESURGENCE OF VAWTS?

HAWTs comprise more than 95% of all utility-grade turbines on the market. In part, this reflects the lower efficiency of VAWTs, and in part, it is a matter of them being developed later, plus the smaller amount of empirical research in perfecting them. The limited research that had been done led to the false conclusion that the maximum coefficient of performance C_P for VAWTs was necessarily less than that for HAWTs under all circumstances. Some newly designed VAWTs, however, have C_P as high as 0.38, nearly the same as the best $C_P = 0.40$ found to date for HAWTs. Moreover, VAWTs may have some properties that make them especially suitable for far offshore applications, where HAWTs are more prone to failure due to the occasional very high wind and where very large VAWTs can better deal with winds that can change direction on a short timescale. Finally, VAWTs require less maintenance, and if they are floating, they have no need for the guy wires that pose a major problem on land (Figure 7.15).

Figure 7.15 Aerogenerator X wind turbine—a proposed 10 MW design measuring 274 m tit to tip intended for offshore use. The British consortium proposing to build them Wind Power claims that they will weigh about half as much as conventional turbines and have smaller towers and lower loads, because of the base mounting of the rotating blades. (Courtesy of Windpower Ltd., Suffolk, UK.)

Figure 7.16 Aerogenerator X wind turbine—a proposed 10 MW design measuring 274 m tit to tip intended for offshore use. The British consortium proposing to build them, Wind power Ltd, claims that they will weigh about half as much as conventional turbines and have smaller towers and lower loads, because of the base mounting of the rotating blades. (Image provided by Windpower Ltd., Suffolk, U.K.)

The lack of need for a VAWT to be oriented to face the wind is not its only advantage, however. At least as important is the elimination of gravity-induced cyclic stresses that occur with each rotation of the blades in a HAWT that can over time be very damaging. There are a number of different designs for VAWTs in addition to the cup anemometer, and one of them, the Darrieus VAWT, is shown in Figure 7.16. One of the chief disadvantages of VAWTs, which close inspection of Figure 7.16 will reveal, is the need for a complex system of guy wires that stabilize the tower. Without these guy wires, the resonances that occur driven by the periodic rotation of the blades would likely cause fatigue over time and eventually a tower collapse in high winds. The need for many guy wires makes it difficult to imagine deploying many such VAWTs in an agricultural setting where the large majority of the land is used to cultivate crops and which is where many wind farms often tend to be located. Normally, in such cases, land may be cultivated right up to the base of the tower for HAWTs. Another VAWT disadvantage is that the torque that turns the rotor periodically varies each rotation, and the result may be unwanted periodicities in the electrical output. Finally, the efficiency of VAWTs tends to be less than HAWTs, which can be most easily understood for a drag device such as the cup anemometer, where the approaching wind acts on one cup that contributes to the rotation and another that retards it.

7.4.4 Number of Turbine Blades, and Solidity

Among turbines with one, two, or three blades, the three-bladed ones are the most common because they tend to have the highest efficiency, have less vibration, and even look cooler! One might naively think that the more blades, the better in terms of efficiency, since the wind passing

Figure 7.17 Many-bladed wind turbine used to pump water.

between the blades is wasted, in generating torque on the blades, but it is not so. The blades affect the entire airstream that passes through the rotor, and the entire airstream in turn reacts back on the blades, causing them to turn. Many-bladed turbines, which of course have high solidity, are commonly used to pump water. Structures such as the wind pump and water tank of Figure 7.17 were and are a familiar sight in many farms and ranches in the regions of the southwest United States and other nations such as Australia and those in Southern Africa that are short of surface water. At their 1930 peak, there were 600,000 wind pumps in the United States alone. Such many-bladed turbines tend to rotate slowly, but they have large torque and start at low wind speed, making them suitable for pumping water.

RISING FROM POVERTY VIA RENEWABLE ENERGY

William Kamkwamba has now received international attention. As a boy, he had to drop out of school since his family could not afford the tuition in Malawi. He studied on his own, reading every book he could lay his hands on. Upon coming across a book about energy, Kamkwamba resolved to build a wind turbine. Since then, he has built a solar-powered water pump that provides his village with its first drinking water, and his family earns

Figure 7.18 Makeshift wind turbine assembled using bicycle parts and materials collected in a local scrap yard by a boy in Malawi. (Courtesy of Erik (HASH) Hersman, http://en.wikipedia.org/wiki/William_Kamkwamba, image is licensed under the Creative Commons Attribution 2.0 Generic license.)

income by using the power to charge cell phones. Kamkwamba's achievements, born out of a determination to raise himself and his family out of poverty, are told by the young man himself in a truly inspiring TED talk. It is a good illustration about how decentralized renewable energy can transform lives throughout the developing world (Figure 7.18).

It is uncommon for turbines to have an even number of blades, since this results in more vibration problems. The essence of the problem is that at the instant the top blade is exactly vertical, the bottom blade is in line with the tower and in its wind shadow, where it experiences little force at the same instance the top blade is pushed backward by the wind. The result is a significant torque at this moment, which has a period equal to the rotation period divided by the number of blades. Turbines having odd numbers of blades do not experience this problem.

7.4.5 Variable and Fixed Rotation Rate Turbines

Although some turbines rotate at a fixed rate, the newer ones have a variable rotation rate that depends on the speed of the wind. The fixed rotation rate is simpler—at least for alternating current (AC) electricity production where it is needed to have the generator driven by the turbine rotate at a frequency that matches that of the grid—typically 50 or 60 cycles per second (3000 or 3600 revolutions per minute), depending on the country. Wind turbines rotate far more slowly than that, but if the rotation has a fixed period, a gear system and a multipole generator could accomplish the interface with some inevitable losses in power. Nevertheless, most turbines rely on variable speed rotation because the interface problem has now been solved, in part with the usage of the induction motor generator that allows a variable speed rotor to induce a constant-frequency voltage—a trick which is discussed in a subsequent section. This was one key technology that allowed turbines to have electrical uses beyond charging batteries. A further advantage of the variable speed turbine is that it allows for greater efficiency, because for any wind speed, there is an optimum rotation rate or an optimum tip–speed ratio.

7.5 CONTROLLING AND OPTIMIZING WIND TURBINE PERFORMANCE

The basic components of a typical HAWT are illustrated in Figure 7.19, but not all of them are present for every turbine. We have already mentioned the anemometer that senses the wind speed and the vane that determines wind direction. These data are used by the control system to orient the turbine to face the wind using the yaw motor and drive—both of which are contained within the nacelle—the streamlined housing that rotates at the top of the tower. The nacelle also contains a low-speed shaft driven by the turbine rotor, which is connected via a gearbox to a high-speed shaft that turns the generator to create electricity, which needs to rotate at a much higher rate. Finally, two other important controls are the brake that can stop the rotation of the blades and a pitch control that can alter their pitch angle.

There are essentially three specific wind speeds that need to be considered in order to properly control and optimize turbine performance, namely,

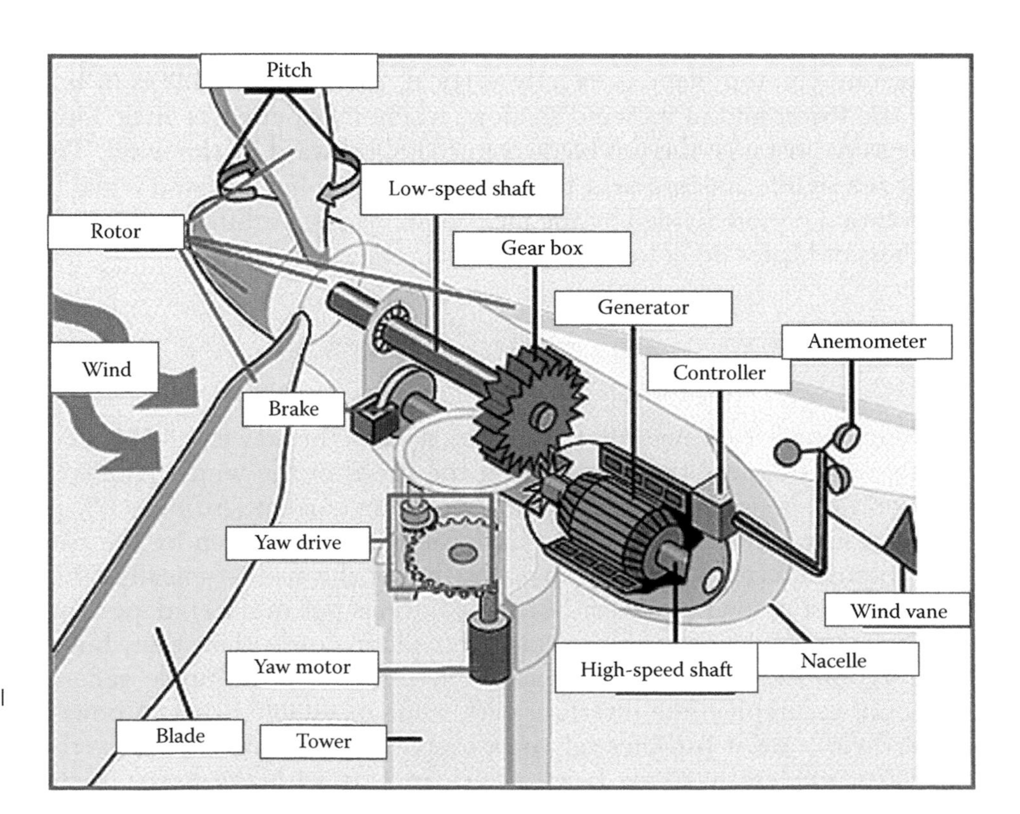

Figure 7.19 Components of a typical wind turbine. (Image by the U.S. Department of Energy, Washington, DC, is in the public domain.)

Table 7.1 Four Important Speeds and Associated Actions for Controlling a Wind Turbine with Their Typical Values

Speed	Typical Value (m/s)	Action Needed
Cut in (startup)	4	Maximize power
Rated (full power)	14	Limit power
Cutout (stop)	25	Apply brakes
Survival	50–65	Hope for best

the cut-in speed, at which it starts to turn, typically around 3 m/s; the rated wind speed, typically 11–14 m/s, at which the turbine generates its full rated power; and the cutout wind speed, typically 25 m/s, at which the turbine needs to shut down to avoid damage. Given the definition of the rated wind speed and rated power of a turbine, it follows that

$$P_{\text{rated}} = \frac{1}{2} C_P \rho A v_{\text{rated}}^3. \tag{7.19}$$

Between the rated and cutout speeds, the power is usually limited to roughly a constant level. Given the v-cubed power in the wind, this requires that the value of C_P (or the turbine efficiency) be reduced as v increases. There will be a loss of power when C_P is reduced, but if it were not limited, the turbine would overheat.

As we have noted earlier, C_P depends on the aerodynamic properties of the turbine for any given wind speed. Many commercial wind turbines adjust C_P by controlling one or more parameters such as blade pitch φ, rotation rate ω, tip–speed ratio λ, or—for at least one unusual case—blade length! The tip–speed ratio is defined by the ratio of the tangential speed of the blade tip to that of the wind v:

$$\omega = \frac{v_{\text{tip}}}{R} = \frac{\lambda v}{R}. \tag{7.20}$$

Thus, the two key variables in determining C_P are φ and λ, so we express the power coefficient C_P to be a function of them: $C_P = C_P(\lambda, \phi)$. The nature of the control is different for fixed and variable rotation speed turbines, which are considered separately. A summary of the actions needed in different wind speed regimes is shown in Table 7.1.

7.5.1 Example 3: Turbine Power and Typical Blade Diameter

What blade diameter d is typically required for a turbine that has a rated power of P kilowatts?

Solution

For a commercial wind turbine, typical values for the power coefficient and rated wind speed are $C_P = 0.4$ and $v_{\text{rated}} = 13$ m/s. Substituting these values and the density of air taken to be $\rho = 1.25\ \text{kg/m}^3$ into Equation 7.19, we find $P_{rated} = 1/2C_P\rho A v_{rated}^3 = 1/2 0.4 \times 1.23 \times (\pi d^2/4) \times 14^3 = 431 d^2$ W. Thus, if P is expressed in kilowatts rather than watts, we find that the typical diameter in meters is $d = \sqrt{P/0.431} = 1.52\sqrt{P}$. For example, we would estimate a 10 MW (10,000 kW) turbine to have a blade diameter of 152 m, which is fairly close to the actual value of 144 m for the new Norwegian 10 MW turbine.

7.5.2 Maximizing Power below the Rated Wind Speed

In this section, we consider the specific ways to control a turbine so as to (1) maximize its power below the rated wind speed and (2) limit its power above that speed. A turbine that rotates at a fixed rotational speed also, of course, has a fixed tip speed ratio λ, so the main ways to control it involve choosing the best values for λ and the blade pitch φ—for those turbines that allow active pitch control. Typically, a carefully designed HAWT with n blades has an optimum tip–speed ratio given by the empirical relation

$$\lambda_{opt} \approx \frac{18}{n}. \tag{7.21}$$

The reason there is an optimum choice of λ is that for too small a value, the blades rotate too slowly to affect most of the air that passes between them, while for too large a value, the rapid rotation induces turbulence which wastes energy. The optimum choice is where the time it takes a blade to rotate through $2\pi/n$ (the angle between blades) equals the time it takes the passing airstream to travel a distance over which it is "strongly affected by" the turbine. This condition explains the inverse dependency of optimum λ on the number of blades—but it is a condition that is only approximate and applies only to lift-dominated turbines. The situation is different for a turbine driven by drag forces, such as the cup anemometer or a Darrieus turbine (Figure 7.16). In this case, the optimum λ must be less than 1.0, since the blade tip speed cannot exceed the wind speed. Do you see why?

In order to optimally control a turbine, we cannot rely on an approximate relation like Equation 7.21, but rather, we need to learn how C_P empirically varies with λ from wind tunnel tests for a particular turbine design. Figure 7.20 shows one typical result for a three-blade HAWT. In this case, in order to maximize the power below the rated wind speed, we would opt for a choice $\lambda = 8$ and $\varphi = 0$, which gives a respectable value of $C_P = 0.45$. What does the choice $\lambda = 8$ imply about the required turbine rotation speed in this case? From Equation 7.20, we see that for a turbine of blade radius R, the proper choice of rotation rate depends on what wind speed v is present. However, since ω is fixed and v varies, C_P can be optimized only for one wind speed. One might be tempted to choose a

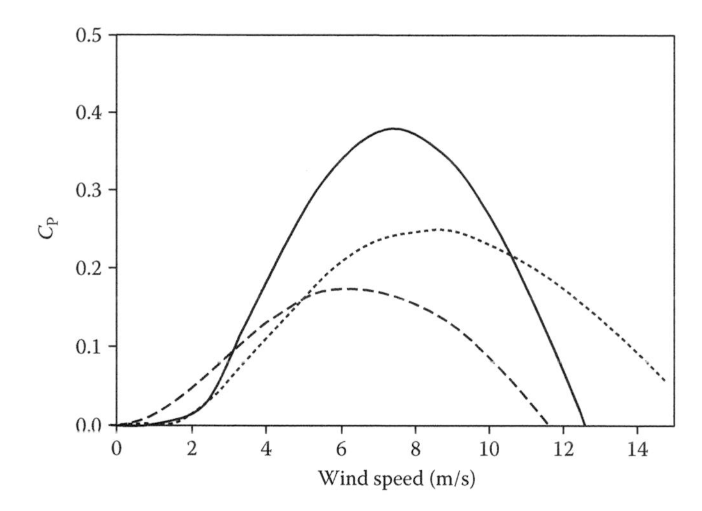

Figure 7.20 Typical curves of power coefficient $C_P = C_P(\lambda, \varphi)$ versus tip speed ratio λ for three values of the blade pitch angle: $\varphi = 0$ (*solid curve*), $\varphi = 5°$ (*dotted curve*), and $\varphi = 10°$ (*dashed curve*).

wind speed equal to the average of the cut-in and rated wind speed, but that fails to take into account that the power in the wind varies as the cube of the wind speed. Thus, the best choice would be closer to the rated wind speed.

For a variable rotation rate turbine—the more common type—an additional parameter can be made to vary (the rotation rate), so as to achieve even higher efficiency levels for speeds below the rated speed. However, there is a fairly limited range (about 10%) for the extent of this variation without suffering efficiency losses.

7.5.3 Limiting Power above the Rated Wind Speed

Above the rated wind speed, there are two basic methods to limit the power and prevent overheating—one passive and one active. Passive control (also known as stall control) applies to turbines having blades that are locked in place and, hence, lack any active pitch control. In such turbines, if the wind speed should increase beyond the rated value, the angle of attack increases, and at some point, it increases sufficiently that the blade stalls, losing all its lift with a concurrent drop in power. The situation is quite analogous to what can happen with an airplane wing, where the result of a stall can be catastrophic—rather than beneficial as in the case of a turbine!

The active method of limiting turbine power above the rated speed (known as pitch control) applies to those turbines for which the blade pitch can change, which is the case for all standard modern turbines. For large commercial-grade turbines, a computer-controlled interface senses that the wind speed is above the rated value and varies the pitch angle φ to result in the desired drop in C_P (and the power produced) so as to keep the power at the rated value. In smaller turbines, the active pitch control is sometimes accomplished by centrifugal forces acting on spring-loaded

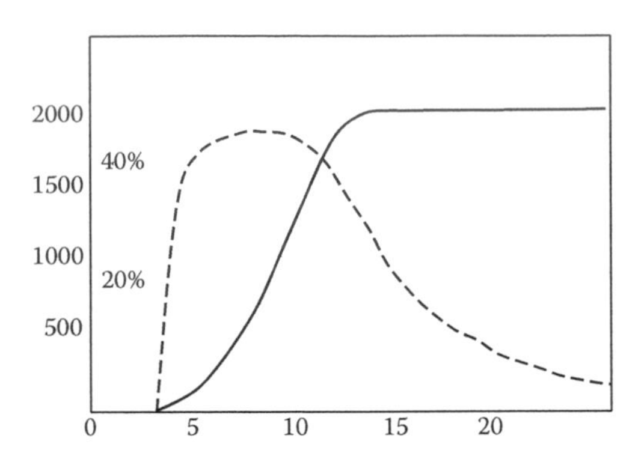

Figure 7.21 Measured wind turbine power (kW) and efficiency (%) or CP versus wind speed for a Vestas 2000 kW wind turbine. The power curve shown solid remains nearly constant above the rated wind speed of about 12 m/s, but the efficiency shown dashed dramatically drops. (Data from a pamphlet on Vestas wind turbine, http://www.wind-power-program.com/large_turbines.htm.)

blades. The advantage of passive control is that it happens automatically. This method is less expensive and mechanically simpler, but it results in a lower efficiency, and less power is generated.

A typical situation is shown in Figure 7.21 for a particular turbine. We can see how power is both (1) maximized (below the rated speed) and (2) kept limited to a constant value (above that speed). While these results apply only to a particular turbine, its cut-in speed (3 m/s) and rated speed (14 m/s) are quite typical. Note how the speed at which C_P maximizes in Figure 7.21 (see dashed curve in Figure 7.21) is not too far below the rated wind speed at about 9 m/s, and that above the rated wind speed, C_P needs to be made to significantly drop so as to prevent excessive power and overheating. The high degree of constancy of the power above the rated wind speed (see solid curve) indicates that the more efficient active pitch control is the method used to limit power in this case.

7.6 ELECTRICAL ASPECTS AND GRID INTEGRATION

A major issue involving both wind-generated and solar-generated electricity that is grid connected is their significant variations with time. The grid allows a sharing of electricity between different regions as individual supplies and demands vary, but there must be an overall supply–demand balance, which is a problem when it comes to highly intermittent renewable sources. As long as the percentage of electricity generation by wind and solar is small, such intermittent supply is not a major issue, but as its degree of penetration rises, it could become a major one that would require both major upgrades to the grid and much more extensive energy storage (such as pumped storage hydro) than what exists now. Some observers believe that without such major changes, the fraction of grid-connected electricity by solar and wind power would probably be limited to around 20–30% for many nations. At a 20% level, the cost of modifications required to the grid would greatly vary from country to country—being estimated at only 0.3 euro cents for Norway (which has a considerable amount of hydropower that can be easily dispatched to offset wind power variations) to as much as 10 times that for Germany which is far more densely populated and lacks substantial hydro and already has a significant solar commitment.

Turbines that are not grid connected need to satisfy far less stringent electrical requirements than those that are, and they may even generate wild AC—whose frequency dramatically varies depending on the wind speed. Such wild AC when passed through a rectifier to produce direct current (DC) is quite suitable to charge batteries. If it is also desired to connect the turbine output to the grid, there are two possible approaches. One involves passing the DC so generated through an inverter to produce AC that is grid compatible. A second more efficient approach would be to avoid wild AC in the first place and generate fixed-frequency current that also needs to satisfy several other requirements to be grid connected. These involve safety, e.g., automatic disconnect from the grid in case of a power failure (to avoid electrocuting linemen), and the quality of the power produced, in terms of tight limits on variations in voltage and frequency. The detailed requirements vary from country to country.

7.6.1 Asynchronous Generator

The requirement of a constant-frequency AC can be met by using either a turbine whose rotation rate is fixed regardless of wind speed or, more conveniently, one that manages to generate a fixed-frequency current even when its rotation rate changes. The asynchronous generator invented by Nikola Tesla is the device that manages this clever trick. Also called the induction generator, the device was first invented to serve as a motor, but it works just as well as a generator. Induction motors and generators are inexpensive, reliable, and efficient, as they lack the brushes or slip rings normally required to electrically connect the rotating coil with the stationary components of the device. Induction motors consume a third of the entire world's electricity in connection with a wide variety of applications, including pumps, fans, compressors, and elevators. One of the more common types of induction motor generators, the squirrel cage type, is shown in Figure 7.22a. The squirrel cage is the rotating part with copper or aluminum bars—the figure has 12 of them—connecting the front and back rings of the rotor. Rather than being driven by a rodent running in place (hence the name), the device responds to a rotating magnetic field

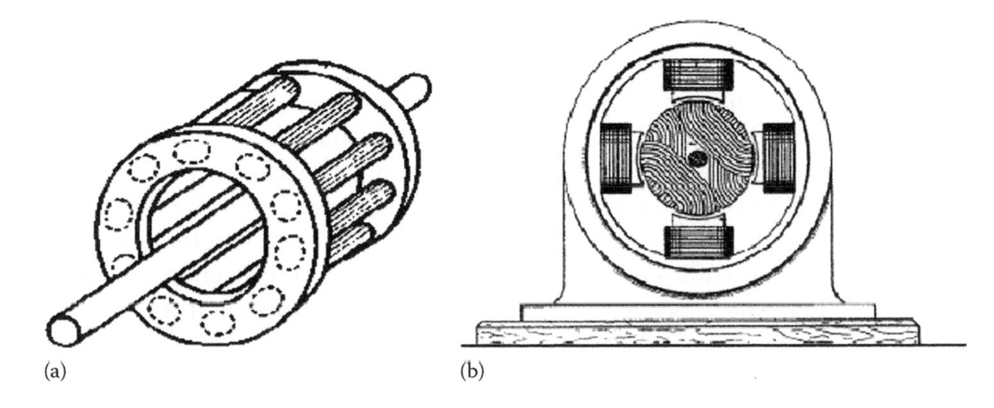

Figure 7.22 (a) Rotor component of a squirrel cage induction motor generator. (Courtesy of Meggar, Creative Commons Attribution ShareAlike 3.0 License.) (b) Drawing from Tesla's 1889 patent of a four-pole induction motor is in the public domain.

that is created by feeding electric currents into the windings of the four electromagnets surrounding the squirrel cage in the right figure. In what follows, we refer to the rotating component of the device as the rotor and the surrounding stationary part as the *stator.*

In order to better understand the concept of a rotating magnetic field, it is simpler to consider the four-pole version of the device illustrated in Tesla's original patent—Figure 7.22b. Suppose the vertically oriented pole pairs are fed by a current that is 90° (a quarter of a cycle) out of phase with that connected to the two horizontal poles. In other words, the magnetic fields generated by the two pole pairs can be written as

$$B_x = B_0 \cos \omega_s t \quad \text{and} \quad B_y = B_0 \sin \omega_s t. \tag{7.22}$$

Together, these represent the x and y components of a magnetic field vector that rotates with an angular speed ω_s. The reason for the subscript s on the angular frequency is because the rotating field is generated by currents in the stator, which is itself not physically rotating. Exactly the same situation (a rotating magnetic field) occurs in the three-pole design, the only difference being that the windings on the three poles must be fed with ACs that are not a quarter of a cycle out of phase but a third of a cycle. Hence, this version is often referred to as a three-phase induction motor generator (Figure 7.23).

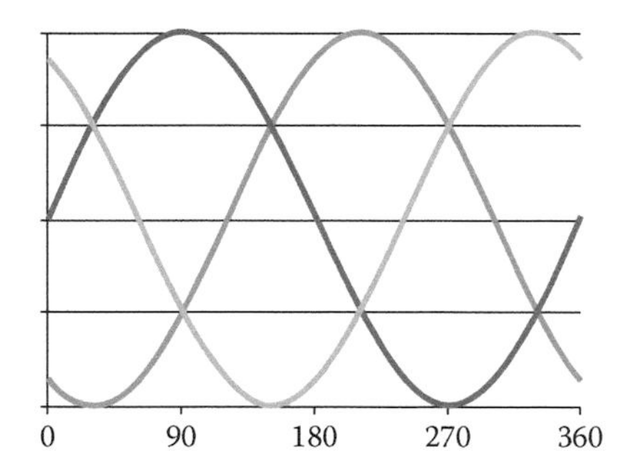

Figure 7.23 Three voltage waveforms (120° out of phase) created in three-phase electric power from the voltage versus time across each of the three electromagnets.

You might wonder how the device "knows" when to be a motor and when to be a generator? It all depends on the relative magnitudes of two rotation rates: that of the stator magnetic field ω_s and the physical rotation of the rotor ω_r. The percentage by which the stator field rotation rate exceeds the rotor physical rotation rate is known as the *slip*, defined as

$$s = \frac{\omega_s - \omega_r}{\omega_s}. \tag{7.23}$$

For a positive slip, the field rotates faster than the rotor, while for a negative slip, it rotates slower. If the slip is positive (ω_r is slower than ω_s), the device acts like a motor, and if the slip is negative, it is a generator. In the case of zero slip, it is neither, which requires that the stator be initially magnetized by current from the grid. Rather than simply memorize a rule, we can easily understand these conditions based on simple physics. When the slip is positive (ω_r is slower than ω_s), the faster rotating magnetic field drives the rotor (gives it mechanical power) and the device is a motor. When the slip is negative, the slower rotating magnetic field pulls back on the rotor (takes away its mechanical power), but energy is again conserved because positive electrical energy is created when mechanical energy is lost.

NIKOLA TESLA: AN OVERLOOKED GENIUS

Nikola Tesla was the inventor of the induction motor generator making use of a rotating magnetic field. Outside the community of electrical engineers, Tesla was one of history's underrated electrical pioneers. Tesla initially worked as a lowly paid assistant to the far better known Thomas Edison, but he lacked Edison's flair for promotion or his ego that often stood in the way of Edison recognizing the merit of others' ideas. The two geniuses parted company over Edison's many refusals to grant Tesla a raise and his disinclination to listen to Tesla's well-reasoned support for AC in the AC/DC "war of the currents." While not quite as prolific as Edison in terms of sheer numbers of inventions, the idealistic Tesla was far more interested in discoveries that might eventually lead humankind to a better life than in profiting from his inventions. Tesla's altruism led him to voluntarily give up royalties worth nearly $20 million (in nineteenth-century dollars!) in order to help save George Westinghouse's fledgling company when it was threatened with bankruptcy. Among Tesla's discoveries were the fluorescent lamp, the spark plug, remote control via radio, the principles of radar, and—of special relevance to the topic of this book—ocean thermal energy conversion). Perhaps his greatest discovery was the radio—having initially been granted a patent for the latter in 1897, 3 years before Guglielmo Marconi. In a shameful episode, the US Patent Office reversed itself on Tesla's patent, perhaps succumbing to political influence, and it awarded the patent to Marconi in 1904. It was then Marconi who won the Nobel Prize for the discovery in 1911. Some measure of justice was eventually restored during Tesla's lifetime when the US Supreme Court declared (as a result of a Marconi lawsuit) that Tesla was the real inventor of the radio.

The way the asynchronous generator works as part of a grid-connected wind turbine is shown in Figure 7.24.

The system is called *doubly fed* because both the rotor and the stator can transfer power to the grid, although the power transferred from the rotor is either positive or negative, depending on whether it is functioning as a generator or a motor. As the wind speed changes and influences the rotor speed, the device functions as either a generator or a motor. Aside from

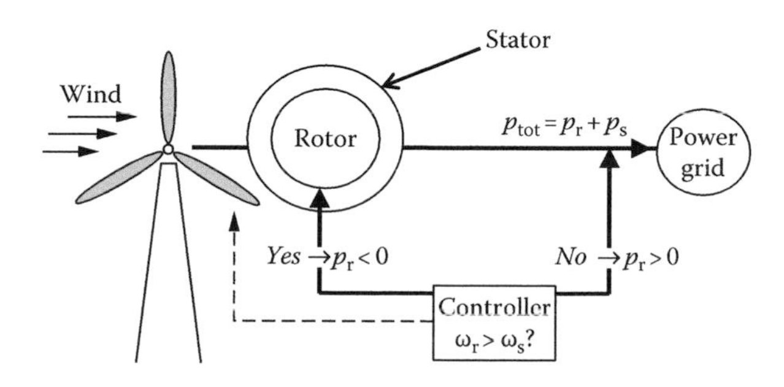

Figure 7.24 Doubly fed asynchronous generator.

having high efficiency, being reliable, and being able to cope with varying rotation speeds of the rotor (up to around ±10%), the asynchronous generator is also very useful in capturing the significant energy present in very brief wind gusts in a way that also cushions changes in the rotor speed that could otherwise cause damage to a generator and gearbox. The cushioning occurs because during a sudden wind gust that tries to accelerate the rotor (negative slip), the device acts like a generator, and the result is a torque that retards the rotor speed. The limit of 10% depends on the restoring torque remaining proportional to the slip over about this range.

7.7 SMALL WIND

Small wind refers to wind turbines for homeowners, ranchers, or small businesses, and they can have rated powers as low as 50 W or as high as 50 kW. For much of the early years of the twentieth century, small wind turbines provided the only source of electricity in rural areas of the United States. Today, they are purchased for various reasons including battery charging and reducing one's dependence on the grid or one's carbon footprint. The economic feasibility and installation of a small wind turbine can be far more problematic than solar panels, in view of the great variability in wind potential with location, even on a particular property, where one must pay careful attention to the terrain, vegetation, height, and placement of obstacles (Figure 7.25). Densely populated areas, in particular, represent among the poorest locations. For some locations, the average wind speed will be far short of what is needed to generate the rated power more than a very small percentage of the time. Nothing can be more vexing to the naive buyer than to purchase a turbine rated at 3 kW only to find that it generates that power 1% of the time! Moreover, until fairly recently, different manufacturers could rate their turbines at different wind speeds, making a direct comparison very difficult.

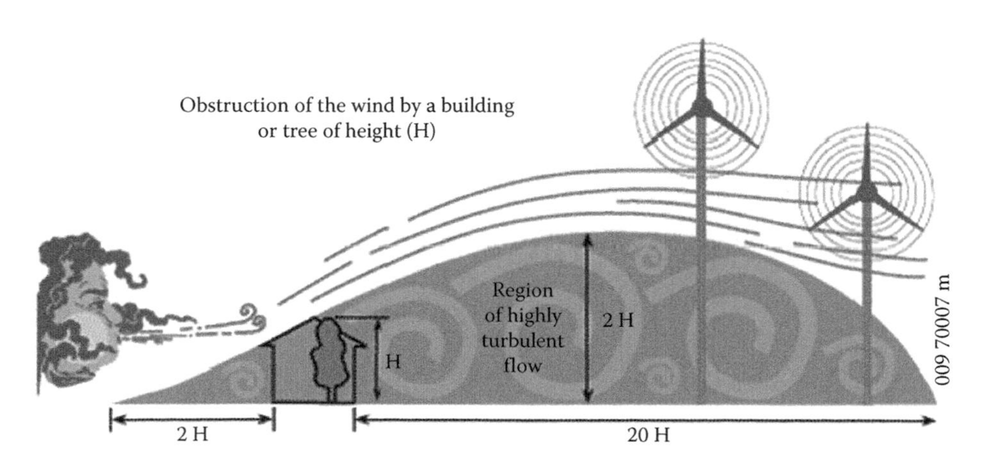

Figure 7.25 Required minimum spacings from obstacles to avoid turbulence for small wind turbines. (Image courtesy of the U.S. Department of Energy.)

7.7.1 Example 4: Comparing Two Turbines

Turbine manufacturer A rates its turbines at an assumed wind speed of 12 m/s and sells a 2 kW turbine for \$5000. Manufacturer B rates its turbines at an assumed wind speed of 15 m/s and sells a 2.5 kW turbine for \$4000. Which produces more power per dollar at a wind speed of 10 m/s?

Solution

For turbine A, we have power/cost = $(2000/5000) \times (10/12)^3 = 0.289$ W/\$, while for B, power/cost = $(2500/5000) \times (10/15)^3 = 0.185$ W/\$. Thus, A is the better buy even though *B* is cheaper and has a higher rated power!

Another complication is the local zoning restrictions that may make a wind turbine impractical—since the maximum tower height might be restricted to values that are insufficient to avoid turbulence. Lastly, there is great variability in the cost, quality, and reliability in the wind turbines made by different manufacturers. Fortunately, impartial organizations exist that can guide the potential buyer of a small wind turbine through all the decisions that need to be made. In the United States, for example, the DOE Wind Powering America program can help the consumer evaluate whether wind power makes sense for them, what turbine make and model is most appropriate, how the economics would work out (either for grid connection or not), and where they can get help in installing the system. Typically, a small turbine is affixed at the top of a tower before the tower is erected. Part of the difficulty in erecting the tower plus turbine is that not only it is top heavy, but also the weight of the tower itself is often substantial since it needs to be strong enough to prevent vibrations at its top in high winds. Installing a tall tower with a wind turbine on top is definitely not a do-it-yourself project, and it can actually be quite hazardous! A good friend required four attempts to erect his wind turbine top a 30 ft. pole, which collapsed on one attempt, fortunately not injuring anyone.

7.8 OFFSHORE WIND

Offshore wind is thought to be one of the most promising frontiers in large-scale commercial wind in light of the faster and steadier winds there, the avoidance of land acquisition or leasing problems, and the lack of opposition from people objecting to wind farms on the grounds of noise or unsightliness. Of course, offshore wind does pose its unique set of problems, namely, the greater difficulty of erecting and maintaining floating towers and the problem of getting the electricity generated back to shore. Denmark was the first to install an offshore wind farm in 1991, and since then at least, 10 others have followed, including eight in Northern Europe, Japan, and China. Offshore wind technology is, however, less mature and more costly than onshore wind. The cost differential between offshore and onshore wind is about a factor of 2—about \$90/MWh generated versus \$50/MWh. As of 2010, only 3.16 GW has

Figure 7.26 Offshore wind turbines near Norfolk coast.

been developed worldwide (about 3% as much as onshore), but it is projected that this total will rise perhaps 25-fold by 2020, with major contributions expected from the United States and China. Interestingly, the largest (7 MW) offshore wind turbine built as of 2015 is located off the coast of Fukushima, Japan, cite of the 2011 nuclear disaster.

The United States, however, has been slow to develop its large potential in offshore wind. After a developer began to install a wind farm off the New England Coast, regulatory confusion, political battles, and NIMBYism have all conspired to put the project on hold. Nevertheless, given the size of the potential resource, the out-of-sight aspect of wind farms that are further offshore appears to have great political appeal (Figure 7.26).

7.9 ENVIRONMENTAL IMPACTS

Wind power, which has no CO_2 or any other emissions, is a relatively benign technology, although it does have its environmental issues, most importantly, its impact on birds and bats, and the noise associated with turbines that some people find extremely annoying. There is also the matter of their appearance, which some people find beautiful, but others find irksome, especially if they happen to be located near offshore and their beachfront home previously had an unobstructed view.

Some studies on bird fatalities have blamed 10,000–40,000 deaths per year on collisions with wind turbines. However, if one considers all causes of human-caused bird fatalities, collisions with buildings and windows account for a majority of them; cats are in second place, while wind turbines account for less than 0.01% Additionally, it is important to note that fossil fuel-generated electricity causes about 20 times as many birds to die as wind power on a per kilowatt-hour basis through pollution. Bat fatalities appear to be a more serious problem with wind power than birds, especially during times of bat migration. On the other hand, since such migrations tend to occur at times of low wind speeds, when very little power is generated anyway, turbine blades could be greatly slowed at such times with little impact on average power output. It appears that such bat deaths are less the result of collisions than the sudden pressure drop that occurs near the rotating blades that harms the bats' sensitive lungs.

One of the most serious issues for some people living near wind turbines is that of noise. Noise sensitivity, of course, dramatically varies by individual, and it would be impossible to be sure that no one would be annoyed by the sound of a large wind turbine. Noise is measured in decibels (dB), which may be defined in terms of the sound intensity I in watts per square meter:

$$dB = 10\log_{10}\left(\frac{I}{I_0}\right), \tag{7.24}$$

and the reference level $I_0 = 10^{-12}$ W/m^2 is the threshold of hearing for an average person.

We can put the noise problem in perspective with the following comparisons:

Note that according to Table 7.2, at a half mile, the noise level of a turbine is about the same as a quiet bedroom. Still some people located a half mile from a commercial turbine are apparently quite disturbed (especially rural folks who are accustomed to quiet), but complaints are quite low at a mile distance. One surprising fact about turbine noise is that the overall level happens to vary as the fifth power of the wind speed as seen from the moving blade tip! Thus, one way to dramatically reduce noise in environments of great noise sensitivity is to limit the blade tip speed to about 60 m/s—although this would lead to a modest reduction in average power generated. The improved design of turbines has led to greater efficiency and less noise, but no complete solution to the noise problem that will satisfy the most sensitive person is likely to be forthcoming. Moreover, noise complaints are far from the norm, since among the new wind farms that have come online in the United States in recent years, only about 5% have generated significant noise complaints—a reduction from wind farms using older model turbines.

Table 7.2 Typical Noise Levels in Decibels from Various Sources

Setting	dB
Rural nighttime background	30
Quiet bedroom	35
Wind farm at 350 m	40
Car at 40 mph at 100 m	55
Busy general office	60

WIND TURBINE SYNDROME

The reality of sensitivity to sound is a complex business. For example, while there are no significant reports of annoyance to sound levels below 45 dB from aircraft, road traffic, or trains, at that same level, nearly everyone in a community would be highly annoyed by the sound of wind turbines. Part of the problem may be associated with the production of infrasound ($f < 20$ Hz) that has been measured from wind turbines, and which while it cannot be heard as sound can affect the vestibular organs—the principal organs associated with balance, motion, and position sense. It has been suggested that such signals can create sleeplessness, vertigo, and nausea in some people. On the other hand, there are also reasons for some skepticism. Research shows that two factors repeatedly came up as dramatically increasing the annoyance level in surveys of people living near wind farms. The first is being able to see the turbines, while the second factor is whether people derive income from hosting turbines. Miraculously, this factor appears to be a highly effective antidote to feelings of symptoms (Chapman, 2011).

A final environmental issue for wind turbines concerns their relatively large footprint in terms of acreage per kilowatt-hour generated—which, like noise, is a nonissue for offshore wind. Even though very large turbines generate far more energy than smaller ones—they do need to be placed further apart, so the proper measure of efficiency is in the energy generated per acre of land. Most current wind farms have an average turbine spacing of seven-blade diameters, although recent research suggests an optimum, more cost-effective spacing is perhaps $15d$ (Meyers, 2011). In either case, since the power generated by each turbine scales as d^2 and the number of turbines that can optimally be placed on a given land area scales as $1/d^2$, the total energy per unit area (ignoring the increase of wind speed with height) is roughly independent of d.

The DOE has studied data on existing wind farms and found that on the average, 34 ha of land are used to generate each megawatt, but with a large variation: (34 ± 22 ha/MW). This energy generation per unit area is significantly worse than is required for either solar thermal or photovoltaic generation, which tend to lie in the range 2–6 ha/MW, or geothermal, which lie in the range 0.4–3 ha/MW. However, mitigating the very poor land utilization rate is the fact that wind, unlike either solar or geothermal electricity generation, allows much of the land to be used for other purposes—often agricultural. According to the DOE study, the amount of land that was directly impacted, and could not be used for other purposes, was 1 ± 0.7 ha/MW, which potentially makes land utilization of wind power better than solar and on a par with geothermal, and it is significantly better than electricity produced from fossil fuels.

Of course, land utilization may not be the best metric to judge between competing alternatives, and the cost per megawatt (taking land costs into account) is probably a far better one. Since wind power is one of the fastest-growing contributors to power in the United States, it would appear that the marketplace has decided in favor of wind.

7.10 UNUSUAL DESIGNS AND APPLICATIONS

7.10.1 Airborne Turbines

Placing turbines aloft and tethering them to the ground rather than placing them on a tower defines the field of airborne wind energy (AWE)—essentially flying wind farms. AWE offers several advantages, notably access to the steadier and higher speed winds at high altitudes and avoidance of the expense of building a robust tower, which, for large turbines, is the biggest single expense. There are, of course, some significant problems associated with AWE, most notably the requirement of a very long tether, bad weather problems, and possibly an aircraft exclusion zone. As of 2010, no airborne turbines are commercially operating, although proponents of the technology suggest that the costs could be as little as \$10–\$20 per megawatt-hour (Figure 7.27).

7.10.2 Wind-Powered Vehicles

Although wind power is well established as a means of propulsion for sailing ships, its usage for vehicles on land is less well known. Of course, one should not expect this to be a feasible alternative for everyday usage, since very special conditions are required, such as a good wind potential, a very smooth flat surface, and a single passenger. There is even a speed record for such vehicles, with the Greenbird, the current land speed record holder (as of 2009) at 126.2 mph. Remarkably, that record is three to five times the real wind speed at the time (Figure 7.28).

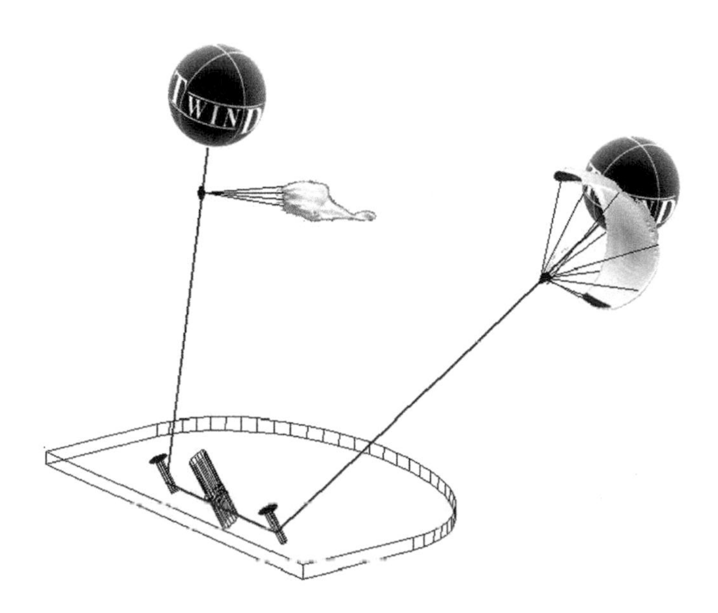

Figure 7.27 How tethered floating wind turbines generate power. In this design, the different tensions in the two cables turn a generator. (Courtesy of Evavitali, http://en.wikipedia.org/wiki/File:TWIND3.JPG.)

Figure 7.28 Greenbird wind-powered vehicle sets land-wind-powered speed limit in Lake Ivanpah, United States. The vertical wooden structure is a stiff sail. (Photo by Peter Lyons.)

7.10.3 Directly Downwind Faster-than-the-Wind

The ability of a wind-powered vehicle to travel faster than the wind itself should not come as a surprise to sailors. However, what would be surprising is a vehicle that travels directly downwind exceeding the speed of the wind. Despite the doubts of many scientists and engineers, such a counterintuitive feat has actually been demonstrated. As shown in Figure 7.29, the vehicle initially gets its propulsive force from the large propeller mounted on it. Now suppose it is traveling exactly at the wind speed, what could possibly make it travel any faster? The answer lies in its rapidly rotating wheels connected by a drive train to the turbine. Even though the device is at rest (in a reference frame moving at the wind speed), the force of the ground on those rapidly

Figure 7.29 Vehicle travels directly downwind faster than the wind. (Photo of Rick Cavallaro's Blackbird; courtesy of Steve Morris.)

turning wheels keeps the propeller turning. In one run (posted on the web), the vehicle built by Rick Cavallaro and John Borton can be seen allegedly traveling nearly three times faster than the wind. Apparently, this run was witnessed by officials of the North American Land Sailing Association and set an official record of 2.8 times the wind speed.

PROBLEMS

1. Investigate how much error is made in the approximation (Equation 7.7) using a Weibul distribution for a number of values of the shape parameter. Do this using a numerical integration of the quantity $v^3W(v)$ in an Excel spreadsheet and plot the percentage error in the difference between the right- and left-hand sides of Equation 7.7 versus the shape parameter.
2. For a wind speed distribution following the Rayleigh distribution, show that the most probable wind speed is 80% of the average wind speed.
3. Using the approximation of Equation 7.7, see if the ranges in power listed for each range of wind speeds in Figure 7.5 is correct for the various wind classes.
4. Consider the definitions of wind classes 3 through 7 in Figure 7.5. For each class find, what fraction of the time the wind speed is greater than the rated wind speed based on a rated wind speed of 14 m/s and assuming the wind speed follows a Rayleigh distribution.
5. Suppose the yaw control on a turbine results in a misalignment of 20° half the time. What power reduction would result?
6. What is a typical blade diameter for a 100 kW turbine?
7. How long does it take a typical 1 MW three-bladed turbine to complete each revolution? Assume that the turbine is operating at the usual rated wind speed and has the optimum tip–speed ratio. How fast is the tip of each blade moving in meters per second?
8. Determine the optimum tip–speed ratio (to obtain the largest C_P) for a cup anemometer. Hint: Using the drag force on each cup, write an expression for the power generated, which can then be expressed as a function of the tip–speed ratio. Be careful of signs.
9. In an effort to reduce wind turbine noise, the turbine tip speed is limited to 60 m/s. Find the typical tip speed for a HAWT functioning at its rated power and find how much of a reduction in audible noise would result from this tip speed limit.
10. A wind farm is placed in a rural setting. Using Table 7.2, find how far away you would need to be from it for the sound level in decibels to be equal to that of a typical rural nighttime background level. Approximate the wind farm as a point source and assume that sound spreads out equally in all directions, so that its intensity drops according to the usual inverse square law.
11. Estimate the number of 500 kW turbines having a typical spacing of seven blade diameters that you could place on 1 km^2 of land.
12. The capacity factor of wind power is the ratio of the amount generated over some extended period of time divided by the

theoretical maximum. Typically, wind farms have capacity factors of 20–40%, with the latter for especially favorable locations. How many megawatt-hours would a wind farm of 100 turbines each rated at 500 kW generate in a year in a very favorable location?

13. It is estimated that a 16 ft diameter wind pump could lift 1600 gal of water by 100 ft when an 18 mph wind blows. If the efficiency of the wind pump is 7%, how long would it take to pump this amount of water?
14. In Example 4, turbine A was the better buy for a wind speed of 10 m/s. For what wind speed would turbine B be the better buy in terms of power generated per dollar?
15. Based on Table 7.2, what would be the probable sound level you would hear if your house were 0.7 km from a typical wind farm in the quiet countryside. Assume that the wind farm is a point source and include both the background sound intensity and that due to the wind farm itself.
16. At a certain location, it is found that at a height of 10 m, the wind speeds during the year follow a Weibul distribution of the form $\phi_u = Cu^k e^{-(0.1u)^k}$, where u is the wind speed in meters per second and the shape parameter $k = 2.5$. Assume further that a wind turbine is erected, and it has a rated power of 150 kW at a rated wind speed of 12 m/s, above which the power generated is constant up to the cutout speed. The cut-in and cutout speeds are 5 and 40 m/s, respectively. Create a spreadsheet so as to find (a) the constant C so that the speed distribution is normalized over the interval $0 < u < 50$ m/s; (b) the average wind speed over this range of speeds; (c) the power the turbine produces at this location in each of 50 wind speed intervals: 1–2, 2–3, 3–4, ..., 49–50 m/s; (d) a graph showing both the normalized distribution of wind speeds ϕ_u and the fraction of the power produced at each wind speed interval.
17. What is the power density for a 22 mph (10 m/s) wind?
18. What is the power density for a 44 mph (20 m/s) wind?
19. Consider a 15 m diameter wind turbine. How much power could be delivered by the turbine if you had 10 m/s wind and you are operating at 40% efficiency? What is the maximum energy density possible from such a system?

REFERENCES

Bloomberg, M. (2011) *Global Renewable Energy Market Outlook*, Bloomberg New Energy Finance, New York.

Chapman, S. (2011) Much angst over wind turbines is just hot air. *The Sydney Morning Herald*. December 21, 2011, http://www.smh.com.au/opinion/politics/much-angst-over-wind turbines-is-just-hot-air-20111220-1p3sb.html.

Meyers, J. and C. Meneveau (2011) Optimal turbine spacing in fully developed wind farm boundary layers, *Wind Energy*.

Robert, W. R. (1996) *Wind Energy in America: A History*, University of Oklahoma Press, Norman, OK.

Shahan, Z. (2011) New wind turbines 300x more powerful than in 1996. *Clean Technica*. August 9, 2011, http://cleantechnica.com/2011/08/09/new-wind-turbines-300x-more -powerful-than-in-1996-top-wind-power-stories/?utm_source=feedburner&utm _mediumfeed&utm_campaign=Feed%3A+IM-cleantechnica+%28CleanTechnica%29.

Hydropower

8.1 INTRODUCTION TO HYDROPOWER

Hydropower, in which the motion of water is harnessed to do useful work, is essentially a form of solar energy that drives the planet's water cycle and accounts for the flow of rivers and ocean waves. Certainly hydro is one of the earliest forms of renewable energy to be harnessed by humans. Its origins go back to usage for irrigation with the development of agriculture that first arose in 6000 BC in Mesopotamia (now Iraq)—the "cradle of civilization." Hydropower has continued to play a significant role over the centuries, and it fueled another major advance in civilization, namely, the industrial revolution. Until steam power was invented and came into widespread use, it was hydro that powered the many useful mechanical devices and machines behind that revolution. A third major technological advance—the large-scale usage of electricity around the turn of the twentieth century—also saw hydropower play a major role. In 1920, e.g., 40% of the electricity in the United States was generated by hydropower—a figure that has now dropped to around 10% with the advent of greater reliance on other methods. However, hydropower now accounts for about 88% of electricity from all renewable energy sources in the world, and in several nations (Norway and Paraguay), it accounts for all or nearly all of their electricity.

Hydropower can usefully generate energy over a vast range of scales, ranging from the largest power plants of any type in the world—the Three Gorges dam and 22,500 MW power station in China down (Figure 8.1) to the nanoscale hydropower plant of a mere 100 W—a factor of 20 million times smaller. Hydro is also an extremely versatile renewable energy source in terms of its wide variety of forms. The most important usage of hydro is the conventional or impoundment power plant hydro, where a dam is constructed to create a reservoir. When the water it contains is allowed to flow out in a controlled manner, it can be harnessed to create electricity. The basic components of a conventional hydroelectric power station are illustrated in Figure 8.2. Power is generated as long as the gate in front of the penstock (the sloped channel running down to the turbine) is opened so that the downhill flow of water drives the turbine and the electric generator connected to it.

Not all types of hydroelectric generating plants require a large dam, as in the case of a run-of-the-river application, where the energy of the flowing water can be directly captured. Hydropower plants are also sometimes used as a vehicle for storing energy in so-called pumped storage applications that may be combined with a conventional power plant. Still other hydropower forms rely on the oceans rather than rivers and streams. Three types in particular are considered in this chapter, namely, wave, tidal, and ocean thermal power, which relies on temperature differences between surface waters and the colder water below.

CONTENTS

Figure 8.1 The Three Gorges Dam is the largest operating hydroelectric power stations at an installed capacity of 22,500 MW.

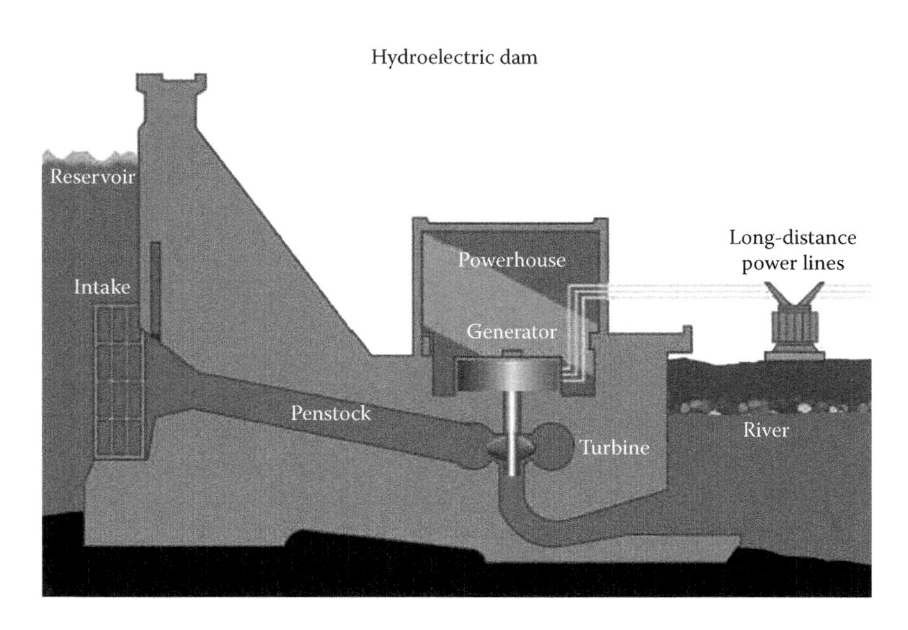

Figure 8.2 Basic components of a conventional hydropower plant.

8.1.1 Advantages of Hydropower

Hydropower, like all renewable energy sources, provides free fuel that never runs out and is more environmentally benign than fossil fuel sources. As we have noted earlier, hydro comprises some of the largest-scale power plants in the world, and hence, it offers economies of scale. Even though it is the largest renewable energy source for producing electricity, much undeveloped potential still exists—even in nations such as the United States, where the best sites have already been exploited. However, the very best opportunities are now probably in the developing nations, especially in Africa, where the amount that is technically feasible to exploit is roughly 20 times what currently exists. Moreover, in the emerging economic powerhouse that is China, the amount that can now be economically exploited is nearly seven times what now exists. Thus, not surprisingly, a large majority of the very large-scale hydro projects now under construction are in China. In fact, of the 12 largest hydropower plants coming on line before 2017, 10 are in China, which will supply that nation with an additional 75,000 MW—an amount that roughly equals the total now present in the United States. Given China's enormous energy needs, this is good news for the environment because hydro accounts for less CO_2 emissions than any other energy source. Small hydro usage also has a great potential, particularly in rural areas of some developing nations, and it lacks the environmental problems of large facilities.

Hydro is also a robust and mature technology, and the power plants last a very long time—50–100 years in some cases. Since the main cost is in their initial construction, averaged over their lifetime, they provide electricity at extremely low cost. Moreover, the electricity that hydropower plants produce has the very useful character of *dispatchability*, meaning that it can be turned on and off on short notice, which is very important in supplying power at times it is most needed or to make up for the inherent variability in nearly all other renewable energy sources. Hydroelectricity production is also extremely efficient—certainly more so than any renewable or fossil fuel plant where heat is converted to mechanical work. Moreover, the dams that accompany conventional hydropower offer some advantages as serving a variety of other purposes as well, including flood control, recreation, and irrigation. Of course, those same dams are also the main source of problems associated with hydro to be discussed later.

8.1.2 Basic Energy Conversion and Conservation Principles

The amount of power that can be generated at a particular location is surprisingly easy to quantify, based on just two quantities—the water flow rate and the head, which is the vertical distance that the water drops. A mass of water that descends a vertical distance h loses potential energy mgh, so that the power p in the moving water can be found from

$$p = \frac{mgh}{t} = \dot{m}gh = \rho\dot{V}gh, \tag{8.1}$$

where $\dot{m}$ and $\dot{V}$ are, respectively, the water flow rates by mass (kg/s) and volume (m^3/s) and ρ is its density (1000 kg/m^3).

Thus, if there were a stream or river that descended over a waterfall of height h, and the volume flow rate were ascertained either by direct measurement or by some estimation process, it would be a simple matter to compute the maximum power of a hydropower plant built at that site using Equation 8.1. There are, however, several caveats. First, if the water descent is a gradual one, as it often is, mechanical energy will be lost and one needs to use some equivalent head h, not the actual vertical descent—with the reduction factor dependent on the gradualness of the descent and the smoothness of the surface over which the river flows. On the other hand, by creating a dam to collect water into a reservoir and eliminating the sections of the river with a gradual descent, losses can be greatly reduced when the water descends a distance h below the level of the reservoir via a smooth bottom penstock.

The main difficulty in using Equation 8.1 in estimating the available power at a particular potential site lies in determining an accurate value for the volume flow rate. In cases of a small stream, it may be possible to block the stream with an improvised dam and then actually measure how much water flows through an exit hole in the dam (and into a large container) in a given amount of time. This, of course, is not feasible in the case of a large flowing river. One approach in this case is to measure the varying depth at fixed intervals across the river and use those to calculate its area of cross section A. One can then estimate the speed v of the flowing water by watching a floating object move past at various points and use the average velocity to find the volume flow rate $\dot{V}$ based on

$$\dot{V} = vA. \tag{8.2}$$

It is common to ignore viscosity when treating water flow, in which case, we may apply the conservation of mechanical energy (Bernouilli's equation) to connect the pressure, speed, and vertical height for any two points in a fluid to obtain

$$P_1 + \frac{1}{2}\rho v_1^2 + \rho g h_1 = P_2 + \frac{1}{2}\rho v_2^2 + \rho g h_2. \tag{8.3}$$

As one application of Equation 8.3, consider a reservoir whose surface is a height h above a discharge pipe out of which water flows and exits the pipe at atmospheric pressure. To find the exit speed of the water out of the pipe, we use $P_1 = P_2$, $h_1 = 0$, $h_2 = -h$, and $v_1 = 0$, so as to find

$$v = \sqrt{2gh}. \tag{8.4}$$

Interestingly, this is exactly the same as the speed an object would attain after falling from rest through a distance h. This result is exactly what is required for a conservative force such as gravity.

8.1.3 Impulse Turbines

Turbines convert the kinetic energy from a flowing fluid (nearly always water) into useful work. The earliest type of turbine design is the impulse turbine, which is turned by the impact of a water stream. Impulse turbines are descendants of the water wheel, which has been used since antiquity to supply the power used to grind grain and for various agricultural and industrial purposes. Since the advent of hydropower to generate electricity, more efficient designs of impulse turbines have been developed. One of these is the Pelton turbine, in which the pressure of a head of water produces a high-speed water stream out of a nozzle that is aimed at a series of buckets arranged around the turbine wheel or runner.

Pelton turbines are well suited for sites having large heads (greater than 10 m), but care must be taken in the choosing of the radius and angular speed of the turbine runner for optimum performance. To achieve high efficiency, it is important that as much of the energy of the water stream as possible be transferred to a rotating wheel. In the ideal case, the water after striking the wheel will be at rest, so that nearly all its energy will be transferred to the rotating wheel, provided this is done with as little splashing as possible to minimize the loss of energy. In the case of the Pelton wheel, the shape of the twin buckets ensures that a properly aimed water stream is smoothly deflected and almost comes back along its original direction—see Figure 8.3. Without the double bucket design, a stream directed at the bucket center would undergo significant splashing.

It can easily be shown that for maximum transfer of energy from the water stream to the wheel, the tangential velocity of the rotating wheel should equal half the velocity of the water stream incident onto it. To prove this, we first note that for maximum energy transfer, the water stream with velocity v should ideally be at rest after striking the wheel, which requires that it have zero momentum after impact, and therefore, the force the water exerts on the wheel is $F = d/dt(mv) = \dot{m}\dot{V}$. The wheel exerts an equal and opposite force on the water and changes its kinetic energy by an amount satisfying $\Delta K = F\Delta x = \dot{m}v\Delta x = 1/2mv^2$, where Δx is the distance the wheel moves while the force of a mass of water m acts

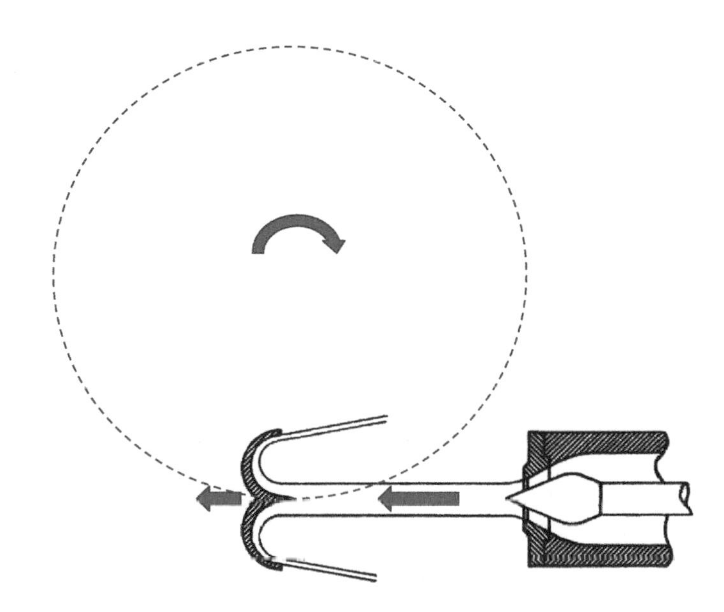

Figure 8.3 Detail of idealized water stream from a nozzle striking the center of a double bucket on a Pelton wheel and reflecting nearly backward (in two separated streams) without any splashing.

on it. Dividing both sides by the time Δt it takes the wheel to move a distance Δx gives $\dot{m}v\Delta x/\Delta t = 1/2mv^2/\Delta t$, which proves that the tangential velocity of the wheel $\Delta x/\Delta t = 1/2v$ for maximum energy transfer. Note that a final water velocity of zero means that it transfers all its energy to the wheel and it then falls straight down. This requirement of maximum energy transfer together with certain other criteria can be used to determine the optimum design of a Pelton turbine.

8.1.4 Design Criteria for Optimum Performance

In Figure 8.4b, the Pelton wheel diameter is roughly eight times that of the 15 buckets, which is probably typical. It would be undesirable if the radius of the jet impacting the buckets were more than 8–10% than that of the wheel because, otherwise, the cups would need to be so large that they would interfere with the flow onto other cups, so a reasonable lower limit to the ratio of the wheel and nozzle radii is perhaps $R/r \approx 12$. While there is no specific upper limit to R/r, we want to keep this ratio as small as possible, or else, we will find that the size of the Pelton wheel becomes too large and unwieldy, and the cost of the installation would rise. Unlike Figure 8.3 where only one jet is aimed at the wheel, typically as many as $n = 6$ are, but it would be difficult to fit more than six nozzles spaced around the wheel. Moreover, it is advantageous in terms of cost and efficiency to have as many nozzles as can easily fit rather than just one, so we shall assume that $n = 6$ is the optimum. A final previously discussed design requirement is that the tangential velocity of a point on the wheel be half that of the water stream incident on it for optimum energy transfer to the wheel, i.e.,

$$v_t = 0.5v = \omega R, \tag{8.5}$$

yielding

$$\omega = \frac{v}{2R}. \tag{8.6}$$

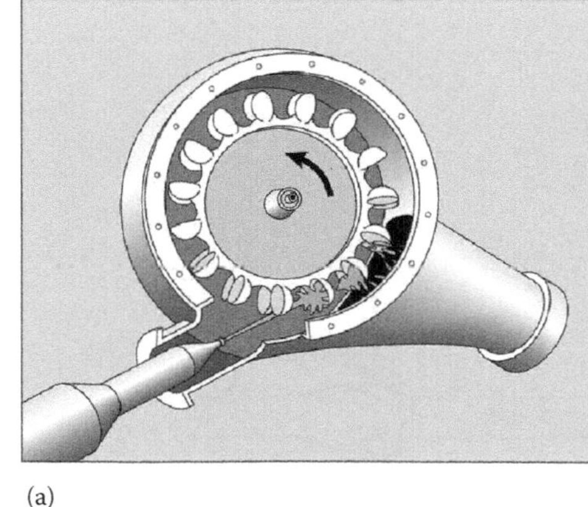

(a)

(b)

Figure 8.4 (a) Drawing of a Pelton turbine having 15 double buckets around the wheel being turned by a water stream. (Courtesy of DOE, Washington, DC.) (b) Assembly of a Pelton wheel at Walchensee Power Plant in Germany. (Courtesy of Voith Siemens Hydro Power, http://en.wikipedia.org/wiki/File:Walchenseewerk_Pelton_120.jpg, image licensed under the Creative Commons Attribution-Share Alike 3.0 Unported license.)

The three design constraints, (1) $R/r \approx 12$, (2) $n = 6$, and (3) $\omega = \frac{v}{2R}$, are sufficient to determine several key properties of a Pelton turbine for a particular site—although not whether a Pelton turbine is in fact the best choice for a given site!

8.1.5 Example 1: Designing a Pelton Turbine

Consider a watershed that has a head and flow rate sufficient to install a hydropower station having a power *P*. Let *e* be the efficiency or the ratio of the electrical power generated to the mechanical power of the water. Calculate the optimum nozzle radii, wheel radius, and rotational speed of a Pelton wheel, assuming $e = 0.9$, two values for both the head *h* (10 and 100 m) and the electrical power *P* (1 and 0.1 MW).

Solution

In order to produce an electrical power *P*, the mechanical power of the water stream at the bottom of a penstock a distance *h* below the water surface is given by P/e. The volume flow rate of the water with velocity *v* out of *n* nozzles, each having a cross-sectional area *a*, is *anv*; thus, we have

$$\frac{P}{e} = \dot{V}\rho gh = anv\rho gh = \pi r^2 nv\rho gh, \tag{8.7}$$

from which we can solve for the required nozzle radii as

$$r = \sqrt{\frac{P}{\pi env\rho gh}}. \tag{8.8}$$

Additionally, based on Bernouilli's equation (Equation 8.3), the water after vertically descending by a distance *h* will exit the nozzles at a speed *v* given by

$$v = \sqrt{2gh}. \tag{8.9}$$

Finally, given the design requirements, we can find the wheel radius in terms of the nozzle radii using $R = 12r$, and solving for the angular velocity from $\omega = v/2R$, we obtain the results shown in Table 8.1 when using

Table 8.1 Optimum Design Parameters for a Six-Nozzle Pelton Turbine Having Several Values for the Power (1 and 0.1 MW) and the Head (10 and 100 m)

Power (MW)	1.0	0.1	1.0	0.1
Head (m)	10	10	100	100
Jet speed (m/s)	14	14	44.3	44.3
Nozzle radius (m)	0.207	0.066	0.037	0.012
Wheel radius (m)	2.49	0.79	0.44	0.14
Rotation speed (rpm)	27	85	478	1512

the two specified heads and powers with $e = 0.9$, $n = 6$, $\rho = 1000$ kg/m^3, and $g = 9.8$ m/s^2.

8.1.6 Reaction Turbines

As we have noted, the Pelton turbine and the more primitive water wheel are two types of impulse turbines. The other category of turbines is known as reaction turbines. The defining characteristic of a reaction turbine is that it is completely submerged in the water, and it spins in reaction to the changing pressure of the water as it flows over its surface. Most commonly, reaction turbines have a vertical axis unlike impulse turbines, which more often have horizontal axes. Reaction turbines are more difficult to mathematically analyze than impulse turbines, and their optimization relies on sophisticated fluid dynamics software in order to find the fluid flow over their surface. The most common turbine in use today, the Francis turbine (Figure 8.5), is one example of a reaction turbine, and it is very efficient. In the Francis turbine, static vanes placed around the runner direct the flowing water tangentially to the turbine wheel, so that as the water encounters the vanes on the runner, it causes it to spin. The static guide vanes also known as a wicket gate is adjustable in angle and spacing so as to allow for changing amounts of incoming flow. The Francis turbine is an inward flow turbine where the water gives up its energy to the turbine and exits out the bottom at low velocity and pressure.

Another type of reaction turbine is essentially a propeller that is run backward. With the propeller of a ship, mechanical energy is supplied to the propeller causing it to turn and drive water backward and thus

Figure 8.5 Francis turbine runner, rated at nearly 1 million hp (750 MW), being installed at the Grand Coulee Dam, United States. (Courtesy of US Bureau of Reclamation, Washington, DC, http://en.wikipedia.org/wiki/Water_turbine.)

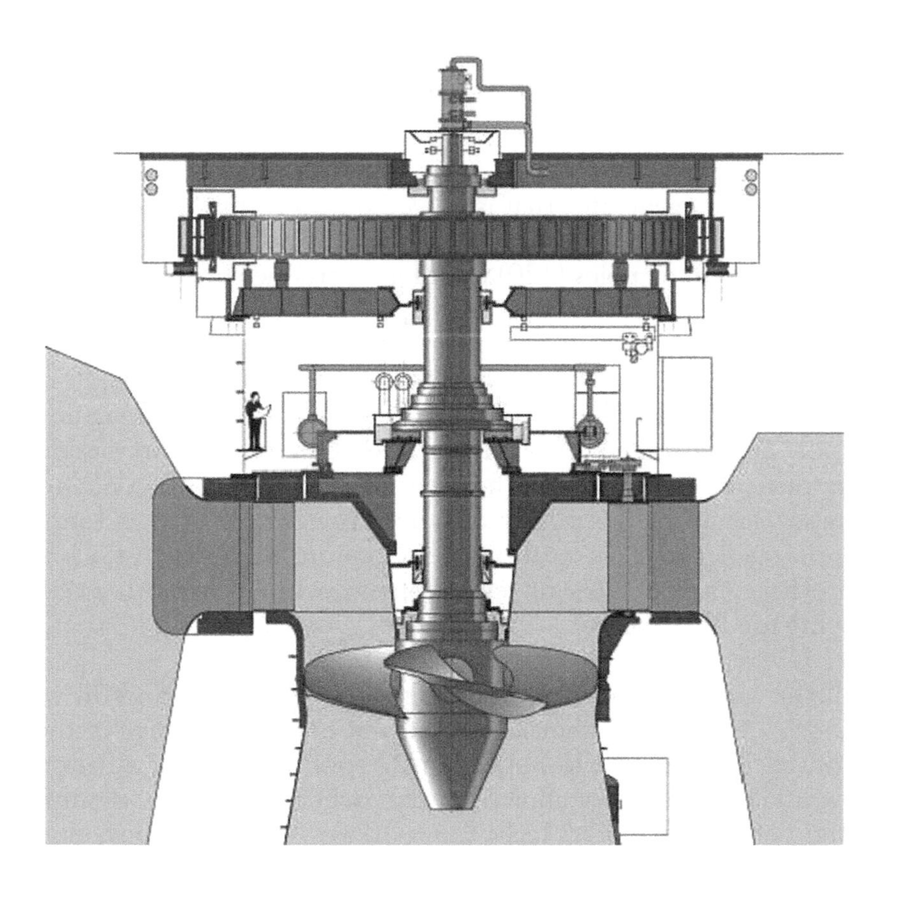

Figure 8.6 Cutaway drawing of a Kaplan turbine, another type of reaction turbine, with a tiny figure of a man on the left to show the scale. The dark parts of the image are static, the black parts rotate, and the light parts represent water, which enters the turbine housing radially and exits at the bottom after passing through the propeller. (Courtesy of Voith-Siemens Hydro Power Generation, http://en.wikipedia.org/wiki/Kaplan_turbine, image licensed under the Creative Commons Attribution-Share Alike 3.0 Unported license.)

move the ship forward. Here in the reverse situation, the motion of water through the propeller causes it to acquire mechanical energy of rotation that can be harnessed to produce electricity. Although many propeller turbines have nonadjustable blades, the Kaplan turbine blades do have an adjustable pitch angle, allowing it to cope with varying flow rates. In addition to being highly efficient, Kaplan turbines, by decreasing the blade angle, can cope with extremely low heads—as little as a meter and still have a sufficiently large ω (Figure 8.6).

The main disadvantage of reaction turbines is the high cost to fabricate and maintain them. The high cost is a consequence of the need to entirely enclose the runner and shaft within a chamber filled with high-pressure fluid, and it explains why they tend to be mainly used for very large hydropower installations.

8.1.7 Turbine Speed and Turbine Selection

The rotational speeds of hydro turbines tend to be lower than either those of gas or steam turbines, typically being in the range 60–720 revolutions per minute (rpm), although it is not unusual to have them as high as 1500 rpm. Recall that the frequency of AC is 60 Hz (in the United States) and 50 Hz (in Europe). Thus, if the hydro turbine is being used to power a generator that produces electricity for the power grid, too low a turbine speed can be a major problem. You might think that a generator producing electricity having a 60 Hz frequency would need to be spinning at 60 revolutions/s = 3600 rpm, but that assumes that the coil in the generator has just two magnetic poles. If many magnets are used (having a total of p poles), the coil could spin at a slower speed of $7200/p$ rpm. However, a generator that is powered by a turbine spinning at only 60 rpm would be problematic since it would need an extremely large number of magnetic poles (120) in order to produce 60 Hz AC.

An alternate way to deal with the problem of too slow a turbine rotation rate is through a gearbox that connects the turbine to the generator. One might also use a combined approach. Thus, e.g., one might imagine using a 12-pole generator and a gear ratio of 10 between the turbine and the generator to allow a turbine spinning at 60 rpm to power a synchronous generator that yielded a 60 Hz current. However, high gear ratios and many pole generators do result in greater complexity and cost, as well as lower efficiency, which is why few turbines operate below 60 rpm if they are being used to generate AC electricity.

Recall the earlier example of designing a Pelton turbine, and the results in Table 8.1 for the two choices of head and power. As shown in the first column of the table, the low speed of 27 rpm implies that a Pelton turbine would be a very poor choice for a site where the head was only 10 m and the power was 1 MW. In fact, h = 10 m is generally considered the minimum head for using a Pelton turbine. However, the more acceptable speed of 85 rpm (column 2) implies that Pelton turbines can be appropriate at such heads (and even lower) when the required turbine power is very low—see Section 8.1.10.

The earlier discussion of the need to avoid very low values of the rotational speed of a turbine primarily applies to the dominant usage of hydroelectricity generation, but this would not be relevant if the turbine is used for other purposes, such as pumping water or for producing a DC. For AC hydroelectricity, even though an asynchronous generator can tolerate small changes in speed, it is important that these changes stay within bounds, or else, the power output of the generator will significantly vary. Therefore, most hydropower installations have some kind of governor mechanism, so that even though there may be varying amounts of water flow rate, the turbine will turn at constant speed. There are a number of ways to accomplish this, but all rely on a form of negative feedback wherein when the device senses a reduction in flow, it takes appropriate actions to counter its effect, such as changing the angle of the wicket gate, opening the penstock entrance wider, or opening the nozzles wider (in the case of a Pelton wheel). Whatever action the governor takes, it is essential that there be as little delay as possible, because, otherwise, speed oscillations may be the result. Such oscillations in speed can be very destructive because of the torques they cause that can weaken the turbine. It is also important that a runaway speed be avoided, which is the maximum turbine speed when the generator is suddenly disconnected from the grid. This dangerous situation is similar to the extreme racing of the motor of your car when it is shifted out of drive and into neutral. Turbines are built to withstand runaway speeds for a short time, but their persistence can be very destructive.

8.1.8 Specific Speed

Engineers are quite fond of defining dimensionless numbers that can be quite useful in many cases where closed-form simple solutions are not available. One such dimensionless number is the specific speed ω_S of a turbine, which is a function of its shape, and is sometimes called the *shape number*. Specific speed ω_S is defined in terms of the turbine power P, the available head h, and the actual rotational speed ω as follows:

- **English units** (rpm, hp, ft)

$$\omega_S = \omega P^{1/2} h^{-5/4} \tag{8.10}$$

- **Metric units** (rpm, kW, and m)

$$\omega_S = 0.2626 \omega P^{1/2} h^{-5/4} \tag{8.11}$$

Of course, Equation 8.11 directly follows from Equation 8.10 with just the conversion of a unit. Thus, by Equation 8.10, a 1 hp turbine having a head of 1 ft has a specific speed ω_S equal to the actual speed ω in revolutions per minute, and a 1 kW turbine operating with a head of 1 m would have a specific speed that is 26.26% of its actual speed. For any given turbine, ω_S refers to the point of maximum efficiency for that type of turbine, and this number can be obtained from the turbine manufacturer. The specific speed allows a useful comparison to be made between different types of turbines and can be a guide as to which is best suited to particular applications. Finally, it can be used to scale an existing design of known performance to a new size and predict its performance. Here are typical ranges of specific speeds for various turbines, some of which are the impulse (I) and some the reaction (R) type: Pelton (I) 10–30, Turgo (I) 20–70, Crossflow 20–200 (I), Francis (R) 30–400, Propeller (R) 200–1000, and Kaplan (R) 200–1000.

Since Pelton turbines tend to have the lowest specific speed of any type, they will therefore have the lowest actual speed (at maximum efficiency) for a specified head and power. As a result, they are unsuitable in applications having a small head (unless the required power is also very small) because their rotational speed would then be too low. Conversely, Kaplan and propeller turbines, which have among the highest specific speeds, are best suited for operating in situations where the head is very small. The range of heads and flow rates for which different types of turbines are best suited can be discerned from the turbine application chart in Figure 8.7. Note that it is a log–log plot, so that the turbine power being the product of head and flow rate is constant along lines having a slope of −1.0.

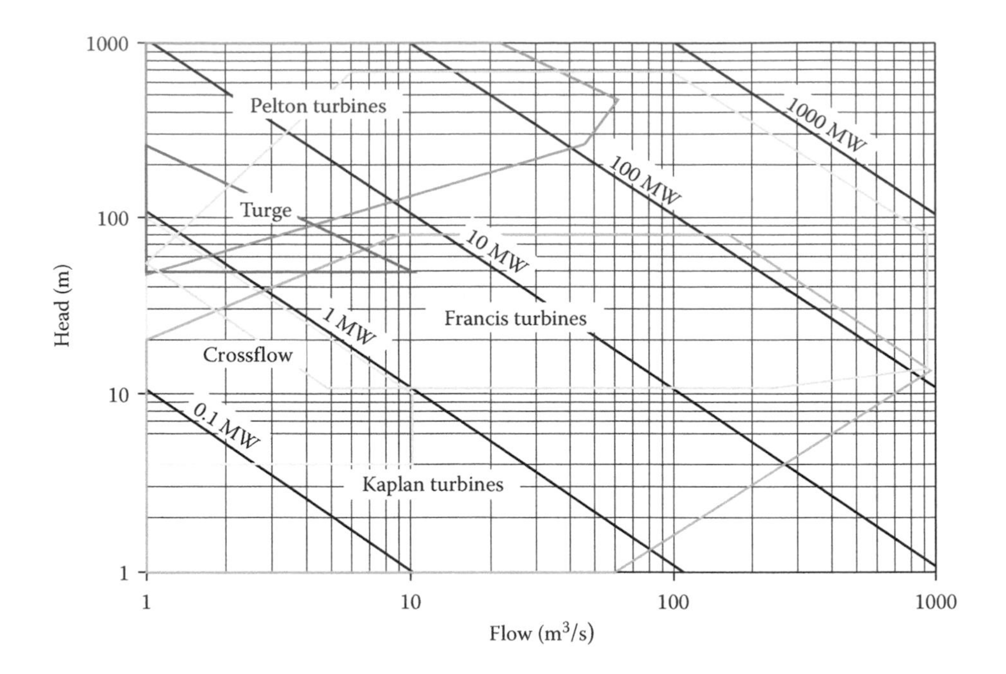

Figure 8.7 Water turbine application chart. (Courtesy of Teratornis, http://commons.wikimedia.org/wiki/File:Water_Turbine_Chart.png, image licensed under the Creative Commons Attribution-Share Alike 3.0 Unported license.)

Some attentive readers may have noticed that the specific speed as defined (Equations 8.10 and 8.11) is not truly dimensionless as asserted earlier. However, if two constants g and ρ are added to the expression by making the substitutions $h \to gh$ and $P \to P/\rho$, then the equation for the specific speed is truly dimensionless, and the result (Equation 8.12) agrees with Equations 8.10 and 8.11, where C is an appropriately chosen dimensionless constant:

$$\omega_S = C\omega\left(\frac{P}{\rho}\right)^{1/2}(gh)^{-5/4}. \tag{8.12}$$

In fact, if one writes the specific speed as

$$\omega_S = C\omega\left(\frac{P}{\rho}\right)^{a}(gh)^{b}, \tag{8.13}$$

it is a simple matter to show that ω_S will be dimensionless only if $a = 1/2$ and $b = -5/4$.

8.1.9 Pumped Storage Hydroelectricity

One great advantage of hydro is its use for pumped storage applications, whereby energy generated at times of low demand can be used to pump water to a higher elevation (effectively stored potential energy) and then released at times of the day when there is very high demand. Pumped storage can be done either as a stand-alone operation or as part of a river-fed hydropower plant. The suitability of hydro for pumped storage follows from the fact that water turbines of the reaction type are reversible devices that can either generate power when high-pressure water flows through them and gives up its gravitational potential energy or store energy when they are rotated backward and used to pump water to a higher elevation. Despite the high efficiency of hydropower, there are, of course, energy losses during the process, so pumped storage is a net consumer of electricity. Nevertheless, the price differential between peak and off-peak electricity prices makes it very worthwhile.

In many places, there is no cost differential as far as the consumer is concerned for electricity produced at different times of the day, but the electric utility does pay more at times of peak demand when higher-cost of supplementary sources need to be relied on. At times of extremely heavy demand, the cost rise can be dramatic. For example, during the electricity crisis in California in 2000–2001, it has been estimated that there could have been a 50% reduction in overall electricity prices if only storage could have supplied 5% of the needed energy during peak times—or equivalently if demand could have been cut by 5%. Equivalently, we could say that the last 5% of the energy produced during that crisis cost as much as the other 95%! Typical daily variation is not that extreme, but it is not unusual for the price to vary by a factor of 2–5 during the course of a day.

Pumped storage is currently the most widely used and cost-effective way of storing large quantities of electrical energy. In the United States, in 2009, pumped storage hydro amounted to 21.5 GW, which was 2.5% of the electricity produced, while in the EU, the corresponding figure is 5%. Either of these numbers, however, is a small fraction of the daily variation in demand—typically almost a factor of 2 from times of minimum to maximum consumer demand. The smallness of these percentages greatly understates the ability of hydro to help deal with the daily fluctuations in electricity demand. Even without pumped storage stations, some of the smoothing out of electric power production to match demand can be accomplished by so-called conventional hydropower stations that lack pumped hydro capability. All that needs to be done in this case is to simply produce power only at times when the demand is high, which can be as simple as opening or closing the penstock or wicket gates. This capability to rapidly switch on and off power production at will is something neither nuclear nor coal-fired plants share. One final use of pumped storage is in connection with highly intermittent renewable sources such as wind or solar power. Thus, rather than put the electricity generated by a wind turbine out on to the grid, which exacerbates the problem of demand variations, it can be used instead to power pumps so that the stored energy can be released at times of peak demand.

8.1.10 Small Hydro

Although the main focus of this section has been on large-scale hydropower, as we have seen, the spectrum of scales for the production of hydroelectricity is vast, and for some applications, hydro on a very small scale may be quite useful—either in remote rural communities in developing nations or for individual homeowners. In fact, on a worldwide basis, the contribution of small hydro is substantial (85 GW) with over 70% of that in China alone. *Small hydro* is usually defined as $P < 10–50$ MW, but it is also sometimes further subdivided into minihydro ($P < 1$ MW), microhydro ($P < 100$ kW), picohydro ($P < 5$ kW), and even nanohydro ($P < 200$ W). A turbine in the pico category might be suitable either for one American home or as many as 50 homes in a remote rural community in a developing nation where one or two fluorescent light bulbs and a radio might otherwise not be possible. One way to generate electricity on the nanoscale makes use of the simple Pelton turbine. This may seem surprising, since earlier, it was noted that Pelton turbines were unsuited to sites with small values of the head. However, very low-power applications are the exception to that rule. For example, using the same design restrictions that were used to generate the results in Table 8.1, it is easy to show that for a nanohydro application where we wish to generate 100 W, and a site where the head is a mere 1 m, we find perfectly acceptable values for the turbine speed (151 rpm) and radius (0.14 m).

The manner in which small-scale hydro is produced varies. For example, some companies make compact combined turbine generators that can simply be submerged in a fast-flowing stream without having to build a dam or install any piping. One such UK-made device capable of producing 100 W looks very much like a half meter-long submarine (Figure 8.8). Although such a compact submersible turbine plus generator may be an elegant way to produce a small amount of power, it captures only a small fraction of the power of the stream based on the fraction of the stream cross section that encounters the turbine propeller. The more common way of harnessing nearly all the power of a stream is using a so-called run-of-the-river hydro, whereby the elevation drop of a river is used to

Figure 8.8 Photo of UW-100 pico turbine that generates 100 W. (Image courtesy of Ampair the manufacturer, Johannesburg, South Africa.)

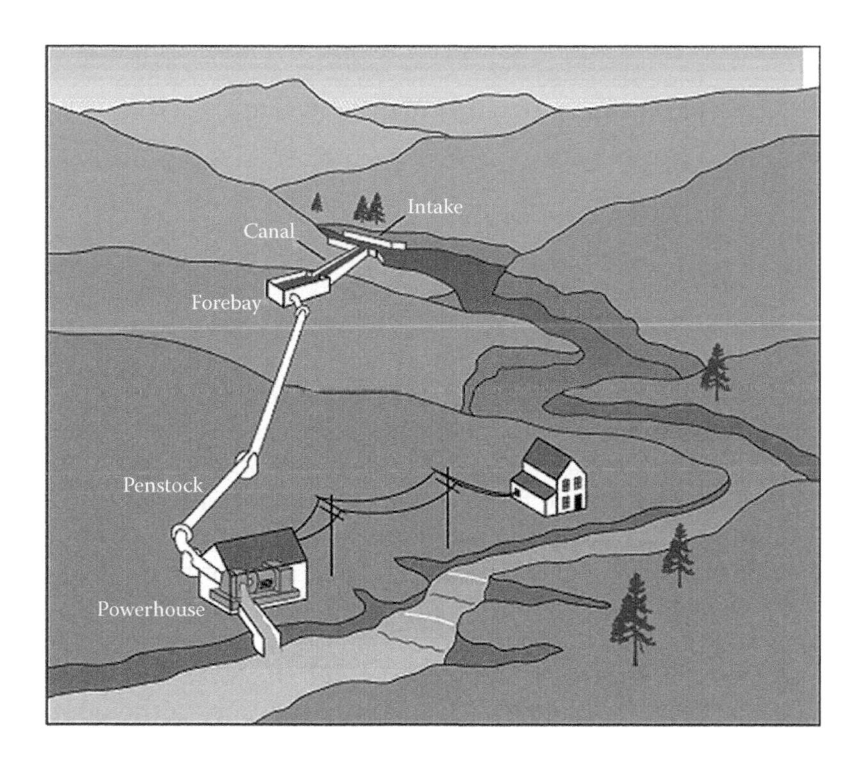

Figure 8.9 Drawing of a micro hydro installation of the run-of-the-river type. (Image courtesy of the U.S. Department of Energy, Washington, DC, is in the public domain.)

generate electricity (Figure 8.9). It is common in this case to divert the bulk of the flow through pipes or a tunnel and then allow the water to return to the river after passing through a turbine. Many small run-of-the-river hydro projects have zero or minimal environmental impact because they do not require a large dam. This cannot be said for large-scale conventional hydropower, but the subject of environmental impact will be deferred until we have looked at the other ways waterpower can produce useful work or electricity.

8.2 WAVE, TIDAL, AND OCEAN THERMAL POWER RESOURCES

Aside from freshwater hydropower, there are many ways to harness the power present in the oceans, but we shall discuss three here, which are probably among the more promising: wave power, tidal power, and ocean thermal energy conversion (OTEC). Ocean waves are ubiquitous and are caused by the interaction of the wind with the water surface. The amount of power in waves worldwide is enormous—probably greater than for freshwater hydro, but very little by way of wave power devices have been installed to date because the technology is not nearly as mature as for conventional freshwater hydro—see Table 8.2. OTEC is in even greater abundance worldwide, but has been even less installed to date for much the same reason. The only one of the three alternatives to freshwater hydro that has a nonnegligible amount installed (at just a single power plant in France) is tidal power. The conventional wisdom on tidal power is that it suffers from having a very limited number of locations where

Table 8.2 **Relative Sizes of Some Various Hydro Resources**

Energy Source	Potential (GW)	Practical (GW)	To Date
Freshwater hydro	4,000	1,000	654
Waves	1,000–10,000	500–2,000	2.5
Tides	2,500	1,000	59
OTEC	200,000	10,000	0

Source: Tester, J. W. et al., *Sustainable Energy: Choosing among Options*, MIT Press, Cambridge, Massachusetts, 2005.

it is practical and that the total amount of additional power worldwide that could practically be exploited is relatively small. As we shall see, this conventional wisdom may be wrong owing to some recent technological advances. In fact, of the three ocean-based alternatives to freshwater hydro (tidal, wave, and OTEC), tidal power or possibly wave power may be the most promising of the three, although none of them are likely to ever come close to the amount of hydroelectricity that freshwater hydro now generates. A large number of practical devices have been developed with respect to both wave and tidal power, and the total amount of power that could be feasibly exploited could be significant—although as with any new technology, the economics might prove a daunting challenge.

8.2.1 Wave Motion and Wave Energy and Power

Sizable ocean waves can be generated when the wind acts for a sustained period and interacts with the surface of the water. The height of the waves depends on the wind speed, how long a time it has been blowing, and various other factors. In introductory physics classes, water waves are sometimes described as being an example of a transverse wave where the motion of a water molecule is simply up and down as the wave goes by. However, water waves are in fact neither transverse nor longitudinal, but both combined. A molecule of the water has both up and down (transverse) motions as well as back and forth (longitudinal) motions as the wave passes. The combination of these two motions out of phase by 90° results in either a circle or an ellipse, depending on whether the water is deep or shallow—with *deep* meaning that the depth is large compared to the wavelength (see Figure 8.10). Note that real water waves have a decidedly nonsinusoidal shape, and in fact, they certainly lack the symmetry of the idealized one shown in Figure 8.11.

Waves, of course, transmit energy, and we can derive an expression for the power transmitted by a wave that shows it is proportional to the square of the wave height and the wave period.

Even though waves are usually nonsinusoidal, they can be decomposed into a series of sinusoidal functions by a Fourier analysis. Let us find the power that passes a given point x along its direction of travel. We are actually finding the power per unit width of the wave along a perpendicular

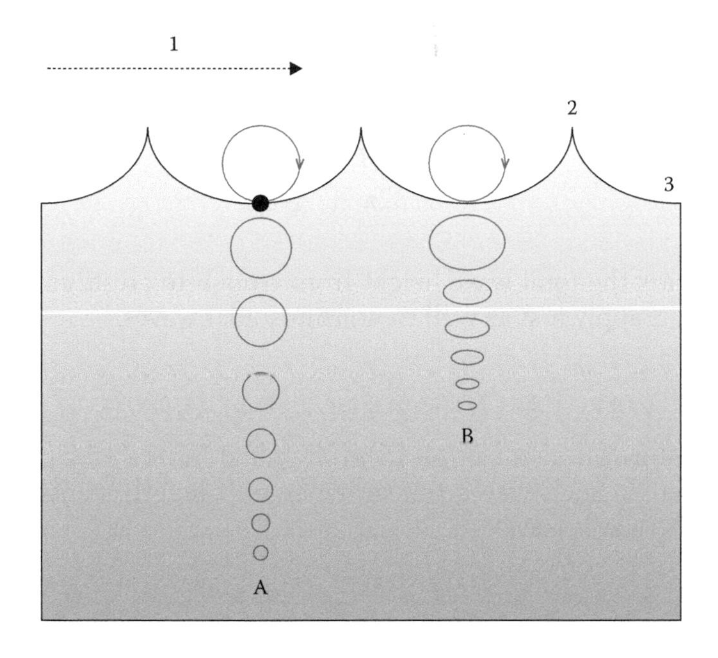

Figure 8.10 Simplified motion of a water molecule at various depths as a wave passes. Case A is for deep water and case B is for shallow water.

to the plane shown. The mass of the column shown is $dm = \rho y dx$, so its gravitational potential energy is

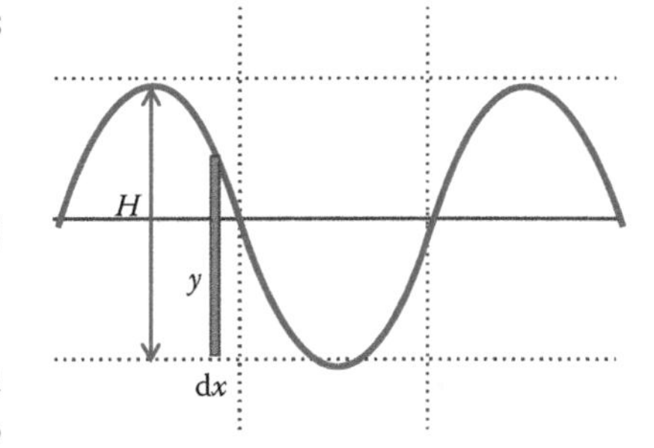

Figure 8.11 Wave of height H. y shows the height of a column of water of thickness dx.

$$PE = \frac{1}{2}mgy = \frac{1}{2}\rho g y^2\, dx, \tag{8.14}$$

and the power passing point x associated with the potential energy is given by $P = PE/t$. To find the average power over time, we use $y_{avg}^2 = H^2/2$ so as to obtain

$$P_{avg} = \frac{1}{4}\rho g v H^2. \tag{8.15}$$

But recall that this is only the power associated with the potential energy. For a mass moving in simple harmonic motion, the kinetic energy adds an equal contribution, which doubles the previous result. Power (and energy) is transported by a wave by its group velocity, which, for deepwater ocean waves, is given by

$$v = \frac{gT}{4\pi}. \tag{8.16}$$

Using Equations 8.14 through 8.16, we obtain for the average total power (kinetic plus potential):

$$P_{avg} = \frac{\rho g^2 T H^2}{32\pi}, \tag{8.17}$$

which can alternatively be expressed in terms of the wavelength rather than the wave period as

$$P_{avg} = \frac{\rho g^2 H^2}{32\pi}\left(\frac{2\pi\lambda}{g}\right)^{1/2}. \tag{8.18}$$

Note that H is the total wave height from trough to crest, and Equations 8.17 and 8.18 apply just as well to nonsinusoidal waves.

8.2.2 Example 2: Power of a Wave

In a major storm, waves can be 15 m high and have a 15 s period. How much power do such waves transmit per unit length transverse to the direction the waves travel?

Solution

Using Equation 8.17 with a seawater density ρ = 1025 kg/m^3 gives P = 13.2 MW.

8.2.3 Devices for Capturing Wave Power

Many devices have been developed to harness wave power, dating all the way back to 1799 and well over 300 patents filed in the United Kingdom alone. A spurt of interest in this possibility occurred in the early 1970s, and one promising device from that period, Salter's duck, was proposed by Stephen Salter in the United Kingdom—see Figure 8.12.

One of the greatest drawbacks to wave power, and one which existing devices have not been able to overcome, is their vulnerability to being destroyed by severe storms. They are unlike wind turbines in this respect, as those devices are capable of withstanding hurricane winds without damage. It is this vulnerability of wave power devices plus the high cost of power generated that probably accounts for their lack of widespread use to date. One intriguing device, the Anaconda, may remedy that drawback (Figure 8.13).

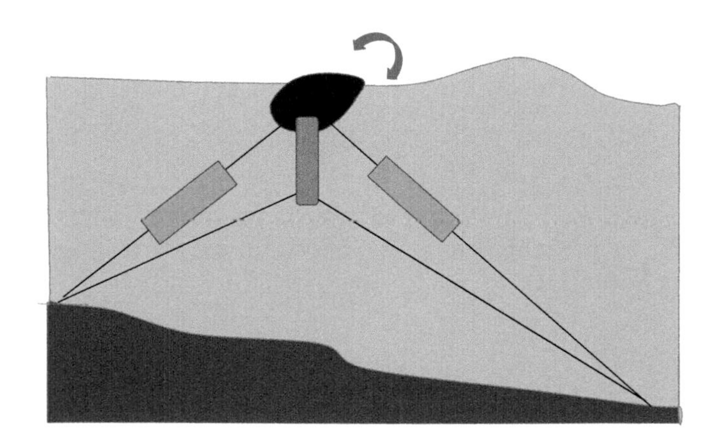

Figure 8.12 The duck's body (shown in *black*), which is moored to the seabed, oscillates as a wave passes it, causing the lines on each side to do work on the two rectangular devices on each mooring line. Remarkably, this device was found to extract 90% of the energy in the wave in tests, but a full-sized version has never been deployed.

Figure 8.13 As a wave passes the Anaconda made by Checkmate Sea Energy, it creates a pressure pulse (see the bulge) that travels down the length of it. (Courtesy of Francis J. M. Farley.)

CAN GIANT RUBBER SNAKES HELP SAVE THE WORLD?

The robust Anaconda wave power generator should survive severe storms, being made of rubber and having few moving parts. Although still under development, and not ready for commercial use before 2014, this device is said to be among the next generation of wave machines. The device responds to passing waves that cause a bulge that propagates along its length and drives a turbine at its end. Although the prototype shown is only 9 m long, a full-scale Anaconda capable of producing a megawatt of power would be 200 m in length and would cost an estimated US$2.8 billion to build. Given the very low maintenance required, Anacondas could produce electricity at around 9 cents/kWh—far less than previous wave generators at least according to the company that makes them.

8.3 INTRODUCTION TO TIDAL POWER AND THE CAUSE OF TIDES

At some locations on Earth, the daily variation of the tides has been harnessed to generate electricity and for other useful purposes. To date, however, tidal power has not been widely used, and it has proven quite expensive compared to some other renewable sources such as wind. In fact, the first and so far only one of a handful of tidal power stations (240 MW power) was completed at La Rance, France, in 1966. Tidal power is often considered to be suitable only to certain restricted locations having unusually high tides, and thus, its total potential that can be practically exploited around the globe is generally thought to be limited. Where conditions for its exploitation are right, tidal power does offer some advantages over some other renewable energy sources, however, in view of the predictability of the tides, compared to highly intermittent sources such as wind. Thus, if the availability and cost issues can be overcome, it could prove to be a very feasible means of generating electricity.

Most people are aware that the moon is primarily responsible for the tides, although the sun is also partly responsible. The basic mechanism is fairly easy to understand, and it was first mathematically explained by the great Isaac Newton, as an application of his law of universal gravitation. Although Newton's formulation has been further refined to take into account the Earth's rotation and other effects, it is a useful way to understand the basics. Consider the net gravitational force on the Earth due to the moon, which follows the usual inverse square form:

$$F = \frac{GMm}{r^2}. \tag{8.19}$$

Now consider the gravitational forces on a small part of the Earth nearest to and furthest from the moon (the right and left sides in Figure 8.14), which may be written respectively as

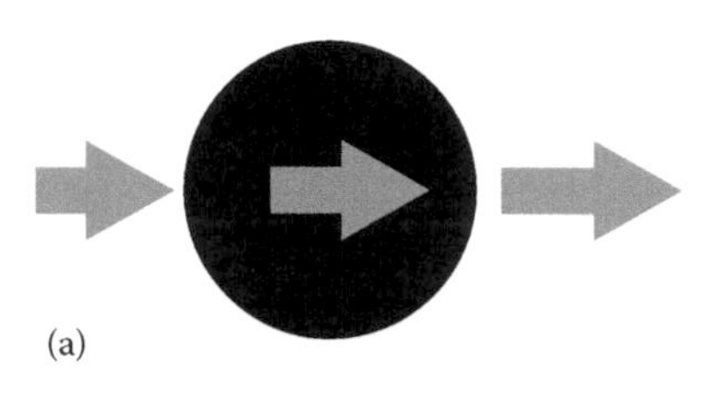

$$F_R = \frac{GMm}{(r-R)^2} \quad \text{and} \quad F_L = \frac{GMm}{(r+R)^2}, \tag{8.20}$$

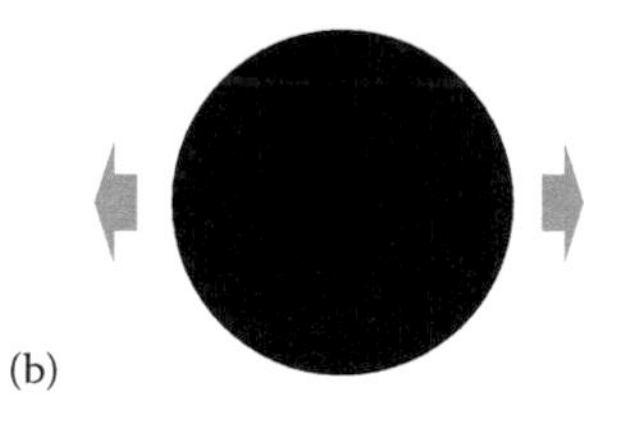

Figure 8.14 (a) Gravitational forces acting on three parts of the Earth due to the moon assumed to lie off to the right. As a result, the force on the left side of the Earth is greater than that on the center and that on the right side is less than that on the center. (b) Forces on the left and right sides of the Earth after subtracting the force on the Earth as a whole. These residual tidal forces act to stretch the Earth horizontally.

where R is the Earth's radius, and the moon is assumed to lie a distance r to the right from the center of the Earth. As long as $R \ll r$, we may use the first-order approximation $(1 \pm R/r)^2 \approx (1 \pm 2R/r)$ to find that

$$F_R \approx \frac{GMm}{r^2}\left(1+\frac{2R}{r}\right) \quad \text{and} \quad F_L \approx \frac{GMm}{r^2}\left(1-\frac{2R}{r}\right). \tag{8.21}$$

In an Earth-based accelerating reference frame, we may subtract the force of the moon on the Earth as a whole, which essentially cancels the one inside both parentheses. We therefore find that for the residual tidal forces on the right and left sides of the Earth, forces are the same magnitude but opposite directions, and they act to stretch the Earth along the line toward the moon—along the x-axis in Figure 8.14:

$$F_x \approx \pm\frac{2GMmR}{r^3}. \tag{8.22}$$

Note that tidal forces vary as the inverse cube or r, not the inverse square. A similar analysis can be done for the gravitational forces acting on the top and bottom of the Earth, and the result is a downward tidal force on the top and an upward force on the bottom of magnitude one quarter that of the horizontal forces, i.e.,

$$F_y \approx \pm\frac{GMmR}{2r^3}. \tag{8.23}$$

Thus, the action of these four forces, two pushing the Earth inward along the vertical and two stretching it horizontally, act to deform the Earth and convert a sphere into an ellipsoid, with the extent of the deformation greater for the water than the more rigid solid Earth. We see that there are two tidal water bulges of water on opposite sides of the Earth held in place by the moon's attraction, and the solid Earth rotates daily underneath them. It is for this reason that most places on Earth experience two high and two low tides each day.

There are, of course, a couple of points that need to be made about this simplified treatment of the tides. First, we need to consider the effect of the sun as well as the moon, even though the former is only 46% of the latter. Whether the tidal effects of sun and moon tend to cancel or add entirely depends on the relative positions of these two bodies in the sky. At those times of the month when they lie along the same straight line (at times of either the full or new moon), the solar and lunar effects reinforce each other, and the resulting tides are called *spring tides*. The opposite situation known as *neap tides* is depicted in Figure 8.15. Note that at such times, the directions of sun and moon are 90° apart (the first and last quarter moons). During the neap tide while the moon stretches the Earth vertically in Figure 8.15, the sun stretches it horizontally (by a lesser amount), so that the net distortion is less than that due to the moon acting alone, and the difference between high and low tide is at its minimum.

The second caveat concerning the analysis so far is that the orientation of the tidal bulges due to the moon and sun are only on the line toward those bodies in the case of an imaginary nonrotating Earth. In the actual situation, the line connecting the bulges on each side of the Earth is 10° ahead of the direction of the moon in the direction of the Earth's rotation. As a result, the gravitational forces on the two bulges exert a torque on the spinning Earth that causes the Earth to gradually slow down over

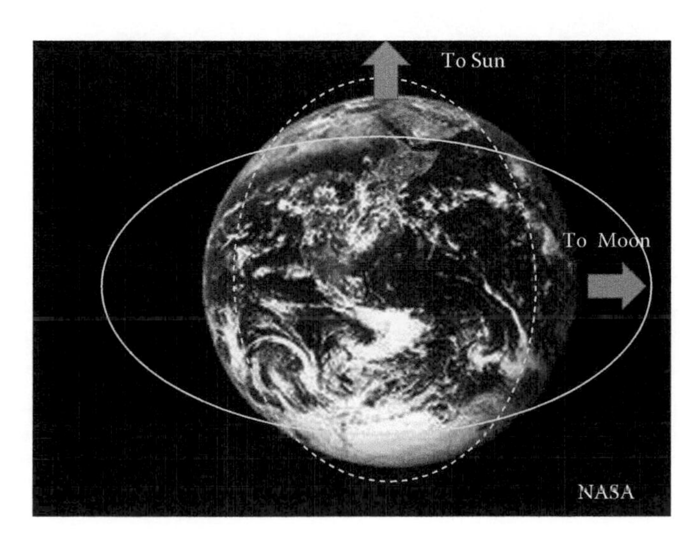

Figure 8.15 Neap tides occur when the direction to the moon and sun are at right angles. In this case, the distortions of the Earth's shape due to the tidal effects of the moon (*solid ellipse*) and the sun (*dotted ellipse*) tend to cancel. Obviously, the magnitude of the distortions is greatly exaggerated in this figure, but their relative magnitudes are about right. (Courtesy of National Aeronautics and Space Administration, Washington, DC.)

the course of the millennia. In fact, 400 million years ago, the day was only 22 h long.

The effect of the tidal forces causes different tidal currents that dramatically vary depending on the local geography of the sea floor and nearby coastlines. On average, the difference between high and low tides is about a meter, but there are places on Earth, such as the Bay of Fundy in Canada, where it reaches an incredible 16 m—the highest on Earth. In fact, there is a small (20 MW) tidal power station operating there—currently the only one in North America. The reason for the exceptional tides in Fundy has to do with the unique funnel shape and immense depth of the Bay, which gives rise to a resonance in the tidal flows where the frequency of water sloshing in and out of the Bay just matches the driving frequency of between 12 and 13 h. Note that the time between successive high tides is not 12 h (half a day) because the moon's position in the sky changes from each day. In fact, because of the two motions (rotation of the Earth and orbit of the moon), the moon returns to the same point in the sky every 24 h and 50 min, and the time between successive high or low tides is half that value.

8.3.1 Tidal Current Power

Accompanying the variations in sea level, the daily tides produce oscillating currents known as tidal streams. The tidal current, of course, is not simply along the water surface, and the underwater currents can be harnessed to power turbines, as illustrated in Figure 8.16. Compared to the

Figure 8.16 Commercial tidal current generator made by Seagen installed in Strangford Lough, Northern Ireland. The underwater tidal currents drive turbines. Note the visible wake indicating the strength of the current. (Courtesy of Fundy, http://en.wikipedia.org/wiki/File:SeaGen_installed.jpg, image licensed under the Creative Commons Attribution-Share Alike 3.0 Unported license.)

two other ways of harnessing tidal power, the amount generated is fairly small, since the propeller blades intercept only a small fractional area of the entire tidal flow. However, there are compensating advantages: the units are self-contained and low cost and have less environmental impact than other methods requiring a large dam. Given the modularity of these units, new ones can be readily added. The first commercial tidal current turbines are being built in 2012 in the gulf of Maine by the company Ocean Renewable Power, although the amount of power will be quite modest—only enough to power 20–25 homes.

8.3.2 Impoundment (Barrage) Tidal Power

Impoundment tidal power also termed *tidal barrage* has been the technique used in some of the few largest existing and planned tidal power plants. A 254 MW plant has recently been completed in South Korea, and one for over 1300 MW is also being planned there. In this approach, a dam is built across the width of a tidal estuary, so that as the tide comes in and goes out, a head develops across the dam. Channels in the dam allow water to flow through driven by the head, and this flow powers turbines and the generators to which they are connected—see Figure 8.17.

One alternative to barrage tidal power, which is built across an estuary and, hence, has negative impacts on habitat, is known as the tidal lagoon. In this case, a completely self-contained structure is built, so that when the tide rises, the lagoon fills with water, and when the tide falls, it empties. In both cases, the flowing water passes through turbines placed in holes in the structure. Tidal lagoons can generate more power, as well as be more environmentally benign.

8.3.3 Dynamic Tidal Power

This newest method of harnessing tidal power was patented by Dutch engineers Kees Hulsbergen and Rob Steijn in 1997. Dynamic tidal power

Figure 8.17 Basic idea of the impoundment method of harnessing tidal power. With high tide on the right instead of the left, the water flow through the dam would be in the reverse direction from that depicted, so it is important that the turbines placed in the dam can run in either direction.

(DTP) promises to greatly expand the areas where tidal power could be profitably exploited and, hence, the total amount of tidal power exploitable worldwide. In addition, power plants relying on this method could generate far more power than what has been the case to date with either the tidal current or impoundment methods. The basic idea of DTP involves building a very long dam from the coast directly out to sea. A typical dam length might be 30–60 km, and it would have a T shape at its end as illustrated in Figure 8.18. Note that such a dam does not enclose anything, but rather, it interferes with the twice daily oscillating tidal currents that run parallel to continental shelves, and those oscillations create a head or water level difference across the dam. Finally, electrical power could be generated in the usual manner using hundreds or even thousands of low head turbines placed all along the dam.

Fluid dynamic simulations show that the presence of such a dam would result in a head whose magnitude is proportional to the length of the dam. In the simulation used to create the image of Figure 8.18, the tidal current of the coast oscillates east–west, and at maximum, the head across the dam reached 1.3 m.

Here is a simple way to understand how a head develops across the dam as a result of the oscillating tidal current. One way to visualize the tidal forces on the ocean is to introduce a small horizontal component to gravity that oscillates in a horizontal direction as shown in Figure 8.19. Simply imagine water in a tray with a partition in the middle and the tray was slowly shaken side to side. The impact of this oscillation would be to cause the water surface to be deformed as shown in Figure 8.19 during one time in the oscillation and to be deformed in an opposite manner a half cycle later.

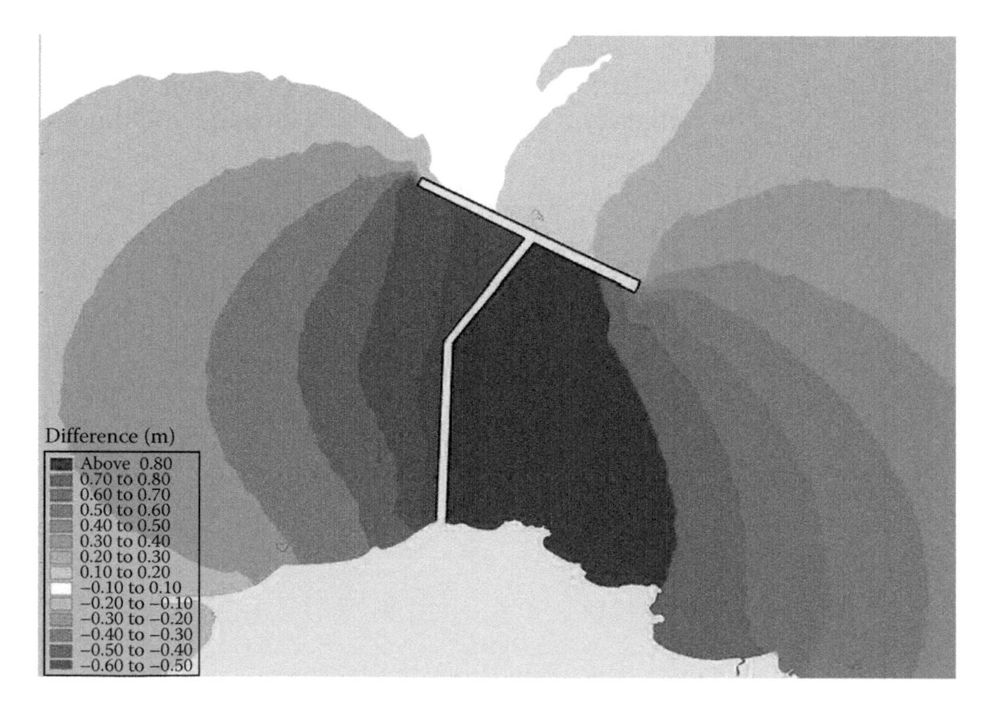

Figure 8.18 Top-down view of a DTP T-shaped dam. The blue and dark red colors indicate low and high tides, respectively. Note that the level of the tide is both high and low simultaneously at opposite sides of the dam. (Courtesy of Wikimedia Foundation, San Francisco, California, http://en.wikipedia.org/wiki/Dynamic_tidal_power.)

Note that using bidirectional turbines, power can be generated whichever side of the dam has the higher water level. The amount of power that could be harnessed by such a dam is said to be huge—perhaps 8000 MW. Most importantly, DTP does not require much of a tidal range, so the list of potential coastal areas where it could be profitably installed is far greater than for the other types of tidal power generation. In China alone, it has been estimated that between 80 and 150 GW of electricity could be generated using DTP. As of 2010, however, no DTP plants are under construction, and the concept needs a pilot test of its feasibility. One problem in implementing DTP in a pilot project is that as noted, the magnitude of the predicted head is proportional to the dam length, and so the power generated scales with the square of the dam length—assuming the head was even large enough to generate power.

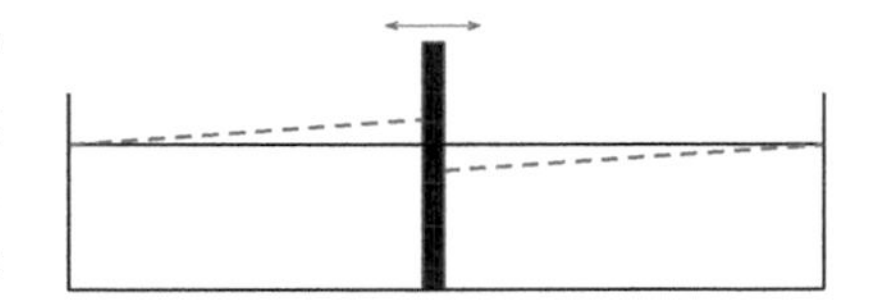

Figure 8.19 Simple model of how the head across the dam arises from oscillating tidal currents, based on a model of water in a tray that oscillates horizontally. The dashed lines show a possible deformation of the water surface on both sides of the dam.

Thus, were one to build a 1 km long pilot dam (1/30 as long as the one in Figure 8.18), the predicted head would be at most 1.3 m/30, or a trivial 4 cm at the point of maximum amplitude (or 2 cm on average)—far less than what is needed to generate even any power and possibly not even detectable, given realistic wave conditions in the ocean. Thus, obtaining funding to build a full-scale plant could prove very difficult without the ability to demonstrate power production in a small-scale facility. In addition, even with the full 1.3 m maximum head in the simulation in Figure 8.18, this translates into an average head of only 0.65 m. While turbines do exist that claim to generate power with heads under a meter, the ability to generate useful amounts of power under such a miniscule head would be marginal. Thus, one would probably want to use a dam length and/or location that would yield a head of 2–3 m, which would require a 60 km long dam!

Aside from the preceding difficulties, there are also issues of economics, since building a 30 km long dam would also be extremely challenging and very costly. One study done for a proposed DTP dam at Ijmuiden, Holland, estimated that a 30 km long dam there would cost $4.2 billion and might generate perhaps 1000 MW. The $4.2 billion cost is for the dam alone and does not include the hundreds or even thousands of turbines and generators, regular maintenance, storm damage, and other costs. When all costs are included, even allowing for green subsidies such as lower interest rates on capital construction, it has been estimated that the project would generate electricity with an estimated 30-year payback time and that it would cost perhaps 21 cents/ kW/h—far more than many other electricity-generating methods.

8.4 OCEAN THERMAL ENERGY CONVERSION

OTEC systems use the very small temperature differences between ocean surface layers and the water at significant depth to run a heat engine that can be used to generate electricity or for other purposes. The first proposal to harness this energy source was made by French physicist Jacques Arsene in 1881, and his student actually built a pilot plant many years later (1930) in Cuba. Of course, the efficiency of any heat engine is limited by Carnot's theorem, so that the smaller the temperature differential, the smaller the maximum possible efficiency, which is around 6–7%, given typical ocean temperatures. Actual efficiencies based on a Rankine cycle have turned out to be typically less than half the maximum at 1–3%, but newer designs have come closer to the theoretical limit. The attraction of OTEC is that the amount of power worldwide that could theoretically be produced dwarfs any other form of hydropower—see Table 8.1—however, the largest system built to date (by India) generated 1 MW. The primary problem with OTEC plants is of course one of economics, and some systems have been abandoned before being completed. The high cost of such systems arises because of both the initial capital investment and the energy costs needed to continually pump and circulate water through the system.

8.5 SOCIAL AND ENVIRONMENTAL IMPACTS OF HYDROPOWER

The environmental impact of the various forms of hydropower are most obvious in the case of freshwater conventional hydropower probably because it accounts for nearly all hydropower currently deployed. Those negative impacts stem from the consequences of building a large dam and the reservoir it creates. Such reservoirs flood the downstream areas, destroy natural habitats and agricultural land, and displace many people. According to a 2008 estimate, upward of 40–80 million people worldwide have been so displaced over the years. In fact, the Three Gorges dam and power plant project in China has alone displaced 1.2 million Chinese peasants from their homes. This project has also endangered the remaining residents of the area, causing landslides, a rise in waterborne diseases, and a decline in biodiversity. The greatest fear, however, is that the Three Gorges Dam may trigger severe earthquakes because the reservoir sits on two major faults. Nevertheless, it should also be noted that the dam did provide many positive benefits, even for many of those who were displaced, including financial compensation and, in some cases, better land.

One of the major ecosystem disruptions such large dams cause concerns fish, especially salmon and trout, which migrate upstream to spawn. In some cases, steps have been taken to alleviate this problem with the installation of fish-friendly turbines or fish ladders that allow fish to find their way around a dam, but even with such measures, some residual impact will remain. Another negative environmental impact that has been cited in connection with large dams and the reservoirs they create is the GHGs emitted from the decaying vegetation that is created when the upstream side is flooded. Although this onetime impact is real, one must also consider the GHG emissions avoided by using hydropower (the least carbon-intensive way to generate electricity), which lasts for the entire life of the dam, as well as the very low cost of hydropower compared to alternatives.

A final negative impact we consider here, and perhaps one of the most serious, is the consequences of a catastrophic dam collapse. Some of these events are among the most catastrophic of all disasters in history in terms of the loss of life. For example, the Banqiao dam collapse in China caused the immediate death of 26,000 Chinese plus many among another 145,000 who later died from epidemics. In fact, in the United States alone, more than 4000 older dams are said to be at high risk of imminent failure, according to the Association of State Dam Safety officials. Dam collapses, of course, are not solely due to poor construction methods or aging, as some past collapses were caused deliberately as the result of wartime attack and sabotage. Moreover, in an era when concerns over terrorism are a high priority, a successful attack on a well-chosen dam could well be more deadly than some other high-visibility targets, such as a nuclear power plant.

Small hydro projects and particularly run-of-the-river hydro do not require a large dam and are therefore much more environmentally benign. Likewise, while wave power may have an impact on commercial and recreational fishing, it is a far cry from the negative impacts of building a large dam. On the other hand, one reason that many modular technologies such as wave power turbines are considered more environmentally benign than large-scale projects is that they just do not generate very much power. Thus, for a collection of small (200 kW) wave or tidal stream generators to match the 20 GW Three Gorges dam power station, it would require a hundred thousand of them. While one wave power generator would have a miniscule environmental impact that is probably not the case for a collection of a hundred thousand.

8.6 SUMMARY

Hydropower comes in many forms, and this chapter considered the four most important among them: conventional freshwater hydro, wave power, tidal power, and OTEC. The first of these is by far the most important in terms of its exploitation to date and its future potential—especially in parts of the world that have so far not exploited a large fraction of their potential. Conventional hydro also has a number of highly desirable features compared to nearly all other renewable energy sources, given its dispatchability. Of the remaining three forms of hydro, very little power has been produced to date, and the potential for the future hinges on the technical and economic feasibilities of several new developments. Each of the forms of hydropower is not without its environmental problems, but those problems are very probably less serious than generating the same amount of electricity from nonrenewable sources.

PROBLEMS

1. Based on the four speeds and heads for a Pelton turbine tabulated in Table 8.1, which combinations of h and P fall in regions of the

turbine application chart that is appropriate for a Pelton turbine to be used? What type of turbine should be used in those cases where the Pelton is inappropriate?
2. Explain exactly why must lines of constant power fall on lines of slope −1 in Figure 8.7.
3. Show that the constant in Equation 8.12 is actually 0.2626.
4. Explain why impulse turbines are not used for pumped storage applications.
5. Explain why it is unimportant if the turbine speed remains constant for a pumped storage hydropower application.
6. Develop an Excel spreadsheet to reproduce Table 8.1.
7. Kaplan turbines can have a specific speed as high as 1000 and can be used with heads as low as a meter. How high could the turbine power be and still have it rotate above 60 rpm?
8. Show that the only possible choices for a and b in Equation 8.13 are the values indicated.
9. Verify the assertion made about a nano-Pelton turbine in Section 8.1.10.
10. While wind turbines have a maximum theoretical efficiency of 59% (the Betz limit), hydro turbines can have much higher efficiencies. Explain the difference, which at first thought seems strange since they both depend on extracting power from a flowing fluid.
11. Show that the tidal compression forces on the top and bottom of the Earth due to the moon are, as Equation 8.23 indicates, a quarter of those of the two stretching forces (Equation 8.22).
12. Show that spring tides are 20% higher than average and neap tides 20% lower than average. Hint: See Equations 8.22 and 8.23 and think about the sun–moon arrangements in the two cases.
13. Prove that the tides due to the sun are 46% of those due to the moon. How much does the magnitude of solar tides vary during the course of a year due to varying Earth–sun distances.
14. Suppose that tidal power began to generate 1000 GW—a vast increase from the present. Assuming that the Earth and moon form a closed system, estimate by what percentage this would reduce the Earth's rotational kinetic energy each year. At this rate, how long would it take for the length of the day to increase by 2 h? Note that due to tidal friction, the length of the day has spontaneously increased by about 2 h during the last 620 million years.
15. Consider an ocean wave in deep water whose wavelength and wave height are, respectively, 50 and 2.0 m. Find the wave power along a 100 m length transverse to the direction the waves travel.
16. How much energy does it take to raise 1 g of water to the top of the troposphere? (Hint: Assume that troposphere top is 10,000 m).

REFERENCE

Tester, J. W. et al. (2005) *Sustainable Energy: Choosing among Options*, MIT Press, Cambridge, MA.

Solar Radiation and Earth's Climate

9.1 INTRODUCTION

The Earth is bathed by solar radiation, which is responsible for the existence of virtually all life on the surface of the planet. In total, the amount of radiation is enormous and is far more than any other renewable energy source, many of which are in the end also driven by the incoming solar radiation. Thus, the subject of this chapter is important because it underlies the way in which solar energy can be harvested on Earth, and it can help decide where, when, and how it can be optimally done or whether it even makes sense at all in some cases. An additional theme of this chapter concerns the energy balance of the Earth and how that balance is shifted by the presence of GHGs in the Earth's atmosphere—an issue of considerable importance in connection with the habitability of the planet. Approximately 1000 W or 1 kW (or, more precisely, 0.865 kW) of solar power reaches each square meter of the surface when the sun is directly overhead, assuming clouds or atmospheric pollutants are not in the path of the sun's direct rays, although the amount that is incident on the top of the atmosphere is 1366 W/m^2, with the remainder being blocked by atmospheric absorption or reflection. That number (1366 W/m^2) is referred to as the solar *constant*, even though it is not truly constant. Let us briefly consider three timescales over which it varies.

On an annual basis, given that the Earth travels in an elliptical orbit rather than a perfect circle, the solar constant varies by around ±0.3% due to the varying distance to the sun. Interestingly, the Earth is closest to the sun in January rather than a summer month. On a bit longer timescale, there is also a variation of ±0.04% as a consequence of the 11-year sunspot cycle. Finally, the sun has dramatically evolved during the course of its life that began an estimated 4.5 billion years ago. Its brightness has drastically changed over that time and will do so again as it nears the end of its life when it approaches the red giant stage in perhaps another 4.5 billion years. It is believed, however, that the solar output has been relatively constant throughout the last 2000 years, with variations of no more than 0.1–0.2%.

The amount of solar power per unit area of the Earth's surface is known as solar irradiance, and it varies both spatially and temporally. Irradiance depends on the angle of the sun relative to the plane of the surface on which the incident power falls. It also depends on whether the radiation is diffuse (coming from all directions), as it does on a cloudy day, or direct, when it is straight from the sun. In general, there may be a mixture of the two—both direct and diffuse, since even when it is not cloudy, there is diffuse radiation reaching a point on the ground where the light came from sources other than the sun such as the sky. Were this not true, the

CONTENTS

sky would appear black, as it is on the surface of the moon, where there is no light reaching the ground from the airless sky. During midday, the total irradiance (direct and diffuse) reaching the ground can widely vary by as much as a factor of 10 or more, depending on how overcast the sky is. Finally, the direct portion of the irradiance depends on the angle θ that the sun's rays make with the normal to the surface on which they fall in accordance with the relationship

$$G_S = G^* \cos\theta, \tag{9.1}$$

and G* is 865 W/m² or whatever else the direct irradiance is found to be for normal incidence.

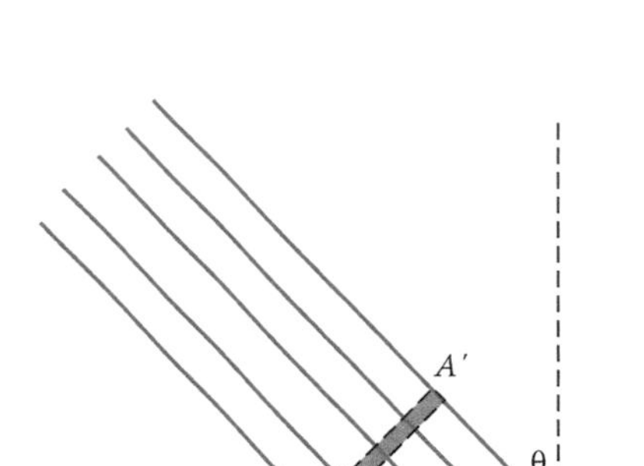

Figure 9.1 Basis for the cos θ factor in the irradiance falling on a surface *A*.

The appearance of the cos θ is a simple projection effect where the solar power is spread out over a larger area when the angle is other than 0°. Thus, in Figure 9.1, we see that the same irradiance falling on an imaginary surface *A*′ falls on the actual surface *A*, which is larger by the factor of 1/cos θ, so the power per unit area for *A* is, therefore, reduced by the factor cos θ.

9.2 ELECTROMAGNETIC RADIATION

When we speak of solar radiation, we are, of course, referring to radiation that is electromagnetic in character unlike some of the particulate radiation emitted from the atomic nucleus. There is, however, one overlap between the two types of radiation, i.e., the gamma rays. Gamma radiation, which is emitted from nuclei, has the highest frequency *f* or shortest wavelength λ of any type of electromagnetic radiation, the two quantities being reciprocally related through the equation

$$f\lambda = c, \tag{9.2}$$

where $c = 3.0 \times 10^8$ m/s is the speed of light in vacuum, which is shared by all types of electromagnetic radiation. In fact, many physicists use the word *light* to refer to all parts of the electromagnetic spectrum, as distinct from visible light.

The various types of electromagnetic radiation are all shown in Figure 9.2. Thus, on the opposite end of the spectrum from gamma radiation are radio waves, which have the longest wavelengths and lowest frequencies, and right in the middle of the spectrum is visible light. Figure 9.2 also indicates whether each type of radiation penetrates the atmosphere with the gray shading, obviously meaning that there is partial penetration. Note the strange horizontal scale that is neither linear nor logarithmic, but which allows each type of radiation to have an equal horizontal space. The two end categories, however, are unbounded off the ends of

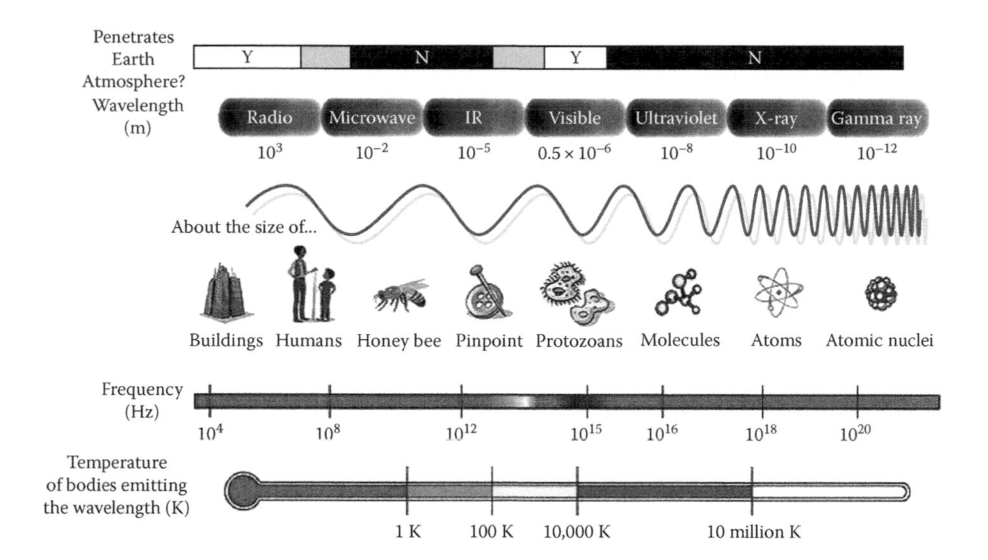

Figure 9.2 Electromagnetic spectrum. (Image courtesy of NASA, Washington, DC, in the public domain.)

the chart. Thus, gamma rays may have frequencies of unlimited size, while radio waves may have arbitrarily large wavelengths.

9.3 TYPES OF SPECTRA

When electromagnetic radiation of any sort is separated out according to wavelength, we create a spectrum, which essentially shows how much radiation is present in each wavelength interval $\Delta\lambda$. Spectra can be categorized according to whether they are continuous or discrete, with the latter also called line spectra. The presence of spectral lines having specific frequencies is the way it is possible to detect the presence of various elements or molecules in the source that is emitting the light. Thus, on this basis, we know just what elements the stars are made of.

9.3.1 Blackbody Spectrum

One important type of continuous spectrum known as the blackbody spectrum was first explained by the physicist Max Planck. A blackbody spectrum is emitted by a body that is a "perfect" absorber of light at all wavelengths. Furthermore, a blackbody can also be defined as one that emits light only from thermal processes, and hence, it neither reflects any light nor emits light from atomic or molecular internal transitions. A blackbody is only an approximation in the real world, but it is often an excellent one. One example of a near-ideal blackbody would be a small viewing hole in a heated oven. Clearly, if the hole is very small, nearly all the radiation incident on the hole from the outside will get trapped inside the oven. The spectrum of light out of the hole had been carefully measured over a century ago, but explaining its shape eluded all physicists before Max Planck. Planck realized that the radiation emitted by the internal heated walls of an oven resulted from oscillating atoms, which you might visualize as being interconnected by springs and forming a lattice. In 1900, Planck obtained an excellent fit to the measured spectrum by making a radical assumption for which he had no justification. He assumed that when the oscillating atoms lost some energy by emitting light, the lost energy was always quantized, i.e., it was an integral multiple of some energy E that is proportional to the light frequency or inversely proportional to the wavelength:

$$E = hf = \frac{hc}{\lambda}. \tag{9.3}$$

In Equation 9.3, $h = 6.63 \times 10^{-34}$ J s is a universal constant now known as Planck's constant. Based on this ad hoc assumption, Planck was then able to derive a formula for the intensity of light as a function of wavelength and temperature:

$$I(\lambda, T) = \frac{2hc}{\lambda^3} \frac{1}{e^{hc/\lambda kT} - 1}. \tag{9.4}$$

Planck's formula is displayed as a function of wavelength for five different temperatures T in Figure 9.3. Since Planck had no justification for his assumption of quantized energy other than it giving an excellent fit to the data, the value of h was simply chosen to give the best fit. Nevertheless, as a result of subsequent work by Einstein and others, our understanding of blackbody radiation in terms of photons having quantized energies is now completely solid, and Planck has been credited with essentially starting the quantum revolution in physics.

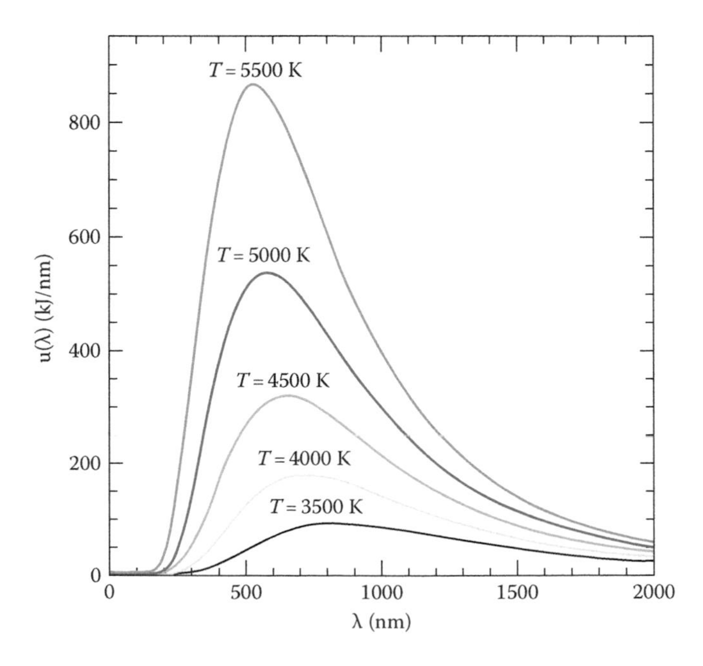

Figure 9.3 Five blackbody spectra corresponding to the indicated temperatures. (Courtesy of Geophysical Fluid Dynamics Laboratory, http://en.wikipedia.org/wiki/File:Wiens_law.svg, image licensed under the Creative Commons Attribution-Share Alike 3.0 Unported license.)

LIGHT: IS IT A WAVE OR A STREAM OF PARTICLES?

So far, we have consistently referred to light as being a wave. However, building on Planck's work, Albert Einstein in 1905 showed that it can also be thought of as consisting of a stream of particles known as photons. It was for this work (not relativity) that Einstein received the Nobel Prize. According to Einstein, the energy of a single photon is given by Equation 9.3, which Planck had discovered in his fit of the blackbody spectrum. Thus, if one had a source of intensity I, we could conclude that it emitted N photons per second, where

$$N = \frac{I}{hf}. \tag{9.5}$$

Based on this relation and typical values of I, h, and f, it is clear that observing the granularity of light (detecting individual photons) is not a commonplace event—although there are very sensitive detectors capable of it. Remarkably, the rod detectors in the eye are capable of registering a single photon, but the eye needs perhaps nine photons to trigger vision in the human eye.

There are two especially noteworthy features of the blackbody spectra shown in Figure 9.3. First, it is clear that the total amount of radiation emitted for a blackbody at a certain temperature (the area under the curve) dramatically increases with the absolute (kelvin) temperature; in fact, if we integrate Equation 9.4 over all wavelengths, we find that the power radiated per unit area of the body obeys the Stefan–Boltzmann law:

$$q = \frac{P}{A} = \varepsilon\sigma T^4, \tag{9.6}$$

where

$$\sigma = 5.67\times10^{-8}\ \mathrm{W/m^2\ K^4}.$$

The emissivity ε that ranges from 0 to 1 is a measure of how close the surface is to being a perfect blackbody.

Second, as seen in Figure 9.3, the peak of the spectrum shifts toward shorter wavelengths the hotter the emitting blackbody—a relation known as Wien's law:

$$\lambda_{max} = \frac{0.002898\ \mathrm{m\,K}}{T}. \tag{9.7}$$

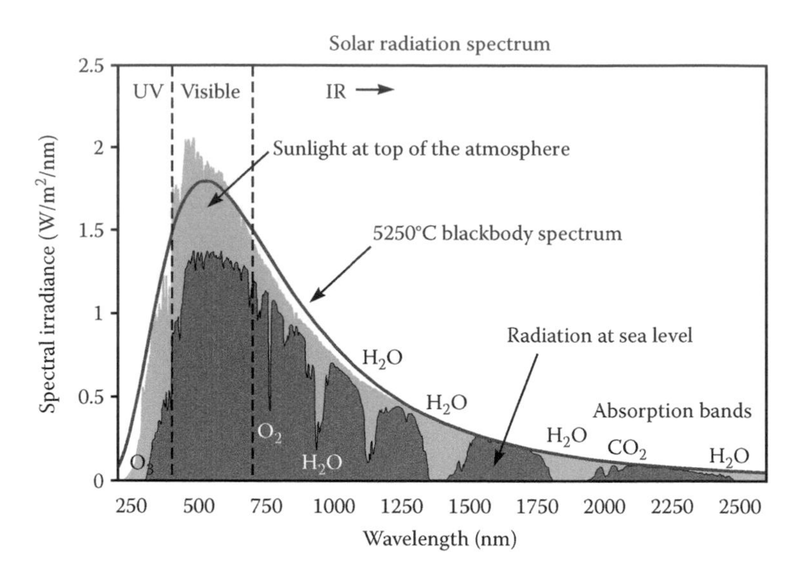

Figure 9.4 Solar spectra at sea level (*gray shading*) and the top of the atmosphere (*light green shading*) and the spectrum for a 5260°C blackbody. (Courtesy of Robert A. Rohde, http://en.wikipedia.org/wiki/File:Solar_Spectrum.png, image prepared as part of the Global Warming Art project and licensed under the Creative Commons Attribution-Share Alike 3.0 Unported license.)

Interestingly, most stars including our sun have spectra that fit the blackbody shape fairly well, which allows us to deduce their surface temperature—note the agreement in Figure 9.4 between the idealized blackbody curve and the actual spectrum shown in light shade, which is what is observed above the Earth's atmosphere looking sunward.

The dark-shaded fraction of the spectrum shows what survives after the solar radiation passes through the Earth's atmosphere and reaches ground level, with portions of the spectrum having been depleted by absorption by various indicated atmospheric constituents. The figure also indicates the dividing lines between three components of the spectrum: infrared (IR) (which accounts for 52%), visible (which accounts for 43%), and ultraviolet (which accounts for 5%). Note that gamma rays and X-rays constitute $\ll$1% of the spectrum.

Recall that the irradiance at ground level (the total area of all gray-shaded portions in Figure 9.4, which amounts to about 1000 W/m^2) assumes normal incidence of direct sunlight unobscured by clouds. However, the actual amount of sunlight reaching a surface at ground level having a fixed orientation during an entire day depends on many factors, most importantly, the position of the sun in the sky and the path it appears to take during the course of the day.

9.4 APPARENT MOTION OF THE SUN IN THE SKY

In order to understand the apparent motion of the sun across the sky, which depends on both the time of year and your latitude, we first need to consider some of the basics of the Earth's motion around the sun. During its orbit around the sun, the axis of the Earth's spin makes a fixed angle of 23.45° with respect to the normal of its orbital plane, and

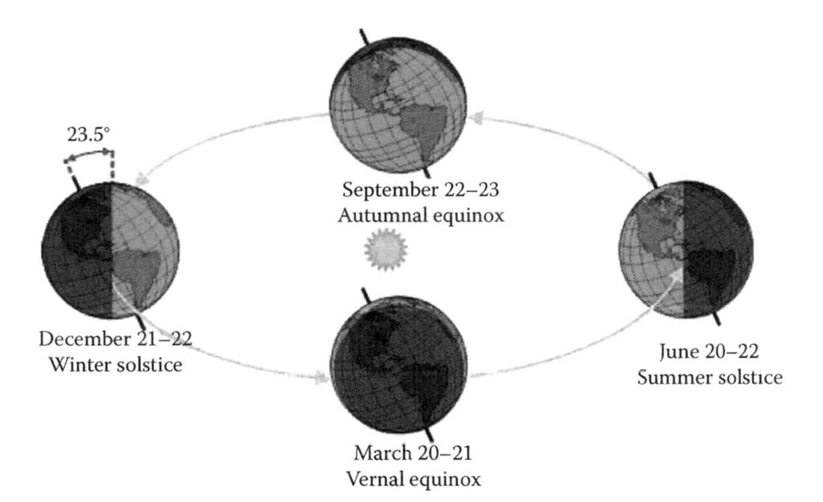

Figure 9.5 Motion of Earth in its orbit shown at the times of the two equinoxes and two solstices. (Image courtesy of NOAA, Washington, DC, is in the public domain.)

it points very nearly toward the North Star. There are four special times of the year, as indicated in Figure 9.5—the two equinoxes and the two solstices. During the spring and fall equinoxes, the Earth's axis is tilted neither toward nor away from the sun, and as a consequence, there are very nearly 12 h of daylight during a 24 h rotation. The other two special days are the summer and winter solstices corresponding to the Earth's axis tilted toward or away from the sun by a maximum of ±23.45°. For the summer solstice (at least summer in the Northern Hemisphere), the axis tilts toward the sun by this amount and the longest day of the year occurs—in terms of the fraction that is daylight. The exact date at which these four special days occur does vary a bit owing to the fact that the year is not an integral number of days long.

Now let us consider not how the picture of the Earth–sun system might look to some imaginary observer outside the solar system (Figure 9.5), but rather how it might look to an actual person standing somewhere on Earth watching the apparent motion of the sun across the sky during the course of a day as the Earth rotates on its axis (Figure 9.6). The ellipses in Figure 9.6 show the path of the sun on the same four special days we have

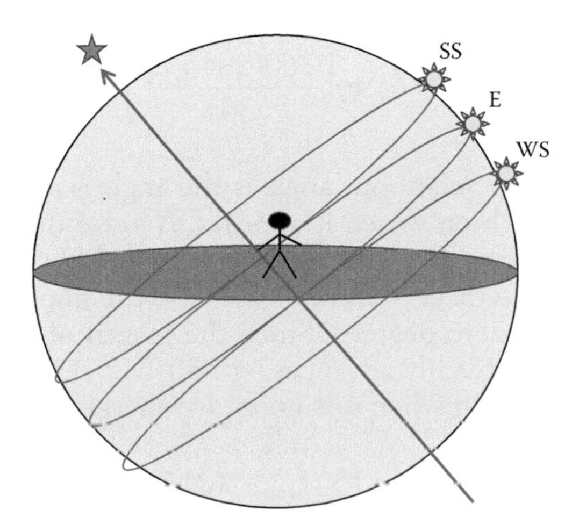

Figure 9.6 Apparent motion of the sun as seen by an observer on Earth.

just considered. There are only three ellipses, not four, shown because the vernal and autumnal equinoxes yield the same (middle) ellipse. Also, the actual path of the sun is very close to being circular, and the elliptical paths are just there for perspective. Finally, a portion of each ellipse lies below the horizon indicating nighttime hours. There are many other features of interest in this diagram, which bears careful study, including the location of the North Star, which is directly above the North Pole lying an angle above the horizon equal to the latitude of the observer, and the fact that the sun rises in the east and sets in the west only on the dates of the equinoxes. In the summer (when the sun is highest in the sky), the sun rises north of east and sets north of west, while in the winter (when it is lowest), it rises south of east and sets south of west.

The three suns show the position of the sun at noon on (SS) the date of the summer solstice (when it is highest in the sky), (E) the dates of the spring and fall equinoxes, and (WS) the date of the winter solstice (when it is lowest in the sky).

Examining Figure 9.6 further, it is clear that specifying the position of the sun in the sky at any given time and place requires two angles, which are known as the declination and the hour angle. The exact definitions of these two angles are as follows:

- *Solar declination*: Solar declination is the angle δ between the position of the sun at noon on a given day and its position at noon on the date of the equinoxes at the same location. Thus, from Figure 9.6, we see that on the dates of the summer and winter solstices, we have $\delta = \delta_0 = \pm 23.44°$, and on the dates of the two equinoxes, we have $\delta = 0°$. Furthermore, it can easily be seen that the elevation angle above the northern horizon of solar noon on any given day equals $\varphi + \delta$, where φ is your latitude. How can we find the solar declination on days other than the four special days of the year? It can be shown that Equation 9.8 will serve that purpose, where $n = 1, 2, 3, \ldots, 365$ is the day of the year. You should easily be able to verify that Equation 9.8 does give the correct answer on the four special days.

$$\delta = \delta_0 \sin\left[\frac{360(284 + n)}{365}\right]. \tag{9.8}$$

- *Solar hour angle*: Solar hour angle is the angle ω of the sun around the circular path on which it appears to move during the course of a day, measured from the last solar noon. Thus, at the instance depicted for the three suns in Figure 9.6, the hour angle on those days would be zero degrees. Since the length of the day is 24 h, the hour angle steadily advances by 360/24 = 15°/h in solar time, or since $t_{solar} = 12$ h when it is noon, we therefore have

$$\omega = 15°(t_{solar} - 12\text{h}). \tag{9.9}$$

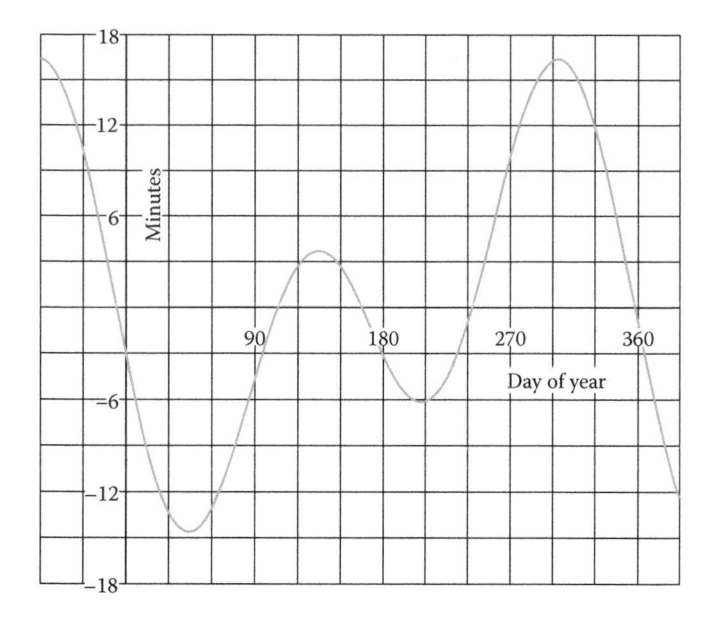

Figure 9.7 Equation of time. The number of minutes that the sun is running fast, or slow, relative to standard time versus day of the year. Negative values means it is running slow.

Unfortunately, the local time you measure on clocks is not exactly the same as solar time based on the apparent motion of the sun for two reasons. The first reason is the presence of time zones, so you need to correct the solar time based on how far you are in longitude from the western edge of your time zone, i.e., $(\psi - \psi_{zone})$. The second correction is based on something with the poetic name: the *equation of time* ω_{Eq}. The rotation of the Earth from one noon to the next takes exactly 24 h on average, but there are small variations due to the fact that the sun advances in the sky at a nonuniform rate each day due to the changes in the speed of the Earth in its elliptical orbit around the sun. These small variations are depicted in Figure 9.7, which is a plot of the equation of time.

Taking both corrections to solar time into account (the time zone and equation of time), we may write Equation 9.10 for the hour angle in degrees:

$$\omega = 15°/\text{h}\,(t_{solar} - 12\,\text{h} + \omega_{Eq}) + (\psi - \psi_{zone}). \tag{9.10}$$

9.4.1 Example 1: Finding the Solar Declination

Find the solar declination on October 4 in Washington, DC, and the hour angle of the sun at 3 p.m. on that day and location.

Solution

October 4 is the 277th day of the year; thus, Equation 9.8 with $n = 277$ yields $\delta = -18°$. The negative declination makes sense since the date is past the autumnal equinox. On the 277th day of the year, the correction due to ω_{Eq} based on Figure 9.7 is about +12 min = +0.2 h. Also, Washington, DC, is at west longitude 76.0°, and it lies about 7° west of the edge of the nearest time zone; thus, at 3 p.m. (when $t_{zone} = 15$ h), we have, using Equation 9.10,

$$\omega = 15°/h(15h - 12h + 0.2h) + 7° = 55°.$$

9.5 AVAILABILITY OF SOLAR RADIATION ON EARTH

The amount of solar energy that can be harvested at any given location depends on many factors, including the degree of cloudiness, the extent of atmospheric absorption, and especially the number of hours of daylight N at a given time of year. Deducing the last of these factors involves a bit of complicated three-dimensional geometry, but the result can be expressed in terms of two variables: the latitude φ and the solar declination δ on that particular day of the year:

$$N = \frac{2}{15}\cos^{-1}(-\tan\varphi\tan\delta). \tag{9.11}$$

9.5.1 Example 2: Finding the Number of Daylight Hours

Find the number of daylight hours on October 4 in Washington, DC.

Solution

On October 4, we have seen that the declination is −18°. Since the latitude of Washington, DC, is approximately 39°, we therefore have by Equation 9.11 $N = 9.1$ h. A value less than 12 makes sense since the date is past the autumnal equinox.

You might want to check for yourself that Equation 9.11 gives the right answer (i.e., $N = 12$ h) for the cases when either $\varphi = 0$ or $\delta = 0$. Figure 9.8 shows how the results from Equation 9.11 depend on arbitrary choices of the latitude and the time of year (which determines δ based on Equation 9.8). Note that there are certain latitude regions where the sun never rises nor sets during some times of the year.

As previously noted, the availability of solar radiation depends on local conditions including the degree of cloudiness, which itself will depend on the nature of the biome. Thus, desert regions are bound to be less cloudy than rainforests, and hence, the availability of solar radiation is not a simple function of latitude, as indicated in Figure 9.9.

Solar insolation (not insulation) is a measure of the amount of solar radiation received on a given surface area in a given time or, equivalently, the

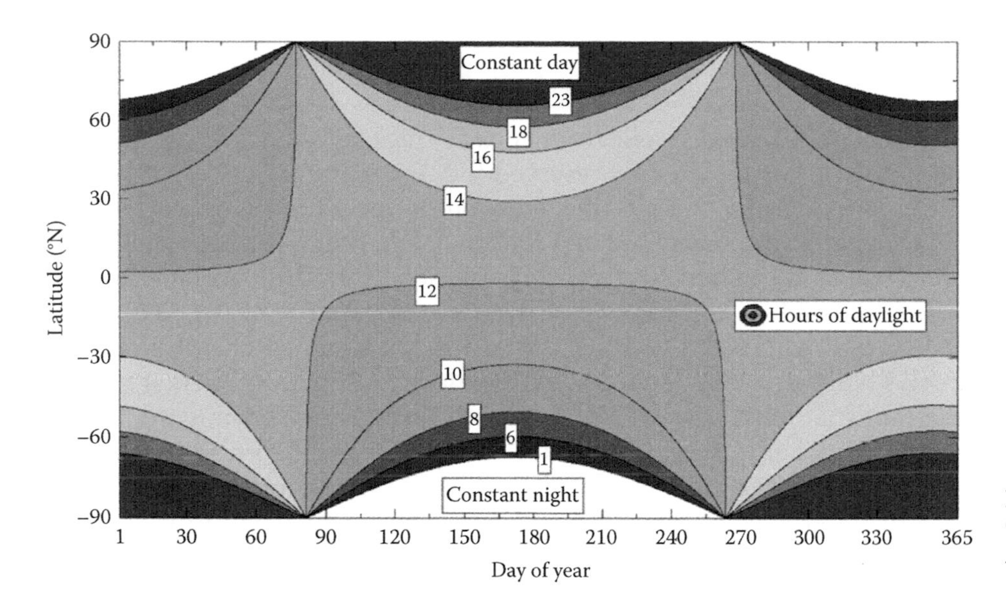

Figure 9.8 Number of hours of daylight *N* versus latitude and time of year. (Courtesy of Jalanpalmer, http://en.wikipedia.org/wiki File:Hours_of_daylight_vs_latitude_vs_day_of_year.png.)

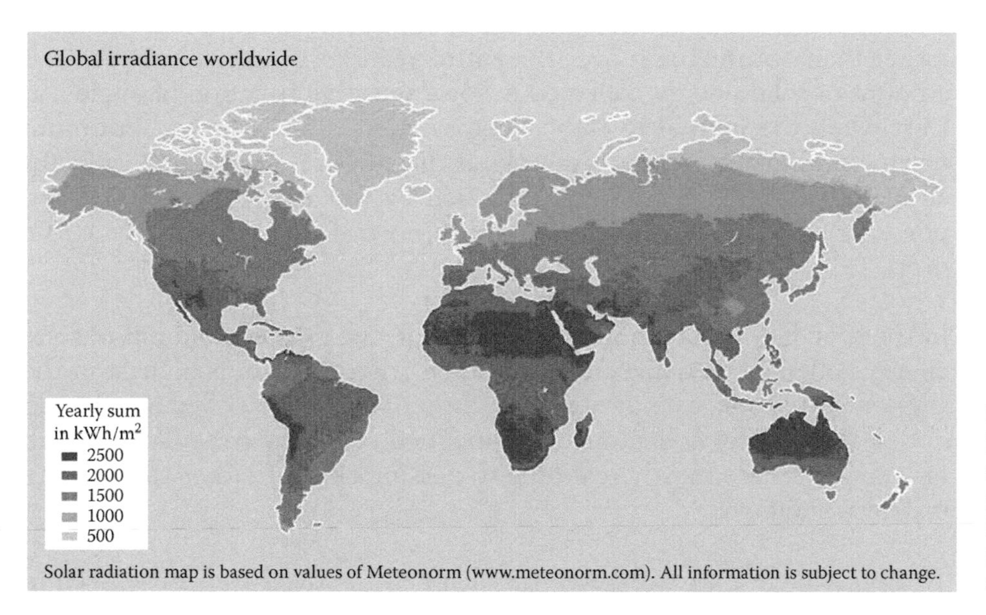

Figure 9.9 Yearly average solar insolation in units of kilowatt hours per square meter on the surface of the planet. The map is based on values of Meteonorm (http://www.meteonoem). (Courtesy of CHH Solutions, http://www.creativhandz.co.za/solar.php.)

integral of the irradiance over time. It may appear from Figure 9.9 that there is a pronounced asymmetry between the hemispheres, but that is only because the equator is not in the middle of the map, but rather passes through Brazil. Note how much greater the insolation is in known deserts, such as the Sahara and the American Southwest, compared to other places having the same latitude. Obviously, places such as deserts are excellent places to harvest solar energy not only because the insolation there is very favorable, but also because there are less other productive uses for the land, such as agriculture or human habitation—although even for deserts, some group will surely find reasons (perhaps legitimate ones) solar collectors should not be placed there.

9.6 OPTIMUM COLLECTOR ORIENTATION AND TILT

In certain cases, a solar collector needs to track the sun as it moves across the sky, but this is generally reserved for concentrating collectors, where the amount of solar energy collected would precipitously drop otherwise. Concentrators are discussed in a follow-up chapter. Normally, collectors are oriented at some fixed tilt angle and orientation throughout the entire year or possibly adjusted once. Obviously, if they are being placed on a sloped roof, which is the typical situation for a homeowner, if you lived in the Northern Hemisphere, you would choose the particular part of the roof that most nearly faces the south. In terms of optimizing the tilt angle, many homeowners prefer for aesthetic reasons to place the collector flush against the incline of the roof rather than propping them up a bit for greater efficiency. If instead you should seek the maximum efficiency averaged over the year, the proper slope of the collector would equal your latitude, so that at solar noon, on the day of the equinoxes, the sun's rays would strike the collector at normal incidence. Other choices are also possible, however. You might wish to optimize the solar energy collected either during winter or during summer, depending on your local climate, rather than optimizing it over the entire year. Fortunately, however, the amount of solar energy collected is not a sensitive function of angle—at least when the normal to the collector differs a bit from the optimum normal incidence case. For example, if the sun's rays strike the collector at an angle 30° off the normal direction by Equation 9.1, the collected power is 86.6% of that at the optimum normal incidence—only a 13% drop.

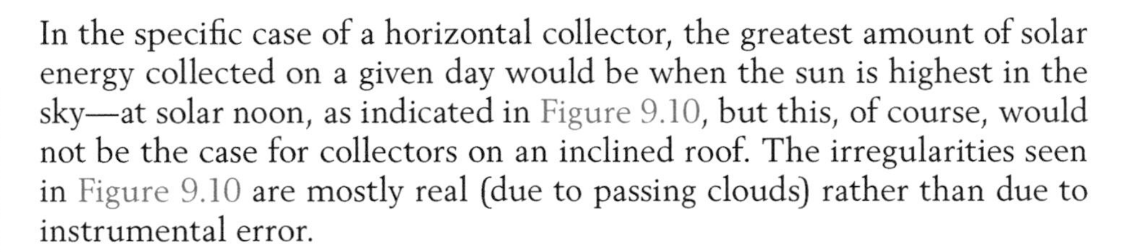

In the specific case of a horizontal collector, the greatest amount of solar energy collected on a given day would be when the sun is highest in the sky—at solar noon, as indicated in Figure 9.10, but this, of course, would not be the case for collectors on an inclined roof. The irregularities seen in Figure 9.10 are mostly real (due to passing clouds) rather than due to instrumental error.

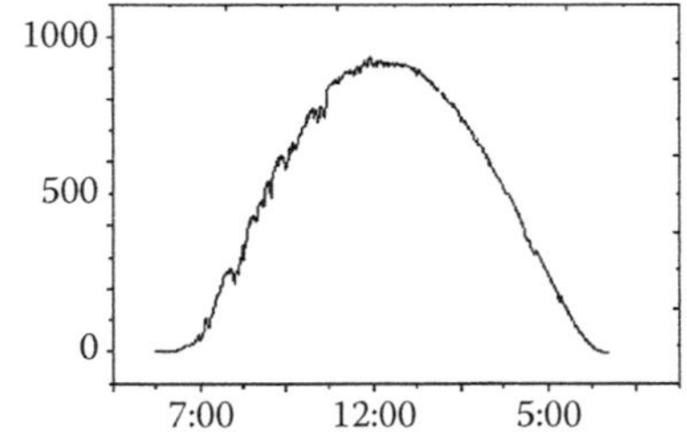

Figure 9.10 Measured solar irradiance in watts per square meter versus time at a location in Thailand (latitude 15° north, longitude 102° east) on February 21, 2008. Aside from the small irregularities, the data are well described by a sine function, indicating this was a very clear day. The maximum irradiance was 930 W/m^2 and the total energy for the day was 6192 Wh/m^2. (Courtesy of Wichit Sirichote.)

To a reasonable approximation, the shape of the measured curve seen in Figure 9.10 (aside from the irregularities) is a simple sine function of the time t since sunrise:

$$G(t) = G_{max}\sin\left(\frac{\pi t}{N}\right), \tag{9.12}$$

where N is the number of hours of daylight in the day and G_{max} is the irradiance at noon, but recall that this only holds for a horizontal surface.

In many realistic situations, the homeowner seeking help with planning a solar installation has a variety of online tools that can help figure out what can be achieved at any given place. For example, there are online tools that

1. Allow you to optimize the tilt angle of the collector for any given location under various assumptions and report back how much kilowatts per square meter per day of solar insolation is available.
2. Allow you to digitize the four corners of your roof (using a Google Earth photograph) and obtain a report back how many kilowatts the solar collectors placed on your entire roof would actually generate.
3. Use a panoramic picture taken from your roof to see what the effect of shading (due to trees and other houses) would be for any given time of the year. There is even an application of this sort that works using your smart phone.

9.7 GREENHOUSE EFFECT

Given that solar power continually bathes much of the Earth's surface, might one expect the Earth's temperature to steadily rise? Of course not, because we must take into account that the Earth sheds heat to space. In fact, by balancing the incoming and outgoing energy flows, it is possible to estimate the average temperature of the Earth because the surface temperature of the globe is what determines how much energy the Earth radiates to space using the Stefan–Boltzmann law (Equation 9.6).

9.7.1 Expected Average Surface Temperature of the Planet

The calculation of the expected surface temperature is quite straightforward. Light from the sun always illuminates at most half the Earth's surface at one time (see Figure 9.11). To find the total irradiance over the Earth, we need to do an integral over that hemisphere so as to take into account the varying values of $\cos\vartheta$ over the illuminated hemisphere.

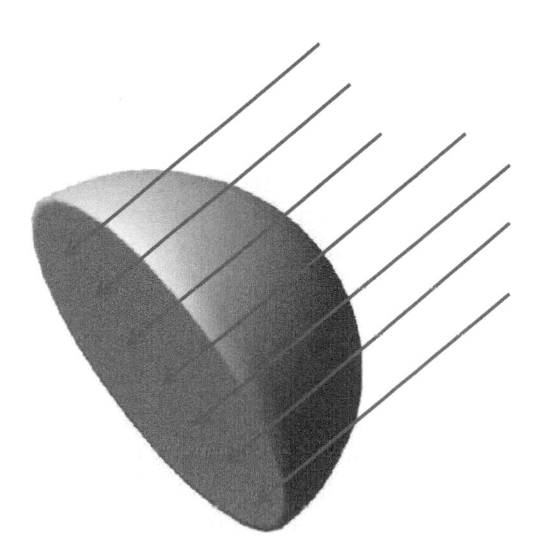

Figure 9.11 The total solar radiation incident on the hemisphere facing the sun may be calculated using radiation incident normally on a disk having the Earth's radius.

However, a much easier way is to realize that all the sun's rays that fall on that hemisphere land with normal incidence on a disk whose radius is that of the Earth.

Thus, if the incoming solar irradiance at ground level is 865 W/m^2, at ground level, we simply multiply that figure by the area of the Earth disk to find the incoming total irradiance over the entire Earth or

$$P_{\text{in}} = 865\pi r^2. \tag{9.13}$$

The outgoing radiant power due to the Earth being warmed to some absolute temperature T is radiated outward by the Earth's spherical surface not a disk, so that we have

$$P_{\text{out}} = \varepsilon\sigma A T^4 = 4\pi r^2 \varepsilon\sigma T^4. \tag{9.14}$$

Equating the incoming and outgoing powers, and taking the emissivity ε to be one for simplicity, yields

$$T = \left(\frac{865}{4\sigma}\right)^{1/4} = 248\,\text{K} = -25°\text{C}, \tag{9.15}$$

which is far colder than the actual average surface temperature of the Earth. Clearly, something must have been left out of this simple calculation. One possibility would be that the true emissivity of the Earth is less than one, but since T depends on ¼ power of the quantity in parentheses, our result would not change much if a correct figure were used. The likely missing piece to the calculation that makes the surface of our planet a much more comfortable average temperature of +13°C, and allows life as we know it to exist, is the natural greenhouse effect.

9.7.2 Natural Greenhouse Effect

Certain gases such as carbon dioxide, water vapor, and methane are present in the atmosphere (some only in trace amounts), and they have a dramatically different effect on the incoming and outgoing radiation. As with the incoming solar radiation, the outgoing radiation from the heated Earth also has a spectrum shape approximating a blackbody, but the temperature is close to 258 K rather than to the 5500 K temperature of the solar surface. Thus, by Wien's law, the peak for the outgoing or upgoing radiation is shifted to much longer wavelengths—into the IR region of the spectrum—see curves labeled "outgoing thermal radiation" for three different temperatures in Figure 9.12. The dark-shaded fraction of the outgoing radiation is all that survives its upward journey through the atmosphere. The remainder of the outgoing

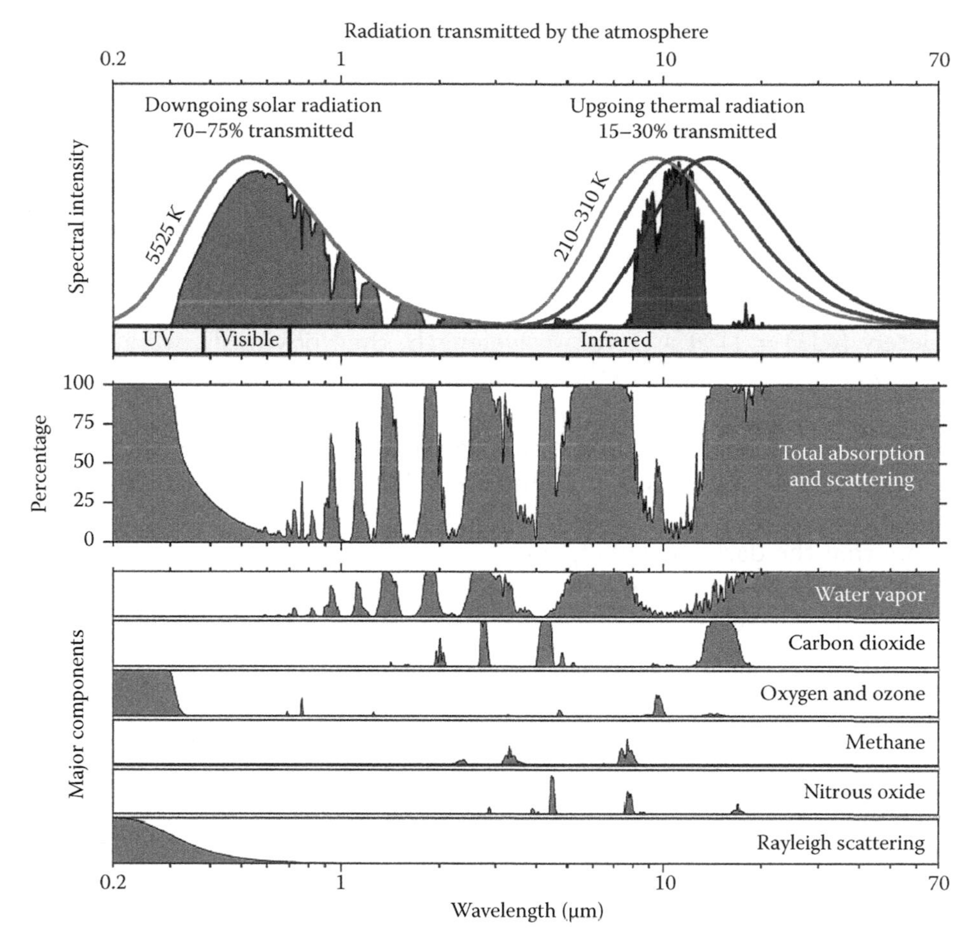

Figure 9.12 Top graph shows the intensity of radiation transmitted through the atmosphere—both incoming (*black*) and outgoing (*gray*). The bottom graphs show the total atmospheric absorption coefficient and the separate contributions from five sources: four GHGs plus Rayleigh scattering. (Courtesy of Robert A. Rohde, http://en.wikipedia.org/wiki/File:Atmospheric_Transmission.png, image prepared for the Global Warming Art project and licensed under the Creative Commons Attribution-Share Alike 3.0 Unported license.)

radiation is absorbed by the atmosphere, which acts like a trapping blanket around the Earth. In equilibrium, the amount of radiation incident on the Earth's surface must equal the amount leaving it. However, from Figure 9.12, we see that while a large fraction (around 70–75%) of the incoming radiation is transmitted through the atmosphere and reaches the planet's surface, only a small fraction (around 15–30%) of the upward radiation is transmitted through the atmosphere, with the rest heating it.

We can understand why so much more outgoing than incoming radiation is blocked by considering specific GHGs and the extent that they absorb radiation of different wavelengths. The extent of absorption for any particular gas can be characterized by its absorption coefficient $\alpha(\lambda)$, which is the fraction of the incident radiation that it scatters or absorbs as a function of wavelength. The absorption coefficient is restricted to lie between 0 and 1, with the latter meaning 100% absorption and zero transmission. Note that in general, the transmission coefficient is simply given by $\tau(\lambda) = 1 - \alpha(\lambda)$.

How is the total absorption curve found from the separate contributions? First, we must find the total transmitted through each of the five gases based on the product of the individual transmission coefficients, $\tau_{total}(\lambda)=\prod_j \tau_j=\prod_j (1-\alpha_j)$, which immediately leads to

$$\alpha_{total}(\lambda)=1-\prod_j \left(1-\alpha_j(\lambda)\right). \tag{9.16}$$

Note that according to Equation 9.16, if any one gas blocks the light completely ($\alpha_j(\lambda)=1$) at a certain wavelength, then obviously, we also have $\alpha_{total}(\lambda)=1$. It is clear from the total absorption coefficient graph in Figure 9.12 that there is only a fairly narrow window centered on 10 μm in where light from the surface of the planet can escape through the atmosphere. The rest of the (IR radiation is blocked mainly by absorption from water vapor and CO_2 at wavelengths much above or below 10 μm. Note that the dark-shaded region in Figure 9.12 showing the transmitted light is exactly the inverse of the window centered on 10 μm in the total absorption curve.

The relative contributions of each of the atmospheric GHGs in blocking the escape of heat does of course partly depend on how much of the gas is present in the atmosphere. Thus, just as you get warmer when you add an extra blanket, the planet gets warmer when more GHGs are present. A glance at Figure 9.12 will reveal that based on their absorption coefficients, the most important of the GHGs in the atmosphere is actually water vapor. However, water vapor is unlike all the other GHGs in that its maximum equilibrium concentration in the atmosphere is strictly based on surface temperature. Thus, additional water vapor above the saturation concentration added to the atmosphere will not stay there, but will condense out.

9.7.3 Climate Change Feedbacks

Water vapor is also fundamentally different from other GHGs, because when its concentration rises with increasing temperature, it can serve as a source of positive feedback. This means that when rising levels of other GHGs raise the planet's temperature, this raises the amount of water vapor in the atmosphere due to evaporation from the oceans, which increases the temperature still further. It is believed that the effect of water vapor in the atmosphere could actually double any increase caused by CO_2 and other GHGs. The situation, however, is still more complicated because water vapor can also result in negative feedback should it form low-lying clouds rather than remain in the atmosphere as vapor. Low-lying opaque clouds tend to block incoming sunlight, and they act to cool the planet, thus counteracting the effect of increasing levels of other GHGs. Although most atmospheric scientists believe that the positive feedbacks of water vapor outweigh the negative ones, the situation with respect to clouds is an important source of uncertainty—especially in regard to the effect of aerosols in fostering cloud formation. Part of the complexity of making a reliable model of how the atmosphere responds to rising GHGs is the large number of feedbacks that exist—some positive and some negative, each with a magnitude that may be highly uncertain (NAS, 2003). The following is a partial list of the more important feedbacks.

9.7.3.1 Positive Feedbacks

- *Water vapor* (already discussed).
- *High-altitude clouds*: If the extra water vapor from higher temperatures results in more thin high altitude clouds, these will block outgoing IR radiation more than they block incoming visible light, with the result being a trapping of the outgoing heat radiation and a further temperature rise.
- *Dissolved gas release*: Various GHGs, especially methane, exist trapped in very substantial quantities in a number of separate reservoirs, including the oceans, methane hydrates (also known as clathrates), and peat bogs. The largest of the bogs includes the permafrost bog in a million square kilometers of western Siberia. A rise in global temperatures would accelerate the melting of the permafrost and release large quantities of methane that would further contribute to a warming trend. Similarly, the oceans contain huge quantities of dissolved CO_2, whose equilibrium concentration decreases with increasing temperatures. Thus, as global temperatures rise, more CO_2 is released, and the temperature further rises.
- *Ice-albedo feedback*: The albedo of a surface is the reflection coefficient or the ratio of reflected radiation from the surface to incident radiation upon it. Of course, any surface must have an albedo in the range of 0 (perfect absorber) to 1.0 (perfect reflector). In the case of the Earth as a whole, the value of its albedo in relation to incoming solar radiation is 0.367. Since white ice obviously has a greater reflectivity than an ice-free surface, if rising temperatures should reduce the amount of the surface that is ice covered and decrease its albedo, the effect is to increase the absorption of the incoming radiation, which will increase the temperature still further.
- *Rainforest drying, fires, and desertification*: According to climate models, continental interiors in a warmer world should experience less rainfall and more droughts. A drying of the rain forests would lead to more fires and eventually their possible destruction and a spreading of the boundaries of the world's deserts. This would put immense stores of carbon (now tied up in vegetation) into the atmosphere and accelerate the warming.

9.7.3.2 Negative Feedbacks As we have noted, while positive feedbacks tend to drive the climate system further away from equilibrium and accelerate any warming trend, negative feedbacks have the opposite effect.

- *Blackbody radiation*: This source of negative feedback was alluded to earlier, although it was not specifically identified as such. It refers to the greater extent of outgoing radiation from a hotter Earth (according to the T^4 law), which acts to reduce the size of the temperature rise from what it would have been in the absence of any such increase in outgoing radiation.
- *Low altitude clouds*: If the extra water vapor from higher temperatures results in more low altitude clouds, these will block incoming visible radiation—reflecting some and absorbing some. The reflected radiation never reaches the Earth's surface, while the absorbed radiation causes the clouds to heat up a bit. The hotter clouds radiate the excess heat half downward and half upward, resulting in more radiation escaping to space than that reaching the surface, which reduces the temperature rise.
- *Fertilizing effect of* CO_2: Extra CO_2 in the atmosphere could stimulate plant growth, which would absorb the CO_2 and limit the extent of warming. Of course, the impact of rising CO_2 levels needs to be considered in conjunction with rising temperatures to decide if the net effect on plant growth is a positive one or, as many scientists believe, a negative one.
- *Spontaneous removal of* CO_2 *from atmosphere*: Various natural processes should lead to greater CO_2 absorption by the oceans as the atmospheric CO_2 levels increase. One such process is chemical weathering of rocks, and biological processes such as greater shell formation in the oceans is the other. However, both of these sources of negative feedback operate on extremely long timescales and are quite limited in their ability to restrain rising temperatures on a decadal or century-long timescale.

There are other positive and negative feedbacks in addition to the ones noted earlier, which should give you some idea of why modeling climate is very complex. Clearly, it is possible to overstate or understate the likely impact of the probable rise in future global temperatures if one does not do an objective assessment of the relative magnitudes of the positive and negative feedbacks.

9.7.4 Four Greenhouse Gases

Table 9.1 illustrates the extent to which four GHGs of concern in relation to anthropogenic climate change have increased over time. Even though, as noted, water vapor is the most important GHG of all, it has not been included in the table because its concentration is driven by the temperature, so it is not considered a source of anthropogenic warming.

Table 9.1 Four Important Greenhouse Gases and Their Atmospheric Concentrations in Parts per Million, Parts per Billion, or Parts per Trillion and the RF due to the Increase in Each One

GHG	Preindustrial	Current	Increase	Lifetime (Years)	RF (W/m^2)
CO_2	280 ppm	387 ppm	107 ppm	100 s	1.46
Methane	700 ppb	1745 ppb	1045 ppb	12	0.48
NO_2	270 ppb	314 ppb	44 ppb	114	0.15
CFC-12	0	533 ppt	533 ppt	10	0.17
Total					2.26

Note: ppb, parts per billion; ppm, parts per million; ppt, parts per trillion.

CONFIRMATION BIAS

When matters of judgment are involved, objectivity can be very tricky. Psychologists have confirmed that most people (scientists included) tend to favor information that confirms their preconceptions or hypotheses regardless of whether the information is true and look much more skeptically at contradictory information. Thus, when evaluating the relative importance of various feedbacks, positive and negative, it should not be surprising if climate change skeptics arrive at much lower estimates for projected future temperature rises by emphasizing the negative feedbacks, while minimizing the positive ones—while at the same time, some climate scientists, whom skeptics deride as global warming alarmists, might be guilty of doing the reverse.

Several points are worth mentioning in connection with this table. First, all but one of the gases were originally present before the industrial revolution began and before humans were on the scene, but they have significantly increased since that time, mainly due to not only combustion of fossil fuels, but also animal husbandry, agriculture, and deforestation. The most important of these GHGs in terms of its impact on climate is CO_2. The level of atmospheric CO_2 has steadily risen since the beginning of the industrial revolution, and it has risen even more steeply in recent decades—see Figure 9.13. Superimposed on the steady rise is an annual cycle due to times of year when the level falls due to photosynthesis and then later rises due to the decay of leaves and other debris. As shown in Table 9.1, not only is the CO_2 concentration greatest among all the GHGs listed, but the degree of its radiative forcing (RF) is also greatest—this being the extent of the energy imbalance it causes. Thus, an RF of 1.46 W/m^2 which the extra atmospheric CO_2 gives rise to is equivalent to having an extra 1.46 W/m^2 coming in from the sun on top of the 1000 W/m^2 now reaching the surface. Based on the RFs in Table 9.1, we see that CO_2 accounts for about 3/4 of the global warming potential (GWP) of all the GHGs. It is therefore worrisome that the concentration of this gas in the atmosphere continues to steadily rise.

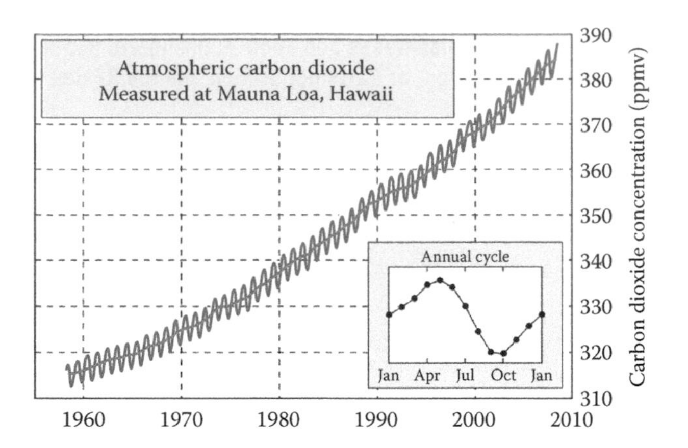

Figure 9.13 Atmospheric carbon dioxide concentrations as directly measured at Mauna Loa, Hawaii. This curve is known as the Keeling curve, after Charles David Keeling, the scientist who began making the measurements in 1960 and who first brought the problem of rising CO_2 levels and their contribution to climate change to the world's attention. (Courtesy of Sémhur, http://en.wikipedia.org/wiki/File:Mauna_Loa_Carbon_Dioxide-en.svg, image licensed under the Creative Commons Attribution-Share Alike 3.0 Unported, 2.5 Generic 2.0 Generic and 1.0 Generic license.)

9.7.5 Global Temperature Variation and Its Causes

The large majority of atmospheric scientists who have studied the matter are convinced that the rising trend in mean global surface temperatures seen since around 1960 is primarily due to the increased level of GHGs in the atmosphere, although for the variations seen in earlier decades, natural causes might be more important. We see that during the ensuing 30 years, the Earth's surface temperature has risen about 0.6°C or 0.2°C per decade on average. Were that trend to continue over the course of the twenty-first century, the result would be a rise in average global temperature rise of 2°C or 3.2°F (Figure 9.14).

What is the evidence that rising levels of GHGs are the main culprit for the warming, rather than some natural phenomenon? Climate scientists are able to fit the temperature variations using models that take into account a mix of natural and anthropogenic forcings. As can be seen in Figure 9.15. The model results fit the data reasonably well, and we see

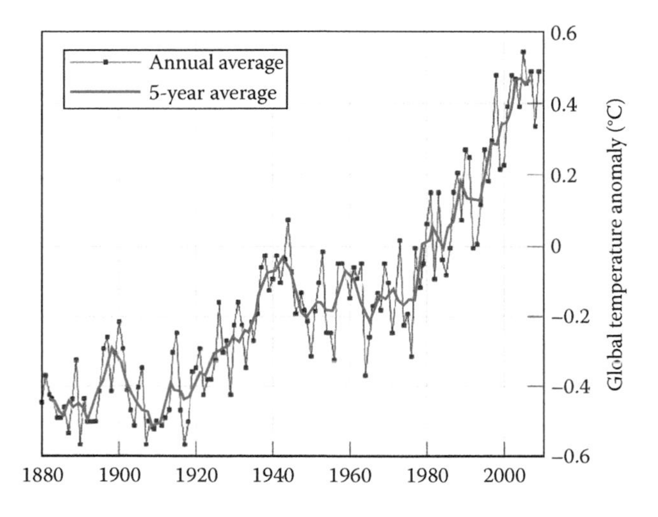

Figure 9.14 Mean global surface temperature (gray) and its 5 year moving average (green) over time. (Image courtesy of NASA, Washington, DC, is in the public domain.)

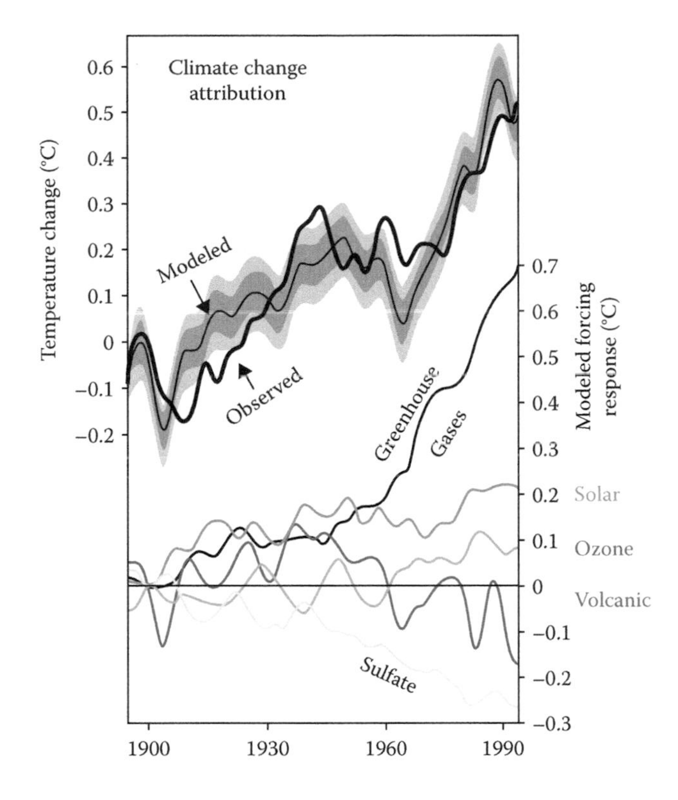

Figure 9.15 Observed and modeled temperature changes since 1900 with separate RFs shown from various causes, both human-made and natural. The gray range for the modeled temperatures indicates their range of variation. (Courtesy of Robert A. Rohde, http://en.wikipedia.org/wiki/File:Climate_Change_Attribution.png, image incorporated into the Global Warming Art project and licensed under the Creative Commons Attribution-Share Alike 3.0 Unported license.)

that the dominant forcings since around 1960 have been human caused, namely, rising levels of GHGs and sulfate aerosols, which have the effect of suppressing some of the warming, due to their impact on cloud formation and blocking of incoming solar radiation.

However, the evidence for the greenhouse effect being the cause of the warming rests not merely on models giving a reasonable fit to the observed data, but on distinct patterns to the warming. Specifically, warming due to the greenhouse effect is expected to cause more warming at night rather than daytime, more warming in winter than summer, and more warming in cold rather than hot areas of the globe—essentially greenhouse warming tends to even out temperature variations temporally and spatially. It is also expected to cause stratospheric cooling at the same time it causes the lower atmosphere to warm. These patterns—essentially a fingerprint of greenhouse warming—have all been confirmed by observation, and they would not occur if the warming were due instead to other causes such as an increase in the solar output.

9.7.6 Climate Projections for the Coming Century

What is likely to be the temperature increase over the course of the twenty-first century? Making such predictions can be hazardous,

MORE EXTREME WEATHER?

When "freakish" weather occurs, it can be very tempting to attribute it to the effects of climate change, which may or may not be appropriate. Climate scientists have, in fact, carefully studied the record of extreme weather since about 1950, and they find that some changes are well established, while others are much less so. For example, according to the IPCC (2012), it is "very likely" (>90%) that there have been fewer cold days and nights, but they have only "moderate confidence" (50%) that there have been more heat waves, droughts, and floods and "low confidence" (20%) that there have been more severe storms.

because we have no way of knowing how human behavior with respect to moving away from fossil fuels may be modified by a continued rise in global temperatures. The first decade of the twenty-first century, however, offers little cause for complacency that actions will replace rhetoric. In addition, it is quite possible that the actual rise in temperatures will be far worse than what a simple linear extrapolation over the last three decades suggests. In fact, according to a business-as-usual scenario for emissions over the course of the coming century, different groups of atmospheric scientists (each with their own sophisticated computer model) have produced results as indicated in Figure 9.16, and a 2°C rise in temperatures would be at the extreme lower range of those predictions.

> Making predictions is hard—especially about the future.
>
> Yoggi Berra, former baseball player and manager noted for his malapropisms

The best estimate range of predictions by the IPCC that has compiled such projections is between 1.8°C and 4.0°C for the coming century, although their full range is 1.1–6.4°C (2.0–11.5°F). The full range includes not only the range due to different models, but also varying assumptions of whether CO_2 emissions will be high or low. No one

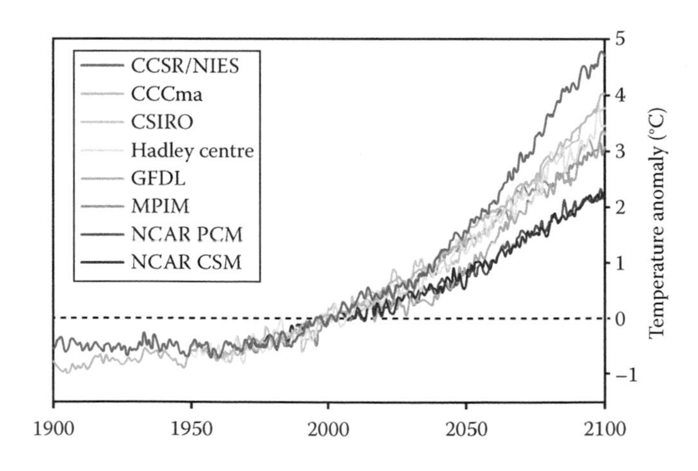

Figure 9.16 Projected rise in mean global temperatures under a "business as usual" emissions scenario, according to eight different computer models from different sources. (Courtesy of Wikimedia Foundation, San Francisco, California, http://en.wikipedia.org/wiki/File:Global_Warming_Predictions.png, image created for Global Warming Art and licensed under the Creative Commons Attribution-Share Alike 3.0 Unported license.)

knows exactly how catastrophic the results will actually be. A rise closer to the lower extreme of that range might not be a matter of grave concern and might even be beneficial on balance, but one closer to the upper limit would certainly be. Since many positive feedback processes exist, it is also possible that potential tipping points may exist in the climate system, and these have the potential to cause abrupt and catastrophic climate change.

9.7.7 Tipping Points in the Climate System

The concept of a tipping point is intuitively clear to most people. It refers to a situation where small disturbances to a system initially have only small effects, but eventually a point is reached where the system tips over to a different state—perhaps a radically different one. A very simple example would be that of a tall rectangular block resting on a horizontal surface, which is shaken back and forth. For some amplitude of the shaking, the surface the block will begin to rock on its base—dotted rectangle in Figure 9.17a—but it does not yet tip over. If the amplitude of the shaking increases further, a tipping point will be reached where the equilibrium state of the block is entirely different, as it has tipped over and is lying on its side.

One way a tipping point can occur is through positive feedbacks that drive the system further and further from its initial equilibrium state when an initial disturbance occurs. Gravity supplies this feedback for the block once the shaking amplitude has reached some critical point. (Before that point, the feedback of gravity was negative when the block merely rocked back and forth.) However, while positive feedback is necessary in order for a system have a tipping point, it is not a sufficient condition, since it is possible that other (negative) feedbacks might occur at some point that restore equilibrium and prevent the system from reaching a radically altered state. Moreover, the existence of a tipping point need not mean that the system undergoes a runaway effect that leads to a

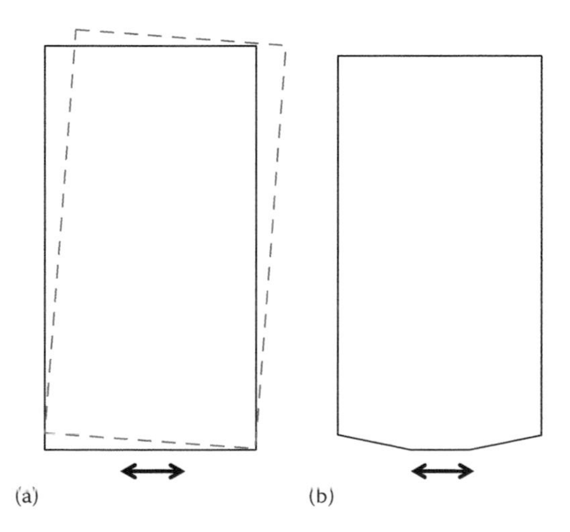

Figure 9.17 (a) Tall rectangular block resting on a horizontal surface, which is oscillating back and forth. (b) Tall block whose base is beveled on an oscillating surface.

catastrophic change in its state. For example, in Figure 9.17b, due to the beveled base of the block, it is possible that shaking might lead it to tip on to one of the two bevels (while remaining nearly upright) rather than tipping over entirely.

In the block example, it is possible to explicitly calculate, given its height-to-width ratio and the frequency of shaking, exactly what the critical amplitude of the shaking would need to be for it to tip over. It will not surprise you to learn that no such explicit calculation can be made for the climate system, which is far more complex than the simple block example. Nevertheless, even for the climate system, one can make some guesstimates over the kinds of possible tipping points that might exist based on the known positive feedbacks—bearing in mind that in a complex system like climate, there may also be other important unknown sources of positive feedback that could also be operating. One possible tipping point in the climate system that would be especially worrisome would be the melting of the Greenland and West Antarctic ice sheets. Were this to occur at a faster rate than is currently happening, i.e., in a warmer world, the climate could reach a point of no return. According to two 2014 studies, the melting of the West Antarctic ice sheet may have already past the point of no return. If these studies are correct, there will be an eventual sea level rise of 10 ft in the coming centuries.

RUNAWAY GREENHOUSE EFFECT?

Runaway greenhouse effects are not just hypothetical possibilities. Our neighboring planet Venus has an atmosphere of almost entirely CO_2, which is 92 times as dense as Earth. The planet is only 28% closer to the sun than Earth—certainly not enough of a difference to explain why temperatures on its surface are hot enough to melt lead (457°C). Scientists believe that at some point in its past, Venus experienced a runaway greenhouse effect that radically altered its atmosphere and boiled away any water that may have been present on its surface. Some climate scientists are convinced that this sort of runaway change has virtually no chance of being induced by anthropogenic activities, since the positive feedback effect from water vapor is well below what is needed to boil away Earth's oceans (Houghton, 2005). Other climate scientists disagree, but put the catastrophe so far in the future (2.5 billion years) that it is of no real concern (Kasting and Ackerman, 1986).

9.7.8 Categories of Positions in the Global Warming Debate

It is probably too simplistic to pretend that there are just two sides to the question of human-caused global warming; a more nuanced division would be in four categories: catastrophists, realists, skeptics, and deniers, with the preponderance of climate scientists falling in one of the first two categories. Realists are differentiated from catastrophists in that while they acknowledge that the consequences of climate change largely driven by human actions could be very dire, they are less certain that the worst-case scenarios will occur and that the imposition of immediate large cuts in emissions will make a significant difference in the outcome. The distinction between the skeptics and the deniers is that while the former believe that the extent of the warming and the evidence for it are overstated, the latter group believes it is all a hoax essentially manufactured by politically motivated grant-seeking scientists. A more refined characterization of the American public positions on global warming has been suggested by experts on the field of climate change communication (Leiserowitz, 2011). According to this 2011 study, 12% of citizens are alarmed, 27% are concerned, 25% are cautious, 10% are disengaged, 15% are doubtful, and 10% are dismissive.

9.7.9 Arguments of Global Warming Skeptics and Deniers

Those who are skeptics and deniers (or doubtful and dismissive) have raised numerous arguments for their positions. Here, we consider 10 of them that are not of an *ad hominem* nature (Table 9.2).

1. *Urban heat island effect*: The urban heat island effect is that more urbanized areas (where many weather stations are located) are expected to have higher temperatures, so that as time goes on and areas surrounding weather stations in these areas becomes more urbanized, there will be a rise in temperature in the record that is simply an artifact of where the weather stations are located. However, the scientists who compile the data around the globe take great pains to correct for this effect, which, in any case, is considered to be small. For example, according to a paper by

Table 9.2 Ten Arguments Raised by Skeptics and Deniers against Either Human-Caused Global Warming or the Validity of Global Warming Itself—Human-Caused or Otherwise

1. The rise in temperatures is due to the urban heat island effect.
2. Satellite data show cooling contradicting land-based measurements.
3. At midcentury, global temperatures fell even though CO_2 was increasing.
4. Global warming stopped in 1998.
5. There is no scientific consensus on the matter.
6. We cannot trust computer models.
7. We cannot even predict the weather next week.
8. Current global warming is just part of a natural cycle.
9. Ice core data show CO_2 levels following not leading temperature changes.
10. Warmer weather and higher CO_2 levels are beneficial.

Thomas Peterson (2003), "Contrary to generally accepted wisdom, no statistically significant impact of urbanization could be found in annual temperatures."

2. *Satellite data show cooling contradicting land-based measurements*: Unlike land-based measurements, which when suitably averaged can give a mean global temperature, satellite measurements based on the position of the peak of the blackbody spectrum cannot be influenced by the urban heat island effect or other selection biases, since they take a whole-Earth temperature at once. On the other hand, satellite measurements are also subject to a host of corrections, including combining the data from a host of different instruments, each of which may have their own calibration biases. At one time, there was a discrepancy between the satellite and land-based measurements, with the former indeed showing cooling. However, after errors were corrected, the satellite and land-based data turn out to be remarkably consistent in the degree of warming they show (Figure 9.18).
3. *At midcentury, global temperatures fell even though CO_2 was increasing*: As seen in Figure 9.14, mean global temperatures did indeed fall rather slightly during the four decades following 1940. Moreover, this was a period during which CO_2 levels rose. The explanation of this behavior hinges on recognizing that there are a variety of causes for global temperature changes, as recognized in Figure 9.15. In that figure, it can be clearly seen that while rising levels of GHGs did contribute a forcing term that caused temperatures to rise, another factor increasing levels of sulfate aerosols (due to greater emissions from burning fossil fuels) led to the opposite effect, since sulfate aerosols contribute both to cloud formation and to the phenomenon of global dimming—a reduction in the level of solar radiation reaching the surface. Apparently, the magnitude of the aerosol effect was slightly greater than the GHG effect during this period, which stopped being the case around 1980 when global temperatures resumed their rise.
4. *Global warming stopped in 1998*: As can be seen in Figure 9.18, the year 1998 does appear to have been the warmest year on record.

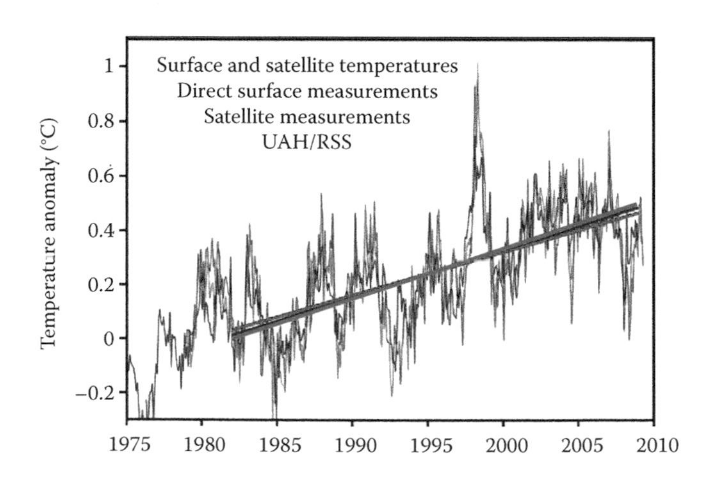

Figure 9.18 Comparison of land-based and satellite data on mean global temperature over a 35-year period, with trend lines for each data set. (Courtesy of Wikimedia Foundation, San Francisco, California, http://en.wikipedia.org/wiki/File:Satellite_Temperatures.png, image created for Global Warming Art and licensed under the Creative Commons Attribution-Share Alike 3.0 Unported license.)

Moreover, if one selected the time interval from say 2003–2010, the data shown would be best fit by a line having negative slope or declining temperatures. However, selecting those 2 years to look for a trend involves cherry picking the data, because they were picked in order to support a position. In general, when dealing with noisy data, one needs to look at a long enough period to see a trend, and over the full time period when satellite data have been available, it is clear that the trend shows rising temperatures as Figure 9.18 shows. In fact, the significant rise in temperatures for the years 2011–2015 (off the graph) leaves little doubt that the trend is a steady rise, with 2015 being so far the warmest year on record, according to National Aeronautics and Space Administration and National Oceanic and Atmospheric Administration data.

5. *There is no scientific consensus on the matter*: In science, *consensus* does not require that a view be unanimously accepted, since many ideas, including, for example, cold fusion, will be adhered to by some advocates long after the mainstream scientific community has come to regard the issue as settled.

   Some aspects of human-caused global warming are known with a very high degree of scientific certainty, such as the importance of rising levels of CO_2 as the chief contributor, while others, such as the matter of tipping points, are less certain. In any case, among climate scientists, surveys show that those in the categories of skeptics or deniers are among a small minority. Furthermore, the main conclusions of the IPCC reports have been endorsed by scientific academies in at least 19 nations, as well as many other prestigious scientific organizations, including the American Physical Society.

6. *We cannot trust computer models*: The computer models used in connection with climate are based on the laws of physics, and they are extremely sophisticated. They do, of course, require input data to run them, and their results can vary from one group to the next, but the variations tend to be fairly small if the inputs are the same. Most importantly, the models can be validated by comparing their predictions with observations. For example, each of the following model predictions has been found to be correct from direct observation:
   a. Stratospheric cooling accompanying surface warming
   b. Small energy imbalance between incoming solar radiation and outgoing IR radiation
   c. Specific short-lived temperature drops following major volcanic eruptions
   d. Larger degree of warming in the arctic region and, in general, in places and times when it is cold

   Although it would be wonderful if we had a number of duplicate Earths on which to do direct experiments on the level of warming that occurs with and without various levels of GHGs, unfortunately the experiment we are running on this planet is the only one we have.

7. *We cannot even predict the weather next week*: Predicting weather is an entirely different proposition from predicting future climate, which may be thought of as the average weather and where the random variations affecting weather have been smoothed out. A useful comparison might be between our relatively good ability to predict population trends in a nation based on demographic drivers, such as fertility rates, level of education of women, and their participation in the workforce versus the poor ability to predict the number of babies born in a particular hospital on a given weekend. Admittedly, however, both long-term climate change and long-term population growth are limited by the same sorts of uncertainties, namely, the impact of future national and international economic growth and government policies.
8. *Current global warming is just part of a natural cycle*: Natural causes, including variations in solar output and the Earth's orbital parameters are certainly part of the driver of climate change, as was the case for the ice ages, and in fact, they are taken into account by climate modelers. Nevertheless, given the ever-increasing levels of GHGs in the atmosphere emitted as a result of human actions, the natural causes have now become a small component of the total. There is no reason to believe that just because natural causes resulted in major climate changes in the distant past, what is happening today must also be natural. On the contrary, skeptics have no serious explanation as to why rising levels of human-caused GHGs, especially CO_2, will not cause further warming. The exact extent of the warming may be disputed, and this is acknowledged by the IPCC reports that provide uncertainty ranges for all their predictions, but it should also be noted that the skeptics rarely provide any uncertainty limits on their prediction of no warming.
9. *Ice core data show CO_2 levels following, not leading, temperature changes*: Atmospheric CO_2 levels can be both the effect and the cause of rising global temperatures. It is a cause through the greenhouse effect, and it is a result based on CO_2 dissolved in the oceans and methane dissolved in the tundra being released as the planet warms. There is, in fact, evidence that in ancient times, global temperature changes occurred naturally and were not driven by GHG variations as already noted, so that in those times, atmospheric CO_2 levels did occur after the temperature changes, not before. Today, however, CO_2 in the atmosphere functions as both a cause and an effect of warming. The fact that it is also an effect today means that oceanic release of more CO_2 is an important positive feedback that contributes to even greater warming driven by atmospheric GHGs.
10. *Warmer weather and higher CO_2 levels are beneficial*: Undoubtedly, there are some areas of the globe where the human inhabitants might welcome a warmer climate, although much depends on the extent of that warming. Nevertheless, given the magnitudes projected, not only for increases in temperature, but also for rising

sea levels, rising levels of ocean acidity, and rising numbers of extreme weather events, it is likely that there will be far more losers than winners as the planet continues to warm. Moreover, many developing nations are expected to be far more vulnerable to these changes than those in the developed world who may be able to more easily adapt to climate change. Finally, natural ecosystems and threatened species are even less able to adapt than humans. While some plant types may benefit from rising levels of CO_2, which has a fertilizing effect, the harm done by warmer temperatures could be more of a detriment to them. In general, while natural ecosystems can evolve to adapt to environmental changes, the timescale for such adaptations is far longer than the timescale over which large changes in climate are anticipated.

9.8 SUMMARY

In this chapter, we examined the nature and availability of solar radiation, which in part depends on the apparent motion of the sun across the sky at different locations and times of year. We also considered other factors that determine the amount of solar radiation that can be harvested, such as the tilt and orientation of a solar collector. In the final section, we examined the energy balance of the Earth and the role that the greenhouse effect plays in shifting that balance.

PROBLEMS

1. Estimate the power output of the sun in watts based on the amount of solar radiation reaching the Earth's surface per square meter, the radius of the Earth, and the distance to the sun.
2. Suppose that the demand for energy were to increase at the steady rate of 2% per year. Estimate how many years would need to elapse before humans used an amount of energy equal to all the solar energy striking the planet.
3. Estimate the number of photons per second that a 100 W bulb emits. Assume a wavelength in the middle of the visible spectrum.
4. Owls have excellent night vision. Assume that their eyes can detect a light intensity as small as 4.5×10^{-13} W/m^2. What is the minimum number of photons per second that an owl eye can detect if its pupil has a radius of $R = 7.5$ mm and the light has a wavelength of 503 nm?
5. Show that Wien's law follows from Equation 9.4.
6. Above what latitude are there 24 h of daylight on August 15?
7. Assume that at a given location and on a clear day, the irradiance incident on a solar panel having an area of 1 m^2 is well described by Equation 9.12 with the peak irradiance given by 800 W/m^2.

How much total solar energy is incident on the panel if there are 10 h of daylight from sunrise to sunset?

8. Equation 9.13 is based on the integrated solar irradiance over a hemisphere being equal to that falling normally on a disk having the radius of the Earth. Do the integration over a hemisphere to prove this to be true.
9. At a certain wavelength, the absorption coefficients of the four GHGs are 0.9, 0.5, 0.1, and 0.3. Find the total absorption coefficient at that wavelength.
10. According to Table 9.1, the RF of all the GHGs added since the start of the industrial revolution is 2.26 W/m^2. Assuming that the average emissivity of the Earth as a whole is 0.64, calculate the expected global rise in temperature associate with that RF, ignoring feedbacks. Hint: You will need to do a first-order Taylor expansion of $(T + \Delta T)^4$.
11. What are three defining characteristics of GHGs?
12. The GWP of a GHG can be defined by comparing the RF it creates per unit mass compared to CO_2, which is arbitrarily defined to have a GWP of 1.0. (a) Find the GWP of the gases listed in Table 9.1. (b) The GWP of any GHG depends on the absorption of IR by its molecules, the spectral location of the absorbing wavelengths, and the atmospheric lifetime of the GHG. Explain why each of these three factors contributes to the GWP.
13. Consider a rectangular block whose width-to-height ratio is x, which is shaken back and forth at a frequency ω. Calculate the minimum amplitude of shaking A for the block to tip over. Hint: In the accelerating reference frame of the oscillating block, a noninertial horizontal force acts on the block at its center of mass, which can be written as $F = m\omega^2 A$ when the block is at either end of its oscillation.
14. Some skeptics of human-caused climate change argue that ancient climates before humans existed changed by far larger amounts than at present, so how do we know that the natural causes responsible for those changes are not responsible for any changes going on today? Discuss the possible flaws in this argument.
15. When scientists examine data on air bubbles trapped in ice cores at various depths, they can obtain data showing the temperature and atmospheric CO_2 concentrations that existed hundreds of thousands of years ago. The two trends in time for the data sets for the two variables tend to be very similar. Explain why this is not convincing evidence that the variations in temperature were driven by changes in the CO_2 levels.
16. Which arguments of the skeptics of global warming, if any, do you think have the greatest validity? Explain.
17. Given the location and time of year listed in Figure 9.10, (a) verify that the number of daylight hours depicted there is roughly correct, (b) write an equation for the irradiance versus time, and

(c) integrate the equation to calculate the total energy incident on the solar collector on that day.

18. Show that above the Northern Hemisphere latitude 90° – δ_0, the sun never rises on the shortest day of the year. Your answer should be based on solving the appropriate formula or reasoning it out using a diagram, rather than using Figure 9.8.
19. Can you suggest a reason for the annual cycle seen in Figure 9.13 for the atmospheric CO_2 concentration? Hint: The fraction of land area in the Northern Hemisphere is far more than that in the Southern Hemisphere.
20. Suppose you are located on the edge of a time zone. If you have a sundial in your backyard on what days of the year would it be most accurate?
21. Explain carefully how satellites are used to determine the Earth's average temperature.
22. If the Earth's albedo were to decrease by 1% (from melting ice), estimate the impact this would have on the average temperature of the Earth.
23. Consider three hypothetical GHGs whose absorption coefficients per gigatons can be expressed as Gaussian functions of the wavelength $\alpha_j(\lambda) = A_j e^{-((\lambda - b_j)/c_j)^2}$, where $A_j = 0.1$, 0.1, and 0.1; $b_j = 10$, 10, and 50 μm; and $c_j = 2.0$, 0.1, and 1.0 for $j = 1$, 2, and 3. Which of these three would be most harmful and which would be least harmful if added to the current atmosphere—see Figure 9.12.
24. Presently, 179 days elapse after the autumnal equinox before the next vernal equinox, but only 186 days after that before the next autumnal equinox. Explain this discrepancy.
25. A major concern about global warming is the associated rise in sea level largely due to thermal expansion of the oceans. It has been established that during the last half century, the sea level has risen about 1.7 mm per year, and during that time, the upper 700 m or so of the oceans has warmed by about 0.1°C per decade. Calculate how the measured sea level rise compares with the rise expected from thermal expansion. What other causes might contribute to sea level rise besides thermal expansion? Why is the melting of floating sea ice not a cause?
26. The solar radiation reaching the top of the Earth's atmosphere is roughly that of a blackbody spectrum corresponding to a temperature of 5500 K. Create an Excel spreadsheet to integrate the solar spectrum, so as to find out what fraction of the solar spectrum lies in the visible region of 450–700 nm.
27. Find some arguments climate change skeptics use to support their position beyond those listed and critically examine them in a one-page essay.
28. What is the average available solar radiant power available to the entire United States? (Hint: The average insolation for the United States is 177/m^2 and, total area 3.615×10^6 mi^2.) About how

much solar energy is available every year? What is the equivalent number of BTUs?

29. Assume that we were to convert 1/500th of the available solar radiant energy of the total US area to usable form assuming 100% efficiency. What state would be closest in area to be required to be completely covered with solar panels? What state area would be required if you had only 15% efficiency?

REFERENCES

Houghton, J. (2005) Global warming, *Rep. Prog. Phys.*, 68(6), 1343–1403, http://www.iop.org/EJ/abstract/0034-4885/68/6/R02/ (Accessed August 26, 2009).

IPCC (Intergovernmental Panel on Climate Change) (2012) *Special Report on Managing the Risks of Extreme Events and Disasters to Advance Climate Change Adaptation*, IPCC, Geneva, http://www.ipcc-wg2.gov/SREX/ (Accessed March 2012).

Kasting, J. F., and T. P. Ackerman (1986) Climatic consequences of very high CO_2 levels in earth's early atmosphere, *Science*, 234, 1383–1385.

Leiserowitz, A., E. Maibach, C. Roser-Renouf, and N. Smith (2011) *Global Warming's Six Americas*, http://environment.yale.edu/climate/files/SixAmericasMay2011.pdf.

NAS (National Academy of Sciences) (2003) *Understanding Climate Change Feedbacks*, National Academies Press, Washington, DC, http://www.nap.edu/catalog/10850.html.

Peterson, T. (2003) Assessment of urban versus rural in situ surface temperatures in the contiguous USA: No difference found, *J. Clim.*, 16, 2941–2959.

Solar Thermal

10.1 INTRODUCTION

The two primary ways of harvesting solar energy use either solar collectors that convert the incident solar radiation into heat or photovoltaic (PV) cells that convert incident solar radiation into electricity. This chapter considers the first of these methods. Solar thermal energy can serve a wide variety of applications from heating homes to cooking food to generating electricity. It is only feasible, however, to use solar thermal energy to generate electricity in a centralized manner in large power stations, whereas solar PV can also be used by individual homeowners and businesses for that purpose. Electricity generation using solar thermal is in fact among the fastest-growing renewable energy applications. Thus, while only 600 MW was in use worldwide in 2009, there are projects under development for an additional 14,000 MW. While solar thermal may not be suitable for use by homeowners to generate electricity, it is ideally suited for use in space heating and hot water heating, which collectively amount to 60% of the average homeowner's energy bill. That figure is for American homes, but it probably applies equally well to other developed nations having a similar climate. Hot water and space heating require that water be heated only to low or moderate temperatures, and this is relatively easy to do using simple solar thermal systems that do not require any concentration of the sun's energy, which is necessary for electricity generation (Figure 10.1).

SOME KEY IDEAS FROM HEAT TRANSFER THEORY

1. Heat never flows spontaneously from a cold object to a hot one, which is one version of the second law of thermodynamics.
2. The maximum possible efficiency for creating useful work in a heat engine (where heat flows between hot and cold temperature reservoirs) is given by the limiting value discovered by Carnot, $e_{Carnot} = 1 - (T_C/T_H)$, which is another version of the second law.
3. The power p associated with heat flow through an object or from a hot object to its cooler surroundings can be described in terms of the object's thermal resistance R and the relevant temperature difference ΔT, according to $p = \Delta T/R$, which is the thermal version of Ohm's law for electricity $i = \Delta V/R$. Like electrical resistance, thermal resistance R is usually a function of temperature, but the variation of R with T may be small in many practical cases.
4. If an object loses heat by several different mechanisms in parallel, e.g., conduction, convection, and radiation, each of these mechanisms has a specific resistance (see the appendix to the chapter for details).

CONTENTS

5. When several parallel mechanisms are involved, the object's net resistance R is found by adding these separate resistances in parallel: $1/R = 1/R_1 + 1/R_2 + 1/R_3$.
6. If an object loses heat that passes through several layers in sequence, their resistances must be added in series: $R = R_1 + R_2 + R_3$.
7. A closely related quantity to R is $r = RA$, where A is the relevant surface area across which the heat flows. Physically r is the thermal resistance of a unit area ($A = 1$), and is known as the r-value of a particular material. A layer's insulating ability is based on its r-value.

10.2 SOLAR WATER-HEATING SYSTEMS

Solar hot water heaters (SHWs) represent a particularly cost-effective means of reducing one's energy costs, and in some nations, the majority of homes use them. For example, SWH is widely used in Greece, Turkey, Israel, Australia, Japan, Austria, and China, but much less so in the United States—a nation that could also greatly benefit from their use. Consider, for example, that the payback time from energy savings or natural gas costs and state-by-state tax incentives is about 4 years. Currently, on a per capita basis, Israel is the world leader in the use of solar thermal hot water systems. In fact, 85% of Israeli households now use solar thermal systems, which constitute 3% of Israeli national energy consumption. In absolute numbers, China is the world leader, and that country uses 80% of all new solar hot water systems coming on the market worldwide—with much room for continued expansion, as only around 30 million Chinese households as yet have one. In part, the popularity of solar hot water heating in China is a consequence of their very low subsidized cost (about $200),

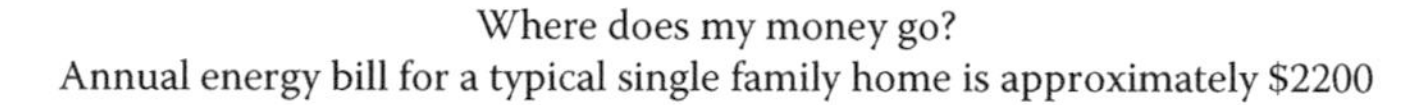

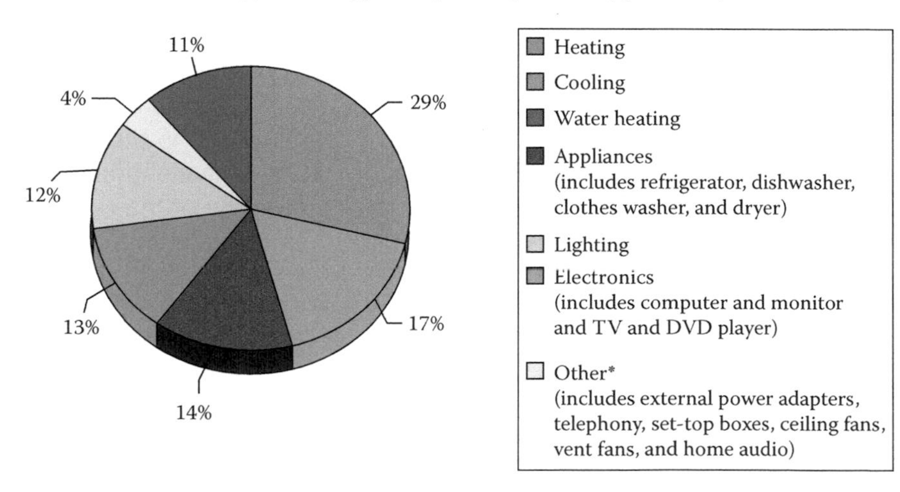

Figure 10.1 Energy expenditures of the average US homeowner. (Courtesy of DOE, Washington, DC.)

which is perhaps a fifth of the cost in the United States where usage of such systems is much lower (0.5% of all new systems worldwide).

SWHs have been around for over a century, and the technology is robust, mature, and relatively straightforward. The degree of complexity and cost of SWH in part depend on a nation's climate, since more sophisticated (costly) systems tend to be needed in colder climates. Fortunately, there are independent certification agencies in the United States and other countries that evaluate the many systems on the market and allow homeowners to assess the virtues of various systems relative to their needs. A decision to replace an existing system with SWH depends on many factors, especially economics. Replacing a hot water heater that has many years of life left with a SWH that saves energy but results in only a modest dollar savings per month would make neither economic nor environmental sense—considering the *embodied energy* represented in the existing unit.

10.3 FLAT-PLATE COLLECTORS

A typical commercial-grade flat-plate collector is shown in Figure 10.2 in a cutaway view. The dark-colored solar-absorbing surface or plate is covered here by a sheet of glass that allows the incident shortwave solar radiation to easily enter. However, the cover tends to trap the longwave radiation emitted from the heated absorber—essentially using the greenhouse effect. The tubing shown, which carries a fluid, needs to be in good thermal contact with the dark-colored metal plate that absorbs the solar radiation. To cut down on thermal losses and achieve high efficiency, the plate rests on a thick insulating layer that reduces heat loss out the underside.

Even though the purpose of the solar collector is to heat water, the fluid piped through the tubing is often not water but antifreeze in the case of typical two-loop systems, where the heat in the primary loop is transferred to a hot water tank without making direct contact with the pure water in the secondary loop. The antifreeze is important to prevent

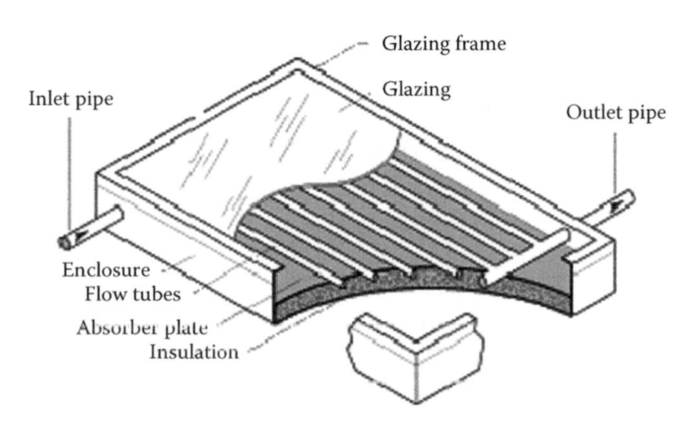

Figure 10.2 Cutaway view of a typical flat-plate solar collector. (Image courtesy of the U.S. Department of Energy, Washington, DC, is in the public domain.)

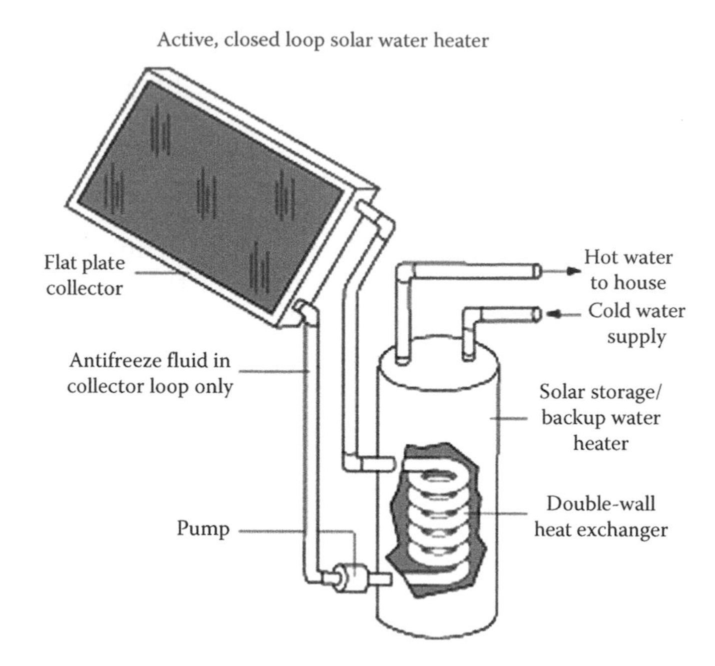

Figure 10.3 Solar collector and other components of a two-loop "active" water heating system.

damage to the collector in climates where temperatures can drop below freezing. As shown in Figure 10.3, the secondary loop carries hot water to the boiler, which serves as a backup source of hot water when there is not enough provided by solar radiation. The complete system also includes a pump to drive the water flow through the system, and a controller that shuts off the flow to the collector when the sun is not shining. Without such a controller, the system would lose heat to its surroundings at night, as the collector sheds heat to its surroundings rather than absorbing it.

Unlike Figure 10.2, some collectors lack a cover, but they are necessarily less efficient because of their greater heat loss, and they cannot heat water to high temperatures. Simple flat-plate collectors with covers can potentially achieve temperatures as high as 180°C—much higher than what is needed for domestic hot water, but the actual temperature can be regulated by having the controller adjust the flow rate through the tubes. In fact, in such collectors, the flow rate of water through the tubes must be adjusted to match the solar insolation in order to avoid overheating and serious damage to the collector.

10.4 EVACUATED COLLECTORS

Should it be desired to reduce thermal losses and achieve high efficiency, a common method is to use evacuated collectors (vacuum between the plate and cover), since this eliminates both conductive and convective heat losses above the heated collector plate. However, evacuated collectors of the flat plate variety can be problematic because their structural ability to withstand a vacuum is poor. Thus, evacuated flat-plate collectors tend to leak (admit outside air) over time. Generally, evacuated collectors have a cylindrical geometry, which has greater structural strength.

An example of an evacuated tube solar collector is shown in Figure 10.4. For this collector, 21 parallel evacuated tubes are heated by the sun and transport heat to the water to a storage tank.

In many cases, the evacuated tubes constitute heat pipes in which heat is transported very rapidly up the tube by means of a phase change. The basic principle of the operation of a heat pipe is illustrated in Figure 10.5—where a highly foreshortened picture of an evacuated tube is depicted. Solar radiation incident on the vacuum-sealed pipe causes a volatile liquid that it contains to vaporize. The hot vapor spontaneously rises to the top of the pipe into the *canula*—a bulb at its upper end—which is in thermal contact with water in the heat exchanger.

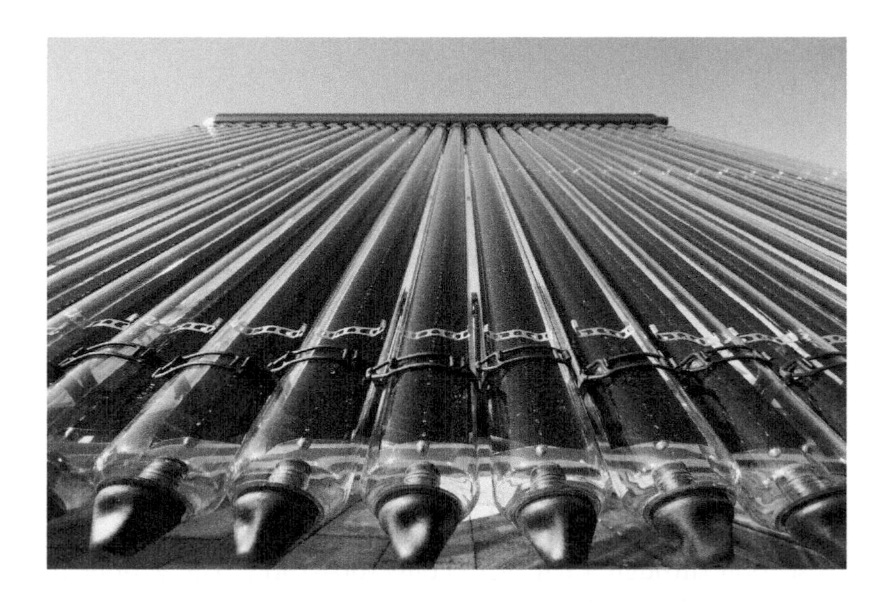

Figure 10.4 An evacuated tube solar water heater.

Figure 10.5 Operation of a heat pipe. A volatile liquid at the bottom is vaporized after absorbing solar heat, and the rising vapor then condenses in the canula at the top when it transfers heat in a heat exchanger. The liquid then flows back down completing the cycle.

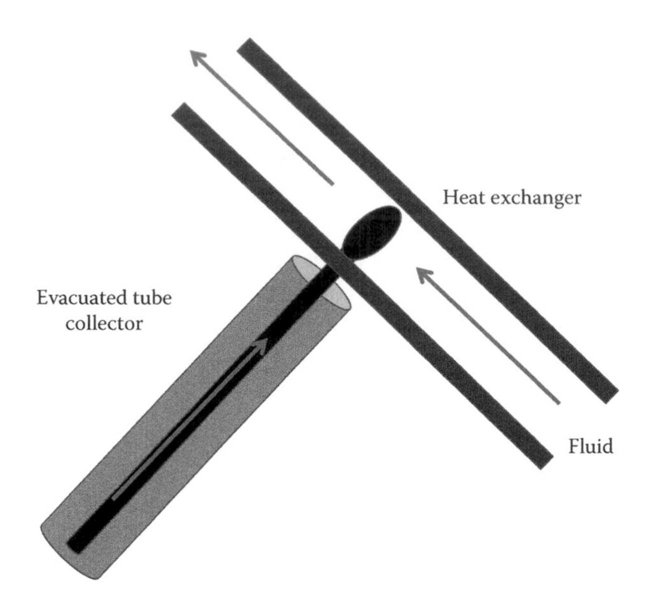

Figure 10.6 Evacuated tube with its canula inside a heat exchanger.

It is in the canula that the vapor condenses back to the volatile liquid. The liquid then descends back down the heat pipe due to gravity completing the cycle. Heat pipes transfer heat more rapidly than any other type of device (Figure 10.6).

10.5 COLLECTOR AND SYSTEM EFFICIENCY

In order to determine the efficiency of a collector, we need to consider the energy balance equation, i.e., power in = power out. Figure 10.7—a cross section through a flat-plate collector—shows the various inflows and outflows of thermal power. There is only one power inflow G_0A, where G_0 is the incident solar power per unit area, of which the fraction ρ is reflected off the front cover. For simplicity, we assume normal incidence. The outflows include the thermal power lost from the collector

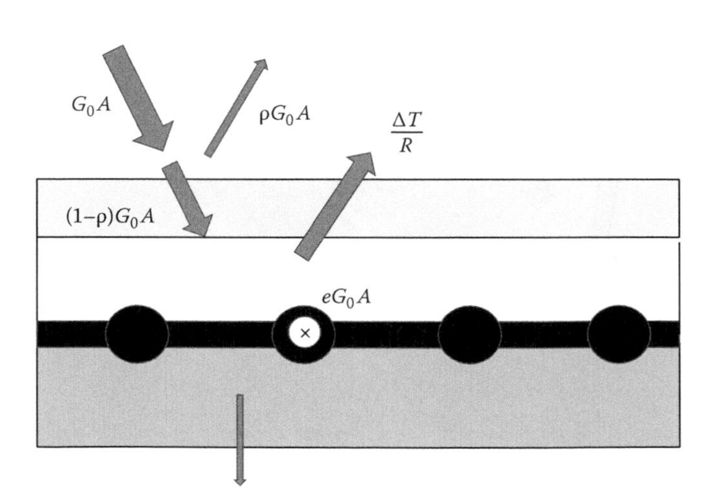

Figure 10.7 Energy flows for a flat plate solar thermal collector.

$\Delta T/R$, where R is its thermal resistance. The loss is primarily out the top cover and is mainly radiative and convective. An additional outflow is the power carried away by the fluid down the tubes—indicated by the X shown in one of the tubes standing for a flow perpendicular to the plane of the page. Since this power is the useful output of the collector, we write it as the input power times the efficiency eG_0A.

Thus, we set the power inflow to the sum of the three outflows to obtain

$$G_0A = eG_0A + \rho G_0A + \frac{\Delta T}{R}, \tag{10.1}$$

where ΔT is the temperature excess above the ambient temperature. Solving for the efficiency, we find

$$e = (1 - \rho) - \frac{\Delta T}{RAG_0}. \tag{10.2}$$

If the collector net thermal resistance is independent of temperature, the graph of efficiency e versus the temperature excess ΔT is linear with y-intercept $1 - \rho$ and slope $-1/RAG_0 = -1/rG_0$, where r is the collector r-value assuming that r does not vary with ΔT.

It should be clear from Equation 10.2 and Figure 10.8 that

1. In areas having higher solar irradiance, the rate of the falloff of efficiency e as the collector temperature rises is more gradual; that is, the curves have a shallower slope.
2. In areas having lower solar irradiance, we need to use more sophisticated (and expensive) collectors having a higher r-value to prevent the falloff in efficiency with rising collector temperature. Thus, as seen in Figure 10.8, only the evacuated tube collector yields a suitable ΔT for the low irradiance case.
3. The stagnation temperature, which is the highest temperature achievable with a collector (the x-intercept in Figure 10.8), increases as either the irradiance or the collector r-value increases.
4. At the stagnation temperature, the efficiency is zero, since the useful power transferred to the plate (and, hence, the fluid) just equals the loss to the environment. The stagnation temperature is also the steady-state operating temperature of a solar collector if the irradiance stays constant.
5. Zero efficiency in the sense used here certainly does not mean that all the power is wasted—only that there is no further rise in plate and fluid temperature if the irradiance stays constant.

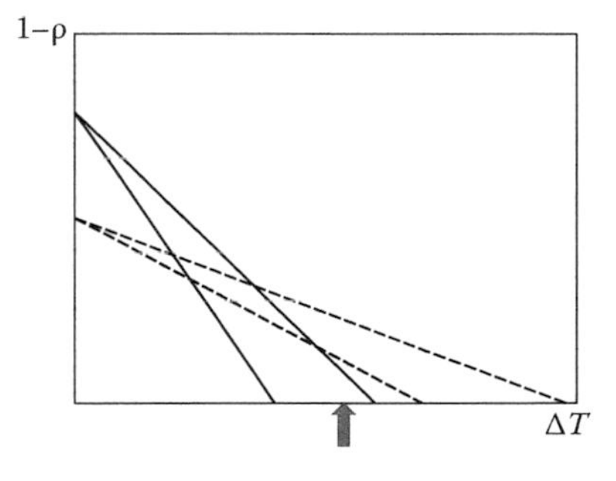

Figure 10.8 Ideal efficiency curves versus temperature excess above ambient temperature ΔT. The two solid curves are for a flat-plate collector and they correspond to a high and low value of the irradiance G_0. The two dashed curves are for a more expensive evacuated tube collector. The black arrow shows the ΔT needed for producing hot water at some desired temperature. We have assumed here that the collector resistance is constant independent of ΔT.

More realistic efficiency curves are nonlinear and take into account a variation in thermal resistance with ΔT. If we only consider variations in R to first order in ΔT, Equation 10.2 could be written as

$$e = (1-\rho) - \frac{\alpha_1 \Delta T + \alpha_2 \Delta T^2}{G_0}, \tag{10.3}$$

where the constants α_1 and α_2 depend on the properties of the particular collector and can be empirically determined by measuring the collector efficiency when it operates at three different temperatures for known G_0 values. Some of the independent agencies that evaluate and certify various solar collectors report values of the constants ρ, α_1, α_2, as well as many other parameters. Figure 10.9 illustrates a comparison of quadratic efficiency curves for two collectors of the flat-plate and evacuated tube types. It can be seen that the flat-plate collector (the dotted line) initially has the higher efficiency. However, the flat-plate efficiency falls more sharply as the temperature excess above the ambient value rises, so that its stagnation temperature is less than half that of the evacuated tube collector. Both of these characteristic differences between the two collector types are easy to understand. The higher initial efficiency for the flat-plate collector arises because this design results in less reflectance (a smaller ρ value)—indicated by its larger y-intercept $1 - \rho$. Flat-plate collectors, however, lose efficiency faster than evacuated tube collectors as T rises since they have greater thermal losses (smaller r-values). Given these two differences, less expensive flat-plate collectors tend to be best in moderate climates with a great deal of solar radiation, while evacuated tube collectors tend to be best in colder regions. Note that the evacuated tube collector in Figure 10.9 has the superior performance only if the temperature excess above ambient exceeds about 75°C.

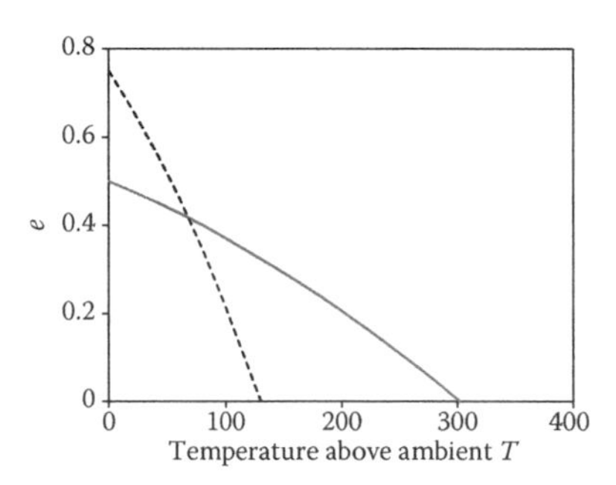

Figure 10.9 Comparison of the efficiency *e* versus temperature for two hypothetical collectors assuming the same solar irradiance; solid line is for an evacuated tube collector and the dotted line for a flat-plate collector.

In order to understand why evacuated tube collectors initially have a much lower efficiency, we need to explain their higher reflectance. As shown in Figure 10.10, when light is incident at glancing angles on a cylindrical surface, a much greater fraction is reflected than when it is closer to being normally incident. Thus, the reflection coefficient is a function of incident angle, i.e., $\rho(\theta)$. The specific function $\rho(\theta)$ depends on the material properties and the polarization of the incident light. Although the details are of no interest here, it should be clear that as incident rays approach grazing incidence $\theta = 90°$, the fraction that is reflected approaches 100% or $\rho(90°) = 1$. It is for this reason that the reflection coefficient ρ averaged over the surface of the tube tends to be greater for an evacuated tube collector than for a flat-plate collector.

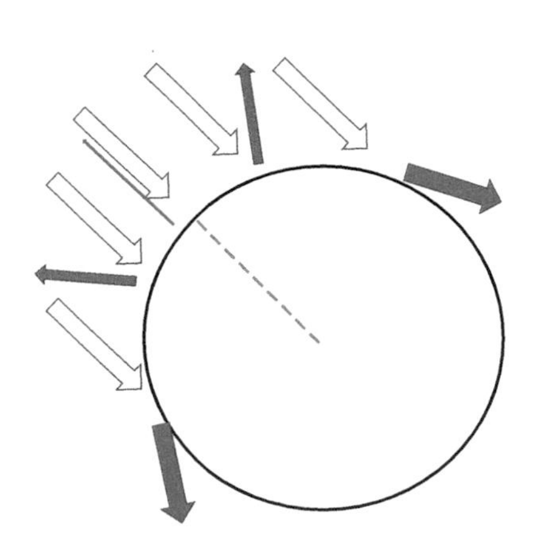

Figure 10.10 Reflected amplitudes when parallel rays from the sun are incident at various points on a tube collector. The reflectance for different incident angles is represented by the width of the dark arrows. Note that more light is reflected, as the angle becomes closer to grazing incidence.

10.5.1 Example 1: A Flat-Plate versus an Evacuated Tube Collector

Consider a flat-plate collector A having the parameters $\rho_A = 0.362$, $\alpha_{1A} = -4.26$, and $\alpha_{2A} = -0.03$ and an evacuated tube collector B with $\rho_B = 0.689$, $\alpha_{1B} = -1.17$, and $\alpha_{2B} = -0.01$, with all values given in SI units. Find the temperature above which the evacuated tube collector has the superior efficiency (a) in a location where the irradiance is 1000 W/m^2 and (b) in a location where it is only 250 W/m^2.

Solution

Based on Equation 10.3, the temperature excess above ambient where the two collectors have the same efficiency is given by

$$e_B - e_A = 0 = (\rho_A - \rho_B) + \frac{(\alpha_{1A} - \alpha_{1B})\Delta T}{G_0} + \frac{(\alpha_{2A} - \alpha_{2B})\Delta T^2}{G_0},$$

so that when the irradiance is 1000 W/m^2 we find

$$0 = -0.327 - 0.00309\Delta T - 0.00002\Delta T^2. \tag{10.4}$$

Solving Equation 10.4 yields two solutions: $\Delta T = 453°C$ and $\Delta T = -144°C$ of which only the first is physically meaningful. In case b where the irradiance is only 250 W/m^2, the quadratic to be solved becomes

$$0 = -0.327 - 0.01236\Delta T - 0.00008\Delta T^2, \tag{10.5}$$

for which the positive solution is $\Delta T = 355°C$. Thus, we see that in case b, the evacuated tube collector has a superior performance at water temperatures that exceed ambient temperature by a significantly smaller amount than that at the higher irradiance location. This dependence on the local irradiance shows the greater importance of using evacuated tube collectors in locations where the irradiance is low, at least when collector efficiency is a concern. On the other hand, it is not until extremely high temperatures (at least in the case of our hypothetical example) that the evacuated tube collector has a superior efficiency, so that for ordinary domestic hot water heating, the less expensive flat-plate collector would be the better choice. A more typical pair of collector parameters would probably give lower values of ΔT more like the example displayed in Figure 10.9.

10.6 THERMAL LOSSES IN PIPES

Although the thermal losses in a solar collector can be significant, the losses in carrying heated water through pipes to a storage tank can be even greater, depending, of course, on the parameters of the pipe and the fluid flow rate. It would hardly make sense to invest in a superefficient collector if the bulk of the heat transported through the pipes is lost before the fluid ever reaches the storage tank. Thus, heat losses through the walls of the pipes are an important factor in judging overall system efficiency.

As a heated fluid flows through a pipe of length L, its temperature steadily drops as long as it is higher than the surrounding ambient temperature T_a. Consider a short length of pipe dx for which the temperature drop is dT. If a unit length of the pipe has a thermal resistance R_1, then the thermal resistance for a length dx is R_1/dx. In this case, the loss of power through the walls of the pipe can be expressed as

$$p = -\frac{T - T_a}{R_1/dx}. \tag{10.6}$$

In the steady-state situation, this loss equals the power loss from the fluid in traveling the length dx, $\rho \dot{V} c \, dT$, yielding upon rearranging terms

$$-\frac{\rho R_1 \dot{V} c \, dT}{T - T_a} = dx. \tag{10.7}$$

Upon integrating both sides over the entire length of the pipe, we obtain an expression for how the temperature excess above ambient varies along the pipe:

$$\Delta T = \Delta T_0 e^{-x/\beta}, \tag{10.8}$$

where $\beta = \rho c R_1 \dot{V}$ is the distance along the pipe for the temperature excess above ambient temperature to decline by the factor $e^{-1} \approx 0.38$ of its initial value when entering the pipe. By the end of the pipe, the decline will be by a factor f, where

$$f = e^{-L/\rho c R_1 \dot{V}}. \tag{10.9}$$

Clearly, based on Equation 10.9, long lengths of uninsulated pipes can have prohibitively high thermal losses if the thermal resistance or if the volume flow rate through them is too low.

10.6.1 Example 2: Thermal Loss in a Pipe

What is the minimum fluid flow rate through a half-inch diameter copper pipe of length 10 m that will result in no more than a 2% drop in the temperature excess above ambient temperature? Given the data available on the web, the thermal resistance of a 1 m length of such pipe is roughly $R_1 \approx 1.0$ m °C/W.

Solution

Substitution in Equation 10.9 yields $0.02 = e^{-L/\rho c R \dot{V}}$, which can be solved for the volume flow rate to find

$$\dot{V} = -L/\rho c R_1 \ln(0.2) = 1.48 \times 10^{-5} \text{ m}^3/\text{s} = 0.23 \text{ gal/min}.$$

10.7 WATER TANKS AND THERMAL CAPACITANCE

In most solar hot water heating applications, the heated water must be stored in a hot water tank for later use. Without a storage tank, the water flowing through the solar collector would need to be heated in a single pass through it, while with a tank, one can circulate the water through the collector multiple times and achieve much higher temperatures. Obviously, in order to achieve significant heating with only a single pass, the water flow rate through the collector would need to be very slow, but then as seen in the previous section, the thermal losses in the pipes would be far too great.

A hot water storage tank is one example of a thermal capacitor, which is quite analogous to an electrical capacitor. Thus, just as electrical capacitance is defined as the charge added divided by the voltage difference across the plates, thermal capacitance is the thermal energy added to a body divided by the temperature difference between the body and its surroundings. Based on this definition, it is clear that thermal capacitance is $C = cm$—the product of the specific heat and mass of the body. Let us now consider the thermal capacitance (also known as *thermal mass*) of a hot water storage tank to investigate its losses.

Suppose that the tank is being heated by a heater supplying a power p_h and that it is losing heat to the environment at a rate of $p_{Loss} = (T - T_a)/R$. The energy balance equation for the water in the tank may be written as $p_h = (d/dt)(mc\Delta T) - p_{Loss}$ or, therefore,

$$p_h = \dot{m}c(T_{in} - T_a) + mc\frac{dT}{dt} - \frac{T - T_a}{R}. \quad (10.10)$$

10.7.1 Example 3: Insulating a Hot Water Tank to Reduce Thermal Losses

Consider a hot water tank whose capacity is 50 US gallons (0.189 m^3), which requires on the average 25 W of electrical power continuously to maintain the water temperature 50°C = 50 K above ambient temperature even when no water is drawn from it, i.e., $\dot{m} = dm/dt = 0$. (a) How much would be gained by adding an insulating blanket around the heater having an *R*-10 insulation? (b) If the heater were turned off at night, how long would it take for the temperature of the water in the tank to drop by 5°C?

Solution

Part a:

Since $\dot{m} - dT/dt = 0$ here, Equation 10.22 becomes $p_h = (T - T_a)/R$, or 25 W = 50 K/R, giving $R = 2.0$ K/W for the thermal resistance of the tank. Insulation having an *r*-value of *R*-10 has $r = 10$ ft^2 °F h/(BTU), which, in the more sensible SI units, is equivalent to $r = 3.51\ m^2\ K/W$. In order to find the thermal resistance of the blanket used to insulate the tank, we can use the relation $R = r/A$, provided we know the surface area of the tank. From commercially available products, we find that a 50 gal tank typically has a height and diameter of 1.51 and 0.56 m, respectively, which yields a lateral surface area of $0.371\ m^2$. If we ignore thermal losses from the top and bottom of the tank, we find that the resistance of the insulating blanket is, therefore, $R = r/A = 3.52/0.371 = 9.48$ K/W. Thus, adding the resistance of the blanket in series with that of the tank itself increases the thermal resistance from $R = 2.0$ K/W to $R = 11.48$ K/W, which means the power drain to maintain the temperature of the water would be reduced nearly sixfold.

Solution

Part b:

Since $\dot{m} = p_b = 0$ here, Equation 10.10 becomes $mc(dT/dt) = -(T - T_a)/R$, which yields as its solution

$$T - T_a = (T_0 - T_a)e^{-t/Rmc}. \tag{10.11}$$

If the initial excess above ambient temperature is 50°C and the water temperature drops by 5°C, we can use Equation 10.11 to solve for the time, giving

$$t = Rmc \ln\left(\frac{T_0 - T_a}{T - T_a}\right) = (2.0)(189)(4186)\ln(50/45) = 1.67 \times 10^5\ \text{s} = 46\ \text{h}.$$

WHEN SHOULD YOU INSULATE YOUR HOT WATER HEATER?

The DOE recommends insulating your hot water tank with an inexpensive precut blanket unless the *r*-value of the heater already exceeds *R*-24. If the manufacturer does not indicate the *r*-value of the unit, a simple test would be to touch the surface of the heater. If it feels warm, then you should insulate it. Typically, standby losses can be reduced by 25–45% by taking this simple action.

10.8 PASSIVE SOLAR HOT WATER SYSTEM

Suppose it were possible to supply a significant fraction of your domestic hot water needs using an inexpensive, easy-to-install solar system that had no moving parts and even functioned during electrical blackouts. Those are the properties of passive SWHs—at least under certain favorable situations. By definition, passive solar heaters circulate the fluid through the system without the aid of pumps and rely instead on differences in fluid density that depend on temperature. One very simple type of passive system is that

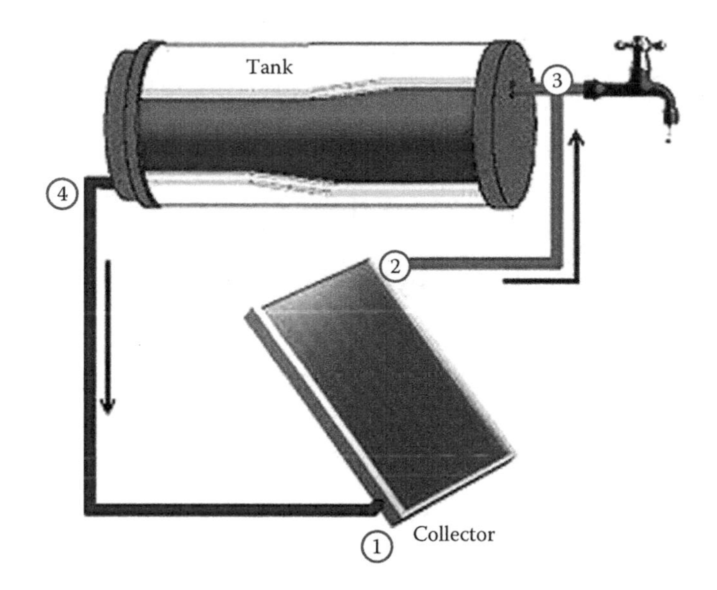

Figure 10.11 A passive solar hot water heater based on the thermosiphon principle. (Image courtesy of the U.S. Department of Energy, Washington, DC, is in the public domain.)

relying on the thermosiphon principle. A simple thermosiphon system was depicted in Figure 10.4, and it is shown schematically in Figure 10.11.

In all passive systems, the water tank must be elevated above the collector. While it is not shown in Figure 10.11, most systems also have a backup tank (which need not be above the collector) to deal with hot water needs when solar heating is insufficient. Note that here the fluid flowing through the collector is in a separate loop from that which is stored in the tank and the two come into contact in a heat exchanger. In this way, one can have antifreeze flow through the collector and potable water in the loop through the tank itself. Four key numbered points in the loop through the solar collector and heat exchanger can be seen in Figure 10.11:

1. Where the cold fluid enters the bottom of the solar collector
2. Where the warmed fluid exits the top of the collector having been heated by the sun
3. Where it enters the top of the tank or the heat exchanger
4. Where cold fluid exits the bottom of the heat exchanger

The flow spontaneously continues in a counterclockwise direction as long as there is sufficient sun, and it is driven by the higher weight of the colder (and denser) fluid prior to entering the solar collector compared to that of the warmer fluid that flows inside it. The pressure driving the flow due to the different weights of the cold and hot waters (the thermosiphon pressure) may be expressed as

$$P_{\text{therm}} = \bar{\rho}_C g\Delta y - \bar{\rho}_H g\Delta y, \qquad (10.12)$$

where Δy is the vertical height difference between points 1 and 3—the highest and lowest points in the loop—and $\bar{\rho}_C$ and $\bar{\rho}_H$ are the average densities of fluid on the cold and hot sides of the loop.

Based on the definition of the average density, Equation 10.12 can be expressed more elegantly as a loop integral performed in a clockwise direction (opposite to actual flow):

$$P_{\text{therm}} = \oint \rho g \, dy. \tag{10.13}$$

Since the variation in density of the fluid is only slight, we may use a linear approximation $\rho = \rho_0(1 + \beta(T - T_0))$, where T_0 is some arbitrary reference temperature and $\beta \equiv (1/\rho_0)(d\rho/dT)$ is the volumetric expansion coefficient. Thus, the thermosiphon pressure may be written as

$$P_{\text{therm}} = g\rho_0 \oint 1 + \beta(T - T_0) \, dy = \beta g \rho_0 \oint T \, dy. \tag{10.14}$$

The second equality in Equation 10.14 is based on the fact that the integral of a constant around any closed path is zero, Another useful quantity known as the thermosiphon head y_{therm} can be defined in terms of the thermosiphon pressure based on $P_{\text{therm}} = \rho_0 g y_{\text{therm}}$, so that

$$y_{\text{therm}} = \beta \oint T \, dy. \tag{10.15}$$

It is the value of the thermosiphon head that will determine the volume flow rate of fluid through a particular passive solar collector. It can be shown, for example, using fluid dynamics that the flow velocity u of a fluid having a dynamic viscosity ν that is propelled through a cylindrical tube of diameter D and length L by a pressure $\Delta P = \rho g y_{\text{therm}}$ is

$$u = \frac{D^2 \Delta P}{32 L \nu} = \frac{\rho g D^2 y_{\text{therm}}}{32 L \nu}. \tag{10.16}$$

10.8.1 Example 4: Finding the Fluid Flow Rate in a Thermosiphon Hot Water Heater

Consider a passive SHW of the type shown in Figure 10.11. Suppose that the points labeled 2 and 4 are elevated 0.5 higher than 1 and point 3 is 0.1 m higher still. Assume that the dynamic viscosity of the fluid has the value of $\nu = 8.9 \times 10^{-4}$ Pa/s and that the hot water tank atop the solar collector holds $V = 100$ L or 0.1 m^3. Further suppose that on entrance to the solar collector, the fluid is at a temperature of 20°C and that it exits the top at 23°C. Obviously, many passes through the system will be required to produce domestic hot water here. (a) Calculate the thermosiphon head, (b) the volume flow rate through the system if the water as it rises through the collector goes through 16 parallel tubes each having an inner diameter of 3.16 cm and a length of 1.0 m, and finally (c) how long it would take for one tankful of water to pass through the collector.

Solution

The volumetric expansion coefficient for water is given by $\beta = 0.000207°C^{-1}$. As shown in Figure 10.12, the closed loop area is simply the area inside the quadrilateral, which is 0.9 m °C, so that from Equation 10.15, we can find that the thermosiphon head is a mere 0.00019 m or 0.19 mm—which would not be expected to drive the flow very rapidly! In fact, from Equation 10.16, we find that $u = 0.065$ m/s = 6.5 cm/s. In order to find the volume flow rate of the water through the 16 tubes, we use $\dot{V} = 16Au = 4\pi D^2 u = 8.2 \times 10^{-4}$ m^3/s. Finally, we can find how long it takes for one tankful of water to pass through the collector using $t = V/\dot{V} - 122s = 2.03$ min. The shortness of this time might seem surprising in light of the slow flow speed, but recall that there are 16 large-diameter pipes feeding the tank.

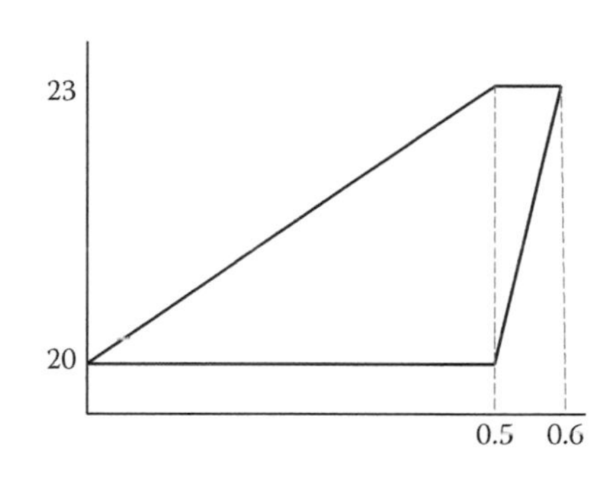

Figure 10.12 Plot of *T* versus *y* for Example 4.

This section on passive solar water heaters began by enumerating some of their advantages, so it is worthwhile to note that they have some limitations or disadvantages as well. As noted earlier, the solar storage tank needs to be at a higher elevation than the collector, which usually means placement of a heavy water-filled tank on the roof that could cause problems. In addition, passive systems tend to be less efficient than active systems, since in an active system, one can control the fluid flow rate through the collector to give the best performance suited for the solar irradiance—a faster flow when there is full sun. In a passive system, given the very slow flow rate (see preceding example), the collector may reach very high temperatures at which the efficiency of the system drops. Finally, contrary to what was stated earlier about no moving parts in passive systems, they generally do include a controller that stops the flow through the collector after the sun goes down. Without this active element, the fluid would circulate through the system in the opposite direction, losing heat to the environment at such times.

10.9 SWIMMING POOL HEATING

One final SWH application we shall consider is that of solar heating of outdoor swimming pools. Its inclusion here is not a reflection of the intrinsic importance of this application, but rather because it represents one of the simplest and most cost-effective ways to use solar energy, if your home happens to have a swimming pool. In fact, the payback time for replacing an electric swimming pool heater with a solar system can be as short as 67 days—less than any other solar application (Figure 10.13).

In the case of a swimming pool, there is no need for a separate hot water storage tank, since the pool itself satisfies this function. Moreover, the thermal losses are primarily from the pool itself, and this is what must be compensated for by the solar collectors, which need not be very sophisticated in view of the relatively low temperature above ambient that is needed. Another factor that makes this solar thermal application relatively easy is that at the time of year when there is least solar heating (winter), homeowners are rather unlikely to want to use their pool, except perhaps for ice skating—not recommended!

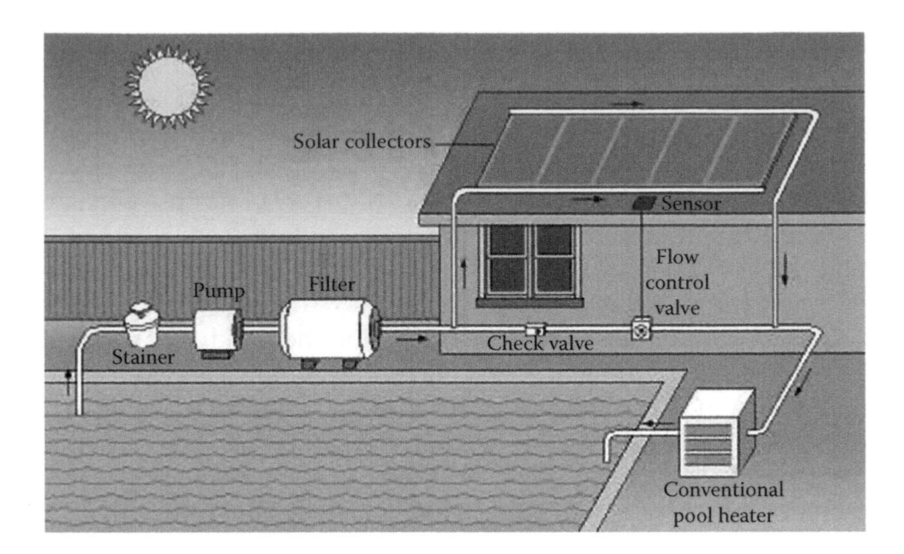

Figure 10.13 A solar swimming pool heating system. (Image courtesy of the U.S. Department of Energy, Washington, DC, is in the public domain.)

The effectiveness of the pool itself as a solar collector depends on the color or darkness of the tiles on the bottom, the depth of the water, and, most importantly, whether the pool is covered at night, which can reduce losses by as much as 50%. Unless the climate is very warm, however, an additional solar collector (besides the pool itself) will be needed whose collection area will strongly depend on climate. Fortunately, online computer programs exist that can be used to size the collector.

10.10 SPACE HEATING AND COOLING

Space heating and cooling of buildings represent the largest single-energy expenditure for the homeowner. For example, in the United States, it represents on average a quarter of the energy used in commercial buildings and nearly half for residences. As with solar hot water heating, space heating and cooling can be of the active or passive type—although it is not uncommon for both types to work in tandem. Passive solar elements are usually incorporated into the original design of a building, although retrofitting is sometimes possible. They do not require the kinds of mechanical or electrical devices that are part of an active system, although they can certainly include solar collectors. Although it is more common to think of solar in connection with home heating, it can also facilitate cooling and ventilation. For example, the solar chimney that has been used since the days of ancient Rome (and still in use in some warm climates) is one such application. When the chimney warms due to the sun, air inside it is heated causing an updraft that generates a continuous flow as the less dense air rises and is replaced by air sucked in from the building, assuming that air can continually flow in through open windows or cracks. Later, we shall consider a distinct application of huge solar chimneys for electric power generation.

A building relying on passive solar heating should incorporate five key elements—other elements may be present, but these five are essential.

According to the DOE, the following five elements constitute a *complete* passive solar home design—each performs a separate function, but all five must work together for the design to be successful:

1. *Aperture (collector)*: The large glass (window) area through which sunlight enters the building. Typically, the aperture(s) should face within no more than 30° of true south and should not be shaded by other buildings or trees from 9 a.m. to 3 p.m. each day during the heating season.
2. *Absorber*: The hard, darkened surface of the storage element. This surface—which could be that of a masonry wall, floor, or partition or that of a water container—sits in the direct path of sunlight. Sunlight hits the surface and is absorbed as heat.
3. *Thermal mass*: The materials that retain or store the heat produced by sunlight. The difference between the absorber and thermal mass, although they often form the same wall or floor, is that the absorber is an exposed surface whereas thermal mass is the material below or behind that surface.
4. *Distribution*: The method by which solar heat circulates from the collection and storage points to different areas of the house. A strictly passive design will only use the three natural heat transfer modes—conduction, convection, and radiation—to circulate heat. In some applications, however, fans, ducts, and blowers may help with the distribution of heat through the house.
5. *Control*: Roof overhangs can be used to shade the aperture area during summer months. Other elements that control underheating and/or overheating include electronic sensing devices, such as a differential thermostat that signals a fan to turn on; operable vents and dampers that allow or restrict heat flow; low-emissivity blinds; and awnings (Figure 10.14).

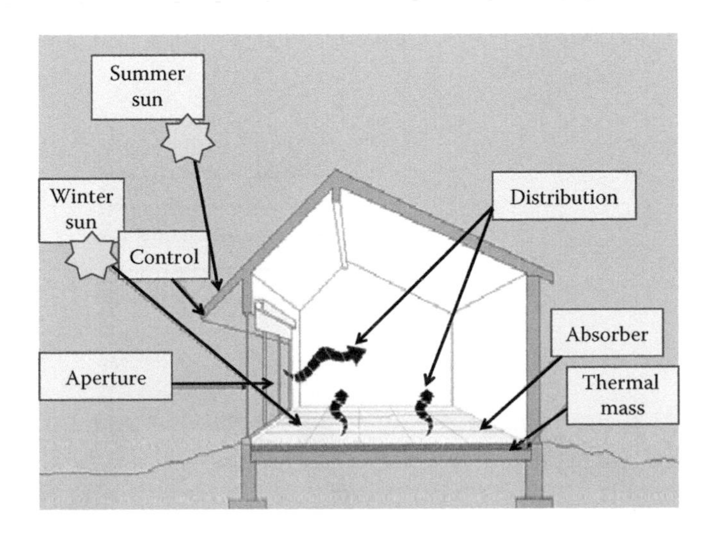

Figure 10.14 Five elements of passive solar design at work in a passive solar building design. (Courtesy of DOE, Washington, DC.)

10.11 THREE APPLICATIONS WELL SUITED FOR DEVELOPING NATIONS

Many forms of renewable energy are well suited for use in developing nations, because of the inadequacy or outright unavailability of electricity and their ability to be implemented on a small scale using locally available materials. However, solar thermal is uniquely well suited because sunshine, unlike hydropower or wind, is ubiquitous, particularly in the tropical regions where many developed nations lie. In addition, the three applications included here all serve to improve the status of human health—especially that of women who are the primary caregivers in developing nations.

10.11.1 Crop Drying

If agricultural products (especially grains) are not dried prior to transport, insects and fungi will make them unusable. Drying must occur within a few days of harvest. On a worldwide basis, more than 70% of all major crop diseases are caused by fungi. In Nigeria, for example, a majority of the crops and grain harvests are lost to fungal and microbial attacks. In many rural areas, conventional sources of energy are absent, so standard drying methods relying on active dryers cannot be used. A passive solar crop dryer usually relies on one of two methods—it either exposes the spread out grains to the solar radiation, where the water they contain is evaporated. Or, alternatively, an airflow is created using a solar chimney type of effect, where the grains are stacked on thin permeable layers so that a solar-driven updraft continuously flows through them and the same purpose is achieved. A simple design of the latter type is illustrated in Figure 10.15. One needs to size the solar collector in order to achieve a particular evaporation rate, since the amount of water evaporated per second cannot exceed $\dot{m} = eG_0A/L$, where e is the solar collector efficiency, A is its area, and L is the latent heat of vaporization. You can easily verify this formula based on the units of all quantities.

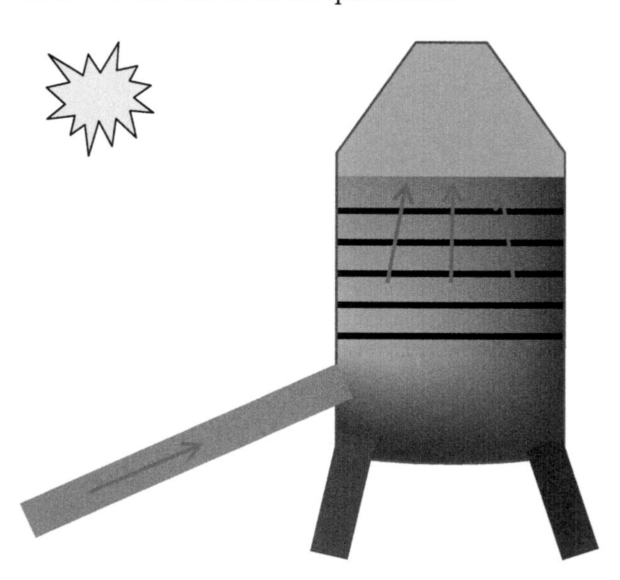

Figure 10.15 A crop dryer relying on the solar chimney principle. Air drawn in through a sloped solar collector flows upward through layers of crops on trays in the silo.

10.11.2 Water Purification

While insect-borne crop diseases may be the bane of much agricultural production in developing nations, unsanitary drinking water is a source of many serious diseases. Simple, low-technology filtration methods can remove much of the problem—especially particulates and bacteria. Solar power, however, can deal with these problems through the simple process of distillation. The solar still is also a useful survival technique to keep in mind the next time you get stranded on a sunny island without access to freshwater (Figure 10.16).

In this simple design, a transparent plastic sheet is stretched over a container or puddle of brackish water with a small container placed in the middle. A rock indents the plastic sheet, and the solar heating under it causes the water to evaporate, condense on the underside of the sheet, and drip into the small container.

Alternatively, there is an even simpler way to use solar energy to disinfect small quantities of water known as the SODIS method (for solar water disinfection), which requires only a clean transparent bottle and sunshine. In this method—discovered by Aftim Acra at the American University of Beirut in the early 1980s, and recommended by the World Health Organization—a sealed bottle containing contaminated water is simply exposed to the sun for 6 h, allowing ultraviolet (UV) radiation to kill diarrhea-generating pathogens. The method does not work, however, if the water is cloudy, if the sky is overcast, or if the bottles are made of glass or plastic that block UV light. Nevertheless, the method has proven so successful at reducing water-borne disease that the Swiss Federal Institute of Aquatic Science and Technology (Eawag) coordinates SODIS projects in 33 developing countries.

(a)

(b)

Figure 10.16 (a) A simple solar still. (b) SODIS project in Indonesia. (Courtesy of SODIS Eawag, http://en.wikipedia.org/wiki/File:Indonesia-sodis-gross.jpg, image licensed under the Creative Commons Attribution 3.0 Unported license.)

Figure 10.17 Woman in rural Ghana standing next to a solar cooker.

10.11.3 Solar Cooking

Normally, biomass is considered a form of renewable energy, because crops may be continually replanted and trees can as well. However, in many parts of the developing world, wood is used for cooking, and there is no replanting, as entire regions are stripped bare, and people (usually women and children) are required to travel longer and longer distances to gather firewood. In some Indian villages, women spent 2 h per trip to gather firewood. After the Indian government adopted forest protection policies to prevent deforestation, the trips decreased from 5 to 2 h (Agarwal, 2001). It is not just a matter of wasted time for many women, since traveling long distances in hostile terrain can wear you out and subject you to many hazards including disease, attack from animals, and rape—especially for women in a refugee camp. In addition, but certainly of less importance, cooking food over an open fire using gathered wood makes use of only 5% of the energy released, which is done by around half the world's population and is a significant waste of energy.

Simple solar cookers can avoid deforestation as well as the waste of time, energy, and lost and impoverished lives associated with long trips to gather firewood. The basic box-type cooker has insulated sides and a transparent cover on top and can reach temperatures of up to 100°C. More complex concentrating versions using mirrors to reflect and concentrate on the solar energy can attain temperatures as high as 350°C (Figure 10.17).

10.12 ELECTRICITY GENERATION

Solar thermal collectors used to generate electricity tend to rely on concentrating the solar radiation using lenses or mirrors. Concentrating collectors are able to heat water to the high temperatures needed to create high-pressure steam used to drive turbines connected to generators.

Moreover, the higher the temperature, the better the achievement, given limits imposed by the Carnot theorem. Therefore, an important consideration is the degree of concentration that can be achieved, as this is closely related to temperature. The two basic geometries for concentrating solar radiation rely on concentrating it either in one dimension or in two dimensions using either mirrors or lenses.

10.12.1 Concentration Ratio and Temperature

Concentrating collectors must track the sun because they can only make use of the direct solar radiation, in contrast to other types of collectors. The solar tracking obviously needs to be done in two dimensions for collectors that concentrate radiation in two dimensions, but only in one dimension if the concentration is done one dimensionally. An example of the latter variety would be solar collectors comprising mirrors in the shape of long parabolic troughs—see Figure 10.18. Here, as long as the angle of the trough is properly tilted to match the sun's elevation angle above the horizon, the sun's rays will be focused onto the long water-filled tube that is located at the focal axis of the parabolic cylindrical mirrors. Solar collectors that concentrate in two dimensions, such as those of the parabolic dish variety, also need to track the hour angle of the sun as well as its elevation angle in order to create an image of the sun near the focal point of the paraboloid. Given that the sun is not a point source, its rays cannot be concentrated into a smaller area than that occupied by its image, which implies the existence of an upper limit to the degree of concentration.

The concentration ratio X of a collector is defined as the ratio of the area of the aperture of the collector itself, i.e., the planar projected area on which

Figure 10.18 Concentrating solar collector of the parabolic trough type. (Image courtesy of the U.S. Department of Energy, Washington, DC, is in the public domain.)

the solar radiation falls, to that of the much smaller area of the image of the sun (formed by a mirror or lens). Thus, in the case of a collector that concentrates in two dimensions, such as a parabolic dish, we have

$$X_{2D} = \left(\frac{d_a}{d_i}\right)^2, \tag{10.17}$$

while for one where the concentration is only in one dimension, such as a parabolic trough, we have the smaller concentration ratio of

$$X_{1D} = \left(\frac{d_a}{d_i}\right). \tag{10.18}$$

Note that in this latter case, there is no true image of the sun formed on the tube—just a long narrow band along its length.

It is useful to find the highest possible values for both the one- and two-dimensional cases, since those relate to the highest temperatures that can be achieved. The solar radiation reaching Earth, a distance r from the sun, originates from the solar surface, which is a sphere of radius r_S. By the time that radiation reaches Earth, the inverse square law implies that it spreads over a sphere whose area is larger by the factor $(r/r_e)^2 = 1/\sin^2\theta_S \approx 1/\theta_S^2$, where θ_S is half the angle subtended by the sun in the sky or approximately 1/213 rad. It is now easy to show that the maximum possible concentration ratio on Earth for two-dimensional concentrators is

$$X_{2Dmax} = \frac{1}{\sin^2\theta_S} = 45{,}300. \tag{10.19}$$

Suppose we imagine that it were possible that by using some clever arrangement of mirrors, we could concentrate the solar radiation reaching Earth by a greater amount and focus that concentrated power onto a perfectly absorbing plate. In that case, the power per unit area radiated by the plate would be greater than that radiated by the surface of the sun itself, and by the Stefan–Boltzmann law, its temperature would have to be higher than the surface of the sun. This situation would be in violation of the second law of thermodynamics, because it would entail a spontaneous flow of radiant energy from the sun to an object at a higher temperature than the imagined surface. Moreover, by comparing Equations 10.17 and 10.18, it is obvious that the maximum concentration ratio for a one-dimensional concentrator must be the square root of the right-hand side of Equation 10.19, or

$$X_{1Dmax} = \frac{1}{\sin\theta_S} = 213. \tag{10.20}$$

In practice, real one- and two-dimensional collectors can attain concentrations of perhaps only half these values. Collectors fall short of the limits due to a variety of reasons, including imperfect optics; tracking errors; reflection coefficients short of 100%; and some component of indirect radiation, due to atmospheric scattering or clouds. Our main performance measure of a concentrating solar collector system is in fact not the concentration ratio anyway, which determines the maximum temperature, but rather the overall system efficiency. The two components of the system efficiency are those associated with the collector itself, e_{Coll} and the efficiency for converting a hot fluid into useful electrical power, with the latter limited to the Carnot efficiency

$$e_{\text{Carnot}} = 1 - \frac{T_C}{T_H}.$$

The collector efficiency may be found by assuming that the concentrated solar power per unit area XG_0 consists of a useful portion of the incident power (that transferred to a heated fluid) plus the wasted power radiated by the tube containing the fluid at temperature T_H. If we assume an emissivity $\varepsilon = 1$ for the wasted radiated part, this condition yields for the power per unit area $XG_0 = e_{\text{Coll}}G_0 + \sigma T_H^4$, which, when solved for the collector efficiency, gives $e_{\text{Coll}} = 1 - (\sigma T_H^4/X)$. Thus, for the overall maximum efficiency, we take the product of the collector and Carnot efficiencies to find

$$e_{\text{Max}} = \left(1 - \frac{\sigma T_H^4}{GX}\right)\left(1 - \frac{T_C}{T_H}\right). \quad (10.21)$$

Let us assume that the cold temperature in Equation 10.21 equals the ambient temperature taken to be 300 K and that we use a range of possible concentration ratios X so as to find the family of curves shown in Figure 10.19 based on Equation 10.21 for efficiency versus hot source temperature.

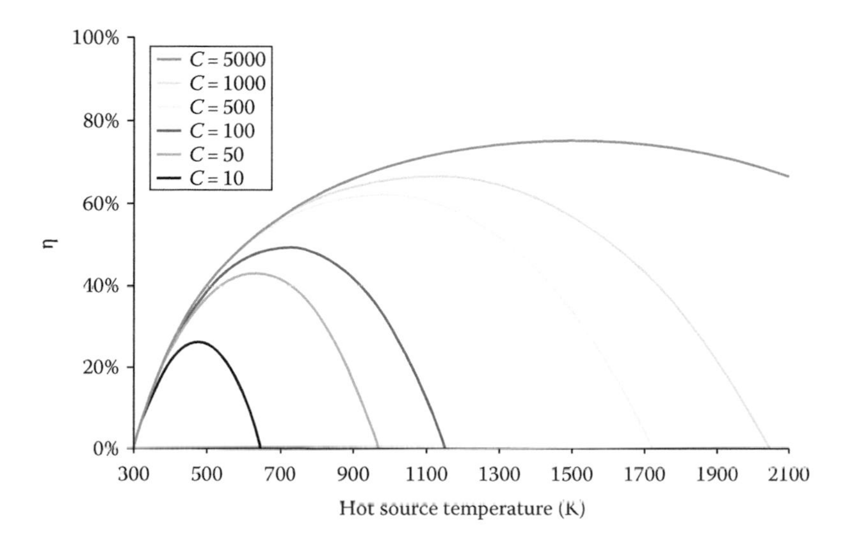

Figure 10.19 Maximum efficiency versus hot source temperature for a variety of concentration ratios from 10 to 5000. (Courtesy of Wikimedia Foundation, San Francisco, California, http://en.wikipedia.org/wiki/File:Solar_concentration_efficiency.png, image licensed under the Creative Commons Attribution-Share Alike 3.0 Unported license.)

It is clear from the form of Equation 10.21 or the plots in Figure 10.19 that there will be an optimum temperature T_H to run a system at for any given concentration ratio, which can be easily found by differentiating Equation 10.21 and setting the derivative to zero. The location of that optimum temperature for highest efficiency results from the competition of the two factors in Equation 10.21—the first, a monotonically declining collector efficiency and, the second, a monotonically rising thermodynamic efficiency for creating work from heat. The temperature of the fluid can be most easily controlled and optimized by varying the fluid flow rate through the tubes for collectors of the parabolic trough variety—the faster the flow rate, the more rapidly heat is carried away, and the lower will be the temperature.

10.12.2 Parabolic Dish Systems and "Power Towers"

One way to concentrate solar power in two dimensions is to use a parabolic dish (paraboloid), which is used in applications such as some large solar cookers. Recall that in this case, it is needed to do solar tracking in two dimensions, which for mechanical reasons makes it impractical to use a very large dish. If we want to use the two-dimensional concentrators to generate electricity by means of heating a fluid, there are several options. One option involves having a large number of parabolic dishes, each of which heats the fluid in small containers placed near the focal point of each dish. However, this method requires piping the heated fluid from the separate collectors to a central storage location before using it to create the steam needed to power a generator. A more elegant option, however, makes use of the power tower such as that depicted in Figure 10.20.

In the power tower design, a large number of mirrors on the ground track the sun so as to reflect the sunlight onto the top portion of the central

Figure 10.20 A power tower solar concentrator system. (Image courtesy of the U.S. Department of Energy, Washington, DC, is in the public domain.)

tower. Inside that top region resides the fluid being heated by the concentrated solar radiation. A high concentration ratio and temperature can be obtained because the concentration is in two dimensions. Rather than use water as the fluid to be heated, many power towers use liquid fluoride molten salts heated to temperatures as high as 800°C. The molten salt is stored in a large underground tank, and its heat is used to power steam-driven generators. This form of energy storage allows the system to generate electrical power day and night, given the high thermal mass and temperatures—thereby avoiding or, at least, reducing the problem of intermittency associated with solar energy.

The technology for solar power towers is less advanced than that for systems relying on trough collectors, but their higher temperatures and efficiencies offer certain advantages. At the moment, the high capital costs of these technologies make them noncompetitive with conventional energy sources. At the rate at which the technology is improving, however, it is likely that their current cost of roughly U.S.$0.25/kWh could be halved in the coming decades. In fact, there are those working in the industry who are far more optimistic, believing that costs of $0.05/kWh are achievable in as little as a few years. If those hopes are realized, the future of concentrated solar power is indeed very bright. Until 2011, Spain was the leader in this technology with 580 MW deployed, but as of 2013, it has been overtaken by the United States, which has two large commercial facilities in place each producing over 400 GWh per year. Although the concentrator designs discussed so far are not suitable for use on a small scale by individual homeowners or businesses, the market is beginning to provide solar concentrators such as the Sunflower made by Energy Innovations that work together with PV systems and can generate electricity on a small scale.

10.12.3 Solar Chimneys

An entirely different way to use solar thermal power to generate electricity relies on the solar chimney or solar updraft tower. The solar chimney discussed in a different context in Section 10.11 is an important exception to the rule that harnessing solar thermal power to produce electricity requires concentrating collectors and high temperatures. In the solar chimney, a large-area solar collector in the form of a large sheet of transparent material open is placed on supports a meter or two above the ground, so that air is able on all sides to enter around the periphery of the circular sheet. Thus, as the air underneath is heated by the heated ground and rises up the chimney, new air rushes in from the periphery to replace it. A continuous flow is generated even after the sun goes down as long as the ground is hotter than the outside air. The speed of the updraft in the chimney can be high enough to drive wind turbines placed at its base, which is how the energy is transformed into electricity.

Although the idea for a solar chimney has been around since 1903 when Isidoro Cabanyes, a Spanish army officer, first proposed it as a way of

Figure 10.21 Solar chimney built in Manzanares, Spain; view through the polyester solar collector. (Courtesy of Widakora, http://en.wikipedia.org/wiki/Solar_updraft_tower, image licensed under the Creative Commons Attribution-Share Alike 3.0 Unported, 2.5 Generic, 2.0 Generic, and 1.0 Generic license.)

generating electricity, very few have been constructed. One small-scale prototype built in Spain, in 1982, generated a peak power of only 50 kW and was decommissioned in 1989. The collector diameter and chimney heights were 244 and 195 m, respectively, but its efficiency was much less than 1%. These facts might not seem to portend a promising future for this technology, especially since the plant was abandoned after the tall chimney collapsed due to high winds when the guy wires anchoring it failed (Figure 10.21).

Despite the unfortunate collapse of the Spanish prototype, the Chinese, in 2010, built a solar updraft tower that has begun producing 200 kW of electric power. When completed, the project is expected to cost $208 million and generate 200 MW using a collector covering 277 ha. Other projects are being planned or considered by various nations including the Australian Company EnviroMission, which has sought funding to build one in the United States.

The EnviroMission 200 MW design envisions a truly colossal chimney having a 1 km tall chimney that would dwarf any human-made structure. The reasons (a) the proposed chimney needs to be so tall, (b) the efficiency of the Spanish prototype was so puny, and (c) some people still believe this all makes good economic sense are all interlinked and tied to the dependence of efficiency on chimney height. The mechanical power in the updraft in the chimney may be expressed as $p_{\text{out}} = (1/2)\,\dot{m}\,v^2$ and the thermal power absorbed by the air under the large transparent sheet is $p_{\text{in}} = c_P\,\dot{m}\,\Delta T$, where c_P is the specific heat of air at constant pressure and ΔT is the amount the air temperature rises above ambient. Thus, if we ignore the losses in driving the wind turbines, we find a maximum theoretical chimney efficiency:

$$e = \frac{p_{\text{out}}}{p_{\text{in}}} = \frac{v^2}{2c_P\Delta T}. \tag{10.22}$$

This, of course, also ignores the collector efficiency. A final step in finding the maximum possible chimney efficiency is to assume that the buoyant air rising up the chimney experiences a modified acceleration due to gravity, which, in the Boussinesq approximation, can be written as $g' = g(\Delta T/T_a)$. Thus, with $v^2 = v_f^2 + 2gy$ and taking the final (exit) air velocity to be small, we obtain the simple expression for the maximum possible efficiency:

$$e = \frac{gy}{c_P T_a}. \tag{10.23}$$

Clearly, this expression yields absurd results for indefinitely tall chimneys, since it ignores heat losses through the walls of the chimney, but it does show a remarkable increase in efficiency proportional to the height of the chimney. Based on Equation 10.23, the theoretical limit for the efficiency of the chimney used in the Spanish prototype was $e = 9.8\ \text{m/s}^2\ (244\ \text{m})/1005\ \text{J/kg K}\ (300\ \text{K}) = 0.008$ (0.8%), whereas a 1 km tall chimney would have a maximum efficiency five times as great (4%)—although the actual figure is likely to be significantly less. Even though the expected efficiency is quite low, it would be enough to generate an appreciable amount of power, given a large collector area. More controversially, the cost per kilowatt-hour might actually be competitive with concentrating solar thermal systems, which have much higher efficiencies. As with concentrating collector systems, the cost of energy from a solar updraft tower is largely determined by the initial construction costs, and these depend on assumptions made about interest rates and years of operation. Estimates range from 5 to 15 Eurocents/kWh under varying assumptions (Schlaich, 2005).

Quite apart from whether the cost estimates for the solar chimney are realistic or competitive, this technology and others like it, where the efficiency scales with their size, suffer from a significant "chicken and egg" problem. Only a large-scale plant will have a decent efficiency and produce an appreciable amount of power, while private investors are unwilling to invest in a large-scale plant without seeing that it is feasible both technically and economically—which cannot be proven with a prototype scale plant. While such considerations may represent a severe impediment for untested technologies to cross the "valley of death" before they make a profit, they are irrelevant to government-supported development, as in the Chinese model. Should it prove economically and technically successful, it could spell good news for similar projects elsewhere.

10.13 SUMMARY

Solar thermal technology can serve a wide variety of applications and is among one of the fastest-growing solar applications. Some uses require only low or moderate temperatures, such as domestic hot water heating, and in this case, simple nonconcentrating collectors of the flat-plate or evacuated tube variety can be used. Hot water heating represents one application for homeowners having a relatively short payback time. To reach high temperatures, concentrating collectors need to be used, and various types of systems have been developed, including parabolic trough collectors and power towers. The latter type of system, unlike the former, concentrates the solar radiation in two dimensions and can be used to attain the highest temperatures. Normally, high temperatures are considered essential for purposes of electricity generation, but the solar chimney concept is one important exception. Using solar thermal for electricity generation may not yet be economically competitive with conventional sources, but the trend is toward lower costs, as the various technologies mature and are more fully deployed. While requiring a large land area per kilowatt, the technologies tend to be relatively benign environmentally. Solar thermal electric systems are often deployed in desert regions unused for agriculture or human habitation—although some objections have been raised by environmental groups concerned about disturbing the desert ecology.

APPENDIX: FOUR HEAT TRANSFER MECHANISMS

The three classical methods of transferring heat are conduction, convection, and radiation, but we shall also discuss mass transport with or without phase changes. In many situations, only one or two of these mechanisms is the dominant one and the analysis can be simplified.

10.A.1 Conduction

Heat conduction occurs when a thermal gradient exists across a slab of material. In this case, the rate of heat flow (thermal power) is proportional to both the thermal gradient and the area of the slab, with the proportionality constant being the heat conductivity k of the material, i.e.,

$$\dot{q} = \frac{dq}{dt} = kA\frac{dT}{dx}. \tag{10.24}$$

For a slab of finite thickness Δx, having a temperature difference ΔT across its faces, this equation obviously becomes

$$\dot{q} = \frac{dq}{dt} = kA\frac{\Delta T}{\Delta x}. \quad (10.25)$$

Three alternative ways to write the thermal power per unit area are

$$\frac{\dot{q}}{A} = h\Delta T = \frac{\Delta T}{r} = \frac{\Delta T}{RA}. \quad (10.26)$$

Equation 10.26 serves to define the three closely related constants: h, the heat transfer coefficient; its reciprocal r, the r-value (a measure of thermal resistance of a unit area); and R, the thermal resistance.

A common situation is that of N slabs of equal surface area in contact with a temperature difference ΔT_j across them. In equilibrium, the same thermal power flows through each slab, so that we can use Equation 10.25 for the temperature across the jth slab (Figure 10.22):

$$\Delta T_j = \frac{\dot{q}\Delta x_j}{k_j A}. \quad (10.27)$$

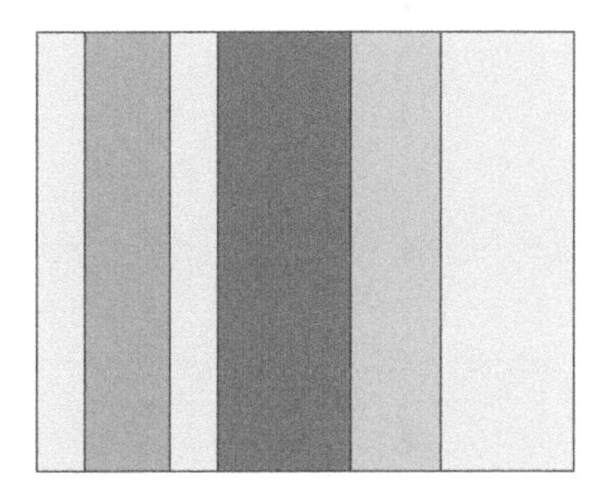

Figure 10.22 Six slabs having different thicknesses and different conductivities (represented by their shading) in contact through which heat flows.

If we add the temperature differences across all the slabs, we obtain the total

$$\Delta T = \sum_j \Delta T_j = \frac{\dot{q}}{A}\sum_j \frac{\Delta x_j}{k_j}. \quad (10.28)$$

Since $r_j = \Delta x_j/k_j$ is the r-value of the jth slab, Equation 10.28 can be rewritten as

$$\Delta T = \frac{\dot{q}}{A}\sum_j r_j = \frac{\dot{q}}{A} r_{\text{equiv}}. \quad (10.29)$$

Thus, we see that the equivalent r-value of the N slabs in series is simply their sum—much as the case with N electrical resistors in series. The SI unit of r-value is meter Kelvin per watt, but in the US building trade, the unit is an awkward combination (ft^2°F h/(BTU in.)). For example, a polystyrene board has an r-value 5.0 ft^2°F h/(BTU in.), implying that a thickness of 1 or 3 in. would have r-values of R-5 and R-15, respectively, using the US units.

10.A.2 Convection

Heat is transferred to or from a surface at temperature T when a fluid at a different temperature T_f flows over it. Convection can be defined as

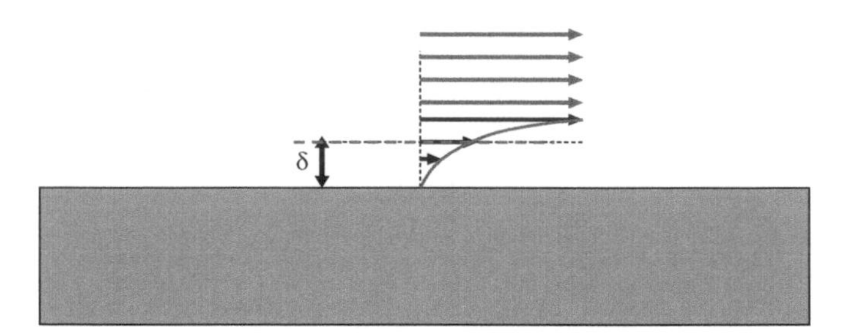

Figure 10.23 Velocity profile for a wind blowing over a slab.

being free (spontaneous) if the fluid flow arises from buoyancy-driven differences in fluid density or forced if the fluid flow is caused by external means, such as a wind or a fan. For example, consider a wind blowing over a horizontal slab of material. The air molecules immediately above the slab will be at rest owing to the strong cohesive forces between them and the slab. As the distance of an air molecule above the slab surface increases, the average molecular velocities will gradually transition to that of the wind—thus the smooth velocity profile shown in Figure 10.23 transitioning to a constant velocity at some distance above the slab.

In the boundary layer approximation, the actual velocity profile above the slab is replaced by a step function. Thus, we assume a layer of stationary air whose thickness is δ. By making this approximation, the problem of heat transfer by convection is transformed into one of conduction through a fixed thickness δ of stationary air, and we may therefore make use of the conduction equation:

$$\frac{\dot{q}}{A} = h\Delta T = \frac{\Delta T}{r}. \tag{10.30}$$

In the convection context, Equation 10.30 is called Newton's law of cooling. ΔT here is the temperature difference across the boundary layer or between the surface of the slab and that of the fluid that flows past it. Being able to treat convection using the same equation as applied to conduction is extremely useful. However, unlike the conduction case, it is not a simple matter to evaluate h or r for convection in terms of a boundary layer of known thickness, since δ varies with the speed of fluid flow and the size, shape, and orientation of the slab. Engineers have developed many empirically based formulas to deal with these complexities, which we shall not consider here.

10.A.3 Radiation

Radiative heat transfer is the direct transmission of electromagnetic energy through space or through transparent media. It is, thus, the only heat transfer mechanism that can occur in vacuum. If a body at absolute (kelvin) temperature T is placed in surroundings at a uniform ambient temperature T_a, the net radiant power per unit area leaving the body is given by

$$\frac{\dot{q}}{A} = \sigma\varepsilon\left(T^4 - T_a^4\right), \qquad (10.31)$$

where the second term accounts for power absorbed by the body from the surroundings and ε is the emissivity of the body and lies in the range of 0–1. Note that the factoring of emissivity ε outside the parentheses implies that the same value applies equally to emission and absorption, so that bodies that are good emitters are necessarily good absorbers. In order to treat radiation in the same manner as conduction and convection, it would be nice if an equation of the form of Equation 10.26 could be found, which at first sight seems incompatible with the form of Equation 10.31. However, given that $\Delta T = T - T_a$ is often small compared to T, we can do a Taylor expansion of the expression $(T + \Delta T)^4 - T_a^4$. Thus, to second order in ΔT

$$\frac{\dot{q}}{A} = \sigma\varepsilon\left((T + \Delta T)^4 - T_a^4\right) \approx 4\sigma\varepsilon T_a^3 \Delta T + 6\sigma\varepsilon T_a^2 \Delta T^2 + \ldots \qquad (10.32)$$

If we retain only the first-order term in ΔT in Equation 10.32, we see that in this approximation, $\dot{q}/A = h\Delta T = \Delta T/r$ remains valid with the heat transfer coefficient given by $h = 4\sigma\varepsilon T_a^3$ and the thermal resistance ($R = rA$) would also be regarded as being constant independent of ΔT. However, if we were to include the second-order term, we find that $h = 4\sigma\varepsilon T_a^3\left(1 + (3/2)(\Delta T/T_a)\right)$. Thus, it is clear that as the temperature of an object rises, and as ΔT increases, the first-order approximation becomes poorer. This breakdown begins to show itself as a linear variation of h with ΔT or equivalently as a quadratic dependence of $\dot{q}/A$ on ΔT (based on the second order in the expansion in Equation 10.32).

10.A.4 Example 5: Validity of the First-Order Approximation

Suppose a plate that is at a temperature T radiates heat to its surroundings at an ambient temperature of 27°C. Up to what temperature T could the radiative resistance be considered constant to an accuracy of 10%?

Solution

An approximate answer can be most easily found by requiring that the second-order term in the Taylor expansion in Equation 10.8 be no more than 10% of the first-order term, i.e., that $6\Delta T_a^2 \Delta T^2 \leq 4\Delta T_a^3 \Delta T$, which requires that $\Delta T \leq (2/3)T_a = 200$ K = 200°C. Here we have just reminded ourselves that when dealing with changes in temperature rather than temperatures themselves, it is, of course, immaterial whether we write them in kelvins or degrees Celsius. It is interesting that the approximation of a constant resistance holds to better than 10% even if the plate temperature rises 200 K or 200°C above ambient temperature. Thus, in many SWH applications, the assumption of a nearly constant resistance applies fairly well.

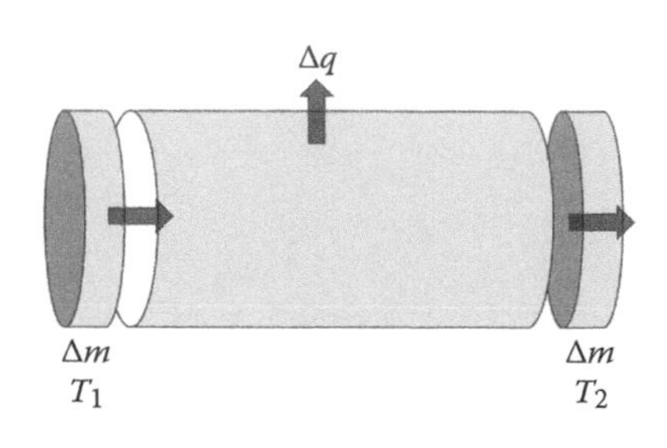

Figure 10.24 Heat transfer by means of mass transport of a fluid through a cylindrical pipe.

10.A.5 Mass Transport

A final heat transport mechanism we shall consider is via mass transport. Convection can certainly be understood in these terms, but here we are using the phrase in referring to heat transfer by transporting a heated fluid through a pipe, which frequently occurs in water heaters of the solar or nonsolar varieties. Let us consider a cylindrical pipe for which fluid enters one end at temperature T_1 and exits the other at temperature T_2. Let us imagine that in a short time interval Δt, a small element of fluid mass Δm shown as a cylindrical slab enters one end of the pipe and an equal mass exits the other (Figure 10.24).

The energy balance equation for this situation can be expressed as

$$\Delta q = \Delta mcT = mc\Delta\bar{T} + \Delta mc(T_2 - T_1), \tag{10.33}$$

where $\bar{T} = (1/2)(T_1 + T_2)$ and the mass flow rate $\dot{m}$ can be expressed in terms of a volume flow rate times the fluid density $\dot{m} = \rho\dot{V}$. Thus, after dividing Equation 10.33 by a short time interval Δt, we obtain for the thermal power

$$\dot{q} = mc\frac{d\bar{T}}{dt} + \rho\dot{V}c(T_2 - T_1). \tag{10.34}$$

Three interesting special cases of this equation would include the following

1. A static mass of water, i.e., where $\dot{V} = 0$
2. The situation where the energy input just balances the thermal losses, i.e., where $\dot{q} = 0$
3. The situation where there is a net energy inflow or outflow, but where the average temperature of the water in the pipe remains constant, i.e., $d\bar{T}/dt = 0$.

In case 3, we obviously have $\dot{q} = \rho\dot{V}c(T_2 - T_1)$, although for greater generality, we might also consider the possibility of a phase change that releases thermal power of amount $\rho V{\cdot}L$, where L is the latent heat. Thus, if there is a phase change, case 3 would be written as

$$\dot{q} = \rho\dot{V}\left[L + c(T_2 - T_1)\right]. \tag{10.35}$$

PROBLEMS

1. Derive the loop integral Equation 10.13 from\ Equation 10.12.
2. Find the thermal conductivity of a slab 1 cm thick if when 5 W/m^2 flows through it, the temperature difference across its faces is 10°C.

3. Derive this expression for the thermal resistance for conductive heat loss through the walls of a long cylindrical pipe having a heat conductivity k, length L, and inner and outer radii r_2 and r_1. $R = (1/2\pi kL) \ln (r_2/r_1)$. The total heat loss would of course also need to consider convection and radiation. Hint: Start with a thin cylindrical shell of thickness dr and length L and apply the conduction equation (Equation 10.24).
4. Prove that when three heat loss mechanisms act in parallel, the usual rule for combining resistances in parallel applies, i.e., $1/R = 1/R_1 + 1/R_2 + 1/R_3$. Hint: How is thermal resistance defined?
5. Suppose you seek a SHW that produced water at least 90°F above ambient temperature. You have a choice between solar collector A—an inexpensive flat-plate collector with a reflectance of 0.1 and rated to have a stagnation temperature of 80°F above ambient—and B—a more expensive evacuated tube collector with a reflectance of 0.5 and a rated stagnation temperature of 150°F above ambient. Note that both stagnation temperatures assume a standard average irradiance of 500 W/m^2. Which one would you buy if you lived somewhere where the average irradiance was (a) 400 W/m^2? (b) 600 W/m^2?
6. For what temperature do the two collectors of the previous problem have the same efficiency at a location where the average irradiance is 400 W/m^2?
7. Carefully explain why you would probably use an evacuated tube solar collector for residential water heating in places having low irradiance and a flat-plate collector in places where the irradiance is high. Hint: Think about how the reflectance and resistance differ between the two types of collectors and how they affect the maximum temperature for a given irradiance.
8. Suppose you had an evacuated tube solar collector whose reflectance was 0.5 and whose r-value was not constant, but rather $r = RA = (1 - 0.004T)$ Km^2/W, where T is the temperature above ambient temperature. Determine the stagnation temperature when the incident irradiance is 500 W/m^2.
9. Thermosiphon solar hot water systems are usually not well suited for large systems having more than 10 m^2 of collector surface. Explain this using some realistic choices of relevant parameters.
10. A rule of thumb is that the fluid flow rate through a solar hot water system should be around 200 gal/h for each square meter of collector area. Explain this using some realistic choices of relevant parameters.
11. Derive a formula for the optimum temperature (for maximum efficiency) using a concentrating collector for any given concentration ratio (see Equation 10.21).
12. Show that the maximum temperature that can be attained using a concentrating collector, whose concentration ratio is X, can be written as $T = T_S(X/X_{max})^{1/4}$, where X_{max} is the maximum possible concentration ratio.

13. Use the data provided for the Spanish prototype solar chimney and assume that peak power was generated when the incident solar irradiance was 1000 W/m^2. What was the efficiency of the system? What fraction of the maximum chimney efficiency was achieved?
14. Assume that all dimensions of a solar chimney used to generate electric power are scaled up by a factor *F*. By what factor does the maximum power generated by the solar chimney increase?
15. Suppose that a concentrating collector has a concentration ratio of 100 and the molten salt is heated to 900 K. Find the maximum overall efficiency if heat is rejected to the environment at a temperature 300 K.
16. Calculate the total power output of the sun at the distance of the Earth. Assume a power density at Earth of 1400 W/m^2 and a distance of Earth from the sun as 1.5×10^8 km.
17. Using the preceding calculation for the total power output of the sun, calculate how much solar energy is emitted in 1 year.

REFERENCES

Agarwal, B. (2001) Participatory exclusions, community forestry and gender: An analysis and conceptual framework, *World Dev.*, 29(10), 1623–1648.

Schlaich, J., R. Bergermann, W. Schiel, and G. Weinrebe (2005) Design of solar updraft tower systems—Utilization of solar induced convective flows for power generation, *J. Sol. Energy Eng.*, 127(1), 117.

Photovoltaics

CONTENTS

11.1 INTRODUCTION

Solar cells, also called photovoltaics or PVs, offer a way to directly convert the energy reaching us from the sun into electricity without the need for any intermediate steps, such as a turbine-driven generator. As was noted in Chapter 1, while for most nations, PV still represents a tiny fraction of all the electric power generated (about 0.2% worldwide), it has been exponentially growing over the last 35 years and may continue to do so for some further time, owing to the rapidly declining prices of solar cells. Two types of crystalline solar cells are shown in Figure 11.1, although noncrystalline (amorphous) cells also exist. The polycrystalline solar modules are less efficient than those made from a single crystal, but are simpler and less expensive to manufacture. Over time, the market share of single crystal solar cells has steadily decreased in light of their much higher manufacturing costs.

PV has a number of advantages for electricity generation, in addition to the usual advantages of renewable sources generally. It also, of course, has some challenges. The list in Table 11.1 omits those advantages that it shares with most other renewable sources.

The "diverse range of uses" of PV in Table 11.1 refers to its suitability for either grid-connected or off-grid uses in remote areas, including space, and in developing nations. It also includes its suitability for vastly different scales from 50 W panels in a rural Third World village (Figure 11.2) to the enormous 1000 MW solar installation being planned in China. The scalability of solar means that one can start with a small installation and easily increase the power by adding new solar panels, which is more difficult with some other renewable energy sources such as hydropower or geothermal.

PV solar cells are based on the PV effect in which light incident on a material creates an electric current. This effect was discovered by Alexandre Edmond Becquerel (1839)—at age 19 while working in his father's lab. A diagram of the apparatus described by young Becquerel is shown in Figure 11.3.

The PV effect is similar to the photoelectric effect in which electrons are ejected from a surface that is exposed to electromagnetic radiation of sufficiently short wavelength. The difference between the two effects is that in the PV effect, an intrinsic (internal) electric field is present that maintains the current flow, allowing PV devices in principle to be used for power generation by directly converting incident solar radiation into electricity. Nevertheless, over a century elapsed before PV could be used for power generation, which became feasible with the advent of doped semiconductors. The interesting physics responsible for the internal electric field that can be created using doped semiconductors requires some

(a)

(b)

Figure 11.1 (a) Polycrystalline PV cells laminated to backing material in a module. (b) Solar cell made from a single silicon crystal.

Table 11.1 Advantages and Challenges for Solar PVs

Advantages	Challenges
Enormous total amount available (more than any other)	Costly (but cost keeps dropping)
Available everywhere (even when cloudy—Germany)	Upfront cost—but not if leased
Best places not suited to other uses (deserts, roofs)	Intermittent (clouds, night)
Most environmentally benign renewable source	
Uses no water (unlike many other sources)	
Little maintenance needed once installed	
Extremely diverse range of uses	
Easily scalable as need evolves	
Works fine at low temperatures (unlike wind)	
Mature technology (yet still improving)	

Figure 11.2 Fifty-watt solar panels on the roofs of a rural Columbian village. (Courtesy of the DOE, Washington, DC.)

discussion of how energy bands form in solid materials and the nature of semiconductors—topics that rely on the principles of quantum mechanics. This material is covered in an appendix to this chapter, which is not essential for understanding the chapter itself.

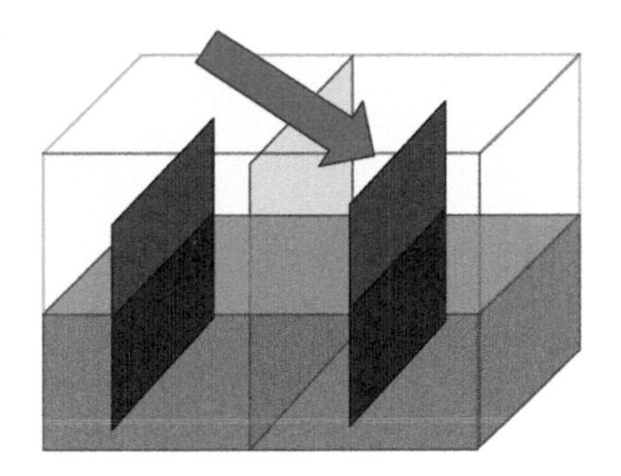

Figure 11.3 Diagram of the apparatus that Becquerel used to discover the PV effect. Light shines on one electrode partly submerged in an acidic solution, causing electrons to be ejected. A thin membrane separates the two halves of the box containing the solution.

11.2 CONDUCTORS, INSULATORS, AND SEMICONDUCTORS

Isolated atoms such as those in a dilute gas have a set of discrete (quantized) energy levels, but atoms in a block of solid matter having N atoms will have energy bands. These bands consist of extremely numerous energy levels ($N < 10^{23}$) that are so closely spaced that the energy might as well be considered continuous within a band. Each of the N energy levels is occupied by at most two electrons (one with spin up and one with spin down) just like the levels in an isolated atom, so the number of electrons filling those levels lying between E and $E + \Delta E$ will be proportional to N, and therefore, it has some maximum value that depends on the number of discrete levels in this energy interval.

Notice that not all the energy levels in a band need be occupied by electrons, and in general, some are filled, some empty and some are partly filled. Equivalently, we say that the "filling fraction" of the levels between E and $E + \Delta E$ always lies in the range of 0–1. A useful distinction is between the valence and conduction bands. The former is the band having the highest range of electron energies where electrons are normally present at absolute zero temperature, while the latter is the next band above it. The region between the bands or the bandgap is a forbidden region, and no electrons can be found having those energies. In cases where the valence and conduction bands overlap one another, there obviously is no gap. Electrical conductors are characterized by this gap-free band structure. If a gap is present between the valence and conduction bands, the material will be either an insulator or a semiconductor, depending on the size of the gap, with 4 eV being the arbitrary dividing line between the two types of materials (Figure 11.4).

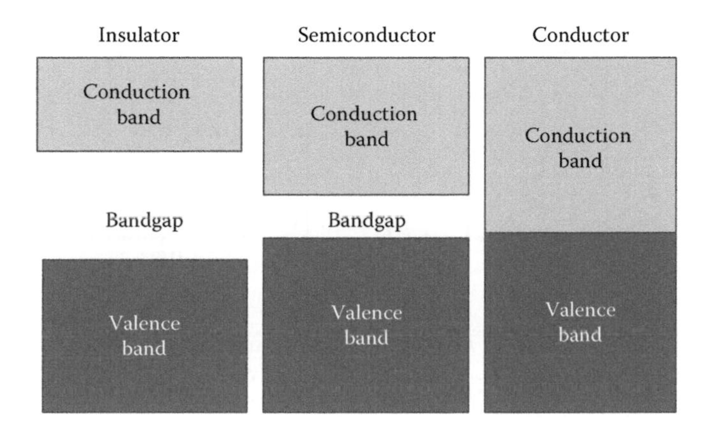

Figure 11.4 Energy bands in conductors, semiconductors, and insulators illustrating the difference in their band structures.

In order to understand the difference between these three types of materials (conductors, semiconductors, and insulators), we need to consider how the filling fraction $f(E, T)$ depends on both energy and temperature. You might expect that electrons fill the levels starting with the lowest ones available and continue filling them in order of increasing energy up to some value, depending on how many electrons there are, but that is only what happens if the material is at absolute zero temperature. Formally, we may say that when $T = 0$ K, the filling fraction is a step function, i.e., $f(E, 0) = 1.0$ for $E < E_F$ and $f(E, 0) = 0.0$ for $E > E_F$. The energy where the step occurs E_F is known as the Fermi energy.

Figure 11.5 shows the filling fraction $f(E, T)$ for both $T = 0$ in Figure 11.5a and $T > 0$ for a conductor in Figure 11.5b and for a semiconductor in Figure 11.5c, where the valence band appears below the gap and the conduction band above it. Note that $f(E, T)$ depicted as a curve is the abscissa and E is the ordinate here. Thus, for $T = 0$, the filling fraction is 1.0 up to the Fermi energy, and it suddenly drops to zero above it, while for $T > 0$, the variation in the filling fraction with increasing energy is less abrupt. It should be clear from these figures that the number of electrons occupying the conduction band depends on both the size of the bandgap and the temperature (which affects the gradualness of the step). Unlike the figure, however, the number of electrons making it up to the conduction band is normally very tiny for a semiconductor at room temperature—see Example 2.

It is only those electrons in the conduction band (not those in the valence band) that can serve as charge carriers in an electric current when a voltage is present. The reason is that if acted on by an electric field, conduction

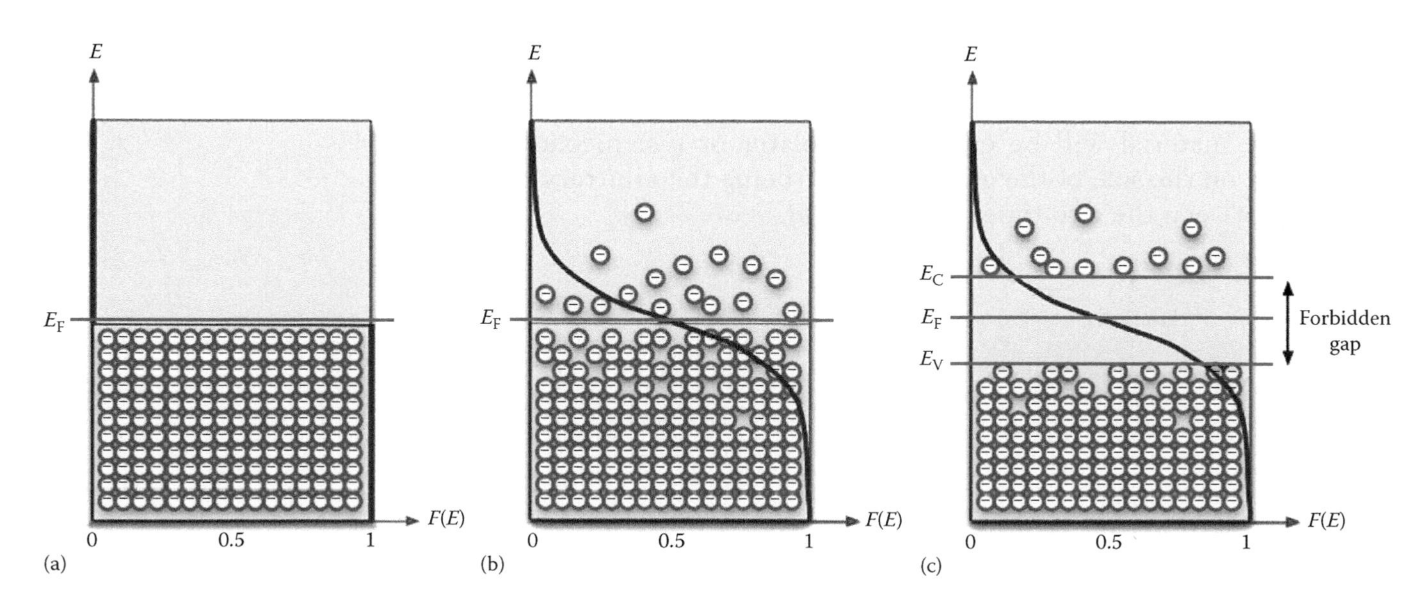

Figure 11.5 (a) $f(E, T)$ for $T = 0$ (absolute zero temperature) and (b) $f(E, T)$ for $T > 0$. The abruptness of the transition from all filled ($f(E, T) = 1$) to all empty ($f(E, T) = 0$) depends on the temperature. (c) $f(E, T)$ for $T > 0$ when there is a bandgap for which no levels are occupied. (Courtesy of Ed Woodward.)

band electrons can accept energy (move to a very nearby higher unfilled energy level), whereas valence band electrons would need to receive enough energy to jump the forbidden bandgap, and this is not possible unless the electric field is so high that the material breaks down and begins to conduct. Conduction band electrons are also known as free electrons, since they are not confined to one atom and can move freely through the material.

Enrico Fermi and Paul Dirac independently discovered how identical particles such as electrons populate energy levels, and the formula they derived for the filling fraction, also sometimes known as the Fermi distribution, is

$$f(E,T)=\frac{1}{e^x+1}, \quad \text{where } x=\frac{E-E_F}{k_B T}, \tag{11.1}$$

where $k_B = 8.62 \times 10^{-5}$ eV/K is the Boltzmann constant. Notice that $f(E, T)$ plotted versus E in Figure 11.6 for various temperatures is a step function only when $T = 0$ K, and the vertical step becomes increasingly rounded as T increases. Also note that at the Fermi energy $f(E, T) = 0.5$ for any temperature $T > 0$ K.

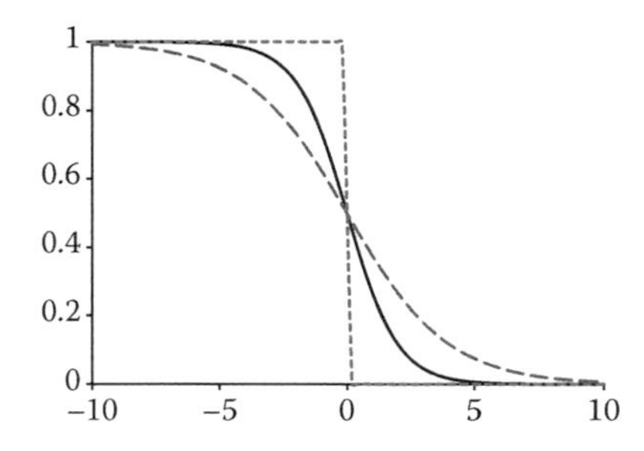

Figure 11.6 Fermi–Dirac distribution versus $x = (E - E_F)/k_B T$ for three values of kT: (a) *solid curve*, $kT = 1$; (b) *long-dashed curve*, $kT = 2$, and (c) *short-dashed curve*, $kT = 0$.

11.2.1 Example 1: Using the Fermi Distribution

Find the fraction of levels that are filled at the bottom of the conduction band for silicon at room temperature.

Solution

Room temperature is about 300 K, and for silicon, the bandgap is 1.11 eV. At the bottom of the conduction band, we obviously have $E - E_F = (1/2)E_g = 0.555$ eV. Thus, $x = (E - E_F)/k_B T = 0.555/(8.62 \times 10^{-5}(300)) = 21.2$, and $f(E, T) = 1/(e^{21.2} + 1) = 6.21 \times 10^{-10}$. This means that only 0.0000000621% of the available spaces are filled with electrons—a tiny fraction indeed.

One way to develop some intuition about the Fermi–Dirac distribution without a derivation is through an analogy with a parking garage. Imagine a garage with very many levels that almost always tends to be half filled, presumably due to bad planning on someone's part! Assume realistically that everyone tends to take the first vacant spot they find, and the garage entrance is on level one. How does the fraction of taken spots vary as a function of level? The answer depends on the ratio R of the time between cars arriving at the garage and the time that the average car remains parked. If R is a very large number (cars arriving rarely, and people remaining parked a short time), then virtually all entrants to the garage will find spots in the bottom half of the garage, and the filling fraction by garage level will be a step function—just as was the case for the Fermi function, where the garage level is a proxy for energy level. Now suppose that R is a small number (cars arriving frequently and people remaining parked a long time). In this case, many arrivals will not find spots on the lower levels and will need to find something on the lowest available upper level. The filling fraction distribution will no longer be a step function and will become more rounded—the more so as R decreases. Thus, R is the reciprocal of temperature in our parking garage analogy. But, enough of parking garages (for now); let us return to semiconductors.

Four properties of semiconductors are worth noting:

- *Two charge carriers.* Current flow in semiconductors can occur via the movement of both electrons and positively charged holes, as described in the next section.
- *Conductivity.* Semiconductors have a much lower conductivity than conductors (given the far smaller number of conduction band electrons)—see Example 1. The conductivities of conductors, semiconductors, and insulators are many orders of magnitude apart: conductors: $<10^8$, semiconductors: $<10^{-6}$–10^5, and insulators $<10^{-14}/\Omega/m$.
- *Bandgap dependence.* The electrical conductivity of semiconductors sensitively depends on the size of the bandgap. For example, silicon with a bandgap of 1.1 eV has a conductivity of 0.00043/Ω/m, while that of germanium whose bandgap is 0.67 eV is 1.7/Ω/m.
- *Temperature dependence.* The resistivity and its reciprocal conductivity of semiconductors vary very dramatically with temperature. In fact, they have smaller resistivity, and the higher conductivity, the higher the temperature because the Fermi distribution has an increasingly gradual transition the higher the temperature, leading to many more electrons in the conduction band. A semiconductor decrease in resistivity with temperature is just the opposite from conductors. However, the temperature dependence of the resistance of semiconductors is more complex when they are doped.

11.3 INCREASING THE CONDUCTIVITY OF SEMICONDUCTORS THROUGH DOPING

Doping a semiconductor is the deliberate addition of impurity atoms into the pure semiconductor in order to change its electrical properties, especially its conductivity. Doped semiconductors are also known as extrinsic semiconductors in contrast to pure or intrinsic semiconductors. Typically, the level of doping is quite small, e.g., light doping might entail adding one impurity (dopant) atom per 100 million atoms, while heavy doping might entail one dopant atom per 10,000 atoms. The best doping concentration for silicon solar cells so as to achieve maximum power output has been found to lie in the range of 10^{17}–10^{18} dopant atoms/cm^3, which is equivalent to about one atom in 10^4–10^5 (Iles and Soclof, 1975). Silicon is used for about 95% of solar cells produced, given its low cost and the size of its bandgap, which as we shall see is nearly optimal in terms of efficiency for converting solar radiation into electricity.

Silicon is a semiconductor having valence +4, i.e., it has four electrons outside a closed shell that bind the atom to four of its six neighboring atoms covalently. In pure crystalline form, silicon has its atoms arranged in a cubic lattice. A plane of such atoms would be arranged as shown in Figure 11.7a, with adjacent atoms sharing a pair of electrons that bind it to its nearest neighbors.

The picture changes when a dopant is added, as seen in Figure 11.7b for dopant atoms of valence +5, such as phosphorus, because there is now one extra electron that cannot take place in pair bonds with four neighboring silicon atoms. This unbound electron is free to move through the crystal, thereby enhancing the number of charge carriers participating in conduction and increasing the conductivity of the material. We speak of the dopant phosphorus here being a donor since it donates conduction electrons, and this type of doping is called *n-type* (for negative—the charge of the electron). Now, consider the other type of doping, *p-type* (for positive) shown in Figure 11.7c. Here the dopant atoms are boron (valence +3), so that instead of there being an extra electron involved in a pair bond to a nearest neighbor, there is a deficiency or a hole. Boron is an acceptor, rather than a donor atom. You might think that the hole is shown in the wrong place in the figure, since it should be shown for one of the four bonds to the boron atom. However, like the extra unpaired electron in Figure 11.7b, these holes are free to move throughout the material in just the same way as electrons, and they too add to the electric current, assuming that the electrons and holes travel in opposite directions.

How does doping alter the way that energy levels are filled in a semiconductor? First consider the n-type doping where there are many more

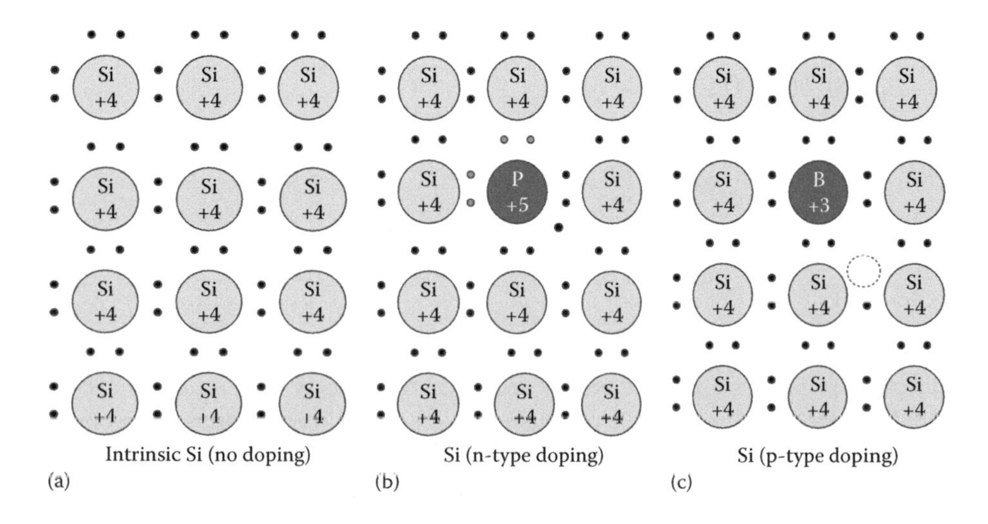

Figure 11.7 (a) Plane of silicon atoms with each bound to its nearest neighbors by covalent bonds (shared pair of electrons), (b) silicon doped with phosphorus (valence +5) and the one unshared electron, and (c) silicon doped with boron (valence +3) and the one hole (dotted circle).

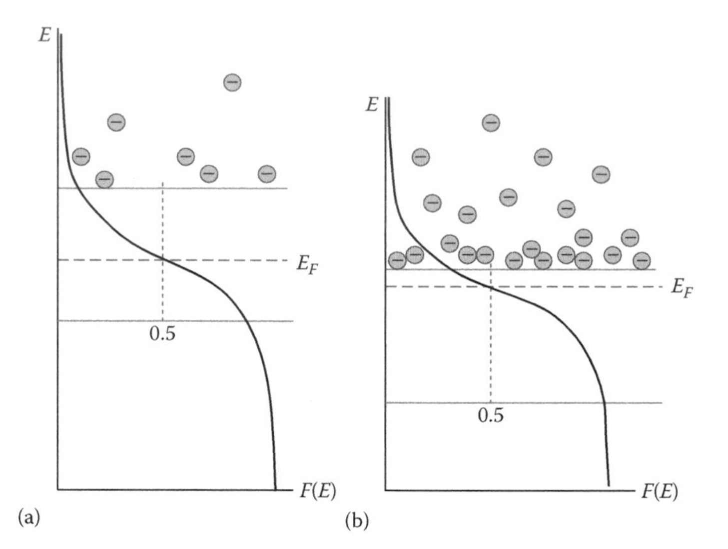

Figure 11.8 Distribution of conduction electrons (a) before doping and (b) after doping. Note the location of the Fermi energy relative to the top and bottom of the bandgap defined by the two solid horizontal lines.

conducting electrons than before. Figure 11.8 shows how the electrons are distributed in energy before doping (left image) and after doping (right image). After doping, there are many more electrons in the conduction band (above the top horizontal line). Unlike Figure 11.7, we have no longer bothered showing electrons in the nearly filled valence band. Essentially, the whole $f(E)$ distribution gets shifted upward, so that the energy level where $f(E) = 0.5$ (the Fermi energy) is no longer in the middle of the bandgap as it was prior to doping but closer to the top. How high it is shifted upward obviously depends on the level of doping.

11.3.1 Example 2: Effect of Doping on Conductivity

How much does the number of conduction electrons change as a result of doping at the bottom of the conduction band, if the result of doping is to raise the Fermi level in silicon up to 0.8 times the bandgap energy? Assume that the room temperature as before is 300 K and $E_g = 1.11$ eV.

Solution

$E - E_F = 0.2E_g = 0.222$ eV. Thus,

$$x = \frac{E - E_F}{k_B T} = \frac{0.222}{8.62 \times 10^{-5}(300)} = 8.58,$$

$$\text{and} \quad f(E,T) = \frac{1}{e^{8.58} + 1} = 1.88 \times 10^{-4},$$

which represents an increase by a factor of $1.88 \times 10^{-4}/6.21 \times 10^{-10} = 303{,}000$ from the value for intrinsic pure silicon. Given that the conductivity is proportional to the number of charge carriers, it will increase by the same factor.

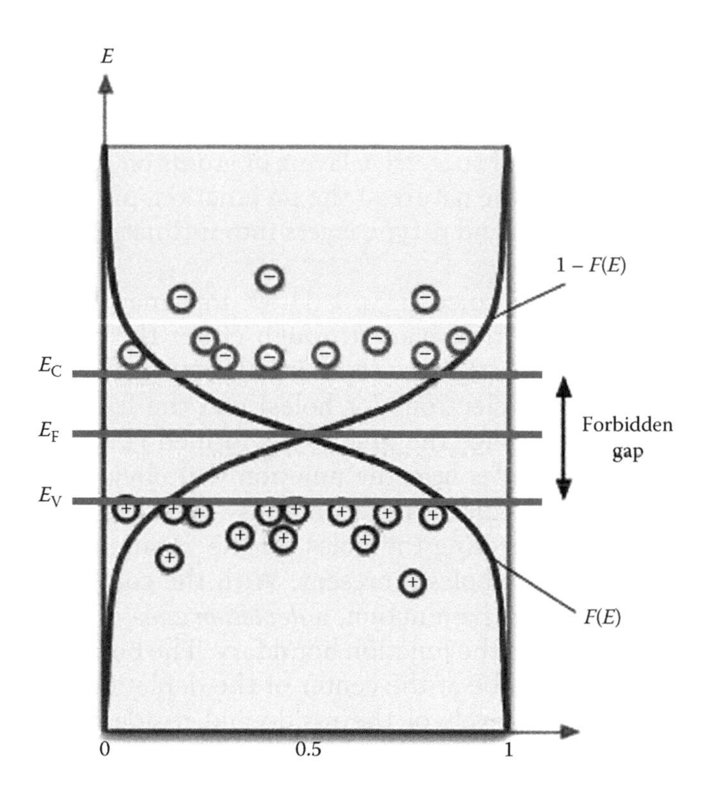

Figure 11.9 Filling fraction distribution for electrons $F(E)$ and holes $(1 - F(E))$. (Image created by Ed Woodward and included with his permission.)

In the parking garage analogy, n-type doping dose is equivalent to adding many more cars to the garage, so that the garage is more than half full on average. Obviously, this would populate higher levels with cars to a greater extent (since lower levels would more often be filled than before, and the distribution of the car filling fraction by garage level would get shifted upward. The parking garage analogy also applies to the energy distribution of holes in the case of a p-type semiconductor, since holes are the equivalent of vacant parking spaces. p-Type doping is the equivalent to having the garage less than half full on average, so that the distribution of vacant parking spaces is now shifted downward rather than upward. Also, with the vacant spots, the highest levels of the garage have the largest population of vacancies, not the lowest. Thus, the distribution of vacant spaces and cars has the same mirror image relationship as the distribution of electrons and holes in doped semiconductors—see Figure 11.9. Notice that holes in a p-type semiconductor populate energy levels in the valence band just below the bottom of the bandgap, not above it.

11.4 pn JUNCTION

pn Junctions are formed at the boundary between p-type and n-type semiconductors. They are the key building block of solar cells as well as a host of other solid-state electronic components, including diodes, transistors, and light-emitting diodes. In fact, solar cells essentially have the same basic structure as semiconductor diodes. In order to create a pn junction, you cannot simply place layers of p-type and n-type materials in contact, because the atoms will not be close enough. Instead, the junction is formed by growing a crystal and abruptly changing the type of doping as a function of depth by various processes, such as epitaxy, which involves the deposition of successive layers of atoms on top of a substrate. In order to understand the nature of the pn junction, pretend that it were possible to bring p-type and n-type layers into intimate contact.

Before the junction is formed (Figure 11.10, top image), no net electric charge exists in any vertical plane through either the p-type (on right) or n-type material (on left), since there is balance between the charge of the mobile conduction electrons (or holes) and the fixed atomic nuclei everywhere. However, once the junction is formed (bottom image), the mobile electrons and holes near the junction will obviously attract each other and upon their collision form photons—alternately, we could say that the electrons in meeting the holes fill the vacancies in the valence band, which is what the holes represent. With the conduction electrons and holes removed near the junction, a *depletion zone* is formed having a width d and centered on the junction boundary. The boundary where the net charge vanishes will be at the center of the depletion zone; however, only when the doping levels of the n-side and p-side are the same. In this depletion zone, the removal of free electrons and holes has exposed vertical layers of residual charges left behind: positive on the n side of the junction and negative on the p side—see Figure 11.10.

These charge layers create an electric field E in the space between them, but none outside—just like a charged parallel plate capacitor. The field is directed from the n-type to the p-type material, not the reverse, since the positive charge layer is on the n-type side. Accompanying the electric field, a built-in voltage develops across the junction of magnitude: $V_{bi} = Ed$ with the n-side having the higher potential.

The presence of this internal voltage serves as a barrier that tends to keep the holes on the p-side and the electrons on the n-side of the junction,

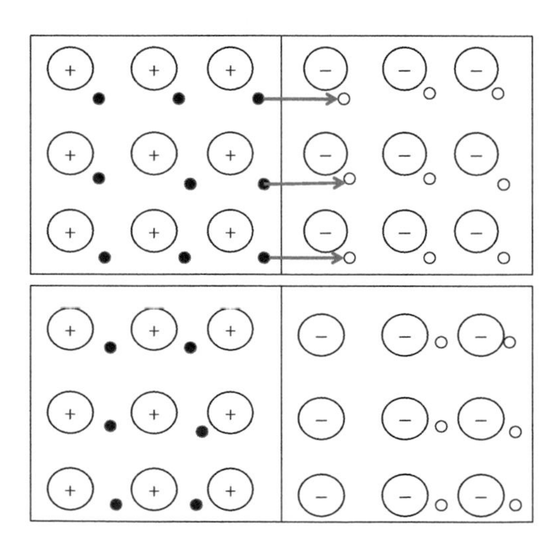

Figure 11.10 Formation of pn junction: top image (before) and bottom image (after). Mobile electrons are represented by small filled circles, and mobile holes by open circles.

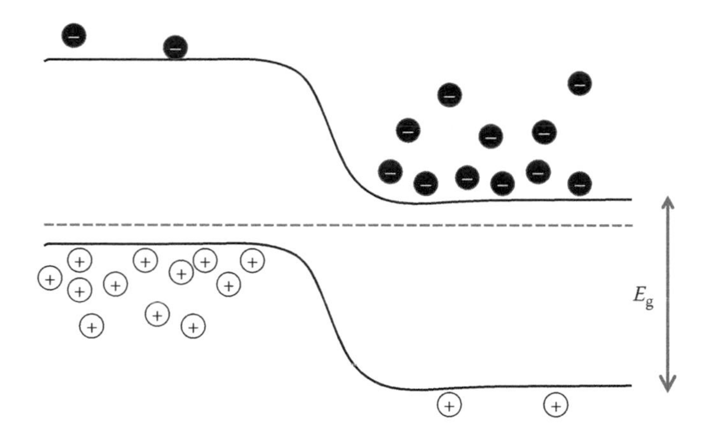

Figure 11.11 Shift in energy levels defining the bandgap across the pn junction (p on left) and the majority and minority carriers in each material. E_g is the bandgap energy, and the dotted line shows the Fermi level.

but a small number of charge carriers of the wrong type will always be found on each side, i.e., holes on the n-side and electrons on the p-side. These minority carriers exist because the equilibrium that established the depletion zone is a dynamic one so that in addition to electrons and holes meeting and creating photons, the reverse occurs at an equal rate, with the creation of new electron–hole pairs at the junction due to thermal excitation. Nevertheless, very few of the holes have enough energy to make it across the depletion zone (uphill) into the n-type material and, similarly, for the electrons, so the ratio of minority to majority carriers is extremely small on the p-side and n-side.

The presence of an internal voltage across a pn junction modifies the energy level diagrams we considered earlier for the n-type and p-type materials separately in an interesting way. The diagram in Figure 11.11 shows how the energy levels and their population by electrons and holes vary as we move across the pn junction.

In Figure 11.1, the p-type material is again on the left and the n-type is on the right. Moving across the junction from left to right results in a drop in potential energy given by qV_{bi}, where q is the electron charge. This drop has the effect of lowering all energy levels as we cross the junction—both the top and bottom of the bandgap as well as the Fermi energy. The conduction electrons in the n-type material and the holes in the p-type material occupy the bands identified earlier, and the minority carriers are also shown here on each side. Notice how, with the energy shift, the Fermi energy is now at the same level on both sides of the pn junction.

11.5 GENERIC PHOTOVOLTAIC CELL

A solar PV cell basically consists of a slab of semiconductor with one or more pn junctions that can be electrically connected to the outside world through metal contacts on the p and n faces (Figure 11.12). As a result of the internal electric field and built-in voltage across the junction, PV cells can generate electricity from incident photons that reach the pn junction provided the photons have sufficient energy to create an electron–hole

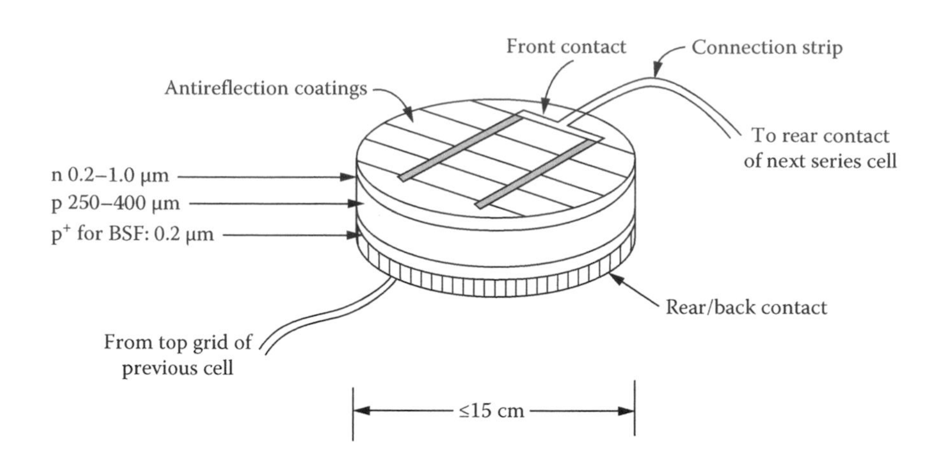

Figure 11.12 Typical solar cell. (From Twidell, J., and Weir, T., *Renewable Energy Resources*, 2nd edn., Taylor & Francis, Boca Raton, Florida, 2006, p. 183. With permission.)

pair there. The junction is typically only a few hundred nanometers below the front surface of the cell, so light can easily penetrate and reach it.

The built-in electric field will drive the freed electrons to the n-side of the junction, and the freed holes to the p-side. Both electrons and holes contribute to an electric current that will persist only as long as the incident radiation is present and so long as a complete circuit is present, i.e., the cell is connected in series with other cells and eventually across a load. No holes actually flow through the cell outside the depletion region, because they combine with electrons once they reach the n-type side. Replacement electrons, however, must enter the other side of the n-type region, and so the current continues throughout the circuit.

11.6 ELECTRICAL PROPERTIES OF A SOLAR CELL

A typical solar cell has a current–voltage curve that looks like the top curve in Figure 11.13. Notice that the current remains nearly quite constant over a range of voltages, which is why we may think of a solar cell as a constant current source, unlike a battery, which is a constant voltage

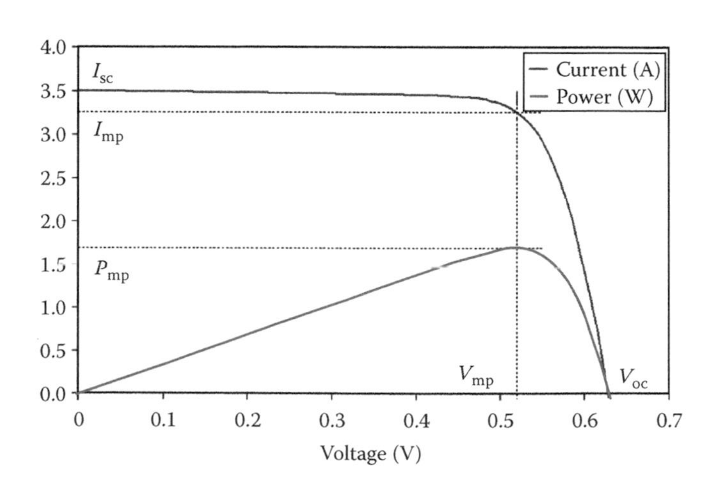

Figure 11.13 Current I and power p of a PV cell versus voltage V. (Courtesy of Eget Arbejde, http://da.wikipedia.org/wiki/Fil:I-V_Curve_MPP.png, image licensed under the Creative Commons Attribution 3.0 Unported license.)

source. The value of the voltage across a solar cell entirely depends on the external load to which it is connected—zero resistance load will obviously give zero voltage across the cell, and the current at that point is I_{SC} (SC for short circuit). When the external load resistance increases to infinity, the current stops flowing ($I = 0$), and the voltage across the cell is V_{OC} (OC for open circuit). The power curve of Figure 11.13 immediately follows from the I versus V curve, since $p = IV$. Do you understand why the power equals zero when $I = I_{SC}$ and when $V = V_{OC}$? Do you see why it linearly increases with voltage over the region where I is constant? The point where the power generated by the cell is a maximum corresponds to the knee of the curve. This is the point where a line of slope −1 is tangent to the I–V curve—remembering that when a slope −1 line is drawn on a graph whose horizontal and vertical axes are on different scales, it will not appear to be at a 45° angle. One other thing to keep in mind about Figure 11.13 is that the current generated by the cell depends on its illumination—obviously, it produces nothing in the dark. In general, the height of the horizontal part of the curve is directly proportional to the normal irradiance incident on the cell as well as its area.

11.7 EFFICIENCY OF SOLAR CELLS AND SOLAR SYSTEMS

The efficiency of a solar cell is defined as the electrical power output divided by the incident power in the electromagnetic wave normally incident on it, or the fraction of incident solar power converted to electrical power. Efficiency depends on four key factors, discussed in the following.

11.7.1 Fill Factor

The fill factor (not the filling fraction) of a solar cell is the ratio of maximum obtainable power to the product of the open-circuit voltage and short-circuit current. The basis of the name of this quantity can be understood as follows. The product of $V_{OC} \times I_{SC}$ is the area of a rectangle whose horizontal dimension is V_{OC} and whose vertical dimension is I_{SC} see Figure 11.13, while the actual maximum power is the area of the dotted rectangle whose horizontal dimension is V_{mp} and whose vertical dimension is I_{mp} (for maximum power). Obviously, the largest possible fill factor (*ff*) is 1.0, for which the I–V graph is a step function, but the fill factor is not the same as the cell efficiency. For the cell depicted in Figure 11.13, it would appear that *ff* is around 0.8 (80%), and in fact, it can be as large as 88% in silicon. Typically, grade A commercial solar cells have $ff \geq 0.7$, while less efficient grade *B* cells have $ff = 0.4 - 0.7$.

One way to understand the importance of a high fill factor is by realizing that solar cells (just like batteries) have an internal resistance that degrades performance. In the case of a solar cell (or a battery), there are actually two kinds of internal resistance: a series resistance and a shunt

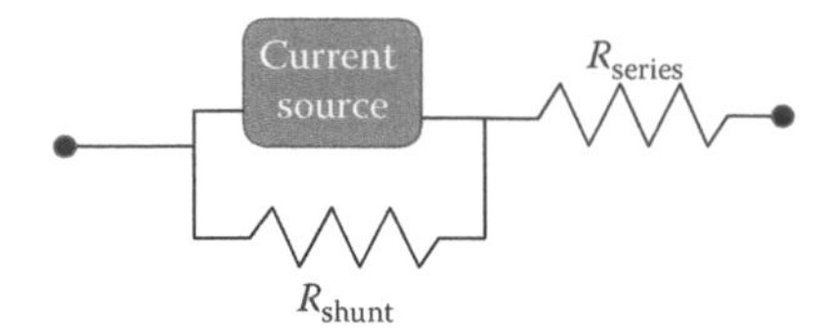

Figure 11.14 Electrical diagram of a PV cell not connected to an external load with its internal series and shunt resistances.

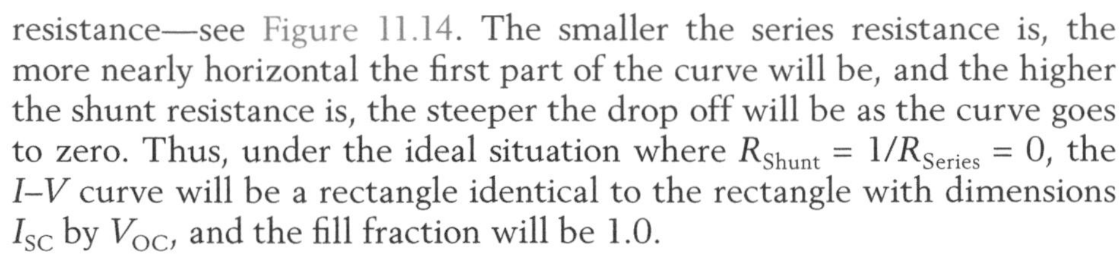

resistance—see Figure 11.14. The smaller the series resistance is, the more nearly horizontal the first part of the curve will be, and the higher the shunt resistance is, the steeper the drop off will be as the curve goes to zero. Thus, under the ideal situation where $R_{Shunt} = 1/R_{Series} = 0$, the I–V curve will be a rectangle identical to the rectangle with dimensions I_{SC} by V_{OC}, and the fill fraction will be 1.0.

It should be clear from this diagram that we want the series resistance to be as low as possible, and the shunt resistance to be as high as possible, so that when the cell is connected to a load, the power wasted is as little as possible. Incidentally, in the case of a rechargeable battery that is unused for a considerable time, its shunt resistance is what causes it to self-discharge.

11.7.2 Temperature Dependence of Efficiency

As with other semiconductor devices, solar cells are affected by temperature, and they become less efficient as the temperature rises. The most temperature-sensitive parameter is the open-circuit voltage V_{OC}, but the short-circuit current I_{SC} is also slightly affected, as Figure 11.15 depicts.

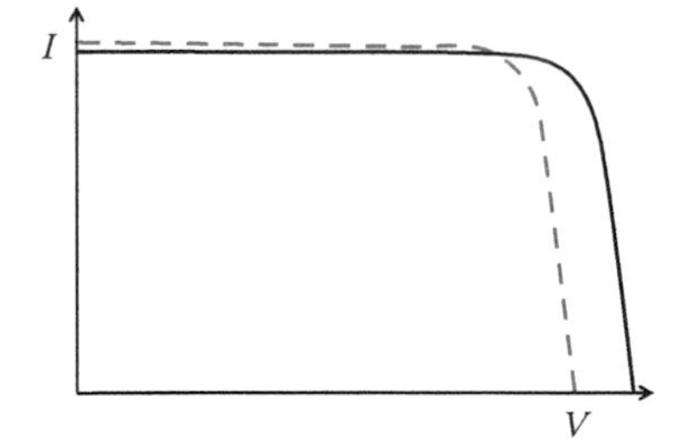

Figure 11.15 Impact of temperature change on the I–V plot for a solar cell. Solid curve is the low-temperature I–V plot, and dashed curve is for the higher temperature. Note that at the higher temperature, the open-circuit voltage drops, and the short-circuit current drops, but by a much smaller amount.

Based on theoretical considerations and empirical data, the maximum power output of a silicon solar cell varies with temperature according to

$$\frac{1}{p_m}\frac{dp_m}{dT} \approx -0.0045/°C. \tag{11.2}$$

11.7.2.1 Example 3: Impact of temperature on efficiency and power output

At a temperature of 20°C, a given solar panel generates 200 W. What would be its output under the same irradiance if the temperature warms by 10°C?

Solution

By Equation 11.2, we have $dp_m/dT \approx -200\ W \times 0.0045/°C = -0.9\ W/°C$. Thus, a rise of 10°C would lead to a drop in output of 9 W—a 4.5% drop.

11.7.3 Spectral Efficiency and Choice of Materials

For any given material such as silicon, the efficiency for radiation of a given wavelength to eject electrons and holes from the depletion zone depends on the relationship between the energy of an incident photon and the size of the bandgap for that material. Figure 11.16 applies in the case of silicon whose bandgap is $E_g = 1.1$ eV (dashed vertical line). The solid curve shows a blackbody spectrum for $T = 6000$ K—the surface temperature of the sun appropriately modified by the Earth's atmosphere. This curve is an approximation to the spectral distribution of solar irradiance at sea level,

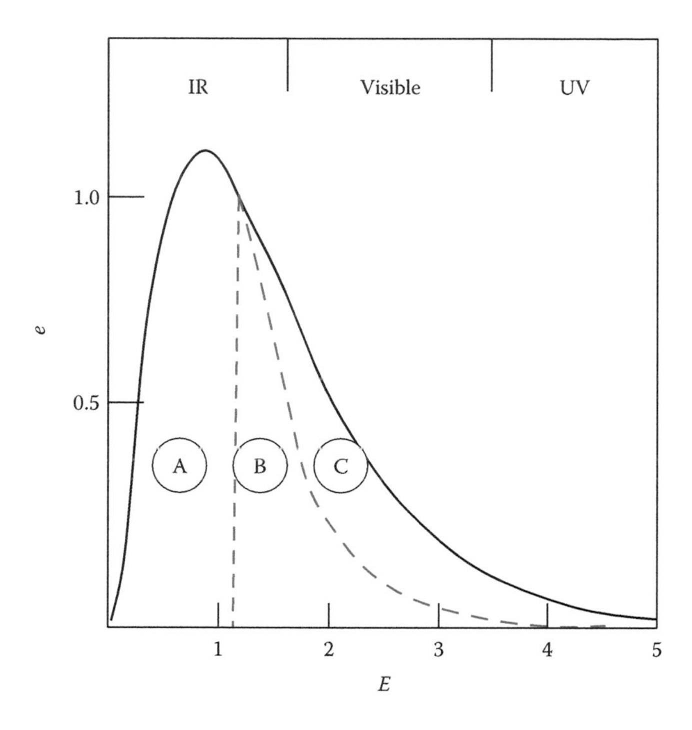

Figure 11.16 Spectral efficiency *e* versus photon energy *E* in electron volts (*dashed curve*) for a silicon solar cell on which solar radiation is incident. Solar spectrum shown by solid curve.

and it is graphed as a function of photon energy, rather than wavelength, but the two quantities are related through

$$E = hf = \frac{hc}{\lambda} = \frac{1240 \text{ eV nm}}{\lambda}. \tag{11.3}$$

The fraction of the light at any given wavelength or photon energy that creates a current is indicated by the dotted curve. Thus, photons of energy $E = E_g$ are 100% efficient because all their energy is used to create electron hole pairs when they are absorbed and an electron is raised from the top of the valence band to vacant energy levels at the bottom of the conduction band and when holes make a jump in the reverse direction. This is the reason why there is a sudden rise in the dashed curve at $E = E_g$ below which photons simply do not have sufficient energy to give to electrons to make the jump. The decline in spectral efficiency for $E > E_g$ is more gradual than the rise was because these photons have more than enough energy to cause electrons to make the jump, but the excess energy is lost as heat to a greater and greater extent as *E* increases. If you try to imagine what the dashed curve might look like for a material with higher or lower bandgap energy than silicon, it will be easy for you to qualitatively understand why silicon's bandgap is quite close to being ideal. A pair of researchers, William Shockley and Hans Queisser (1961), first calculated the Shockley–Quesisser limit in 1961, which gives the maximum possible efficiency as a function of bandgap assuming a 6000 K blackbody spectrum. The efficiency curve they found peaks at around 1.4 eV, but it has not dropped very much by 1.1 eV, where the solar conversion efficiency can be up to 33.7%. Real silicon solar cells cannot achieve efficiencies this high because they have other losses, such as reflection off the front surface and light blockage from the thin wires on its surface. Typically, modern monocrystalline solar cells produce only about 22% conversion efficiency, but this does not include losses external to the cell itself.

11.7.4 Efficiency of Multijunction Cells

The Shockley–Quesisser limit on efficiency of a solar cell applies only to cells having a single pn junction. Significantly higher efficiencies can be achieved with multijunction cells, in which successive layers are chosen having suitable bandgaps, with each layer absorbing photons from a different region of the spectrum (Figure 11.17). It is extremely important in multijunction cells that the layers be arranged starting with the top layer having the highest bandgap and successive layers having progressively lower bandgaps. In this way, incident photons that have an energy $E < E_g$ simply pass through the first layer unabsorbed. If the ordering of layers were reversed, then the most energetic photons would knock electrons into the conduction band for the first layer, but since $E > E_g$, much energy would be lost as heat owing to the fall off in efficiency that occurs in Figure 11.16 in region B. Figure 11.17 shows one possible sequence of layers that satisfies the aforementioned condition and which the region of the spectrum each layer tends to absorb photons in.

In addition to sequencing layers according to their bandgaps, it is also important that the current produced by each layer closely matches one another. If this condition is not satisfied, one has a series of current sources, each with their own internal resistances (see Figure 11.14), which will result in internal current loops that waste energy and lower efficiency. As of 2011, multijunction solar cells have been produced that have efficiencies exceeding 43%, but of course, they are much more expensive than lower efficiency cells.

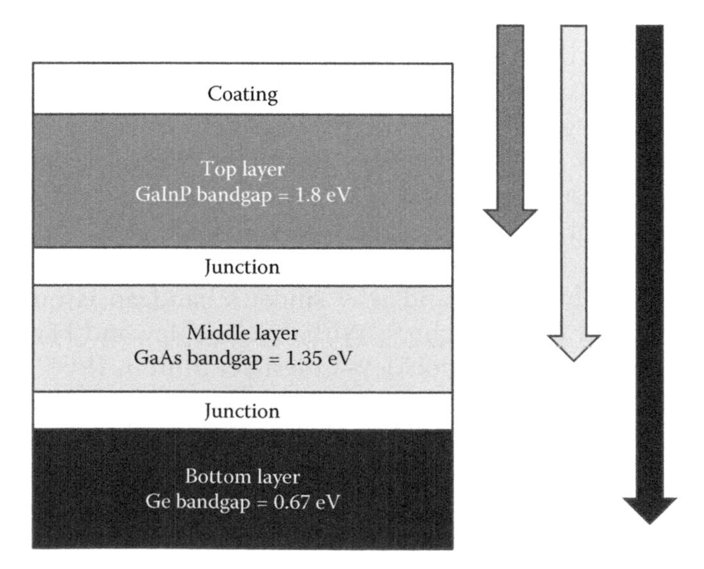

Figure 11.17 Successive layers of p and n materials in a multijunction solar cell.

TABLE 11.2 **Average PV System Component and Total Efficiencies**

Component	(%)	Grid-Tied (%)	Off-Grid AC (%)	Off-Grid DC (%)
PV array	80–85	X	X	X
Inverter (to convert DC to AC)	80–90	X	X	
Wire	98–99	X	X	X
Disconnects and fuses	98–99	X	X	
Batteries (round trip)	65–75	X	X	X
Total		60–75	40–56	49–62

11.8 EFFICIENCY OF SOLAR SYSTEMS

For purposes of power generation, a collection of solar cells need to be electrically connected in series to form a solar module or panel, and then a collection of panels are connected together to form an array. A typical solar panel produces 200 W of power, so the number of panels needed for an array is determined by the application. For home usage, solar system calculators exist online, such as the PV watts program calculator that can help you size a system for your needs. Some of the online applications allow the user to go to Google Earth and digitize the four corners of your roof, so as to see what maximum solar system size could fit on it. These calculators also take into account your latitude, roof slope, roof orientation, and degree of shading and estimate month by month how much electricity the system could generate. In addition to thinking about the efficiency of individual solar cells, one also needs to consider the efficiency of the whole system, which includes the factors in the second column of Table 11.2.

Notice that the total system efficiency in the last row of Table 11.2 is the product of all the separate efficiencies that apply to that type of system; however, these totals do not include the efficiencies of individual solar cells by which they must be multiplied.

11.9 GRID CONNECTION AND INVERTERS

A major step forward leading to the greater use of PV systems has been the inverter invented in the early 1980s. Inverters are the devices that convert the variable DC power from a solar panel or an array of panels into utility-grade AC. One usage of inverters is to allow the system to put power back onto the grid, which requires an accurate match to the grid frequency and phase. Such grid-tied systems must also incorporate an automatic disconnect from the grid in case of a power outage for the safety of linemen making repairs. A disadvantage of such systems is that they do not provide backup power in case of a power outage. Another use of inverters appropriate to remote locations that are not grid connected is to provide AC power by drawing on batteries that are charged by a solar array.

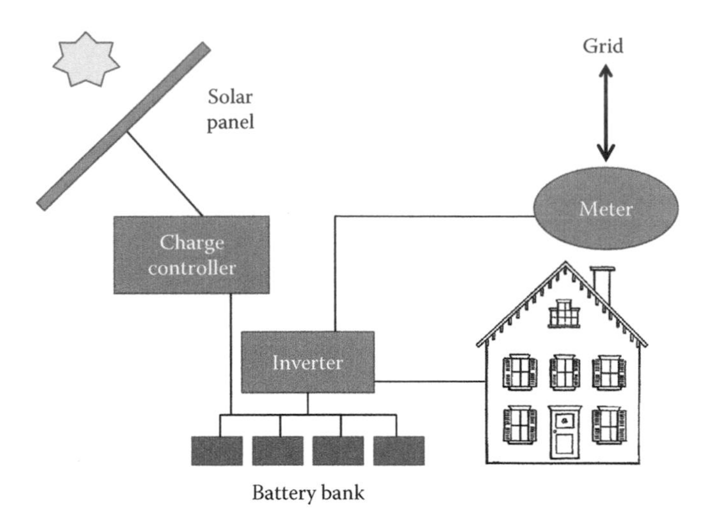

Figure 11.18 Grid-tied system with battery backup. The function of the charge controller is to prevent overcharging of the batteries. Obviously, neither it nor the batteries are present in grid-tied systems lacking battery backup.

The two configurations for using inverters with a solar array are either to combine the DC output of all the panels in the array and send that to a single inverter or, alternately, to have a separate inverter connected to each panel. This latter arrangement, also termed *microinverters*, tends to be used in smaller system. Microinverters allow for one-to-one control of single panels, and they make it much easier to add panels so as to expand the size of the system. A further major advantage is that the failure of a single panel or inverter in a string of panels will not take the system offline, since the arrays of panels are connected in parallel. As a result of their increasing efficiency, and the distributed architecture permitted by microinverters, they have become increasingly popular in recent years. In the future, it can be expected that microinverters will become completely integrated with solar panel modules that produce AC power (Figure 11.18).

11.10 OTHER TYPES OF SOLAR CELLS

11.10.1 Thin Films

Thin-film solar involves the deposition of one or more layers of PV material onto a substrate. They are much easier to fabricate than conventional solar cells and, hence cheaper, but they also have significantly lower efficiency. Although some thin films have been made with cell efficiencies as high as 12–20%, production modules tend to be around 9%. The share of the PV market for thin-film solar has been steadily growing and has currently reached 20.4%, which is comparable to the best conventional solar cell. In applications for home PV systems, besides efficiency and cost, the consumer also needs to consider that the amount of roof area may be limited, so that lower efficiency of thin films may limit the power generated to below the needed or desired amount. Thin-film solar panels can be made from a variety of materials, and manufactured as flexible sheets and even in the form of roof shingles (Figure 11.19).

Figure 11.19 Flexible sheets of thin-film solar cells. (Courtesy of Fieldsken Ken Fields, http://en.wikipedia.org/wiki/Thin_film_solar_cell, image licensed under the Creative Commons Attribution-Share Alike 3.0 Unported license.)

11.10.2 Dye-Sensitized Cells

Dye-sensitized solar cells (DSSCs) are one particular type of organic solar cells that are PV cells made using organic materials (Figure 11.20). Invented in the 1990s, they are currently considered to be the most efficient third-generation solar cells. This is a type of thin-film technology consisting of a semiconductor formed in a layer between a photosensitized anode and an electrolyte. Thus, dye-sensitized cells do not involve a pn junction. Instead, they mimic the process of photosynthesis in separating the roles of harvesting solar energy (absorbed by the dye consisting of titanium dioxide) from that of charge transport through the electrolyte. Efficiencies of 11% have been achieved with dye-sensitized cells, which are not quite as good as the best thin-film cells or conventional silicon panels However, in one 2012 experiment, even higher efficiencies have been achieved when graphene sheets are created by combining titanium dioxide with graphite and baking the resulting paste at high

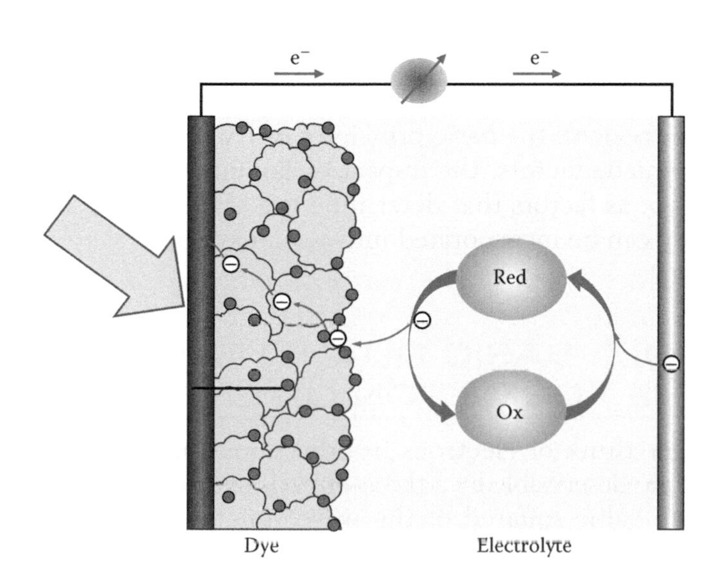

Figure 11.20 Dye-sensitized solar cell. The nanometer-sized dye molecules that capture the light are obviously enlarged here. A layer of these dye molecules exchange electrons with the adjacent electrolyte.

temperatures. Graphene—a wondrous material consisting of a single layer of carbon atoms forming a honeycomb array—has many useful electrical and mechanical properties, including great strength and very high electrical conductivity that make it very well suited to many applications in renewable energy technologies and energy storage—see Chapter 13 for more on graphene.

The biggest disadvantage of conventional solar PV has been cost. DSSCs are very simple to manufacture from low-cost materials. Thus, their power output per dollar ratio measured in units of kilowatt-hour per square meter per dollar (kWh/m^2/$) is expected to make them soon reach grid parity with conventional energy sources.

11.11 ENVIRONMENTAL ISSUES

The environmental impacts of solar PV are almost all positive: reduction of CO_2 emissions, no noise pollution, and their offsetting the need for less benign energy sources. Nevertheless, one can find some environmental issues with any energy technology. One is that the manufacturing process is energy intensive, although the energy that PV systems produce is far greater, with a breakeven point reached in at most about 6 years—about a quarter of their lifetime. There are also issues with the health and safety of workers exposed to dangerous chemicals during the PV manufacturing process and that of workers who install the systems who face hazards of possibly falling off roofs to electrocution. Additionally, there is the disposal problem when systems reach the end of their useful lives, which also poses a moderate environmental hazard. Finally, there is the matter of land use. Even though PV systems often use space not needed for other purposes (roofs) or desert land unsuited to agriculture, there may be groups opposed for environmental reasons (interference with desert wildlife) or on heritage grounds (Native American burial grounds).

11.12 SUMMARY

This chapter reviews the basic principles of PV solar cells, including the nature of semiconductors, the impact of doping, and the pn junction. It considers various factors that determine the efficiency of solar cells, and the way they can be incorporated into a solar system.

APPENDIX: BASIC QUANTUM MECHANICS AND THE FORMATION OF ENERGY BANDS

We no longer think of electrons in orbit about an atomic nucleus, but rather, they are describable by a three-dimensional wave function $\Psi(x, y, z)$. The absolute value squared of the wave function gives the probability density of finding an electron at a particular point in space, thus, the probability of finding an electron in a small volume element dV located at (x, y, z) is

$$\mathrm{d}P = \left|\Psi(x,y,z)\right|^2 \mathrm{d}V. \tag{11.4}$$

The particular wave functions and their associated discrete energy levels for a given system are the result of the confinement of the electrons to some region of space surrounding the nucleus. It is similar to the kind of standing wave patterns found if you confine an ordinary wave—say a wave on a string or a sound wave—to some spatial region. However, unlike those cases, $\Psi(x, y, z)$ has no physical significance apart from the probability density.

11.A.1 Finding Energy Levels and Wave Functions

The mathematical procedure for finding $\Psi(x, y, z)$ in a given situation involves solving the Schrödinger equation with an appropriate choice of potential energy function $U(x, y, z)$:

$$-\frac{\hbar^2}{2m}\nabla^2\Psi = (E-U)\Psi, \tag{11.5}$$

where

$$\nabla^2\Psi = (\partial\Psi/\partial x^2)+(\partial\Psi/\partial y^2)+(\partial\Psi/\partial z^2),$$

E is the total energy of the state

$$U = U(x,y,z),$$

and m is the electron mass

$$\hbar = h/2\pi.$$

Finding the wave functions and their associated energies can be challenging for all but the simplest cases. However, since our interest here is in understanding the basics, we imagine that instead of being confined to a real atom where the potential U is the $1/r$ Coulomb potential, we instead imagine an electron is confined only in one dimension not three. In particular, imagine we had an electron confined to an interval $0 < x < L$, inside which, it experiences no forces, i.e., the potential $U(x)$ is constant (assumed to be zero) in that interval—a situation known as a one-dimensional square well potential (Figure 11.21). Since the electron is absolutely confined, we have implicitly assumed $U(x) = \infty$ outside the well. The use of a simplified square well potential means that the wave

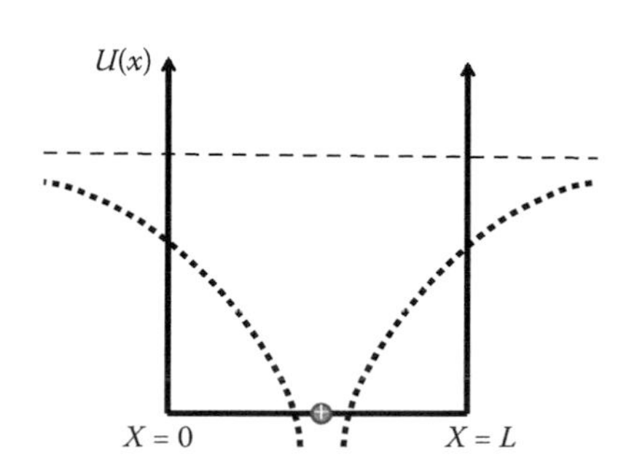

Figure 11.21 Infinite square well potential and $1/r$ Coulomb potential shown for contrast center of atom is identified by the positive nucleus.

functions and associated energy levels we find will have no relation to those obtained using a realistic Coulomb potential, but the point of the discussion here is merely to show how confinement in space leads to quantized energy levels, not to find the energy levels for a real atom.

HOW IS A SQUARE WELL POTENTIAL PHYSICALLY REALIZED?

An obvious choice might be an electron trapped in the space between a pair of very large parallel plates separated by a distance L. However, in that case, the electron, being charged, would induce charges on each plate and would therefore be experiencing forces (a nonconstant potential). Were the electron replaced by an uncharged neutron, we would have a physical situation that approximates a square well potential in which the neutron is confined.

Since the well is one dimensional, we have $\nabla^2\Psi(x) = \partial\Psi(x)/\partial x^2$ and $U(x, y, z) = U(x)$. Thus, Equation 11.5 can be simplified to yield this second-order differential equation:

$$\frac{\partial\Psi(x)}{\partial x^2} = -k^2(x)\Psi(x), \tag{11.6}$$

where the function $k(x)$ is

$$k(x) = \sqrt{\frac{2m|E - U(x)|}{\hbar^2}}. \tag{11.7}$$

Classically, we would describe the electron bouncing back and forth between the walls of the well, but remember that we are instead describing it in terms of its wave function. It is easy to solve Equation 11.6 in this case, since $k(x)$ is a constant inside the well, i.e.,

$$k = \sqrt{\frac{2mE}{\hbar^2}}. \tag{11.8}$$

As you can easily verify, the general solution is

$$\Psi(x) = \pm A\sin kx \pm B\cos kx + C. \tag{11.9}$$

Equation 11.9 is subject to the two boundary conditions $\Psi(0) = \Psi(L) = 0$, since the probability goes to zero at the edges of the well, given that the walls are impenetrable. As you can verify, these conditions require, respectively, that $B = C = 0$ and $kL = n\pi$. Thus, from Equations 11.8 and

11.9, we find that the correct wave functions and associated energies for the infinite square well are

$$\Psi_n(x) = \pm A \sin\frac{n\pi x}{L}, \tag{11.10}$$

$$E_n = \frac{k^2\hbar^2}{2m} = \frac{n^2\pi\hbar^2}{2mL^2}. \tag{11.11}$$

The $n = 1$ state, also known as the ground state, has the lowest energy, and by Equation 11.11, we see that the excited state energies ($n > 1$) can be expressed as $E_n = n^2 E_1$.

If you recall the patterns for standing waves on a string fixed at both ends, you may notice that the wave functions in Figure 11.22 look identical to them. The probability of finding the electron at various places can be easily found as functions of x using $|\Psi(x)|^2$. Solving the problem for a square well having a finite depth is a bit more challenging than the case of the infinite depth well—but the wave functions look similar to those in Figure 11.22, with the difference being that $\Psi(x)$ no longer vanishes at $x = 0$ and $x = L$. Instead, the wave function exponentially decays to zero outside the well boundaries. In addition, $\Psi(x)$ and its first derivative are required to be continuous across the two well boundaries.

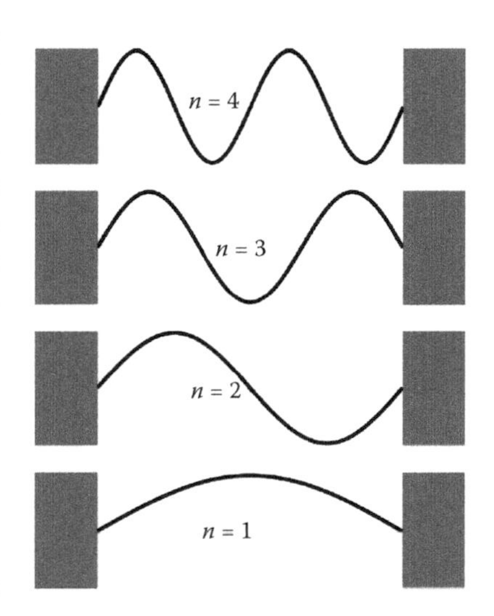

Figure 11.22 First four wave functions for a square well potential that has infinite depth.

11.A.2 Coupled Systems and Formation of Energy Bands

Now suppose rather than a single isolated atom, we had two atoms next to one another—close enough that they influence energy levels of one another to a small extent. In this case, instead of thinking of each atom with its own quantum states, we really need to analyze the combined system as a single system. We again use the square well potential (instead of the true $1/r$ Coulomb potential) for each atom, i.e., we have a double square well potential—but this time with a finite height barrier between the wells. Suppose we try to imagine what the ground state wave function might look like here. A good approximation known as the tight binding model involves taking a superposition of wave functions for individual isolated atoms. The basis of this approximation is that the influence of adjacent atoms on the wave functions of one another is very small.

Thus, the ground state wave function for a double well can be made up from the single well wave functions, which need to be modified only slightly since they do not go to zero at the well edges for the finite well. As a guess for what the ground state looks like, we might try a symmetric combination of the two individual square well wave functions: $\Psi_S(x) - \psi_{1\text{Left}}(x) + \psi_{1\text{Right}}(x)$, i.e., which yields something like the dotted curve in Figure 11.23. On the other hand, an equally good guess might be the

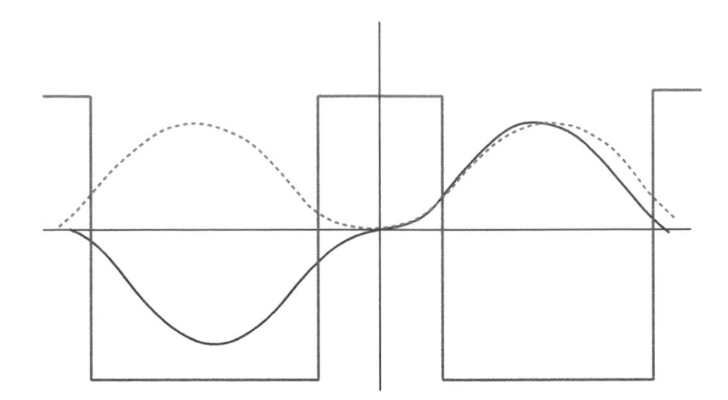

Figure 11.23 Double square well and the two lowest energy wave functions.

antisymmetric combination $\Psi_A(x) = \psi_{1Left}(x) - \psi_{1Right}(x)$—something like the solid curve in Figure 11.23. A proper solution of the Schrödinger equation yields two wave functions quite similar to these *S* and *A* combinations, but having slightly different energies.

Can you tell which of the two solutions *S* or *A* would have the higher energy? Hint: Which of the two has the longer wavelength inside the right well? The procedure to obtain the two wave functions shown in Figure 11.23, which can be somewhat tedious, is to use sine functions for $\Psi(x)$ inside each well and exponentials outside the wells and then require the continuity of $\Psi(x)$ and its first derivative at each boundary. My congratulations if you have successfully solved this problem at some point, but it is not necessary for our purposes to write the exact form of the wave functions and the associated energies.

The only essential point in the preceding discussion is that a single wave function and energy level for the single well gives rise to two wave functions with slightly different energies in the combined (coupled) two-well system. If we were to now consider the case of three nearby square wells (representing three nearby atoms), we would find that each energy level

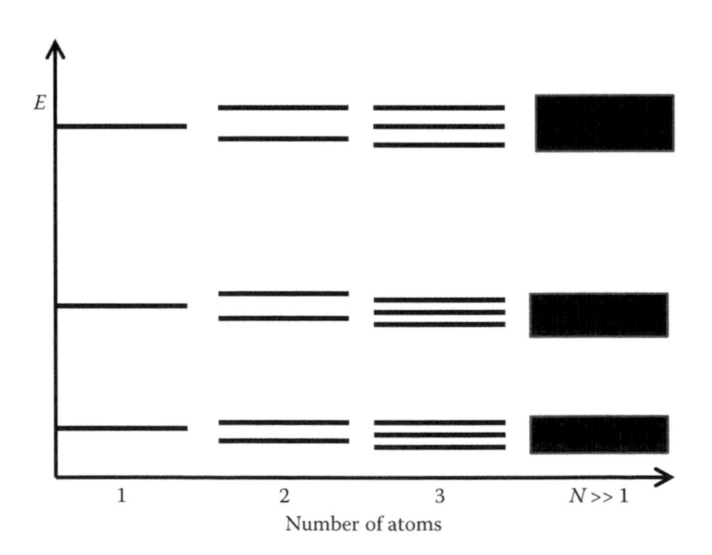

Figure 11.24 Splitting of single atom energy levels when two, three, and $N >> 1$ coupled atoms are present.

for the single atom system now splits into three separate levels, and if we had a row of N atoms, where N is extremely large, each level would split into N levels. This is exactly the situation we have in a crystal, where N is on the order of 10^{23}. Crystals, of course, are three dimensional not one dimensional, but the basic analysis is the same. Given the enormous size of N, the energy levels are so close together, we speak of energy bands—the black rectangles in Figure 11.24.

PROBLEMS

1. Which of the two double-well wave functions shown in Figure 11.23 has the higher energy? Explain.
2. Make a drawing of the three lowest energy wave functions for a triple square well potential.
3. The bandgap of gallium arsenide (GaAs) is 1.4 eV. Find the optimum wavelength of light for PV generation in a GaAs solar cell.
4. Consider a single silicon crystal 1 cm on a side. What is the spacing between the levels in the conduction band, assuming it is 1 eV wide?
5. Find the probability of levels just above the bandgap being filled in undoped GaAs at room temperature (300 K) and compare it with silicon.
6. At what temperature is the number of electrons in some interval ΔE at the bottom of the conduction band of undoped silicon (bandgap 1.1 eV) the same as that in undoped galium arsenide (bandgap 1.4 eV) at room temperature?
7. Prove that the point on an I (current) versus V (voltage) curve where the power is maximum occurs where the straight line of slope -1 is tangent to that curve. Hint: Consider what the power is at that point and at two points a small distance $\pm dV$ on either side of the maximum power point.
8. Solar cells are usually connected in series to form a solar panel. What problem could occur if they were connected in parallel? Hint: Why are batteries normally connected in series?
9. What are the differences between a current source such as a solar panel and a voltage source such as a battery?
10. When we compare the amount of energy that a PV cell can generate on a typical summer and winter day, there are two competing effects: (a) In summer, the days are longer, and (b) the temperatures are higher. Use Equations 9.11, 9.12, and 11.2 to gauge the relative importance of these two effects and determine their relative sizes. Assume we are comparing July 1 and January 1 at latitude 45° north and that the tilt of the solar collector gives the same maximum irradiance at noon on the 2 days. Assume that the average temperatures are 0°C and 25°C in January and July, respectively. Hint: You will first need to find the length of the day for the two named days.

11. Referring to Table 11.2, show that the total system efficiencies (last row in table) follow from the individual component efficiencies for those components that apply in each case.
12. At any distance x from the plane defining a pn junction, the product of the densities of holes and electrons is a constant independent of x. Why must this be so?
13. Consider a p-type material doped with 10^{17} atoms/cm^3 and an n-type material on the other side of the junction with 10^{18} atoms/cm^3. What is the ratio of majority to minority carriers far from the junction on each side?
14. Make a qualitatively correct plot of the density of free electrons and holes as a function of distance x from the junction showing both positive and negative x-values and using a log scale for the two densities.
15. What are the similarities and differences between these three: (a) a battery, (b) a charged capacitor, and (c) a pn junction.
16. Suppose an electron is in the $n = 6$ state in an infinite square well potential well of width L. Find all x values where the probability of finding an electron is a maximum.
17. Assume we were to convert 1/500th of the available solar radiant energy of the total US area to usable form assuming 100% efficiency. (Hint: See the question and your answer in Chapter 9.) What state would be closest in area to be required if you had only 15% efficiency?
18. What is the current available from a typical single solar PV panel (130 W) if the electric potential developed is 17.5 V?
19. Consider a region where it costs an average of $8.00/W of solar panels installed. What is the cost to install a 3 kW solar PV system for a residence?

REFERENCES

Iles, P. A., and S. I. Soclof (1975) Effect of impurity doping concentration on solar cell output, *11th Photovoltaic Specialists Conference*, Institute of Electrical and Electronics Engineers, Phoenix, AZ, May 6–8.

Shockley, W., and H. J. Queisser (1961) Detailed balance limit of efficiency of p-n junction solar cells, *J. Appl. Phys.*, 32, 510–519.

Twidell, J., and T. Weir (2006) *Renewable Energy Resources*, 2nd edn., Taylor & Francis, Boca Raton, FL, p. 183.

Energy Conservation and Efficiency

12.1 INTRODUCTION

Energy efficiency and renewable energy have been called the twin pillars of an energy policy that is sustainable. There are many reasons to pursue energy conservation and efficiency, which have been described as the low hanging fruit in terms of increasing our energy supply—or more properly making the existing supply go further. One 2008 study has placed the potential nontransportation energy savings for the United States from increased conservation and efficiency to be around $1.2 trillion through the year 2020—or 23% of the nation's energy budget (McKinsey & Company, 2008). But economics is only one benefit for increasing conservation and efficiency efforts. Virtually all methods of generating and transmitting energy have environmental consequences—some methods being far more harmful than others. By avoiding or reducing the need to generate and transmit energy, these environmental consequences can be reduced. Energy conservation and efficiency can also prolong the life of equipment, by putting it into a low-power sleep mode automatically if it is not in use. Finally, they can help posterity by leaving more of the world's finite supplies of nonrenewable energy sources for their use (when they will be in even scarcer supply) and help reduce one cause of international conflict—the global competition for those resources.

A physicist's first thought on hearing the phrase "energy conservation" is probably "How could it be otherwise; energy is always conserved!" But of course we are here using the phrase in a different sense of how energy is used by humans and efforts to reduce the amount we use. The two concepts of energy conservation and efficiency are closely related, and both are worthy of support, but they have somewhat different connotations. Efficiency involves using less energy to achieve the same ends, while conservation puts the stress on simply using less energy, even if we have to compromise on the ends. Efficiency generally involves a technical solution, while conservation involves a behavioral one. Thus, we can conserve energy in the home by turning down the thermostat and wearing a sweater or we can use energy more efficiently (and use less of it) by having a programmable thermostat that lowers the heating or cooling system at the times we normally are not at home. Of course, there is in this case every reason to follow both courses—the first action (conservation) requiring a small sacrifice, and the second one (efficiency) requiring neither behavioral change nor sacrifice—except perhaps learning to use the programmable thermostat! Although efficiency and conservation are often mutually reinforcing, they can sometimes be seen as being in conflict, depending on one's worldview.

Examples of the kinds of attitudes and approaches for someone who puts primary stress on either the conservation or efficiency approaches appear

CONTENTS

Table 12.1 Views Held by Those Putting the Primary Stress on Conservation and Efficiency

Conservation Emphasizer	Efficiency Emphasizer
Use less energy	Use energy more efficiently
Emphasize human behavior	Emphasize technology
Educate public on the environment	Educate public on costs and benefits
Slow economic growth if needed	No need to slow growth
Ban wasteful practices and products	Market will choose best products
It is the Third World's turn now	Keep American advantages
Technology is the enemy	Technology is the savior

in Table 12.1. The order of rows in the table is from relatively apolitical to increasingly political. In practice, most of us probably find merit in both the conservation and efficiency approaches and find some of the stark choices listed as being rather simplistic, particularly for items in the bottom half of the table. For example, the idea of always putting the environment ahead of the economy in making energy decisions is just as silly as the converse. Nevertheless, it is also probably true that many of us tend to emphasize one position over the other—even if we add a "yes, but…" to the listed proposition, as in "Public education on costs and benefits is important, but let's be sure to include not only economically quantifiable costs and benefits, but intangible ones as well—such as loss of species."

WHAT ABOUT HUMAN ENERGY?

In thinking about using energy efficiently, it is unclear whether or not to include human energy. Modern technological society has replaced many chores at one time done by human labor with efficient machines, thereby saving much back-breaking work and freeing people for other tasks. When tasks are done by manual labor instead of using these machines, the end result is some extra expenditure of fuel—namely, the additional food intake needed to allow humans to perform the work, as well as whatever energy went in to growing the food and shipping it to market. These considerations are usually not taken into account when evaluating the energy efficiency of a bicycle or other human-powered vehicle—perhaps on the grounds that such vehicles avoid directly consuming any fuel, but perhaps they should be.

Energy, people, and natural resources are the three primary drivers of a nation's economy, with the first two probably being preeminent, as evidenced by those successful nations, such as Japan, that lack many natural resources. The importance of energy in promoting economic well-being can be illustrated through the connection between per capita gross domestic product (GDP) and per capita energy usage for nations across the globe.

The ratio of the two variables plotted in Figure 12.1 is the energy consumption per unit of GDP or the energy intensity of each nation, which is

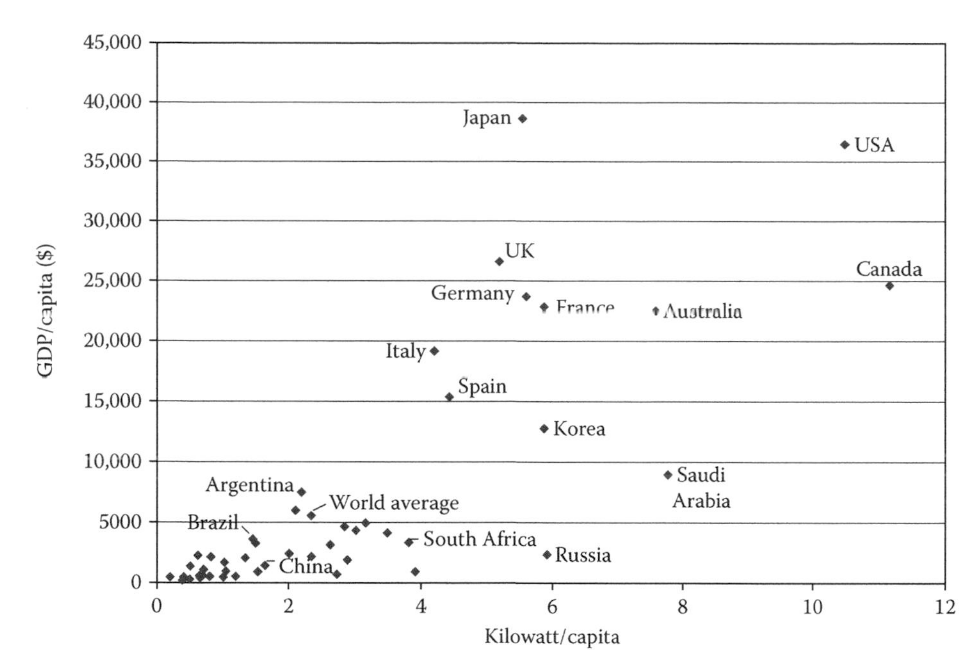

Figure 12.1 Per capita GDP versus per capita energy usage for various nations. (Courtesy of Frank van Mierlois, http://es.wikipedia.org/wiki/Archivo:Energy_consumption_versus_GDP.png, image licensed under the Creative Commons Attribution ShareAlike 3.0 License.)

a measure of how efficiently a nation transforms energy into wealth—with smaller values representing greater efficiency. It can be seen that the two variables in Figure 12.1 are strongly correlated—heavy per capita energy usage tends to be associated with greater GDP per person. Equally interesting to the strong correlation are the numerous outliers that have either far higher energy intensity than average or far less. Nations in the former category may have high energy intensity for a variety of reasons, not just because they use it inefficiently. Other reasons for high energy intensity might include an inhospitable climate, long commutes (dispersed population), more home ownership, or an overabundance of domestic energy sources. Conversely, nations having low energy intensity might benefit from a moderate climate, having a concentrated population, more apartment dwellers, or a scarce domestic energy supply. Clearly, energy intensity involves many factors that are independent of how efficiently the nation uses energy, as the term is normally understood. Even with this caveat, however, it is still true that most of the nations on the top 10 list of lowest energy intensity usage are models of efficient usage of energy: Japan, Denmark, Switzerland, Hong Kong, Ireland, United Kingdom, Israel, Italy, Germany, and Austria. Finally, it is interesting that one nation not on the list (the United States) that is unaccustomed to considering itself as average falls almost exactly on the trend line.

The strong correlation between per capita energy usage with greater GDP per person says nothing about the direction of cause and effect or whether it even exists. For example, does greater energy usage per person promote economic growth or is it the result of it? Or are both the result of some other variable, such as having a high-technology society or a highly educated population? One way to express the relationships depicted in Figure 12.1 is through the equation

$$\frac{\text{GDP}}{\text{population}} = \frac{\text{GDP}}{\text{energy}} \times \frac{\text{energy}}{\text{population}},$$

which can be symbolically written as

$$G_P = E_P \times G_E, \tag{12.1}$$

where G_P is the GDP per person; E_P is the energy used per person; and G_E is the GDP generated per unit of energy.

One key variable that has so far been omitted from these international comparisons is the relative sizes of the nations' populations. One way to include this variable is through the substitution $E_P = E/P$ into Equation 12.1, where E is the total energy used by a nation and P is its total population, giving

$$E = \frac{P \times G_P}{G_E}. \tag{12.2}$$

It can be instructive to see how changes in each variable in Equation 12.2 are interrelated. If we make use of the relation

$$\Delta E = \frac{\partial E}{\partial P}\Delta P + \frac{\partial E}{\partial G_P}\Delta G_P + \frac{\partial E}{\partial G_E}\Delta G_E, \tag{12.3}$$

we find that the fractional change in each variable satisfies

$$\frac{\Delta E}{E} = \frac{\Delta P}{P} + \frac{\Delta G_P}{G_P} - \frac{\Delta G_E}{G_E}. \tag{12.4}$$

12.1.1 Example 1: What Went Wrong?

A nation's government hopes to increase its citizens' standard of living by increasing its energy-generating capacity by 20% over a 10-year period, while simultaneously improving the nation's energy efficiency by 10%. At the end of this time, the nation's leaders are disappointed to find that the GDP per person has only increased by a very modest 3% rather than the expected 30%. What was left out?

Solution

Solving Equation 12.4 for the fractional change in population over 10 years, we find

$$\frac{\Delta P}{P} = \frac{\Delta E}{E} - \frac{\Delta G_P}{G_P} + \frac{\Delta G_E}{G_E} = 0.2 - 0.03 + 0.10 = 0.27.$$

The government forgot to allow for population growth, amounting to 27% during that 10-year period.

12.2 FACTORS BESIDES EFFICIENCY INFLUENCING ENERGY-RELATED CHOICES

The usual definition of energy efficiency is the fraction of the expended energy that produces a desired result. On that basis, for example, the efficiency of an incandescent light bulb is about 2.6%, since 97.4% of the electrical energy goes to invisible IR radiation rather than light. IR radiation will heat your house, which would be useful in winter, but is a detriment in summer, so we leave that consideration out. Consider another example: if electricity is used to heat water, we could say that the efficiency of the process would be virtually 100%, and yet in another sense, electric hot water heating is perhaps only a third as efficient as using natural gas, because it is foolish to neglect the energy losses in producing the electricity in the first place and transmitting it to your home. Clearly, meaningful definitions of efficiency need to consider losses at all steps of a process from generation to end use. Furthermore, the wise use of energy is not just about energy efficiency, as important as that quantity may be—it is also about time spent, safety, convenience, feasibility, cost, environment, and even culture. Here we consider two case studies to illustrate the many factors besides efficiency that go into decisions about energy and how those other factors often assume much greater importance than efficiency.

12.2.1 Biking to Work

Bicycling is an extremely efficient means of transportation, as it uses no energy save your own, and it helps keep you fit as well. In many European cities, a significant fraction of the population commutes to work by bicycle—for example, in Copenhagen, the percentage is estimated at 55%. The practice is far less common in the United States even in places that are more health conscious and bicycle friendly. Among the leading towns and cities in the nation, the numbers are quite small: at 6% Portland, Oregon, tops the nation, followed by Boise at 4%, and Seattle at 3%. America with its vast open spaces is a nation addicted to the automobile, where 90% of the population report spending an hour and a half each day in their car and where the average commute to work is 16 mi. For most US locations, either the commute is too long, or one would need to compete with auto traffic that may not be too keen on giving the right of way to a two-wheeled slow interloper. Still, over 35% of Americans commute less than 5 mi to work, but for most, that 5 mi might as well be 50 mi given the lack of bike paths and the blindness and even outright hostility of some drivers toward bikers.

Biking to work is almost certainly more time consuming than driving, but is it really more dangerous? Statistics show that in the United States, biking comprises around 1% of all trips, but 2% of traffic fatalities. Thus, at least for a nation like the United States, where biking is

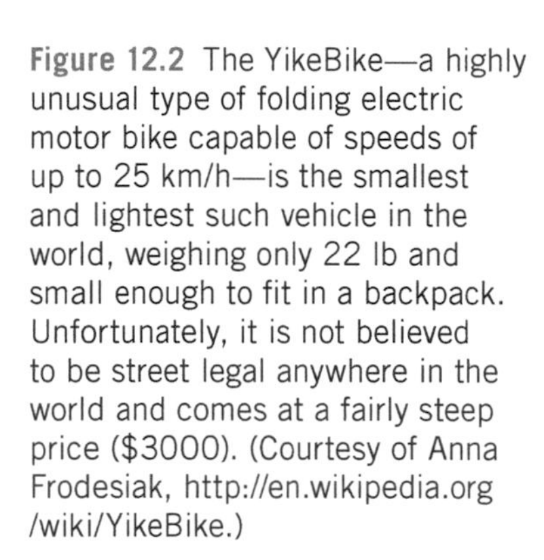

Figure 12.2 The YikeBike—a highly unusual type of folding electric motor bike capable of speeds of up to 25 km/h—is the smallest and lightest such vehicle in the world, weighing only 22 lb and small enough to fit in a backpack. Unfortunately, it is not believed to be street legal anywhere in the world and comes at a fairly steep price ($3000). (Courtesy of Anna Frodesiak, http://en.wikipedia.org/wiki/YikeBike.)

less common, and drivers may not be used to looking out for bikers, cycling to work does put one's safety at greater risk. While biking to work is extremely energy efficient, its safety and its practicality—given many long commutes and many Americans' sedentary lifestyle—make it a poor choice for most commuters. The last consideration is especially ironic, given the amount of money many Americans spend on diets and gym memberships to reduce weight and get in shape. The abstract idea of promoting greater use of bicycles in an era of high-priced gasoline has tremendous appeal, but it would require a sizable cultural transformation ("Copenhagenization?") to become widely adopted in the United States. Still, the more environmentally conscious and fit among Americans may become more used to the idea and perhaps even opt for an electric bike—a mode of transport that could be an ideal way to navigate through traffic-congested cities, except for the unfortunate fact that they have been banned for usage on the streets of some American cities, presumably on safety grounds (Figure 12.2).

12.2.2 More Efficient Solar Collectors

As another example of how energy choices are influenced by many considerations besides efficiency, consider the matter of solar collectors. Most commercial solar PV cells on the market have efficiencies in the vicinity of 15%. In recent years, the push to ever-higher efficiencies using multijunction solar cells have produced cells commercially available with efficiencies about three times that value. However, the higher efficiency comes at a significant increase in cost—typically by a factor of 5. Thus, the power per generated per unit cost is at present less favorable for the higher-efficiency cells. For a homeowner trying to decide what kind of solar installation is best, the situation becomes further muddied when we consider the installation cost and the lifetime of the solar cells. If the installation cost is a significant fraction of the total, it could make sense to go with the higher-efficiency, more expensive panels. At the opposite extreme, one might consider using very inexpensive solar PV panels made from amorphous silicon that is sprayed on a backing, which can even be made into roof tiles or shingles. Yet these inexpensive solar cells have efficiencies of only around 7% and their lifetime is less than crystalline solar cells. Clearly, one would need to weigh all the variables, cost, efficiency, lifetime, total power produced, and more, to determine the optimum choice. In fact, it can be argued that for renewable energy sources where the fuel is abundant and free, efficiency is entirely irrelevant! Perhaps the most important criterion should be cost per kilowatt-hour produced over the lifetime of the system—with two important caveats. First, the capacity of the system needs to be enough to meet your needs, and, second, the upfront cost may be as important, or more important, than the lifetime cost. In fact, the high upfront investment has been the main barrier to a move toward greater usage of renewable energy sources, even when they are cost-effective over their lifetime.

12.3 LOWEST OF THE LOW-HANGING FRUIT

A 2008 study commissioned by the US Energy Information Agency by the McKinsey & Company (2008) attempted to identify the most cost-effective places outside the transportation sector to conserve energy through the year 2020. Although the McKinsey study was specific to the US economy, its conclusions are likely to be applicable to some other developed nations—except that many European and Asian nations have already made many of the changes that are recommended. For example, according to the DOE, while only 20% of homes older than 1980 have adequate insulation, the figure is 80% in the United Kingdom.

12.3.1 Residential Sector

Of the possible cost-effective efficiency improvements identified in the McKinsey study, the residential sector accounts for fully 35% of the savings, with the remainder split between industry (40%) and the commercial sector (25%). The top categories for energy savings in the home identified in the study are listed in Table 12.2.

The order of entries in Table 12.2 is based on their cost–benefit ratio, i.e., the largest cost avoided per dollar expended, and all table entries listed would save at least twice the initial outlay over a decade. The numbers next to each table entry indicate the possible savings as a percentage of the US energy budget based on the analysis in the McKinsey study. That electrical devices should have the largest potential savings should come as no surprise, given the number of these in the home and their energy inefficiency. Just take one ubiquitous electronic device—the computer—the potential future energy savings are very sizable. According to researchers

Table 12.2 Potential Energy Savings Expressed as Percentages of the US Energy Budget They Could Save

	Percentage of Potential Savings of US Energy (%)
Electrical devices	1.48
Lighting	0.25
Programmable thermostat	0.57
Basement insulation	0.72
Duct sealing	1.29
Attic insulation	0.49
HVAC maintenance	0.34
Water heaters	0.38
Windows	0.23
Air sealing	0.68

Note: The order of the items is in terms of the largest savings in home energy usage. Note that the numbers listed were not included in the 2008 McKinsey report but were inferred from a graph that was included.

at the University of California, Berkeley, emerging new technology using magnetic microprocessors instead of silicon-based chips has the potential to consume a million times less energy per operation than existing computers (Lambson et al., 2011).

Energy-saving measures for the home would collectively include half the total upfront investment for the energy improvements for all sectors of the entire economy or about $229 billion. While that is an impressive figure, the lowest of the low hanging fruit in the home would require very little investment.

Many of the specific actions that a homeowner needs to take to save energy involve simple behaviors that involve no sacrifice or initial investment, including the following:

- Turning lights out when leaving a room
- Turning the refrigerator and freezer down, i.e., up in temperature
- Setting the clothes washer to warm or cold, not hot
- Turning down the water heater thermostat
- Running the dishwasher only when full and not using heat in the drying cycle
- Not overheating or overcooling rooms
- Closing vents, drapes, and doors in unused rooms
- Relying more on ceiling fans than air conditioning

Other actions involve very inexpensive items and a minimum of labor, such as the following:

- Cleaning or replacing filters of furnace and air conditioners as recommended
- Wrapping the hot water heater in an inexpensive insulating blanket

- Caulking and weather stripping to plug air leaks
- Insulating ducts
- Adding insulation to attic or basement
- Using energy-efficient lighting (compact fluorescent lamps [CFLs] or light-emitting diodes [LEDs])

There are many good websites that have detailed recommendations on energy-saving measures, and many homeowners have found it useful to have a professional do an energy audit of their home to identify the most cost-effective. Although there are many inexpensive ways to conserve energy, some may require a significant expense. Often, however, this is an expense that the homeowner would eventually need to make anyway, such as buying new, more efficient, major appliances, such as a washer, dryer, refrigerator, or hot water heater. In the last category, many homeowners may wish to look into the solar option, which while more expensive in its initial outlay can pay for itself in several years, depending on climate. There are many good websites that rate appliances based on their energy efficiency, and in the United States an Energy Star label is used to identify those that meet certain standards. Of course, in addition to buying efficient appliances, it is equally important to purchase those that are sized to your needs, rather than the largest or most powerful one available.

VAMPIRES AMONG US

With regard to appliances, there is also the issue of standby power, sometimes called *vampire power*, that they draw even when not in use as long as they are plugged in. By some estimates, vampire power can be as much as 20 W per electrical device and as much as 10% of all residential energy consumption. Some state and national governments have mandated that appliances expend no more than 1 W on standby power, but homeowners can, of course, take their own action by simply unplugging devices when not in use. Since this may be a bit of a chore, it might be worthwhile to identify those devices for which this might be necessary. One of the worst vampires in the home is a plasma TV, if you happen to have one. Plasma TVs can consume as much as 20 W on standby, which might cost around $20 over the course of a year. Another major vampire is the cable box for those who have cable TV channels, and these consume around 10 W. In contrast, modern high-definition liquid crystal display televisions consume far less (about 1 W in standby mode), so they need not be unplugged. In general, any appliance that feels warm to the touch is probably consuming more standby power than is necessary.

The biggest single energy consumer in the home is usually associated with space heating and/or air conditioning, which when combined account for between 50% and 60% of the energy costs. The least expensive ways of cutting losses in this area was noted earlier, i.e., making sure that the home is very well insulated and that any air leaks are sealed. Many air leaks occur around windows, and it is useful (but costly) to replace single-pane windows with double-pane windows that are filled with argon gas rather than air. At some point, it may be necessary to replace the furnace or air conditioner, and here again, efficiency and capacity should be your major considerations in addition to cost. Unlike other major appliances, however, with furnaces, there is the additional matter of choosing the energy source: electric, gas, oil, or geothermal, for example—and the choice has both environmental and economic implications that may not be entirely clear. For example, in Chapter 6, we saw that a high-efficiency gas furnace might involve less CO_2 emissions than a geothermal heat pump, since the latter requires electricity to run the heat pump.

12.3.2 Lighting

Energy-efficient lighting is important both for the home and for the commercial sector, where it is surprisingly the largest single energy consumer—being three times that of air conditioning. In homes and offices combined, it has been estimated that lighting consumes between 20% and 50% of the total energy used. The potential savings in this area is enormous due to the inefficiency of the traditional incandescent light bulb. Although CFLs have been on the scene for some time and represent a significant improvement in energy efficiency, the technology that is likely to overtake them is based on LED lighting. LEDs make use of the same pn junction used in solar cells. Instead of producing an electric current when solar radiation is incident with the LED, the converse process occurs: an electric current through the pn junction causes light to be emitted—whose color or wavelength (in nanometers) is determined by the bandgap energy E of the pn junction according to

$$\lambda\,(\mathrm{nm}) = \frac{hc}{E} = \frac{1240}{E}, \quad (12.5)$$

where E is in units of electron volts (eV). As seen in Figure 12.3, when a sufficient voltage is applied across the pn junction, electrons and holes on either side will flow in opposite directions toward it. When they recombine at the junction, they annihilate and create a photon whose wavelength is given by Equation 12.5. In terms of energy, the electrons jump downward across the bandgap, while the holes jump upward. Alternatively, we can say the electrons are just filling the holes in their downward jumps across the bandgap.

The pace of technological progress in better and cheaper LEDs is unparalleled, outside of that for computer chips. As can be seen in Figure 12.4, for each of the last three decades, the amount of light emitted per LED has increased by a factor of 30—an astonishing rate of improvement known as Haitz's law. Matching the improvement in LED efficiency is an equally impressive exponential decrease in their costs over time—a factor of 10 each decade (Figure 12.5).

Additionally, LEDs have an extremely long lifetime of around 60,000 h, compared to only about 1,200 h for an incandescent bulb, making them

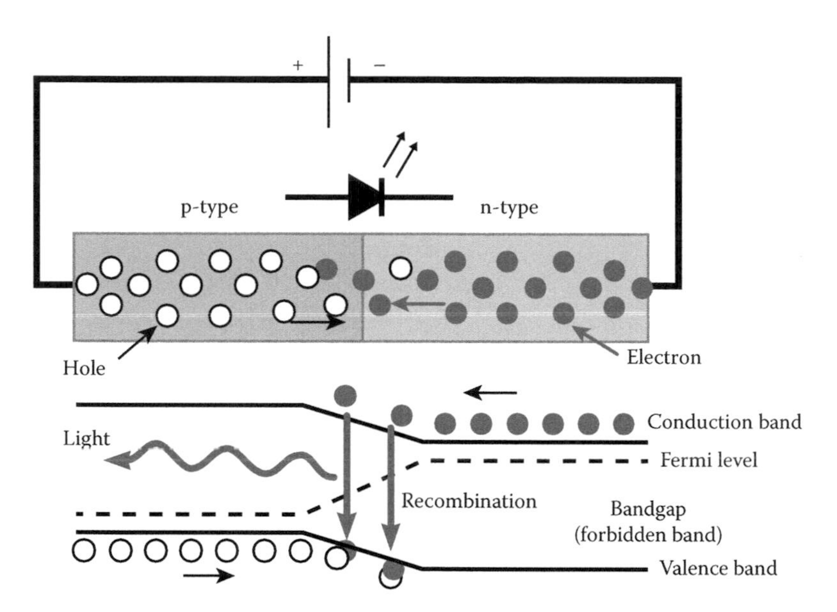

Figure 12.3 Inner workings of an LED—top image shows the flow of electrons and holes across the pn junction; bottom image shows the energy jumps the electrons and holes make across the bandgap.

ideal for use in places where it is inconvenient or difficult to replace bulbs. Moreover, the failure of LEDs unlike most other bulbs is gradual over time, rather than sudden, which usually represents another asset, plus they are much less fragile than other bulbs. The biggest advantage of LEDs is, of course, their high efficiency, as of 2010, reaching 208 lm/W—about 14 times that of a typical 100 W incandescent bulb.

LEDs achieve their high efficiency in one of two ways: either by creating white light from a mixture of red, green, and blue LEDs or else by using a phosphor that transforms the incident light from a blue LED into white light in a similar manner to the way a fluorescent bulb works. In either case, most of the spectrum produced falls inside the region visible to the human eye unlike the case of an incandescent bulb, which is mostly in the IR. The main disadvantage of LEDs is their high initial cost compared to incandescent or CFLs, which, at present, could represent a significant barrier for their residential use, even though over their lifetime, the energy savings much more than outweighs their initial cost, as Example 2 shows.

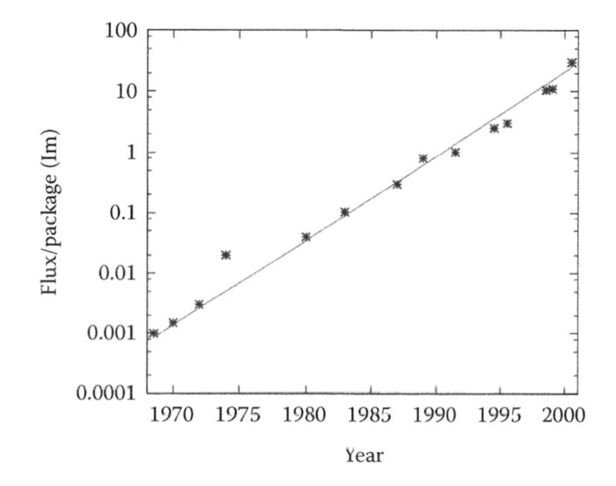

Figure 12.4 Exponential pace of progress in the efficiency of LEDs. (Courtesy of Thorseth, http://en.wikipedia.org/wiki/Haitz%27s_Law, image licensed under the Creative Commons Attribution-Share Alike 3.0 Unported license.)

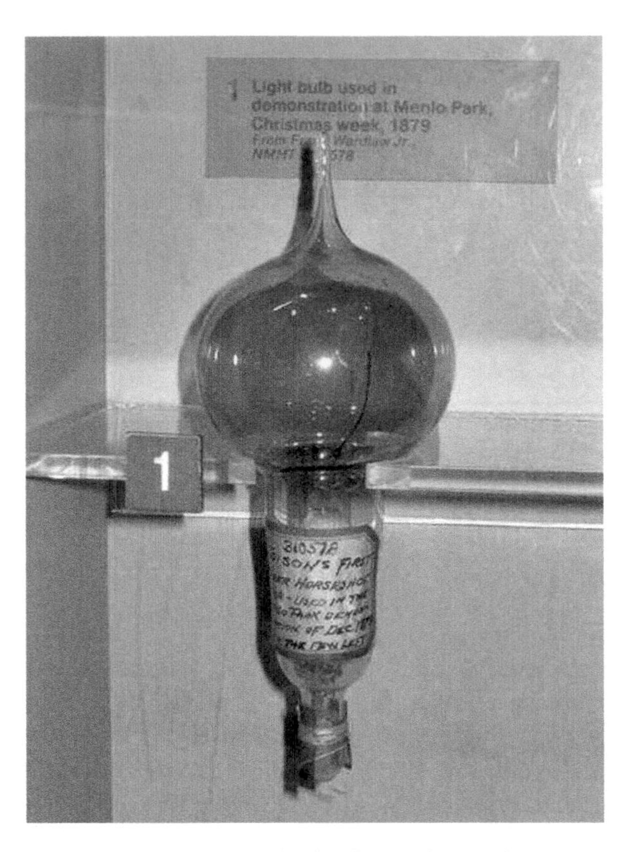

Figure 12.5 Edison's first successful light bulb used in a public demonstration in 1879. (Courtesy of Alkivar, http://en.wikipedia.org/wiki/File:Edison_bulb.jpg.)

Even though the incandescent light bulb is about the worst form of illumination in terms of efficiency compared to its modern alternatives, it is also perhaps the greatest single invention of the last 200 years. That assessment will certainly be disputed by those favoring other more modern candidates, such as personal computers, cell phones, or the Internet. The huge importance of the humble light bulb goes well beyond that one invention, however, because its widespread usage required a generation and distribution system for electricity, and readily available electricity in homes and factories was the stimulus for all the other electrical devices to which we have become accustomed. It is difficult for a person living in the developed world today to fathom what life would be like without access to electricity—although a brief taste is provided during times of electrical blackouts. An extended period with no electricity (including no way to charge batteries) would force the industrial world into a much more primitive existence for which most people would be woefully ill-equipped—even though the half of humanity now living on less than $2.50 per day now finds itself in that predicament.

EDISON AND THE INCANDESCENT LIGHT BULB

The prolific Edison, who held a record 1093 US patents, may not have been the first person to come up with the idea of an electric light bulb, but his persistence in trying any number of materials as bulb filaments finally led to one that did not quickly burn out and could serve as the essential element for his 1880 patent for a practical light bulb. Even though Edison's first successful test of a carbon filament lasted only 40 h, it was a huge improvement over what had come before. Interestingly, Edison never tried tungsten filaments because the technology for creating them in the form of fine wires did not exist at the time (Figure 12.4).

12.3.3 Example 2: Lighting Cost Comparison

Use the data provided in Table 12.3 and calculate (a) the cost savings over a 25-year period for a homeowner who uses LEDs or CFLs instead of incandescent bulbs and (b) the payback time for the initial cost of the LED bulbs over the incandescent bulbs. Assume that a typical home uses the equivalent of 30 incandescent bulbs 60 W each and that the bulbs are on for 3.65 h per day. Assume that the cost of electricity is $0.10/kWh. Different values for the data can be found from many different sources, so this example is only meant as a for-instance case.

Solution

A bulb that is on for 3.65 h/day is on about 60,000 h over the 25-year period, which by no coincidence happens to be the average lifetime of an LED bulb, 6 times that of a CFL and 50 times that of an incandescent. Thus, the cost for the 25-year period is the initial cost of bulbs purchased plus the energy cost or

$$\text{For LEDs:}\quad 30(\$16+\$0.1/\text{kWh}\times 0.006\text{ kW}\times 60{,}000\text{ h})=\$1560;$$

$$\text{For CFLs:}\quad 30(6\times\$3+\$0.1/\text{kWh}\times 0.014\text{ kW}\times 60{,}000\text{ h})=\$3060;$$

$$\text{For incandescent bulbs: } 30(50\times\$1.25+\$0.1/\text{kWh}\times 0.060\text{ kW}\times 60{,}000\text{ h})=\$12{,}675.$$

We can find the payback time for the LEDs over the incandescent bulbs by setting the energy savings over an unknown time t equal to the price differential of the bulbs: $\$0.1/\text{kWh} \times (0.060\text{ kW} - 0.006\text{ kW}) \times t = \$16 - \$1.25$. Solving for the time t, we find 2731 h or 4.0 months. Clearly, based on these data, the upfront investment is justified given the short payback time. But it could be even better in some locations. There are places, such as the state of Hawaii, for example, where electricity prices are triple the figure used in this example. In such a case, by continuing to use incandescent bulbs

Table 12.3 **Comparative Data on Three Types of Bulbs Having the Same Luminosity**

	LED	CFL	Incandescent
Lifetime (h)	60,000	100,000	1,200
Power (W)	6	14	60
Bulb cost ($)	16	3	1.25

in lieu of LEDs, the Hawaiian homeowner would be ignoring an investment with a 6-week payback time and be paying nearly an extra $40,000 over the 25-year period. Based on the rule of 70, a 6-week payback time is equivalent to making a financial investment that increases 600% a year!

12.3.4 Energy Management

Energy management that applies to both the commercial and industrial sectors is the single biggest way to conserve energy. Homeowners can also use the concept, but perhaps in a more informal way. The process of energy management consists of the following four steps:

1. Collecting detailed data on your energy usage
2. Identifying energy savings opportunities and the amounts that can be saved
3. Acting on those opportunities, with the most cost-effective ones first
4. Tracking the impacts of those actions and going back to step one

The author's university (George Mason) has such an energy management plan that has been quite effective, as it has led to significant reductions in energy use and put George Mason 16% below the average for US colleges and universities in terms of energy usage per square foot. Specific actions taken by the university include installing more efficient lighting, installing sensors to turn off lights when rooms are unoccupied, turning off heating, ventilation, and air conditioning (HVAC) after hours, and setting thermostats at reasonable values (not overcooling in summer or overheating in winter). Further, Mason has a fleet of energy-efficient hybrid vehicles, requires that all new buildings be Leadership in Energy and Environmental Design certified, and has a special cost-savings agreement with the electric power company. A large organization that has some electrical energy uses that can be deferred may be able to make an agreement with the power company to shed a certain amount of load (1 MW in Mason's case) on very short notice at times of peak demand in return for a lower rate for its power. The result may not involve a savings of energy for the customer, since the consumption has simply been rescheduled, but it does represent a savings in energy dollars spent and means that the power company can get by without installing new generating capacity, so it is truly a form of conservation. Conceivably, smart meters (see Chapter 13) might allow such deferred electricity consumption to occur with individual homeowners as well as large organizations.

12.3.5 Cogeneration

Although some methods of generating electricity are extremely efficient, such as hydropower, overall, the process of generating electricity in most nations is predominantly done from a combination of fossil and nuclear fuel. Although some plants can have efficiencies as high as 60%, the average efficiency in electricity production is only 33%. Of the remainder of the initial energy, 61% is in the form of heat that is usually expelled to the environment, 3% is used to run the power plant, and another 3% is lost in transmitting the electricity to the consumer. The heat from generating electricity is, in the case of the United States and many other nations, the largest single source of wasted energy.

Cogeneration involves the simultaneous generation of both electricity and heat, where the latter is used for useful purposes, such as heating homes or businesses. Unfortunately, heat, unlike electricity, cannot easily be transported over large distances without significant losses, so that it must be used either close to the power plant or for district heating as is done in some Scandinavian countries and in New York City where 10 million lb of steam is provided by the Con Edison power company to 100,000 buildings. Through cogeneration and recovery of heat energy for useful purposes, the efficiency in electricity-generating plants can be raised as high as 89%. Although usually cogeneration involves the use of heat as a by-product of electricity generation (a so-called topping cycle), it also includes the converse process in which electricity is generated as a by-product of the waste heat recovered in high-temperature industrial heating processes (a bottoming cycle).

Cogeneration is fairly common in Europe where 11% of the electricity is generated using it, with Denmark leading the way at 55%. In fact, Denmark fully heats 60% of its homes through district heating provided by cogeneration. In the United States, until 1978 legislation promoted cogeneration, the practice was much less common, but it has now risen to about 8% up from only 1% in 1980. Ironically, it was in the United States where cogeneration began in Thomas Edison's 1882 first commercial power plant that achieved 50% efficiency because of it. Unfortunately, later developments in the United States involving the construction of centralized power plants managed by regional utilities tended to discourage cogeneration, which is most feasible with smaller power stations that are close to large population concentrations. In fact, the nations that are among the world leaders in cogeneration do get a much higher share of their electric power from decentralized sources. As the world generates more of its electricity from renewable energy, which is more feasible to decentralize than fossil fuel plants, cogeneration should become increasingly feasible. As we shall see in the next section, cogeneration is not the only way to make productive use of heat that is otherwise wasted.

NUCLEAR DESALINATION: AN EXAMPLE OF COGENERATION

Since an estimated 20% of the world's population does not have access to safe drinking water, its lack represents a major and growing public health problem. Although the primary use of cogeneration is for heating homes and businesses, using the heat to produce freshwater is another possibility—for nations having access to seawater or brackish groundwater. Desalination may not be a good reason to build a new nuclear plant, but it is a good application for an existing plant. Using the waste heat from a nuclear reactor has been shown to be a cost-competitive way to desalinate seawater. Desalination usually relies on fossil fuels in the process of reverse osmosis in which brackish water is pumped through membranes—a process that typically requires 6 kWh of electricity per cubic meter of pure water produced. The feasibility of using waste heat from nuclear reactors for desalination has been demonstrated by India and Japan, but it has not been implemented on a large scale. Of course, one can also use renewable energy for desalination—a purpose to which it is well suited given its intermittent nature. In fact, a wind farm in Perth, Australia, yields 130,000 m^3 of freshwater per day. Solar energy is another possibility—especially since many areas of water shortage also have abundant sunshine. Nevertheless, neither wind- nor solar-powered desalination is an example of cogeneration, since unlike nuclear they do not make use of heat that is a by-product of electricity generation.

12.3.6 Thermoelectric Effect: Another Way to Use Cogeneration

The usual way of generating electricity from heat involves using the heat to create high-pressure steam to drive a turbine connected to an electric generator. In contrast, the thermoelectric effect involves the direct conversion of heat into electricity (the Seebeck effect), as well as the converse process of using electricity to create temperature differences (the Peltier effect), in which a temperature difference across two materials creates a voltage across them (Figure 12.6).

According to the thermoelectric effect, when two different conducting materials A and B separately connect a heat source at temperature T_2 to a heat sink at temperature T_1, they will have a different voltage across them. If the top ends are at the same voltage (the meaning of the connecting black bar), then a current can be made to flow between the bottom ends that have a voltage difference ΔV across them for as long as the

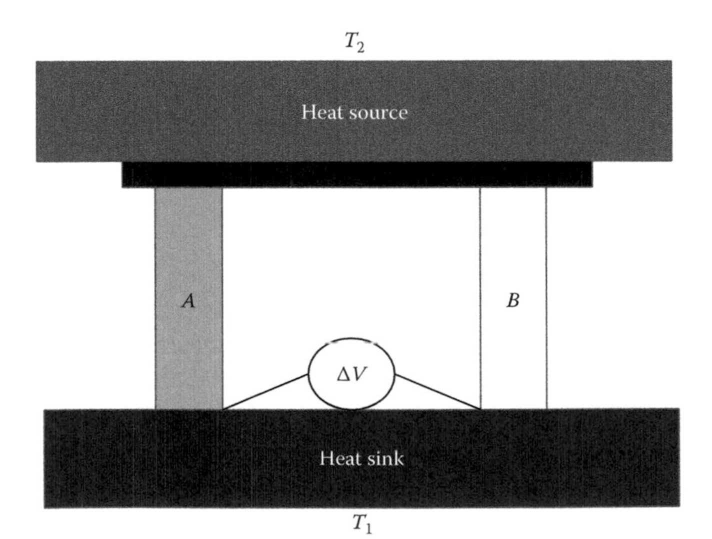

Figure 12.6 Basic idea of the Seebeck effect.

temperature difference $\Delta T = T_2 - T_1$ exists. The parameter that describes the size of the effect is the Seebeck coefficient S, defined as

$$S = \frac{\Delta V}{\Delta T}, \tag{12.6}$$

or how much voltage ΔV is obtained per unit temperature difference ΔT. The suitability of a material for generating electrical energy in the Seebeck effect depends on its electrical and thermal properties, specifically the electrical and thermal conductivities σ and κ, respectively, with the figure of merit Z being defined as

$$Z = \frac{\sigma S^2}{\kappa}. \tag{12.7}$$

Although the basic thermoelectric effect was discovered as far back as 1821 by Thomas Seebeck, its practical feasibility was greatly enhanced in recent decades by making use of semiconducting materials instead of metals for the two materials A and B, as semiconductors have the best ratio of electrical conductivity relative to their thermal conductivities. Nevertheless, the efficiency of generating electricity by this method—typically only 3–7%—cannot compare with traditional methods such as nuclear, geothermal, or burning fossil fuels. To date, there have been two principal uses of the thermoelectric effect: electric power generation using heat from radioactive isotopes in spacecraft and from waste heat in automobiles.

12.3.7 Conservation and Efficiency in Transportation

Roughly 28% of energy generated in the United States is used in the transportation sector, and it is an area that is among the most promising one for conservation given the inefficiencies that exist. In the transportation sector, 75% of the energy is wasted and the figure is even worse (85%) in the specific case of automobiles and light trucks, which accounts for the dominant transportation use. In the United States, for example, 75% of transportation energy is consumed by cars and trucks, and the remaining quarter is divided among all other modes of transport. The transportation efficiency figures are somewhat higher for many other developed nations, where there is less reliance on the automobile, more efficient cars, and shorter average driving distances than in the United States.

Transportation is nearly all (95%) fueled by petroleum, using the internal combustion engine of either the diesel or gas turbine varieties. Although there have been improvements in efficiency in recent decades, currently, automobiles use only about 19–23% of the energy supplied for useful purposes. The main loss occurs in the engine itself, which rejects 70–72%

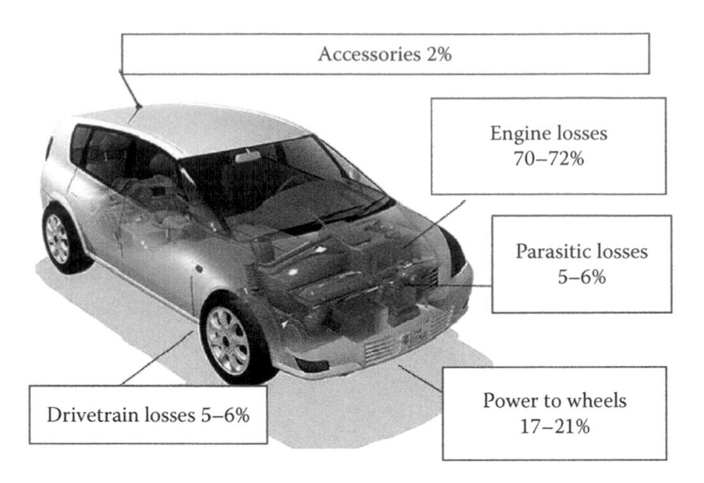

Figure 12.7 Typical losses in an automobile fueled by an internal combustion engine for combined city/highway driving. Parasitic losses include the water pump, alternator, etc. Of the power to the wheels, the majority (8–10%) is dissipated as wind resistance, while 5–6% is dissipated in rolling resistance and 4–5% in braking. (Courtesy of DOE, Washington, DC.)

of the energy as heat, according to US government data for combined city–highway driving.* A further 5–6% loss occurs when the power is transmitted through the drive train to the wheel axles. Of the useful power, 2.2% is used for all the accessories, including lights and radio, leaving only a mere 17–21% for actual propulsion. Of course, even that useful portion eventually winds up as heat—both as a result of air resistance and tire or brake friction (Figure 12.7).

There are various possibilities for improving the overall efficiency, including the following:

- Recovering some of the waste heat by the thermoelectric effect
- Regenerative braking and shock absorbers
- Improving engine efficiency
- Lighter and smaller vehicles
- Alternative fuels
- Alternatives to the internal combustion engine
- Using automobiles and trucks less or more efficiently

12.3.7.1 Thermoelectric energy recovery The exhaust gases from a car are hot enough to melt lead, and they can be used to generate electricity directly through the thermoelectric effect.

The use of automotive thermoelectric generators (ATGs) is an active field of research in the automotive industry.

12.3.7.2 Example 3: What efficiency of a thermoelectric generator is needed? (a) What efficiency e of an ATG would be needed to power all of the accessories of a car? And (b) what improvement would such an ATG mean for gas mileage? Assume values for the amount of energy in the fuel used for all accessories (2.2%), the amount used for propulsion (19%), and the amount wasted as heat (71%).

* The US government data are based on the three publications listed on their website, http://www.fueleconomy.gov/feg/atv.shtml.

Solution

In order to find e, we note that we can write the power needed for all accessories in two ways: $p_{acc} = 0.022p_{fuel}$ and $p_{acc} = 0.71ep_{fuel}$. Combining these relations, we find $e = 0.022/0.71 = 0.031$ (3.1%). Given that the amount of energy available for propulsion is now 19% plus the 2.2% no longer needed for accessories, we have 21.2% of the power in the fuel now available for propulsion. Hence, the improvement in gas mileage would be 2.2/19 = 11.5%—a substantial improvement.

12.3.7.3 Regenerative brakes and shock absorbers In normal braking systems, the kinetic energy of a car is transformed into heat by friction in the brakes as the car is brought to a stop. In contrast, regenerative brakes use the initial kinetic energy of the vehicle to power a generator whose electricity is then stored in the car battery and not wasted. Some hybrid (combined gas and electric) vehicles already use regenerative brakes, which are a major reason for their better gas mileage. Apparently, the key to getting the biggest energy recovery using regenerative braking is to come to a stop very slowly whenever possible, because the current generated depends on how quickly you decelerate, and there is a maximum charging current that the car battery can handle. Regenerative shock absorbers use the up and down motion of the vehicle to generate electricity (and, in the process, cushion the oscillations). Regenerative shocks are another energy-recovery system, but they have much less potential than regenerative brakes—at least for a passenger vehicle on smooth terrain. However, there are applications—including military vehicles traveling on dirt roads—where the fuel efficiency can be improved by up to 10% through their use.

12.3.7.4 Improving engine efficiency Given that the primary energy loss (62.4%) occurs in the engine itself, reducing this percentage offers the most promising of all possibilities. Internal combustion engines operate in a four-stroke cycle consisting of the intake, compression, power, and exhaust strokes. Four-stroke engines can be subdivided according to whether they are described by an Otto cycle or a diesel cycle in which the fuel–air mixture self-ignites without the need for a spark. It can be shown that there is a theoretical maximum engine efficiency given by

$$e = 1 - f(\gamma)/r^{\gamma-1}, \tag{12.8}$$

where r is the compression ratio, i.e., the factor by which the fuel–air mixture is compressed before being ignited and $f(\gamma)$ is a function of the variable $\gamma = C_P/C_V$, which is the ratio of its specific heats at constant pressure and constant volume.

The specific function simply has the value $f(\gamma) = 1$ for an Otto cycle, and it is a more complex function that need not concern us here for the diesel cycle. Note that for all real fuel–air mixtures, γ is confined to the range of 1.00–1.66. As one example of Equation 12.8, if we had an Otto cycle with $r = 10$ and $\gamma = 1.4$, the maximum theoretical efficiency of such an engine would be $e = 1 - 1/10^{0.4} = 0.601$ or 60.1%, but no real engine has an efficiency this high. A close inspection of Equation 12.8 shows that higher theoretical efficiencies can be obtained through increasing either r or γ. Most Otto cycle engines have compression ratios around 10, which cannot be significantly increased without causing engine knock (autoignition of the fuel) that can lead to engine damage. Diesel engines that are designed to operate with autoignition tend to have higher compression ratios, which, in part, explain their higher efficiency—about 30–35% better than Otto cycle engines.

The other alternative way to increase efficiency is to raise γ, which is more achievable in Otto cycle engines. γ can be raised most simply by using a leaner fuel–air mixture (more air and less fuel). However, if the mixture is too lean, the fuel may not even ignite. Thus, one active area of research is to improve the flammability of lean fuel mixtures.

12.3.7.5 Lighter and smaller vehicles A strong inverse correlation exists between vehicle weight and fuel efficiency, since larger heavier vehicles need to have larger more powerful engines. The main reason cars in Europe and Japan have better fuel efficiency than those in the United States is precisely because they are smaller and lighter. There is also the matter of vehicle safety to consider, and in the real world, heavier vehicles do tend to be safer—especially if you live in a society where most drivers have heavier vehicles. This is true even though the law of action and reaction (Newton's third law) requires that equal forces always act on vehicles when they collide regardless of their relative size—which many students refuse to believe! The fact of equal forces, however, by no means implies equal effects on the cars. In a collision, the lighter of the two vehicles will experience much more deformation due to its larger deceleration. Although vehicle weight does matter in terms of safety, much also depends on its design; safety features; and, most importantly, its driver demographics, knowledge, and skills. If almost everyone were driving smaller cars (as in Europe or Japan), the safety penalty to driving a light car would be considerably reduced.

12.3.7.6 Alternate fuels The smallest change of fuels might involve going to diesel engines, which as noted earlier is 30–35% better than gasoline. Environmentally, clean-burning diesel engines would be a step in the right direction, at least on a near-term basis. Other possibilities might involve cars running on biofuels, such as ethanol. Unlike the United States, the nation of Brazil has pioneered the use of engines running on 100% ethanol. Biofuels and hydrogen-fueled vehicles are discussed in other chapters. Here we consider one other possible fuel, namely, natural gas, which is both less expensive (by about a third) and more environmentally friendly, as it emits 29% less CO_2 emissions and 92% less particulates than gasoline. Best of all, for a nation such as the United States, concerned about its access to petroleum, the domestic reserves of natural gas are currently enormous, having greatly expanded over the last two decades. Natural gas vehicles (NGVs) store the fuel as a highly compressed gas (about 3000 psi), which is reduced before it enters the cylinders. The engine works based on much the same process as a conventional internal combustion engine. In case you are concerned about the dangers of very high-pressure flammable gas, the high-pressure tanks are said to be safer than gasoline tanks. It can be expensive to convert car engines to natural gas, but it is much less expensive, of course, to build NGVs in the first place. In European nations, there are around half a million NGVs now on the road, and in some nations, such as Armenia, they constitute as much as 20–30% of all cars, while in the United States, there are only 110,000 NGVs—mostly buses.

12.3.7.7 Alternatives to the internal combustion engine The main alternative to the internal combustion engine is the electric car, which could prove highly popular to those who do not have long commutes to work or a need to take many long trips. For the average driver, the feasibility of the all-electric car is tied to further developments in batteries or fuel cells, which is an issue discussed at length in Chapter 13. Let it simply be said here that in their present state of development, while their debut seems promising, it remains to be seen how much penetration into the market electric vehicles will have, as a result of concerns over vehicle range and recharging issues. Quite possibly, hybrid (gas–electric) vehicles will have greater market appeal, although strong incentives or rising gasoline prices might well change that.

12.3.7.8 Using automobiles and trucks less or using them more efficiently There are many simple ways that cars can be used less—some involving more compromises than others. One simple action would be to always combine errands so that you accomplish the same goals with less driving. For those who can "telework," there is the possibility of exploring the possibility of working from home 1 day per week or seeing if your employer would agree to a 4-day week with extended hours each day. Many employers also encourage carpooling, which can be especially attractive for very large organizations and can have significant savings. In some cities, there are lines of people waiting for someone with whom to carpool—which allows the driver the privilege of driving in certain reserved highway lanes. Finally, most cities and many suburban areas have mass transit possibilities as an alternative to driving. The relative efficiencies of cars and buses, for example, strongly depend on the load factor of the latter, which varies by as much as a factor of 5 from city to city. If mass transit is sparingly used in a city, the fuel efficiency per passenger mile can be significantly worse than for a car that has several occupants, but its efficiency is much higher than cars if it is used to capacity. Of course, even if low intercity ridership on bus routes might make them less efficient than cars, other justifications exist for their support, such as low car ownership among poor city dwellers who rely on mass transit to get to work.

Regarding freight transport, trucks are one of at least four other methods including rail, air, and ship or barge. Trucks are now roughly three times less efficient than rail, but rail is not suitable for some destinations or cargoes, so improving the efficiency of trucks is very important. One short-term improvement for trucks would be to encourage their shift to natural gas through government incentives. It has been suggested that for each 18-wheel truck that is converted to natural gas, the environmental impact would be equivalent to taking 325 cars off the road, given the much poorer efficiency of a truck and their much longer travel distances. Airfreight is of course the most expensive way to transfer cargo, but it is important for time-urgent shipments.

12.4 OBSTACLES TO EFFICIENCY AND CONSERVATION

Given the sizable savings that could result from improvements in energy efficiency and conservation, it is perhaps surprising that so little has been done to date—at least in some nations. Here we consider the obstacles in the way of implementing efficiency and conservation, which are numerous and formidable. A number of these obstacles have already been noted, for example, the competing pressures that influence energy decisions, such as safety, comfort, and convenience, which may legitimately overwhelm considerations of efficiency—especially if the costs of energy are low. Intrinsically, problems such as conservation, which involve many small partial solutions and many different decision-makers (homeowners, companies, utilities, and governments at all levels), are more complex to solve. In one sense, however, the decentralized multifaceted nature of the conservation problem also has its advantages, since individual homeowners, companies, or local governments may try many different courses of action, and the more promising ones may be reported and emulated by others.

Some lack of action on conservation may be due to simple inertia, especially in the face of a plethora of information that may be confusing to the average homeowner or company executive who is unsure where the biggest savings are. Even for a specific application, such as lighting, the various alternatives, including the many types of CFLs and LEDs can easily leave one confused. Homeowners may also have outdated information (such as believing that LEDs are limited to directional lighting) or they may be distrustful of what appears to be an extravagant claim (that a 6 W LED really is as bright as a 60 W incandescent bulb), or they may be unaware of all the existing rebates for some energy efficiency expenditures. In fact, surveys have shown that many citizens have a very poor idea about which courses of action are most effective in saving energy (Attari, 2010).

Other reasons for inaction with respect to energy conservation have to do with one's belief system and the way conservation is presented. If it is portrayed as simply "reducing one's carbon footprint," which is important mainly to those who are very concerned about human-caused climate change, then those who are skeptics (including apparently many members of one of the two political parties in the United States and nearly all its leaders), then conservation will be more limited in its appeal. Quite apart from the climate change issue, there will be those who dislike the moralistic tone associated with promoting conservation, which they may see as an infringement on individual liberty.

The most important barrier to conservation efforts, however, is probably its upfront cost, which, in some cases, can be significant. Many people and companies insist on a fairly short time frame for recouping those initial outlays. The example of the LED light bulb is again instructive, and an average homeowner may wonder what the value is in an expensive product that will last about 25 years, if he/she will probably not be in this house anywhere near that long? Of course, a different conclusion might arise if he/she realizes how short the payback time is in this case—unless she happens to have much higher priorities in a tough economy, such as paying the mortgage to avoid foreclosure!

Another obstacle to efficiency and conservation efforts involves poorly aligned incentives, meaning that the person paying the upfront cost is not the one who reaps the benefits. Some teenagers may be more likely to forget turning out the lights when they leave a room, when it is their parents who pay the electric bill. In the same vein, a property owner might be reluctant to replace energy-inefficient appliances when it is the tenant who pays the electric bills. Companies can also have a similar problem depending on how the budgets of individual departments are allocated and whether energy costs are included in them. Finally, why should a utility company encourage its customers to conserve if that means it sells less energy? All these misaligned incentives do have solutions, especially the last one where the profits of the utility company can be made dependent on its promoting conservation through legislation.

The final obstacle to energy efficiency and conservation we consider is insufficient or misguided government policies. Two examples might include the "cash for clunkers" policy enacted in the United States intended to promote more efficient automobiles—which accomplished no such thing—and the policy of banning electric bicycles from the streets of some major US cities. Presumably, the electric bike bans are on safety-related grounds—which is ironic since they are probably safer than unbanned normal bikes, as they can more easily keep up with the traffic, get away from a dangerous situation on the road faster, and maintain a decent speed uphill.

CASH FOR CLUNKERS

In 2009, the US government initiated a program informally dubbed "cash for clunkers," which was both an effort to stimulate the economy and a way of improving the efficiency of cars on the road. This program allocated $3 billion to offer rebates of $2000 to new car purchasers who were trading in old cars ("clunkers") for new ones that had at least 22 mpg. The net result of this program did indeed improve the gas mileage, since the average mileage of the new cars exceeded those of the trade-ins by 9 mpg, but its impact on the overall US fleet was negligible, given the number of cars involved, and the likelihood that the program probably just made potential purchasers move up the date of their purchase to take advantage of the program. Thus, it is likely that the dip in car sales in the 7 months following the conclusion of the program was a direct result of the stimulated sales during the 2 months of the existence of the program, and one study concluded that the costs of the program outweighed all its benefits by $1.4 billion. Even the impact of the program on the environment is unclear. Thus, while the miles per gallon of cars did improve as a result of the program, that improvement would have occurred anyway if the purchases had taken place a few months later. Moreover, environmentally, there is a cost associated with building a new car—both in terms of needed energy and raw materials. For example, to offset the carbon footprint due to the energy associated with making and shipping a new car by the gain of 9 mpg in mileage, the average driver would need to drive it for 5–9 years.

While governments may go too far with well-intentioned but counterproductive efforts to save energy, they can also not do enough in areas that might really make a difference. One possible example of this kind involves the matter of taxes on gasoline. When Michael Faraday, the inventor of the electric generator, was asked in 1850 by the British Finance Minister about the practical value of electricity, he supposedly replied "One day sir, you may tax it" (MacKay, 1982). Indeed, most nations today are quite well aware of the importance of this source of revenue and how tax rates can affect consumer choices so as to promote energy efficiency. One exception to this general rule is the United States, which has the lowest tax on gasoline of any major nation—in fact, as of 2011 it was five times lower than the next to the lowest nation out of 21 industrialized nations, according to US government data (Alternative Fuels Data Center, 2011). Equally remarkable is the way the federal gasoline tax is calculated in the United States, i.e., as a fixed dollar amount, rather than a percentage. As a result, with no rise in the tax over time, the effective tax rate on gasoline inevitably declines due to inflation—exactly the opposite of what normally happens to income taxes in nations making use of tax brackets. As of 2011, the federal gasoline tax in the United States stands at 18¢, making it in real terms as a percentage of the price only a third as great as it was in 1993—the last time it was raised. Aside from having a gas tax that is exceptionally low and continually decreasing in real terms, the

United States also has a lower average efficiency of its cars whose miles per gallon is about half that of cars in either Europe or Japan, although the Obama administration mandated a doubling of the average automobile mile per gallon standard (to 54.5 mi per gallon [mpg]) by the year 2025.

The price of gasoline clearly has some influence over the motivation to drive energy-efficient vehicles. If expensive gasoline makes people more conscious of the need to conserve, then surely cheap gasoline and low gasoline taxes have the opposite effect. Gasoline taxes at least in the United States have implications that go well beyond incentives to conserve energy; they also relate to the ability of the nation to pay for highways and bridges, which are, in some cases, part of a decaying infrastructure. Consider that in the United States, 25% of bridges have been deemed "structurally deficient" owing to the lack of federal revenue provided by ever-declining gasoline taxes to repair them. Lacking a crystal ball, no one can say whether US lawmakers might, at some point, belatedly raise the exceptionally low (by international standards) tax on gasoline. At the time of this writing, the aversion to new taxes by one political party and lobbying by the energy companies make the prospects appear dim.

12.5 IS ENERGY EFFICIENCY AND CONSERVATION ULTIMATELY FUTILE?

Although most people support efforts to improve efficiency and conserve energy, especially when they are economically viable, as we have seen, there are many reasons for lack of action. The most fundamental obstacle, however, is the belief that the whole enterprise is futile, and so why bother? In this section, we consider a theory in support of this view and assess their credibility.

12.5.1 Jevons's Paradox

Economics has been justifiably called the gloomy science. William Stanley Jevons was a nineteenth-century economist who observed that after James Watt invented his coal-fired steam engine, thereby greatly improving the efficiency of steam power; the result was not a reduction in the use of coal, but just the opposite—a vast expansion as this power source was suddenly found to be much more useful than was hitherto the case. From this observation, Jevons drew the conclusion that technological progress, when it makes a process relying on some resource more efficient, will inevitably lead to greater consumption of that resource rather than less—a counterintuitive idea that is known as Jevons's paradox. In a similar vein, some have argued that all improvements in energy efficiency are ultimately bound to be futile, because they will make the energy source more desirable and, hence, lead to greater consumption. One can indeed find cases besides Jevons's original example that support his hypothesis; however, such supporting examples are most likely to be found in the early stages of a technology when the commodity fueling it has yet to find widespread use and the technology is not mature. It is by no means clear how well Jevons's paradox applies to the situation we find ourselves in today.

The price of gasoline, for example, does matter in affecting how much and how far people drive their cars, particularly in tough economic times, and an increase in miles per gallon (one measure of efficiency) would economically have the same effect on people as a drop in the price of gas. However, economists have observed that the effect of price changes is modest, which they refer to as a price inelasticity (or insensitivity) relative to demand. They conclude that the main impact of a gasoline price rise would be reductions in other spending areas rather than in miles driven. Thus, a hypothetical doubling in the miles per gallon would likewise have little impact (certainly not nearly a factor of 2) on the distances people drive their cars. Nevertheless, such an efficiency increase rise might eventually lead people to be less conscious of the urgency to purchase an energy-efficient vehicle the next time they make such a purchase, so in this sense, this rebound effect would cancel out some of the need for more efficient vehicles and offer partial support to Jevons's hypothesis.

Another example illustrating how well or poorly Jevons's paradox applies to energy efficiency might be the possible replacement of many incandescent bulbs with the 10 times more efficient LEDs. It seems inconceivable that such a replacement would have anything like a 10-fold increase in the use of lighting that would cancel out the efficiency increase—given that we already illuminate virtually everything we wish to illuminate. Thus, rebound effects are bound to be relatively unimportant when the efficiency increase is very large. Moreover, in the area of renewable sources of energy, Jevons's paradox seems especially inappropriate, because the fuel source is free, so that even if efficiency improvements should lead to greater use of, say, solar energy, there could never be a shortage—only a move away from finite nonrenewable resources, which is all to the good. Finally, even if Jevons were correct as far as nonrenewable energy sources are concerned, this is not a valid argument for avoiding efficiency improvements, since while they might make the resource more widely used, they also improve people's lives through greater productivity.

12.6 SUMMARY

This chapter looks at the many benefits that can be obtained through a greater emphasis on energy efficiency and conservation, bearing in mind that many other considerations besides energy efficiency usually guide our energy-related decisions, and some of these other factors may be more important than efficiency. It also considers specific places where conservation and efficiency improvements might have the most impact and be the most cost-effective with particular attention to each of the four sectors of the economy: residential, commercial, industrial, and transportation.

The main focus of the discussion is on the US economy, given that many other developed nations have already implemented many of the changes. The chapter concludes with a discussion of obstacles to action on energy conservation.

PROBLEMS

1. Show that Equation 12.4 follows from Equation 12.3.
2. A nation's population increases by 1% per year over a 10-year period, and its energy-producing capacity and GDP both double in that same time. What would have happened to its energy intensity in the 10-year period?
3. Suppose we wish to choose materials to make an LED that generates white light through a combination of pn junctions that emit red, green, and blue lights. Select approximate values for the three relevant wavelengths and calculate the three bandgap energies required. Search on the web to identify some possible choices of materials to use for the junctions that match these bandgaps.
4. Refer to Figure 12.4, showing the rate of increase in the brightness of LEDs. Based on that figure, estimate the rate of increase per year and the doubling time. How do your results compare with those obtained from the rule of 70?
5. The efficiency of a power plant that generates electricity by heat produced by either fossil or nuclear fuels is limited by the Carnot theorem. Assume that the heat is expelled to the environment at 300 K and the combustion temperature is 800 K; what is the maximum possible efficiency?
6. The compression ratio in an automobile engine is 10, and the ratio of two specific heats for the fuel–air mixture is $\gamma = 1.30$. Suppose that through the use of a leaner mixture (more air and less fuel), the value is raised to $\gamma = 1.32$. How much would this increase the efficiency by assuming it is well approximated by an Otto cycle? What change in the compression ratio would have the same effect on the efficiency?
7. When driving a car with regenerative braking, you want to press the brakes gently, because the power being recovered by charging the battery cannot be absorbed if it exceeds a certain amount. Assume you wish to keep the charging current below 40 A for 12 V battery, and your 1000 kg car is traveling at 20 m/s. If you wanted to bring it to a stop and recover as much of its kinetic energy as possible, what would be the shortest stopping time?
8. Suppose that we wish to use the waste heat of a 1000 MW nuclear reactor for purposes of desalination. If the reactor is 33% efficient and half the waste heat is used to evaporate seawater, how many gallons of freshwater per day could it produce? Find a result using two methods: (a) reverse osmosis and (b) direct evaporation of the water.

9. Explain why each of the three terms defining the figure of merit Z for a thermoelectric appears in the numerator (σ and S) or the denominator (κ).
10. Show that for an appliance that consumes p watts of standby power, the annual dollar savings that can be had by unplugging it when it is not in use is up to around p dollars.
11. It has been suggested that for each 18-wheel truck that is converted to natural gas, the environmental impact would be equivalent to taking 325 cars off the road, given the much poorer efficiency of a truck, and the much longer distances they travel. Find some data on the web for these quantities for both trucks and passenger cars and the relative emissions (of all kinds) when using natural gas and gasoline and see if this number seems reasonable, i.e., if your estimate of the number is within a factor of 2.
12. How far should you be willing to drive out of your way to save 5¢ a gallon on gas, assuming your 15 gal tank has 3 gal left? How about if you could save 10¢ a gallon? Suggest three factors that have been omitted in this estimate that make your calculation an overestimation of the actual driving distances.
13. It has been estimated that extreme obesity contributes to a loss in the gas mileage of your car by 2% if you are 100 lb overweight. Assume that the work done against air resistance is proportional to the weight of a vehicle plus its occupants; make some estimates to calculate whether the preceding claim is roughly correct. Incidentally, in a related vein, the Japanese airline Nippon Airways, as a fuel-saving measure has asked all passengers to visit the restroom before boarding their planes.
14. A heat pump maintains the temperature of a home at 20°C. If the air temperature outdoors is –5°C, what is the maximum COP?
15. A home freezer maintains –5°C. If the home is maintained at 20°C, what is the maximum efficiency in terms of the energy efficiency ratio?

REFERENCES

Alternative Fuels Data Center (2011) Maps and Data, http://www.afdc.energy.gov/afdc/data/index.html#www.afdc.energy.gov/.

Attari, S. Z. (2010) Public perceptions of energy consumption and savings, *Proc. Natl. Acad. Sci. U.S.A.*, 107(37), 16054–16059.

Lambson, B., D. Carlton, and J. Bokor (2011) Exploring the thermodynamic limits of computation in integrated systems: Magnetic memory, nanomagnetic logic, and the Landauer limit, *Phys. Rev. Lett.*, 107(1), 010604.

MacKay, A. L. (1982) *The Harvest of a Quiet Eye: A Selection of Scientific Quotations*.

McKinsey & Company (2008) *Unlocking Energy Efficiency in the US Economy*, New York.

Energy Storage and Transmission

13.1 ENERGY STORAGE

13.1.1 Introduction

The two topics treated in this chapter form a natural pair: energy transmission involves transporting energy spatially, while energy storage involves its temporal movement—whereby energy produced at one time can be stored and delivered at a later time. There are numerous applications of energy storage, each having requirements that are best met by specific technologies. For example, wind-up springs may be fine for toy cars and watches (at least before the advent of miniature batteries), but they would be out of the question for supplying energy to real cars. Likewise, pumped hydro is well suited for storing energy produced in a hydroelectricity plant, but it could scarcely be imagined for use in any applications requiring a portable energy source, such as for vehicle propulsion. Making wise decisions about the relative merits of different technologies for various energy storage applications is a complex subject because there are at least a dozen criteria that can be used. This situation is different from making decisions involving available alternative ways to generate electricity, where two factors are (or should be) of most crucial importance: the cost per kilowatt-hour generated and the harm to the environment.

Energy storage is especially important in connection with electricity generation because the power output of nuclear or coal-fired generating plants cannot be rapidly changed to match the continually varying demand for electricity. Thus, through energy storage, it is possible to produce energy at one time and send it out to the power grid at another—particularly at times of peak demand during the day. This capability of supplying energy at the time of one's choosing is known as dispatchability. It is a property that becomes more valuable as renewable sources with their highly variable output become an ever-larger fraction of generated electricity. One way that energy storage can come to the rescue in providing dispatchable power involves combining pumped storage hydro with a renewable source such as wind power. Wind turbines can be used to pump water up to a reservoir at higher elevation, and at a later time when the water is allowed to flow back down it drives a turbine to create electricity at the desired rate. As we shall see, for storing large quantities of dispatchable energy, pumped storage hydro is far superior economically to other methods (Figure 13.1).

CONTENTS

Figure 13.1 Waste heap in Hamm-Pelkum, Germany, where a power plant supplying 15–20 MW is to be installed by combining wind power with pumped storage. (Courtesy of RWE [Photo by Hans Blossey].)

One of the most crucial reasons to have cheap and reliable energy storage methods is to provide power to vehicles. Short of having long extension cords, in the form of connection to overhead power lines or the third rails of some electric trains, the power source must be onboard the vehicle. Unlike the industrial, commercial, and residential sectors, the transportation sector of the economy in many nations is almost entirely dependent on just one energy source: petroleum. Of course, petroleum or natural gas fuels themselves are a form of energy storage, because their stored chemical energy is released during combustion, but they do have well-known environmental and supply problems, hence the interest in energy storage technologies that are based on clean renewable sources. The largest repository of stored renewable energy on the planet consists of that driven by solar-powered photosynthesis in living plants. Biofuels created from plants and algae were the subject of another chapter and will not be further discussed here, but they are competitors to some methods that will be considered, such as batteries, compressed air, and fuel cells.

There are several ways to characterize the numerous energy storage technologies, one being their physical nature: (1) mechanical (or thermal), (2) electric (or magnetic), or (3) nuclear. An equally important categorization involves the time interval for which devices based on the technology are capable of storing energy or delivering it to a load. We refer to the two processes of storing and delivering energy as *charging* or *discharging* the device. Thus, our classification here is based on how long a time is required to charge and discharge. Charging/discharging time is important because some applications require very short times, i.e., high power, while others do not. One useful division is (1) long charging/discharging times (minutes to many hours), (2) short times (seconds to minutes), and (3) very short times (less than a second). A simpler two-part division would be between power storage applications (short and very short times) and energy storage applications (long times), since energy being delivered on a short time scale typically involves high power. Consider the following example relating to storage for electricity generation at a power plant based on the original three-part division:

- *Energy management (long times)*: Energy management also known as load leveling allows energy generated at a time of day when demand is low to be released hours later at times of peak demand.
- *Supplying bridging power (short times)*: Maintaining continuity of service is essential when switching energy sources. One example would be bridging the gap during the transition from power from the grid to a local diesel generator during a blackout so as to cover the time delay before the generator reaches full power.
- *Maintaining power quality (very short times)*: Monitoring (and correcting) the quality of the power produced at the power plant is important for many end uses. This process of sensing and correcting needs to take place on a time scale much less than 1/60 of a second, which is the time each cycle lasts—based on a US standard of 60 cycles/s.

MAINTAINING POWER QUALITY

Ideally, electric power generated should have a fixed voltage and frequency and a perfect sinusoidal shape, i.e., a single frequency with no harmonics. The quality of power refers to the size of departures from either the standard voltage and frequency values or a sinusoidal voltage waveform. Power quality also involves the extent to which (nonperiodic) transients are present, such as spikes or dips in power, and finally the maintenance of power on a continuous basis with no interruption regardless of the load. Poor-quality power can cause serious problems for many kinds of electrical equipment. If the power at the generating plant has quality problems, it may be treated or conditioned, thereby rapidly sensing any anomalies and automatically self-correcting them. As noted earlier, this requires having a source of stored power that is capable of adding or removing power on a very short time scale. Poor power quality can also be corrected by the end user, such as using a surge protector to protect computers in the event of a surge (higher than normal voltage) or using a battery backup to allow operation to continue through a temporary power failure. Cheaper surge protectors, however, may cause their own set of quality problems, especially in terms of creating harmonics.

13.1.2 Mechanical and Thermal Energy Storage

Many mechanical and thermal storage methods operate on long time scales (e.g., pumped hydro, compressed air, and thermal storage), and we shall consider these first. One indicator of the importance of mechanical storage methods is indicated by pumped storage alone being 5% of all generated electric power worldwide. It is therefore useful to consider the unique advantages that pumped hydro offers as an energy storage method in connection with electricity generation.

13.1.2.1 Pumped hydro As discussed in Chapter 8, pumped hydro that is often an adjunct to a standard hydropower plant stores energy by running the water turbine backward, so that instead of its normal function of using the mechanical power in descending water from a reservoir to generate electricity, the turbine uses electrical energy from the grid or other external source to pump the water up to a higher reservoir, where its potential energy has been increased by mgy. Since the reservoir, dam, turbine, and generator all already exist as part of the power plant, pumped hydro involves minimal additional cost above that of the hydropower plant itself. Of course, considerable incremental costs would be incurred to install pumped hydro if it is used in conjunction with other sources such as wind power. The most significant advantage that pumped hydro offers is probably the large amount of energy that can be stored, as the following example suggests.

13.1.2.1.1 Example 1: Energy stored in a reservoir Consider a lake formed behind a dam in a hydropower plant. The turbine is $y = 20$ m below the surface of the lake whose surface area is $A = 20\ \text{km}^2 = 2 \times 10^6\ \text{m}^2$. Suppose that when the plant pumps water up from the turbine to the reservoir, the water level in the reservoir eventually rises by $d = 0.5$ m. How much energy has been stored in that half-meter thick layer? How many 12 V lead–acid storage batteries rated at 60 Ah would be needed to store the same amount of energy?

Solution

First consider the energy stored in a lead–acid battery. Since $P = iV$ and $E = pt$, we have $E = Vit$. The battery rating of 60 Ah clearly represents the product of current times time, or the *it* in the formula for energy, so that the energy content of a fully charged battery is therefore: 12 V × 60 Ah × 3600 s/h = 2.59 million J. This may sound impressive until we compare it to the potential energy stored in the raised water in the reservoir: $E = mgy = \rho Vgy = \rho Adgy = 1000 \times 20 \times 10^6 \times 0.5 \times 9.8 \times 20 = 1.96 \times 10^{12}$ J. Thus, it would require just under a million lead–acid storage batteries to store the energy of the extra half meter of water.

The million battery alternative would be out of the question economically, given that storage batteries have perhaps only 800 charge and discharge cycles before needing replacement. Thus, the cost of replacing batteries might be around 100 million dollars annually, not counting labor. In addition, the charge–discharge efficiency of batteries can be as low as 50%, meaning that half the charging energy would be lost upon discharge—erasing all the advantage of shifting the electricity to peak times. In contrast, pumped storage is in the 70–85% efficiency range, and it would be far cheaper than batteries. How big is 1.96×10^{12} J? By most standards, it represents a huge amount of energy. However, if the hydroelectricity plant generates 2000 MW, which is not unusual, it would be able to produce full power for the plant for only 18 min.

13.1.2.2 Thermal storage Heat energy can be stored most simply by filling a well-insulated container with a hot liquid, as in the case of many home hot water heaters. Here however, we specifically consider thermal storage in connection with electricity generation. As described in Chapter 10, solar thermal power can be effectively harnessed by mirrors that focus the collected sunlight onto either horizontal tubes or the top of a power tower as was discussed in Chapter 10. In one design, the heat is intense enough to melt a salt compound in the tower that is sent to a storage tank. Salt is a useful material for this application because it is inexpensive, nontoxic, and nonflammable, and it remains liquid at very high temperatures without requiring high pressure. The first such plant was built in Andalusia, Spain, in 2009 and supplies 200,000 Spaniards with their electricity. The well-insulated tanks where molten salt is stored contain enough thermal energy for up to a week and allow the plant to generate electricity even at night. In fact, the use of the molten salt storage allows the plant to double the number of hours it produces electricity at an incremental cost of only 5%.

13.1.2.2.1 Example 2: Molten salt storage tank A storage tank containing molten salt initially at 300°C is used by a solar thermal power plant to generate electricity. The plant requires a minimum salt temperature around 150°C in order to generate steam to power a turbine. If the cylindrical storage tank is 9 m high and 24 m in diameter, how long would the thermal energy in the salt be able to produce electricity in a 100 MW power plant, assumed to be 30% efficient ($e = 0.3$). Assume that the salt has a density of $\rho = 2300$ kg/m^3 and a specific heat of $c = 1800$ J/°C.

Solution

The mass of salt present is given by $m = \rho V = \rho(\pi d^2 h/4) = 2300 \times 3.14 \times 24^2 \times 9 \times 0.25 = 9.36 \times 10^6$ kg, so that the thermal energy it contained above the required minimum temperature is $E = mc\Delta T = 9.36 \times 10^6 \times 1800 \times (300 - 150) = 2.53 \times 10^{12}$ J—enough to run a 30% efficient 100 MW power plant for a time $t = eE/p = 0.3 \times 2.53 \times 10^{12}/10^8 = 7590$ s or just over 2 h. Clearly, higher temperatures or a larger tank would be needed if it is desired to produce power night and day as well as inclement weather. The storage tanks in the Spanish power plant are in fact significantly larger than that in the example, and the plant power is only 50 MW, which allows the turbines to run at full power for close to 8 h on stored heat (Figure 13.2).

13.1.2.3 Compressed air energy storage Storing energy by compressing air is much like compressing a spring. In either case, the stored energy is released when the spring or compressed gas is allowed to expand. The amount of energy that it is practical to store in compressed air energy storage (CAES) can be considerably greater than that stored in a spring because it is much easier to store very large volumes of air and compress it to extremely high pressures than it is to construct a spring that would store the same amount. In fact, while spring-powered cars are unfeasible, cars that run on compressed air are commercially available. Unlike a conventional internal combustion engine that drives pistons by igniting

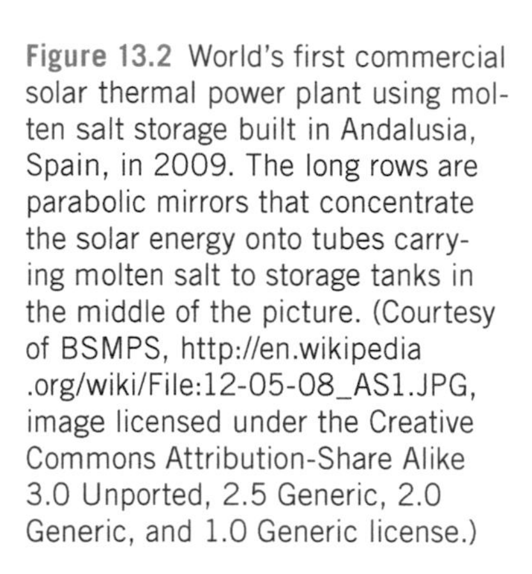

Figure 13.2 World's first commercial solar thermal power plant using molten salt storage built in Andalusia, Spain, in 2009. The long rows are parabolic mirrors that concentrate the solar energy onto tubes carrying molten salt to storage tanks in the middle of the picture. (Courtesy of BSMPS, http://en.wikipedia.org/wiki/File:12-05-08_AS1.JPG, image licensed under the Creative Commons Attribution-Share Alike 3.0 Unported, 2.5 Generic, 2.0 Generic, and 1.0 Generic license.)

gasoline, in a compressed air, car pistons are driven by the expanding air in a way similar to how a steam engine works. There are some advantages to running a car on compressed air, notably the absence of emissions, the relative ease of refueling using a compressor, and readily available fuel.

Unfortunately, despite these advantages, the disadvantages are very serious—the primary one being that of limited driving range on compressed air alone. Although an Indian company (Tata Motors) promised in 2008 to have a car ready for sale in a few years running on compressed air having a 125 mi range, as far as the author is aware, no such vehicle is being commercially sold as of 2014 by Tata. Moreover, a published 2005 study showed that cars running on state-of-the-art lithium batteries outperformed both compressed air cars and cars powered by fuel cells more than threefold (Mazza and Hammerschlag, 2005). Finally, even the emissions issue is misleading. While the car itself produces no emissions, the compressors needed to pressurize the air run on electricity, and emissions are created at the power plant, which on average are more than what an internal combustion car would have emitted. While compressed air cars are technologically feasible, a study by the University of California at Berkeley concluded that "even under highly optimistic assumptions the compressed air car is significantly less efficient than a battery electric vehicle..." (Creutzig et al., 2009).

One application of compressed air energy storage that is well established and less controversial than the compressed air car involves energy storage for improving the efficiency of electric-generating plants by releasing stored energy at times of peak demand. One can use electricity to run a turbine backward to compress air (typically in a large underground cavern or salt dome) and then later remove the pressure and have the air drive the turbine (and generator) at desired times, thereby generating electricity. Round trip efficiencies can be as high as 75%, which is comparable to that for pumped hydro. A number of small storage applications have been developed around the world, and a larger (300 MW) project is being developed in California. The highest efficiency type of compressed air storage involves the use of isothermal (constant temperature) compression and expansion, but this is only feasible for low power levels, since it involves a slow process with efficient transfer of heat to and from the environment.

Another application of compressed air energy storage involves improving the efficiency of gas-fired electric-generating plants (Figure 13.3). These plants are often used to supply electricity at times of peak demand, because unlike coal-fired or nuclear plants, they can quickly change their power output by merely changing the rate of gas flow in the burner. In the CAES version of a gas-fired plant, compressed air is stored in an underground cavern or abandoned mine beneath the power plant. Electric-powered compressors are used to pressurize the air in the reservoir during off-peak hours of electricity demand, while the plant itself generates no electricity until times of greatest (on-peak) demand. During those times when electricity is scarcest and most expensive, the pressurized air is released, mixed with the natural gas, and burned to run the turbines. The end result is that the plant uses 60% less gas for each megawatt of electricity generated than a conventional turbine which uses 2/3 of the fuel to generate the electricity needed to compress the air just before combustion. Of course, in a CAES gas-fired plant, electricity is required to compress the air, but that was done using inexpensive off-peak electricity, which results in lower overall cost of electricity.

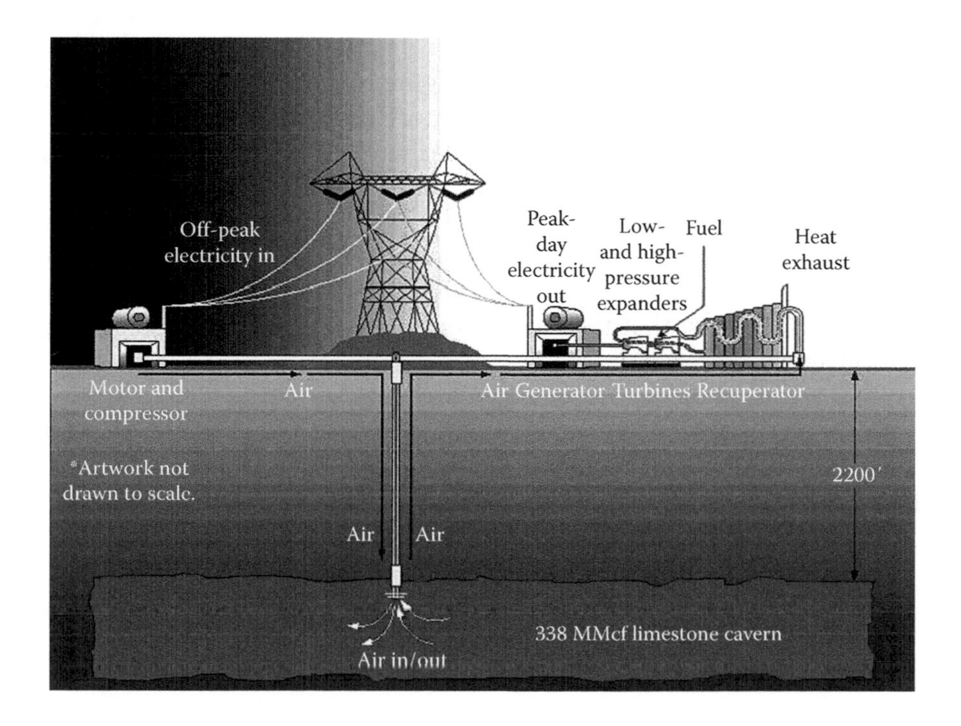

Figure 13.3 Gas-fired power plant using underground CAES. The left portion of the image shows that during night time (off-peak hours), air is compressed and sent into the underground reservoir, while the right portion shows that during daytime (peak hours), it is withdrawn from the reservoir. (Courtesy of DOE, Washington, DC.)

PEAK DEMAND

The term *peak demand*, also called *peak load*, refers to the maximum power requirement of a power plant or an electric utility servicing some region. The demand for electricity varies in a more or less predictable way on all time scales: seasonal, day of the week, hourly, and shorter, with the variation in each case dependent on the climate of a region, extent of air conditioning, presence of heavy industry, and other factors. Apart from relatively predictable variations, there are, of course, other less predictable factors, most importantly the weather. The predictable periods of peak demand tend to be for several hours in the early evening when most people come home from work. In some electricity markets consumers, face real-time pricing (based on the variable wholesale price), but more typically (at least in the United States), they pay a fixed price based on an average annual cost. The unfortunate side effect is that consumers, unlike large commercial and industrial electricity users, have no incentive to reduce demand at times of peak wholesale prices or to shift their demand to nonpeak demand times. If there is a large difference between the minimum amount of electricity needed (the base load) and the maximum peak load owing to unexpectedly heavy demand, the result can be brownouts, blackouts, and/or extremely expensive rates that involve importing power at peak times.

13.1.2.4 Flywheel energy storage Flywheels have an origin going back to antiquity. Originally, pottery was made using long strands of clay that were coiled one atop the other. Around 3500 BC, some clever potter in the Middle East came up with the idea of using a massive slow-turning wheel driven by hand or foot that would allow people to more efficiently shape clay into pottery than could be done using the earlier coiling technique. In a more recent nineteenth-century usage, James Watt, who developed a practical version of the steam engine, included a massive flywheel to keep the rotation rate of the engine more uniform—the same purpose it has in the potter's wheel. In recent years, engineers have literally reinvented the wheel in coming up with flywheels instead intended as energy storage devices. These ultrahigh-speed wheels (in contrast to their low-speed descendants) can rotate at speeds of up to around 100,000 rpm. High-speed rotation is important if a flywheel is to be used as an energy storage device, given that the stored energy depends on the square of the rotation speed.

13.1.2.4.1 Flywheel physics The energy stored in a flywheel rotating at a speed ω (in radians per second) and moment of inertia I (in kilogram-square meters) is given by

$$E = \frac{1}{2} I\omega^2, \tag{13.1}$$

where the moment of inertia or rotational inertia can be written as

$$I \equiv \int_0^R r^2\, dm = kMR^2, \qquad (13.2)$$

where M and R are the mass and radius of the wheel, respectively; $dm = 2\pi r\rho(r)dr$ is the mass contained between r and $r + dr$; and the constant k depends on the shape of the wheel or on $\rho(r)$, i.e., its density (per unit thickness) as a function of r.

In order to find the constant k for any shape, we need to do the integral using some known $\rho(r)$ in Equation 13.2. Two obvious special cases of Equation 13.2 would be where nearly all the mass is at the same fixed distance R from the axis of rotation, i.e., a thin torus, ring, or hoop, giving $k \approx 1$, and another where nearly all the mass is very close to the rotation axis, giving $k \approx 0$. A less obvious, but important, case would be that of a uniform thickness constant density disk of radius R, for which doing the integral in Equation 13.2 would give $k = 1/2$.

For older (slow-speed) flywheels intended for the original purpose of maintaining a constant rotation speed, it was important that they be very large and massive and be made from dense materials such as stone (potter's wheels) or steel (Watt's steam engine). The use of dense materials, however, is not necessarily an advantage when flywheels are used for the purpose of storing the maximum amount of energy where high rotational speed is essential. The use of very dense materials may limit the maximum possible rotational speed, because the larger centrifugal forces they create can cause flywheel disintegration. In fact, the best modern-day high-speed flywheels tend to be made of lighter (less dense) materials having exceptional tensile strength. By making a flywheel from many windings of carbon nanotube fibers, for example, one can create a wheel capable of up to around 100,000 rpm.

CARBON NANOTUBES

The element carbon, so essential for life, can generate some amazingly varied structures ranging from the exceptionally hard and transparent diamond to the soft black graphite used in pencils. Carbon nanotubes are one surprising structure of pure carbon discovered in recent years. Consisting of one-atom-thick sheets of atoms (known as graphene) rolled up into hollow cylinders about a nanometer in diameter, these tubes have extraordinary tensile strength owing to the very strong bonds between the carbon atoms located at the vertices of the hexagons in Figure 13.4b. Individual nanotubes have been created with lengths up to about 10 cm—or 100 million times their diameter. A bundle of nanotubes

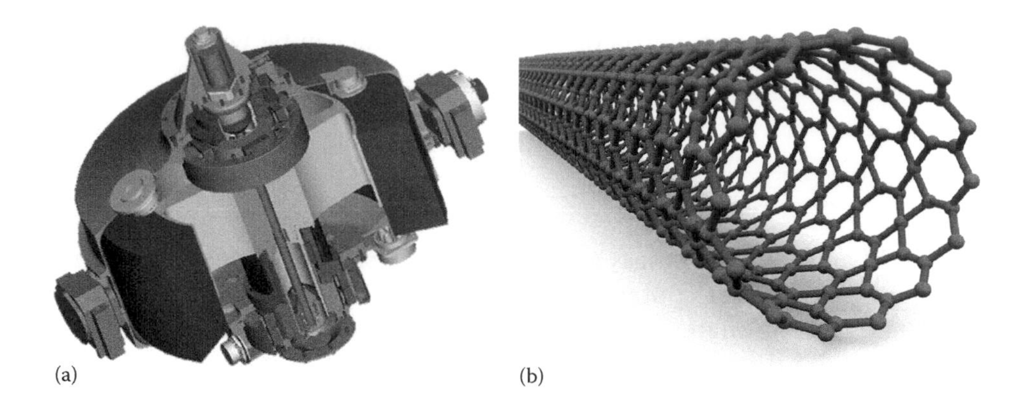

Figure 13.4 (a) Cutaway view of a flywheel rotor (the dark gray parts) and its suspension inside a sealed enclosure used by NASA for space applications. (Courtesy of NASA, Washington, DC.) (b) Drawing of a section of a carbon nanotube.

(which self-adhere) can be grouped together to form ropes or fibers. Various techniques exist for creating carbon nanotube fibers in sizeable quantities, and their cost has precipitously dropped in recent years. Aside from their great tensile strength, they possess a host of other unusual thermal and electrical properties, making them suitable for a number of applications, including some in the areas of energy storage and generation. One other novel application recently devised by MIT scientists involves storing solar energy directly in chemical form, using a technique that should achieve volumetric energy densities as good as the best lithium-ion batteries (Grossman and Kolpak, 2011).

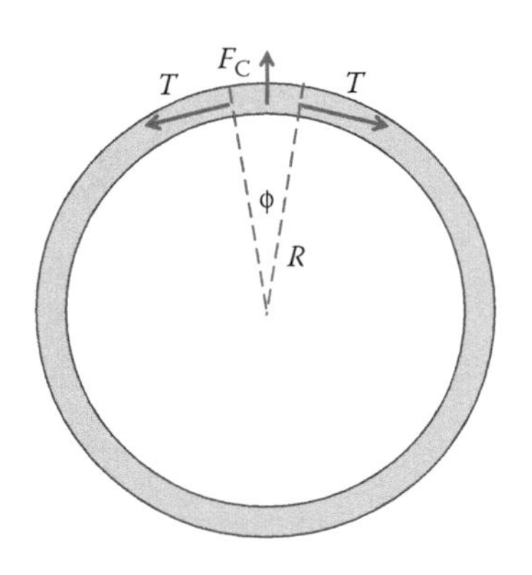

Figure 13.5 Flywheel of radius R with all the mass concentrated close to the edge. Three forces act on a small section of the wheel subtending an angle ϕ. They are the tension T that the left and right sections each exert on it and the upward force F_C, which represents the outward centrifugal force on that small piece of the wheel. The tension forces make an angle $\phi/2$ with the horizontal.

Let us consider exactly how the maximum possible rotation speed of a flywheel depends on the properties of the material of which it is composed, in particular, the density ρ and tensile strength σ, which is the maximum stretching force per unit cross-sectional area before rupture. For simplicity, consider a flywheel having nearly all its mass at a single radius R from the axis, e.g., a torus with a very small cross-sectional area A. Consider a short segment of the torus that subtends an angle ϕ and whose length is $R\phi$—see Figure 13.5. In a reference frame rotating with the flywheel, the three forces acting on this segment are the outward centrifugal force F and the two tension forces T from the rest of the wheel. In order to find the largest possible rotation rate ω_{max}, we set T to its maximum possible value $\sigma A = \sigma \Delta R^2$: Thus

$$T = \sigma \Delta R^2, \tag{13.3}$$

where we have assumed a square cross section of the torus ΔR^2 over which the tension force operates. Since the volume of the small segment is $V = R\phi\Delta R^2$, we can therefore write the centrifugal force as

$$F_C = m\omega_{max}^2 R = \rho V \omega_{max}^2 R = (\rho R\phi\Delta R^2)\omega_{max}^2 R = (\omega_{max} R\Delta R)^2 \rho\phi. \tag{13.4}$$

In a reference frame rotating with the wheel, the net force on the segment vanishes, so we must have for the net y-components of the forces:

$$F_C - 2T\sin\frac{\phi}{2} = 0. \tag{13.5}$$

Substituting Equations 13.4 and 13.5 into 13.6 yields

$$(\omega_{max} R\Delta R)^2 \rho\phi - 2\sigma\Delta R^2 \sin\frac{\phi}{2} = 0. \tag{13.6}$$

Finally, making use of the smallness of the angle ϕ, we have $\sin\phi/2 \approx \phi/2$, and hence, Equation 13.6 becomes

$$\omega_{max}^2 = \left(\frac{\sigma}{\rho}\right)\frac{1}{R^2}. \tag{13.7}$$

This result for the maximum rotation rate (in radians per second) applies only for an unrealistic case where all the mass is nearly the same distance from the rotation axis. For other shapes where the mass is distributed as some function of radial distance, a similar formula would apply with a geometrical constant K multiplying the right-hand side of Equation 13.7. As we can see from Equation 13.7, for a fixed geometry, the maximum rotation speed depends on the properties of the material (the ratio of tensile strength to density) and the radius of the wheel. Choosing the best material out of which to make a flywheel so as to maximize the energy that it can store is a matter of finding one that has the largest value of the ratio σ/ρ.

13.1.2.4.2 Example 3: Power delivered by a flywheel A carbon nanofiber flywheel in the shape of a uniform disk of mass 1.2 kg, radius 0.4 m, and density 2,600 kg/m^3 is spinning at 100,000 rpm (10,500 rad/s). For how long could it deliver 1 kW of power?

Solution

The energy of the spinning wheel is $E = (1/2)I\omega^2 = (1/4)mR^2\omega^2 = (1/4) \times 1.2 \times 0.4^2 \times 10{,}500^2 = 5.29 \times 10^6$ J, which could supply 1 kW of power for a time of $t = E/p = 5.29 \times 10^6$ J/1000 W = 5290s or close to 1.5 h.

13.1.2.4.3 Flywheel technology and applications As a result of the high rotation speeds that can be attained, modern flywheels can store a considerable amount of energy—typically 360–500 kJ/kg of mass—or about three to four times what a lead–acid battery stores per kilogram. Energy can be added or extracted from the flywheel using a combined motor–generator, which need not have physical contact with the spinning wheel—see the description of an induction motor in Chapter 7. Aside from materials that are lightweight and very strong, there are two other features that many modern flywheels possess that allow them to spin at extraordinary speeds: (1) being enclosed in a vacuum compartment (which eliminate air resistance) and (2) having a magnetically levitated suspension so as to eliminate friction that would be present if there were a mechanical suspension with moving surfaces in contact. Flywheels lacking these two features can lose 20–50% of their energy during the course of 2 h, but with them, the loss is only a few percentage. Modern high-speed flywheels, because of the lack of mechanical wear, require virtually no maintenance and have an indefinitely long lifespan—with as many as 10,000 times the number of charge–discharge cycles of a lead–acid battery. This lifetime advantage favors flywheels for applications involving frequent charge–discharge cycles. The flywheel efficiency of 90% energy recovery on discharge is also a significant improvement over lead–acid batteries. In summary, the high energy density, absence of maintenance, and long life make the high-speed flywheel a formidable competitor to batteries for many applications.

- *Vehicle propulsion*: Flywheels have been proposed for use in vehicles, but except for short driving distances, they lack sufficient stored energy. One exception would be buses, and they have actually been used by the Swiss for this purpose in the 1950s (Figure 13.6). Their feasibility for buses depended on the ability to frequently recharge flywheels at passenger stops en route via booms mounted on the roof of the bus. Although flywheels may be inappropriate as the sole means of vehicle propulsion in automobiles, they are useful in providing a power boost to a car at times when fast acceleration is needed. The flywheel can also recover energy during regenerative braking, making recharging possibly unnecessary during a trip. Flywheels have in fact been used for a power boost in some racing cars. Their possible use

Figure 13.6 Only surviving flywheel-powered gyrobus, built in 1955, in a museum in Antwerp, Belgium. (Courtesy of Vitaly Volko, http://en.wikipedia.org/wiki/File:Gyrobus_G3-1.jpg, image licensed under the Creative Commons Attribution 2.0 Generic license.)

in commercial automobiles is under consideration, but there are disadvantages, most importantly, the matter of safety should a flywheel disintegrate either due to weakness or a vehicular collision. In such an event, if the flywheel enclosure is breached, debris could be hurling at extremely high speed in all directions. Nevertheless, there are safety issues with virtually all technologies that have high energy density, and it is unclear if the danger of high-speed flywheels is truly worse than that posed by other technologies—even batteries.

- *Space applications*: Flywheels are ideally suited to many applications aboard spacecraft, in view of their long life, high reliability, low weight, high efficiency, and rapid charge/discharge times. They have the added benefit of providing attitude reference because of the gyroscopic effect of a rapidly rotating wheel. Obviously, in space, flywheels would not need a sealed vacuum enclosure, which would further reduce weight.
- *Grid power interruption and quality correction*: A large collection of advanced flywheels can provide power during a temporary interruption of power from the grid before an emergency power source, such as diesel generators, comes on, which can take as long as 15–20 s. Flywheels are capable of discharging their power quite rapidly—perhaps 100 kW each for that time period. Thus, a farm of 200 such flywheels could deliver 20 MW of power. Flywheels can be useful in correcting the quality of power produced at the generating plant, by shifting the frequency if it differs from the standard before putting the power onto the grid. One of the largest flywheel storage systems in the world capable of delivering 20 MW has been installed in Stephentown, New York, for exactly this purpose.
- *Uninterruptible power supply*: Aside from maintaining quality at the power plant, individual end users of electricity can also guard against power interruptions. For the end user, a flywheel might provide a kilowatt or more for hours (see Example 3).

13.1.3 Electric and Magnetic Energy Storage

In this section, we will consider the ways that energy can be stored electrically and magnetically, including batteries, fuel cells, capacitors, and magnetic fields. Technically, a fuel cell itself does not store energy, because it can only produce energy for as long as reactants, such as hydrogen, are added to it, so it is preferable to think of the hydrogen itself as the repository of stored chemical energy.

13.1.3.1 Batteries Electric batteries were one of the first human-made sources of electricity, having been invented in 1800 by Alessandro Volta. The invention of the battery was well before that of the electric generator and was preceded only by the Leyden jar, essentially a capacitor. Today, batteries are a very prevalent source of electricity useful for a wide range of applications that require either portability or independence from the electric grid. One of these applications that are considered of great importance is the electric battery-powered car, and much research is being conducted on improving battery performance, which hitherto has been the main impediment to an all-electric vehicle that has the range and cost of conventional cars. In general, batteries come in two basic varieties: rechargeable and nonrechargeable. The unhelpful terms of *primary batteries* for nonrechargeable ones and *secondary batteries* for rechargeable ones are also sometimes used. Primary batteries are not rechargeable because the chemical reactions taking place in them are not reversible. Our focus here will be on rechargeable batteries, because one cannot very well consider the nonrechargeable ones cost-effective in the context of large-scale energy storage. Of course, there is still an important role for nonrechargeable batteries for many small-scale storage applications such as flashlights, watches, and portable radios. Before considering the large number of batteries, and their relative merits, it is useful to consider the properties that all batteries share and the way they function.

Batteries contain two electrodes typically made of different metals. The positive electrode is known as the anode and the negative one is the cathode. Inside the battery and between the electrodes is an electrolyte consisting of positive ions and electrons in solution. Some types of batteries actually involve two different electrolytes in each half of a cell. Both electrons (small circles) and ions (large circles) can leave or enter the solution to or from the surfaces of the electrodes, and they can slowly travel through the electrolyte, which is either a liquid (in a wet cell) or a paste (in a dry cell).

Figure 13.7 shows the reactions taking place at each electrode during discharge. A positively charged ion is shown entering the electrolyte from the cathode, while to conserve charge, an electron flows upward through the cathode where it then flows through some device (the load). When the ion eventually reaches the anode, the reverse process happens when an electron combines with the ion on the surface of the anode. For some batteries, the chemical reactions involve ions that are multiply

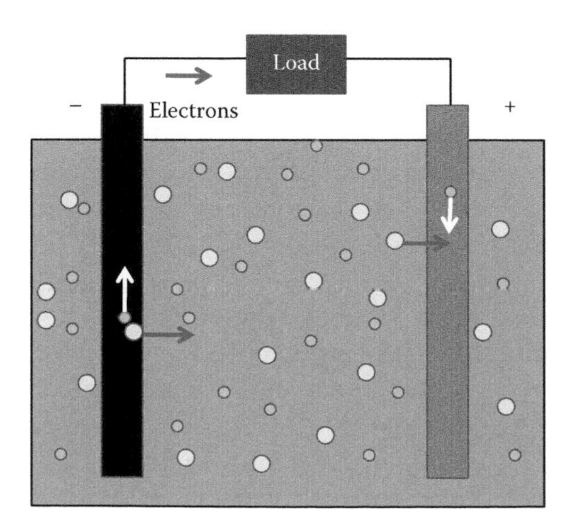

Figure 13.7 Basic elements and functions of a battery during discharge. Small circles represent electrons, and large circles represent ions.

charged, in which case, the reactions at each electrode involve multiple electrons rather than one. The situation depicted in Figure 13.7 is described as one where the electric current is flowing to the left through the load, which is opposite the direction the electrons travel. The only change we would need to make to the diagram to describe a battery being charged (rather than discharged) is to reverse the directions of all arrows and think of the device not as a load, but as a source of energy that drives the current the other way (uphill) through the battery.

The usage of the term *battery* originally referred to a series of separate cells rather than just one, but now, it is commonly used whether there is only a single cell (as depicted in Figure 13.7) or a number of them in series, as there is in the case of a 12 V car battery—with its eight 1.5 V cells in series contained within a single battery. In general, the voltage across each cell of a battery depends on the particular chemical reactions involved in that type of battery, which, in turn, depends on the materials chosen for the two electrodes and the electrolyte. If we call E_A the energy released at the anode (per electron) when ions and electrons combine and E_C the energy consumed at the cathode (per electron) when ions and electrons separate, then the difference $E_A - E_C \equiv \varepsilon$, also called the electromotive force (EMF), is what drives the current flow during discharge. The quantity EMF, which stands for electromotive force, is in fact not a force at all, but it has units of voltage, and it is what a voltmeter would measure across the battery terminals in the limit of no current drawn.

When a current i flows through the battery, the voltage across its terminals or the voltage across the load is given by

$$V = \varepsilon \pm ir, \tag{13.8}$$

where r is the internal resistance that all batteries possess in varying degrees and the sign used depends on whether the battery is charging (+) or discharging (−).

Thus, when a battery sends a current through a load, the terminal voltage is always less than its EMF, and when it is being charged, it is greater than the EMF (Figure 13.7). The current that a battery can supply to a load of resistance R is given by (Figure 13.8)

$$i = \frac{\varepsilon}{r + R}, \tag{13.9}$$

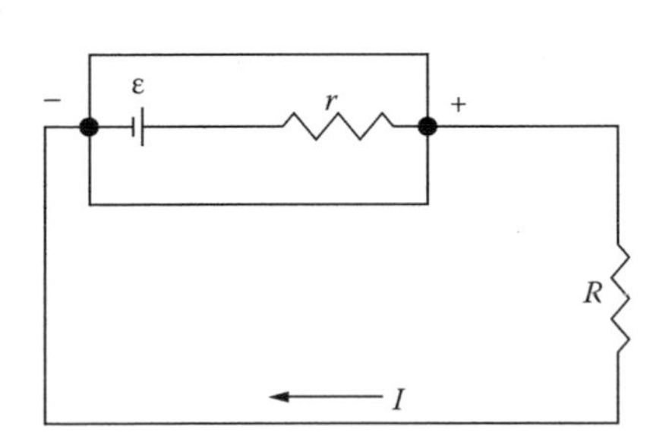

Figure 13.8 Battery connected to a load resistance R, also showing its internal resistance r.

and it has its maximum value ε/r when $R = 0$, i.e., a short circuit. Ideally, the chemical reactions taking place in a battery should be completely reversible for charging and discharging, so that a battery could be charged and discharged indefinitely, but real batteries are far from that ideal, owing to the release of gases, corrosion, and deterioration of the electrodes.

13.1.3.1.1 Example 4: Finding the maximum current from a battery A voltmeter measures a voltage of 12 V across the terminals of a battery, but when a load is connected to the battery that draws 10 A, the voltmeter reading drops to 8.8 V. What is the maximum current that could be drawn from the battery?

Solution

Solving Equation 13.8 (with the minus sign) for the internal resistance, we find $r = (\varepsilon - V)/i = (12.0 - 11.8)/10 = 0.02\ \Omega$. The maximum current possible is when the load resistance R is zero, in which case, $i = \varepsilon/r = 12/0.02 = 240$ A.

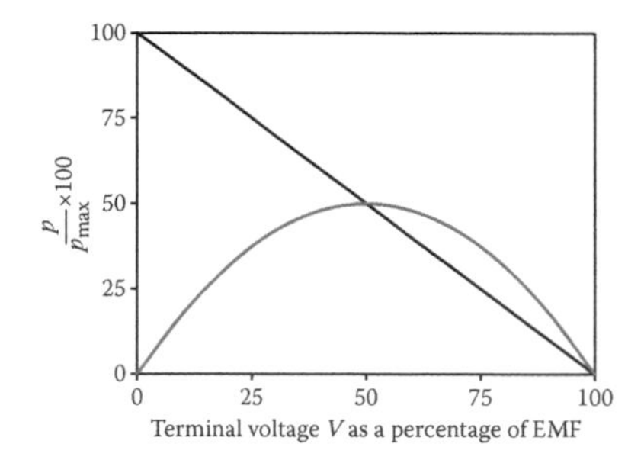

Figure 13.9 Terminal voltage of a battery (as a percentage of the EMF) versus current drawn as a percentage of its maximum value (sloped line). The inverted parabola shows the power delivered to the load. It can easily be shown that the point where the power delivered to the load is a maximum is when $r = R$.

The terminal voltage of a battery V is plotted versus current i in Figure 13.9 (sloped straight line). The voltage here is plotted as a percentage of the EMF, and the current is plotted as a percentage of its maximum value $i = \varepsilon/r$, which is when the external load $R = 0$. Thus, the rate of falloff in voltage with increasing current drawn depends on the size of its internal resistance. This is the reason large batteries, which have a smaller r than small ones, are capable of providing larger currents. The internal resistance of a battery can also increase with age, usage, or low temperature, making it less able to supply high currents. The inverted parabola is the power supplied by the battery. We see that the power is obviously zero when $i = 0$, but it also goes to zero when $V = 0$ (at the maximum current) since $p = iV$.

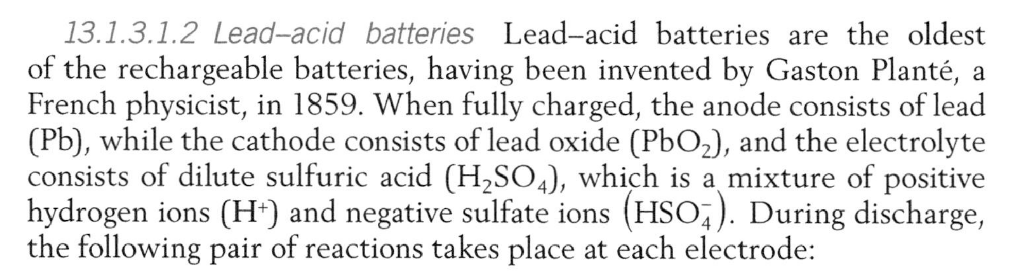

13.1.3.1.2 Lead–acid batteries Lead–acid batteries are the oldest of the rechargeable batteries, having been invented by Gaston Planté, a French physicist, in 1859. When fully charged, the anode consists of lead (Pb), while the cathode consists of lead oxide (PbO_2), and the electrolyte consists of dilute sulfuric acid (H_2SO_4), which is a mixture of positive hydrogen ions (H^+) and negative sulfate ions $\left(HSO_4^-\right)$. During discharge, the following pair of reactions takes place at each electrode:

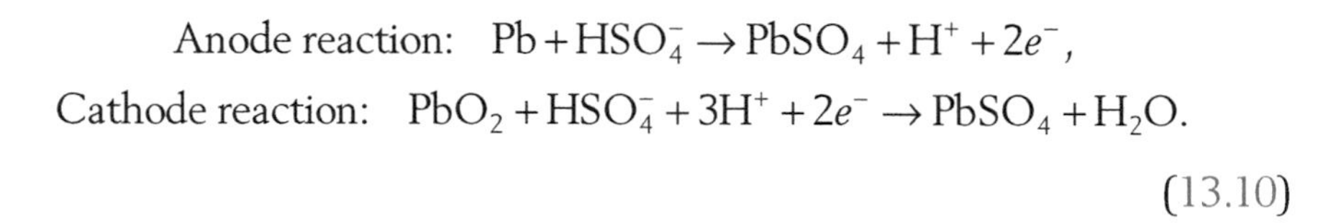

$$\text{Anode reaction:}\quad Pb + HSO_4^- \rightarrow PbSO_4 + H^+ + 2e^-,$$
$$\text{Cathode reaction:}\quad PbO_2 + HSO_4^- + 3H^+ + 2e^- \rightarrow PbSO_4 + H_2O. \tag{13.10}$$

The energy liberated in the anode reaction (per electron released) exceeds the energy consumed at the cathode by 1.5 eV, so the cell has a 1.5 V EMF. During charging, the reactions taking place at each electrode are simply the reverse of Equation 13.10, i.e., just reverse both arrows. When a lead–acid battery discharges, over time, the surfaces of both electrodes gradually become lead sulfate ($PbSO_4$), while the electrolyte becomes more and more diluted (less acid and more water), and during charging. the reverse occurs. The original lead–acid battery did not yield very high current, owing to the limited surface area of each electrode, but improvements in design now using a paste-filled matrix or grid greatly increased the effective surface area and, hence, the reaction rates. Current lead–acid batteries used in cars can produce currents as high as 450 A.

LEAD–ACID BATTERY PARADOX

When lead–acid batteries are new, they are uncharged, and both electrodes are made of the same identical material—lead—which defies the notion that different metals are needed for the battery to have an EMF. During their initial charging, the following pair of reactions occurs at the electrodes, and the result is electrodes consisting of lead and lead oxide, so that an EMF subsequently occurs.

$$\begin{aligned}&\text{Anode reaction: } 2H^+ + 2e \rightarrow H_2\\&\text{Cathode reaction: } Pb + 2H_2O \rightarrow PbO_2 + 2H^+ + 2e\end{aligned} \qquad (13.11)$$

Car batteries can only go through perhaps 800 cycles before needing replacement—although that number depends on the depth of discharge. If the battery is discharged only by a moderate 30%, it can last perhaps 1200 cycles, while a deep discharge of 100% of the energy would shorten its lifetime to perhaps 200 cycles. Deep discharge lead–acid batteries do exist that are intended to be fully discharged, but they cannot deliver a current as high as that needed for car batteries, because their internal resistance is larger. It is their capability of producing very high current (needed for starting the engine) and their relative low cost that accounts for lead–acid batteries being the most common type of battery used in internal combustion vehicles. Lead–acid batteries, however, are not a good choice in the case of an electric car, given two important disadvantages: a low energy density and limited efficiency (50–92%) per charge–discharge cycle. The low energy density, of course, means that for a given weight battery, the energy content (and, Therefore, the vehicle range) is limited.

13.1.3.1.3 Lithium-ion batteries Lithium-ion batteries first proposed in the 1970s are currently the battery of choice for electric vehicles, as well as many consumer electronic devices, in view of their having one of the best energy densities (over three times that of lead–acid batteries), slow loss of charge when not in use, long lifetime, and high efficiency per cycle (80–90%). Inside the cell of the current version of a lithium-ion battery (first developed by Akira Yoshino in 1985), Li^+ ions carry the current across the electrolyte, a lithium salt solution, from a cathode made from one of a number of lithiated metal oxides to an anode usually made of graphite. Essentially, during discharge, lithium ions are extracted from the anode and inserted into the cathode, while the reverse occurs during charging. Despite their ubiquity, owing to their use in cell phones and other common electronic devices, lithium-ion batteries do have some drawbacks, including high cost and safety. Lithium-ion batteries when exposed to high temperature can ignite or even explode. (Recall the earlier comment about the safety of flywheels.) The high cost of lithium-ion batteries may not be much of a barrier to their use in electronics, but their use in an electric vehicle, which requires a battery pack of very large capacity, is another matter entirely, and as of 2011, it can be a sizable fraction of the cost of the car itself.

13.1.3.1.4 Other batteries A number of batteries exist based on chemistries other than lead–acid or lithium ions, including nickel–cadmium (NiCd), nickel metal hydride (NiMH), and nickel–zinc (NiZn), with the last one being a relatively new technology that is not yet widely used commercially. Each of the various types of batteries has its own advantages and disadvantages, which tend to be application dependent. Three battery properties are of particular importance: cost, energy density, and cell voltage (Table 13.1). The importance of cost and energy density is obvious, but that of cell voltage is perhaps not. A high cell voltage allows a battery to deliver greater power for the same current supplied.

A new, extremely novel type of battery invented by an MIT materials chemistry professor, Donald Sadoway, is the liquid metal battery. The electrodes in this battery consist of two metals (magnesium and antimony) that are liquids at the high temperature (700°C) at which the battery operates. When a salt electrolyte solution is added to the mix, the three different liquid densities result in an automatic stratification, with the heaviest layer (antimony) on bottom, the salt solution above it, and the magnesium on top constituting the positive electrode. Its efficient design allows the battery to do away with much of the space in conventional batteries that is used to separate the active materials, and it can provide currents tens of times what conventional batteries produce. Moreover, based on prototypes, the costs of these revolutionary batteries are expected to be extremely low—less than a third that of conventional batteries per kilowatt-hour, because of the use of common materials, simple design, and scalability to very large size. The batteries are also designed to operate at very high temperatures—a major advantage since high operating currents can generate a lot of heat. This novel type of battery offers the possibility of large-scale energy storage for renewable sources that would make them much more compatible to the electric grid (Figure 13.10).

13.1.3.2 Ultracapacitors Ultracapacitors, like ordinary capacitors, store electrical energy by separating positive and negative charges that reside on a pair of plates until the plates are connected across a load, at which point

Table 13.1 Comparison of Various Chemistry Rechargeable Batteries Based on Three Important Criteria

Chemistry	Cell Voltage (V)	Energy Density (MJ/kg)	Cost
Ni Cd	1.2	0.14	$
Lead acid	2.1	0.14	—
Ni MH	1.2	0.36	$
Ni Zn	1.6	0.36	—
Li ion	3.6	0.46	$$$

Note: —, least expensive; $, moderately inexpensive; $$, moderately expensive; $$$, very expensive.

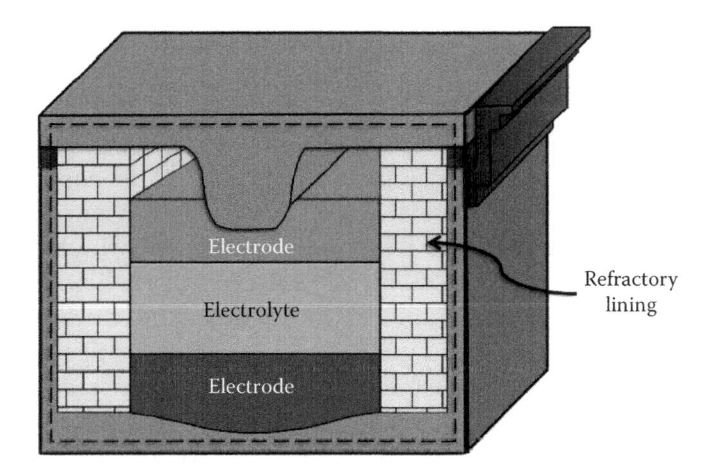

Figure 13.10 Liquid metal battery. In one design, the upper electrode is liquid magnesium (Mg), the lower electrode is a liquid magnesium–antimony (Mg–Sb) alloy, and the electrolyte is a magnesium salt solution. When the battery is discharging, electrons leave the upper negative electrode, go through an external circuit, and then enter the bottom positive electrode, where they then combine with Mg^{++} ions in the electrolyte. The reverse process occurs during charging. (Courtesy of DOE, Washington, DC.)

a current flows through the load and the capacitor discharges. Unlike batteries, no chemical reactions take place during charging or discharging, and hence, there is no conversion of chemical to electrical energy.

Let us briefly review the basics of capacitors, which should be familiar to most readers. In one form of the capacitor, a pair of plates of area A are separated by a distance d, with a dielectric (insulating) material of dielectric constant κ filling the space between the plates. The amount of positive or negative charge q that each plate is capable of holding is proportional to the voltage across them V, so that the capacitance of the capacitor is defined as q/V, which for a parallel plate capacitor can be shown to have the following form:

$$C \equiv \frac{q}{V} = \frac{\kappa \varepsilon_0 A}{d}, \tag{13.12}$$

where C is measured in units of farads (abbreviated F), equal to coulombs per volt, and $\varepsilon_0 = 8.85 \times 10^{-12}$ F/m, a universal constant, is the so-called permittivity of free space.

The amount of energy that a capacitor can store is given by

$$E = \frac{1}{2} CV^2, \tag{13.13}$$

and the energy is usually regarded as residing in the electric field that fills the space between the plates. According to Equation 13.13, the two ways to increase the amount of energy stored involve increasing either the voltage V or the capacitance C. No matter how good the insulating material is between the plates, there will be some maximum voltage that can be sustained between them when they are separated by a given distance d. Air, for example, will break down if the voltage exceeds around 10,000 V per inch separation. Given this limitation, and the typical capacitance of ordinary capacitors, the amount of energy stored tends to be quite low—the energy density being negligible compared to a storage battery, for example.

The ultracapacitor, also known as the supercapacitor or electric double-layer capacitor, was first developed in 1966, and it has dramatically increased the maximum capacitance that a capacitor can attain. As of 2010, C values as large as 5000 F have been achieved, which is thousands of times greater than that of conventional capacitors. The construction of the ultracapacitor—see Figure 13.11—involves the same two plates as in the conventional version; however, the actual electrodes on which the charge resides involves not just the plates, but the porous spongy conducting material between them, and the positive and negative charges are kept apart by the separator between the two halves of the device. Since the cumulative surface area of the tiny nanoscale pores that fill the volume is far greater than that of the plates themselves, by Equations 13.12 and 13.13, the capacitance and the energy stored is hundreds of times greater than that for a conventional capacitor. The improvement in energy density is less than the improvement in capacitance, since the thickness of the separator is less than that of the distance between plates, which results in a lower maximum voltage compared to a conventional capacitor.

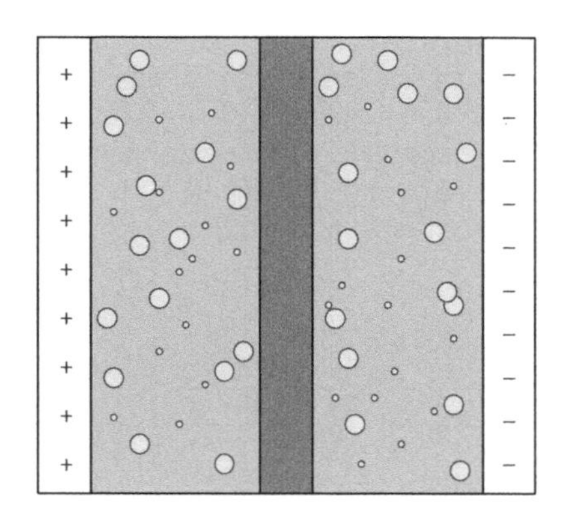

Figure 13.11 One cell of an ultracapacitor. The electrodes where the charge is stored include both the plate and the porous conducting material between them. The gray separator is a dielectric material. The circles represent pores of various sizes.

Even with the greatly enhanced energy density, ultracapacitors still fall short of that of batteries which can generate energy as long as the chemical reactions continue. However, what they lack in energy density, ultracapacitors more than make up for in power density, since unlike batteries where the slow migration of ions limits the rate at which energy can be supplied (the power), there is a much higher limit in the case of a discharging capacitor, just involving the flow of highly mobile electrons. As is well known to most readers, the discharge time for a capacitor is governed by the time constant RC, where R is the resistance of the load across which it discharges. Considering the variety of electrical storage devices, there tends to be an inverse correlation between energy and power densities across devices that chemically store energy (like batteries and fuel cells) and devices like ultracapacitors that do not—a correlation illustrated in the Ragone plot in Figure 13.12.

Since ultracapacitors are capable of delivering a lot of energy in a short time scale, they are used in a variety of high-power applications, such as supplying short-term bridge power in the event of a power failure. They have also proven useful in improving the efficiency of wind turbines by rapidly adjusting the pitch of their blades in response to rapidly changing wind directions and speeds so as to optimize the turbine performance. Batteries could also be used for this purpose, but they lack the ability of ultracapacitors to (1) deliver short bursts of power, (2) do it for many cycles (typically 20,000 charge–discharge cycles), and (3) do it over a wide range of temperatures. Currently, ultracapacitors are more expensive than batteries, but the difference tends to decline over time due to better manufacturing. They also have a lower energy density than batteries (only

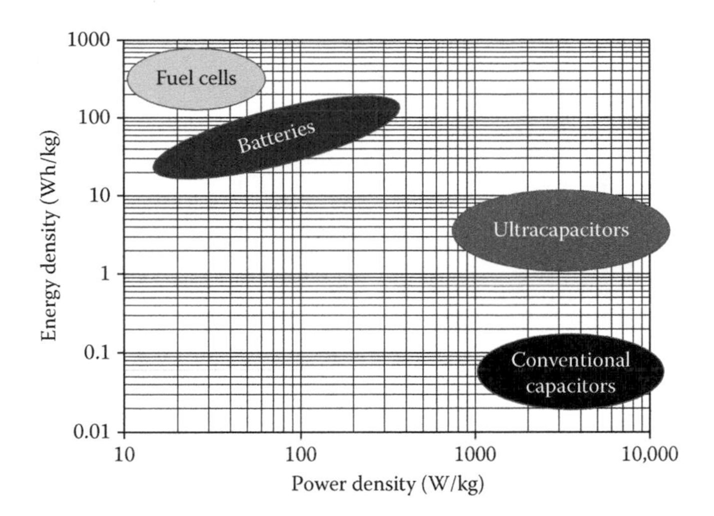

Figure 13.12 Energy densities versus power densities for various electrical devices.

about 5% of Li ion batteries), but that distinction may also be narrowing with the advent of an ultracapacitor–battery hybrid that is an attempt to combine the best of both technologies. Applications of the hybrids would include usage in vehicles—for the purpose of regenerative braking and providing sudden acceleration boosts, much in the manner of flywheels.

13.1.3.2.1 Example 5: Energy density for an ultracapacitor A company sells ultracapacitors whose mass is 0.1 kg and whose capacitance is 5000 F with a maximum voltage of 4 V. How much energy can they store and what is the energy density?

Solution

From Equation 13.13, we find that $E = (1/2)CV^2 = (1/2) \times 5000 \times 4^2 = 4 \times 10^7$ J, and the energy density is 4×10^7 J/0.1 kg = 400 MJ/kg.

13.1.3.3 Fuel cells The basic idea of the fuel cell was discovered by German scientist Christian Friedrich Schönbein in 1838. Fuel cells have some similarities to batteries, in that they are made up of the same three basic elements: anode, cathode, and electrolyte, and they generate electrical energy from chemical reactions and convert chemical into electrical energy. However, unlike batteries, they require a flow of fuel, usually hydrogen gas. from outside the cell, so that the fuel cell itself does not store energy, and we do not speak of charging and discharging it. Electricity is generated in the cell as long as fuel continues to enter it. A block diagram of a fuel cell connected to a load is shown in Figure 13.13.

On entering the fuel cell, hydrogen atoms become ionized by a catalyst at the anode that strips off their one electron. The freed electron goes through a wire to the external load, while the positively charged hydrogen ion migrates through the electrolyte. On reaching the cathode, two hydrogen ions combine there with an oxygen atom in the air and an electron that has travelled through the load to produce water (H_2O).

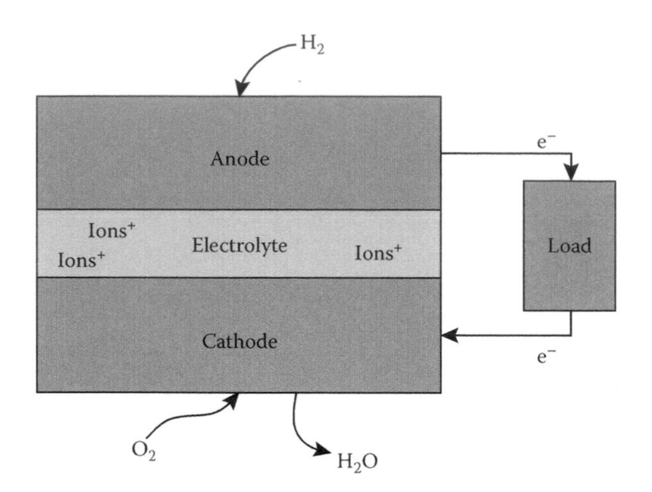

Figure 13.13 Hydrogen-fueled fuel cell connected to a load.

Electrical energy is generated because the energy input to strip electrons off hydrogen atoms at the anode is less than the energy released in the reactions at the cathode. Fuel cells can also work using many other fuels besides hydrogen including natural gas, but hydrogen is a particularly attractive choice since the only end product is water. Of course, hydrogen is an extremely reactive element that is never found in an isolated form in nature. For this reason, we should not really think of hydrogen as a fuel at all, since in order to obtain it, say from electrolysis of water, we need to supply at least as much energy as it later releases in the fuel cell. Thus, hydrogen is really an energy storage medium. Better and more efficient methods to produce hydrogen by splitting the water molecule are an important area of research, and tiny metallic particles less than a nanometer in size appear to be a particularly promising catalyst for the process. If the hydrogen separation can be powered by solar energy, the life cycle of hydrogen production and subsequent electricity generation in fuel cells would be a completely nonpolluting set of processes powered by renewable energy.

13.1.4 Hydrogen Storage and Cars Powered by It

In addition to hydrogen production, there is also the issue of storing hydrogen until it is used, which is of crucial importance in the possibility of using it to power a vehicle using a fuel cell. The three alternatives for storing hydrogen are as follows:

- *Liquefied hydrogen*: The energy density of liquefied hydrogen is extremely high—even better than gasoline on a volumetric basis (see Table 13.2). However, hydrogen does not liquefy until it has been cooled to −253°C, which could pose significant, possibly insurmountable problems for its use in automobile propulsion.
- *Metal hydrides*: Hydrides are metallic compounds with hydrogen atoms stored in them, which can be released as hydrogen gas by heating. Past methods of storing hydrogen in metallic hydrides

Table 13.2 Comparison of Energy Densities by Weight and Volume for Hydrogen and Gasoline

Storage Medium	Energy Density by Volume (Wh/L)	Energy Density by Mass (Wh/kg)
Liquid hydrogen	2600	39,000
Gas (150 atm)	405	39,000
Nickel metal hydride	100	60
Gasoline	9700	12,200

have not been particularly successful, but here too, the world of nanoscience may offer a better way, as scientists have been researching the use of nanoparticles to achieve higher energy densities. Nevertheless, there is a long way to go to achieve energy densities that would be appropriate to vehicular applications. The energy densities on the basis of energy per unit mass for metal hydrides are over two orders of magnitude lower than gasoline, largely because we need to include the mass of the metal hydride itself as well as that of the very light hydrogen.

- *High-pressure gas*: This might be the most promising possibility, but as Table 13.2 indicates, gas stored in tanks pressurized even as high as 150 bar (150 atm) have a volumetric energy density 1/24th that of gasoline, meaning that the tanks would need to be 24 times the size of a standard gas tank to contain the same energy.

The preceding comparison between gasoline and pressurized hydrogen gas storage might seem to relegate the idea of hydrogen fuel cell cars to the distant future, but this is not the case, owing to three neglected factors.

- *Stronger storage tanks*: Originally, storage tanks for hydrogen were made of metal—either aluminum or steel, which can sustain a maximum pressure of about 200 bar, hence the use of 150 bar in Table 13.2. The same carbon fibers that have made faster rotating flywheels a possibility have also allowed the construction of much stronger hydrogen storage tanks that can withstand pressures of 661 bar and up. Using a value of 600 bar, we would get a factor of 4 improvement on the energy that could be stored per unit volume compared to 150 bar.
- *Efficiency of electric vehicles*: In Chapter 12, we learned that on average, automobiles powered by internal combustion engines have an overall efficiency only around 15%—with 62.4% of the energy of the fuel lost in the engine and another 17.2% lost in standby idling. In contrast, fuel cells have an efficiency of about 50%, and the electric motors they would power can have efficiencies above 90%. Using these numbers, we find an overall fuel efficiency for a fuel cell-powered electric car to be about three times that of a car powered by an internal combustion engine in view of the need for frequent refueling stops.

- *Hybrid vehicles*: It is not necessary that a fuel cell be the only power source for a car, particularly for drivers who make a lot of short trips, and infrequent long ones, so that most of their driving could be done powered by a fuel cell. However, unless the vehicle range provided by the fuel cell is significant, it could be an undesirable option for most drivers.

Based on these considerations, given a factor of 4 improvement in energy density (allowed by stronger storage tanks) and a factor of 3 improvement in efficiency of electric vehicles, we find an overall factor of 12 improvement. This means that a hydrogen tank that has the same volume as a standard gasoline tank would be able to provide enough fuel to drive half the range of an internal combustion engine on fuel cell power alone, or perhaps 225 mi—a very respectable distance. Fuel cell-powered cars have now in fact become a reality—witness the fuel cell vehicle made by Toyota with a claimed range of 300 mi with a cost of refueling similar to a gasoline engine.

Readers who are aware of the 1937 Hindenburg catastrophe (Figure 13.14) involving the total destruction by fire of a German passenger airship kept aloft by hydrogen gas might be reluctant to have highly pressurized hydrogen tanks in their cars, but it has been claimed that they are probably safer than standard gasoline tanks in view of their much greater strength. Unfortunately, cost is a sizable negative for fuel-cell powered cars. Although the costs of extracting hydrogen using various energy sources are not out of line with those of producing gasoline, the costs associated with making the car itself are another matter. Finally, in addition to the cost issue, there is also the matter of a current lack of infrastructure in the form of hydrogen refueling stations and a distribution network for them—a problem that is probably far less serious for battery-powered electric vehicles.

Figure 13.14 Hindenburg disaster involving a 1937 hydrogen-filled air ship. (Courtesy of Gus Pasquerella, http://en.wikipedia.org/wiki/Hindenburg_disaster.)

13.1.5 Battery-Powered Electric Cars

There is great current interest in battery-powered vehicles, but their history goes back to the beginning of the automobile era. In fact, in the early years of the twentieth century, electric cars powered by batteries were the preferred choice in view of their being cleaner, quieter, and easier to operate than internal combustion engines, which, in their early years, required a hand crank to start up. However, as internal combustion technology improved, electric battery-powered cars began to fade from the scene, primarily because of the problem of limited driving range—the same issue that has hindered their market acceptance until recently. In recent years, however, there has been a resurgence of interest in electric vehicles motivated by a variety of factors, including the cost of gasoline, concern for the environment, and worries about the overreliance on energy sources from unstable regions. As of 2015, more than 700,000 plug-in electric passenger cars and vans have been sold worldwide, with the United States being the leading market with 41% of global sales since 2008. The main negatives of electric cars are the same now as in the early years of the twentieth century, namely, limited range and high cost, and much revolves around the cost and energy storage capacity of batteries, which continues to improve.

Related to the range anxiety matter, some concerns have been raised about the availability of fast-charging stations. This issue is being addressed in a variety of ways, with some proposing stations that would simply swap batteries in and out of a car in a matter of minutes—although there have been safety concerns expressed about this idea which Mercedes tried in the 1970s (for a bus) and later abandoned due to safety issues. Another way to satisfy the need for readily available fast charging is through home charging stations, and at least one company has introduced compact high-power units for under $1000—although they would do little for drivers finding they needed a recharge while en route to their destination.

THE "LONG TAILPIPE" OF ELECTRIC VEHICLES

Although electric vehicles are often referred to as being emissions free, emissions will have been produced by the power plant that generated the electricity used to charge the batteries, unless the power was generated by nuclear or renewable sources. Such extra power plant emissions may be more or less than those emitted by an internal combustion car, depending on what energy source was used to power the plant. Let us imagine that it were possible to replace most of the cars in the United States by electric vehicles, and consider the electricity source to be the U.S. power grid as a whole with its mix of energy sources. According to an MIT study it would be expected that greenhouse gas emissions would drop more than half if we used off-peak power to charge the batteries of electric vehicles (Kromer and Heywood, 2007). The situation with other kinds of emissions is more mixed: while carbon monoxide and volatile gas emissions would drop over 90%, particulate and SOX emissions

Table 13.3 **Comparison of CO_2 Emissions by Engine Type**

Type of Engine	Annual CO_2 Emissions (t)
Internal combustion	4.98
CNG	2.85
Hybrid electric	2.23
All electric	2.59

Note: CNG, compressed natural gas.

would increase. From an emissions standpoint (including emissions at the power plant), running a fleet of cars on compressed natural gas (CNG) would be nearly as good as an all-electric or hybrid electric car in terms of CO_2 emissions, and considerably better in terms of most other emissions (Yuhnke and Salisbury, 2010) (Table 13.3).

13.1.6 Magnetic Storage

Energy can be directly stored in magnetic fields in a similar way that it is stored in electric fields in a capacitor. The energy per unit volume stored in a magnetic field of magnitude B (in tesla) can be written as

$$\frac{\text{Energy}}{\text{Volume}} = \frac{1}{2\mu_0}B^2, \tag{13.14}$$

where $\mu_0 = 4\pi \times 10^{-7}$ T m/A is a universal constant. If the magnetic field is created by a current in a coil of wire, its magnitude will be proportional to the size of the current. Normally, to sustain a large current in a wire requires a continued input of power owing to resistive losses; however, through the use of a superconducting wire, these can not only be reduced, but also actually brought to zero. A current created in a superconductor (whose resistance is exactly zero) will maintain itself indefinitely without any power input. The coil of wire can be charged by increasing its current with the power supplied from an outside source.

Thus, exceptionally large magnetic fields can be created and stored indefinitely without any concern over losses or coil overheating, since the coil resistance is zero. The discharge process that allows power to be withdrawn from the coil works in the same manner. Superconductivity does not occur unless the coil is cooled below some temperature that depends on the particular material. Superconductivity was discovered in 1911 by Dutch physicist Heike Kamerlingh Onnes when he measured the electrical resistance of mercury as he slowly cooled it to temperatures approaching absolute zero. When he reached the temperature of 4.2° above absolute zero, the slow decline in the resistance of mercury became a precipitous sudden drop to zero! Thus, 4.2 K is known as transition temperature of mercury. Today, there is much research to identify materials having the highest transition temperatures—with the ultimate goal of room-temperature superconductivity, so as to avoid the expense of attaining very low temperatures.

A superconducting magnetic energy storage system (SMES) consists of the coil of wire, a refrigeration unit, and a power conditioning system to convert AC power from an outside source to the DC needed to create the magnetic field. The only losses in the charge and discharge cycle involve the 5% loss in the power conditioning unit, since there are no losses in the superconducting coil, making the overall efficiency per cycle 95%. SMES has some unique advantages over other energy storage systems aside from its high efficiency, including a very short time delay during charge and discharge and no moving parts, meaning that reliability and lifetime are both very good. The main application to date has been to maintain power quality, where fast response times are essential. A large SMES is capable of supplying 20 MWh of energy—say 10 MW for 2 h or more power for shorter times. SMES is probably not a technology suitable for small-scale applications, given the high costs associated with making wire out of superconducting materials, the expensive refrigeration units, and the costs of providing power to them—factors which are mainly responsible for its fairly limited use to date.

13.1.7 Nuclear Batteries

Nuclear batteries are not tiny nuclear reactors, which are not possible—do you know why? Instead, they use the heat from the radioactive decay of some isotope and use it to generate electricity using the thermoelectric effect discussed in Chapter 12. Nuclear batteries have been used for many years, mostly in applications needing very long lifetime and high energy density, such as outer space or underwater military systems. Although most existing nuclear batteries are quite costly, as well as being large and heavy, research is ongoing to create much smaller, lighter, and more efficient versions. One research group at the University of Missouri has created a nuclear battery thinner than a penny and, in the future, hopes to produce a battery thinner than a human hair (Figure 13.15). Although many people might balk at the idea of a nuclear battery powering their pacemaker, the radioactivity is contained within the battery since the emitted radiation lacks the range to escape, and consequently, they have been used for such implanted medical devices. Disposal of used nuclear batteries would add a very minor amount to the existing problem of disposing of low-level radioactive waste. In contrast, disposing of much more numerous car batteries can be a major environmental issue if simply dumped in landfills.

Figure 13.15 Nuclear battery next to a U.S. dime for scale. (Image by Jae Wan Kwon is included with his permission.)

13.1.8 Antimatter

Antimatter would represent the ultimate in energy storage. While not being in the realm of science fiction, the feasibility of using it any time in the foreseeable future for energy storage seems highly remote. Antimatter is a kind of mirror image of ordinary matter, and subatomic particles such as electrons, protons, and neutrons each have their antimatter counterpart, which has the same mass, but opposite sign of electric charge—thus, the antielectron, also called the positron is positive, while the antineutron is neutral like the neutron, but is entirely distinct from it. Antiparticles can even combine to form all the antielements, such as antihydrogen. Antimatter can be created from energy in the very high-energy collisions that take place in a particle accelerator. When antiparticles are created in such collisions, equal numbers of ordinary particles are also created as in the creation of an electron–positron pair in a proton–proton collision: $p + p \rightarrow p + p + e^+ + e^-$. Antimatter is stable as long as it does not come into contact with ordinary matter, but when it does, the result can be complete annihilation, with the mass being entirely converted back to energy as in $e^+ + e^- \rightarrow 2\gamma$ (two gamma ray photons). For this reason, the possibility of finding a mine of antimatter where antimatter could be found is exactly zero. If antimatter created in a high-energy accelerator can be stored in magnetic bottles so that it never physically comes in contact with the walls, one could in principle accumulate a large quantity, as was done in the fictional work *Angels and Demons* (Brown, 2000). If antimatter could ever actually be accumulated in significant quantities, it would be the ultimate in energy storage, as it would have the greatest possible energy density. By Einstein's relation $E = mc^2$, we have the energy density $E/m = c^2 = 9 \times 10^{16}$ J/kg—about 700 billion times better than lead-acid batteries. The obvious application for such fantastic energy densities would be space travel, and some researchers at the National Aeronautics and Space Administration (NASA) have been looking into propulsion schemes that one day might rely on antimatter. However, these possibilities would not be for the foreseeable future. To create 1 kg of antimatter, for example, using known technologies for accelerators using Earth's entire supply of energy each year would require a million years.

13.1.9 Summary

The first half of this chapter considers the topic of energy storage which is important for many reasons, including improving the efficiency and quality of electric power generation, providing backup power in the event of blackouts, providing power to portable devices and vehicles, and making it possible for renewable energy sources to be used more effectively, despite their intermittent nature. There are many technologies for storing energy, and they can be grouped according to their physical nature: mechanical, thermal, electric, magnetic, or nuclear. Which storage device is best suited to a particular application depends on a host of factors, including energy and power densities (on either a mass or a volumetric basis), efficiency, lifetime, reliability, safety, temperature dependence, storage capacity, cost per cycle, and lifetime cost—among others! In some cases, a combination of technologies—one having high power density and another having high energy density—may be the best choice. For example, for vehicles—perhaps one of the most crucial storage applications—one might use an ultracapacitor or flywheel for quick power boosts or regenerative braking (a power application) and batteries for propulsion, given their greater energy density.

13.2 ENERGY TRANSMISSION

13.2.1 Introduction

Different energy sources each have their own technology for distribution to the point where the energy is used, and this issue has been discussed in earlier chapters for some sources. Heat transmission, for example, has been discussed in Chapter 10, while fossil fuel transport was considered in Chapter 2. Our focus here will be on the transmission and distribution of electrical energy. Electricity is, of course, fundamentally different from either heat or fossil fuels, since while they can easily be stored until they are needed, the capability to store large quantities of electricity is much more limited. Although many readers will consider the terms *transmission* and *distribution* to be virtually indistinguishable, within the electric power industry, there is an important difference, as the former refers to the bulk transport of electric power over long distances using transmission lines connecting power plants to area substations, while the latter refers to the distribution of that power beginning from a substation to a surrounding population center.

13.2.1.1 Electricity transmission The transmission network for distributing electricity (the grid) has evolved over time starting in the last few decades of the nineteenth century to the present. Initially, there was no national grid, but only electric power plants serving the needs of some surrounding local area. The grid gradually evolved when these local networks or grids became interconnected to the point where there are over 300,000 mi of interconnected transmission lines (in the United States) operated by 500 different companies. These transmission lines usually provide multiple pathways for electric power from any generating plant to reach a customer. This redundancy provides the system with its fault-tolerant character so that the failure of individual pieces of equipment will (usually) not be fatal.

A remarkable property of the system is the balance between supply and demand. At every moment in time, there is usually a balance between the overall power used and the power generated, so that when you turn on a light switch, a generator somewhere must sense the extra load and "know" to turn a bit harder so as to generate that much more power. Unlike the financial system where borrowing from the future is allowed, the supply and demand of electrical energy must be in balance at all times—ignoring for now any power losses on the lines. The momentary failure of supply to equal demand locally will lead to an import of electricity from connecting grids or, if this is not possible, to load shedding (brownouts) or, in extreme cases, blackouts.

The electrical energy generated is, of course, somewhat greater than that received by customers, because any time that electricity travels through wires having a nonzero resistance, some energy will be converted to heat or be radiated as electromagnetic waves. These losses depend on the length of the wires and other factors to be discussed, but overall, they degrade the system efficiency by only a modest amount. The average losses for the whole system of transmission lines can be estimated by comparing the amount of power generated with that delivered to all customers, which differ by 7% in the United States.

13.2.1.2 Alternating current transmission and distribution In the early years of electricity generation at power plants, there was a rather nasty battle (the "war of the currents") between the proponents of direct versus alternating current (DC vs. AC) over which was the superior choice, and as you are aware, the AC proponents led by Nikola Tesla and George Westinghouse prevailed over Thomas Edison who championed DC. The superiority of AC lies in the ability of AC transmission lines to carry large quantities of power over long distances using very high voltage with low losses. Recall that the loss of power on the line $p_{\text{Loss}} = i^2R$, where R is the line resistance. Thus, in order to minimize the power loss, we want the current in the line to be as small as possible for a given line resistance. Reducing the current is done by stepping up the voltage (using a transformer), prior to feeding the electricity onto the line, since a lossless transformer keeps the product $p = iV$ constant. Transformers are in principle very simple devices for which the ratio of the voltages on the primary and secondary sides equals the reciprocal ratio of the number of turns of wire on each side:

$$\frac{V_S}{V_P} = \frac{N_P}{N_S} = \frac{i_P}{i_S}. \tag{13.15}$$

The basis of Equation 13.15 is Lenz's law for induced EMFs and the requirement that the magnetic flux Φ is entirely contained within the iron core of the transformer. The transformer depicted in Figure 13.16 (for which we obviously have $N_P > N_S$) will function as a step-up transformer that increases voltage so long as the input voltage is connected across the secondary side, while the output fed to the transmission line is connected to the primary side. After the electricity has travelled over some considerable distance on the line, it reaches a substation, where a step-down transformer reduces its voltage before it is distributed to some local area. Exactly the same type of transformer can be used to

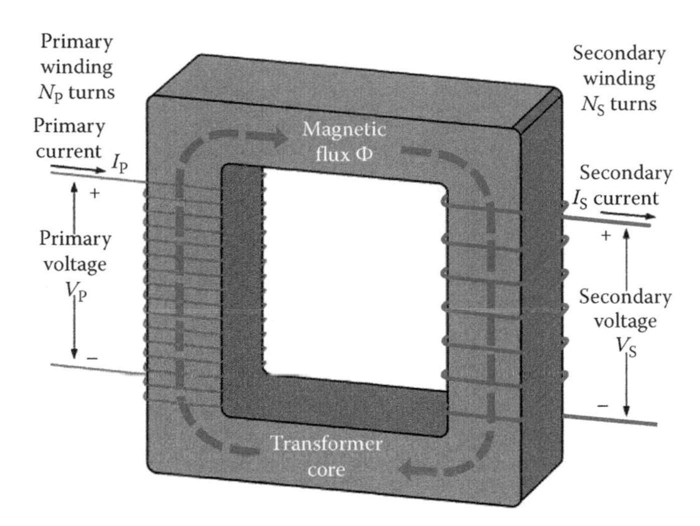

Figure 13.16 A transformer used to increase or decreases the value of an AC voltage.

step up or down voltages, with the roles of the primary and secondary sides reversed in the two cases. As noted earlier, the local lines feeding electricity to some area from a substation are normally referred to as *distribution* rather than *transmission* lines. Historically, the same companies owned and managed both transmission and distribution lines, but in recent years, there has been a separation of the two functions in the United States.

13.2.1.3 Alternating vs. direct current Tesla and Westinghouse prevailed in their promotion of AC over DC generation and transmission mainly because the transformer provided an easy way to increase or decrease voltage in the case of AC, but no such simple device existed at the time for raising or lowering the voltage for DC. As a result, DC power fed onto transmission lines had to be at some fixed voltage needed by particular devices, and if different devices required different voltages, a separate pair of power lines would be required for each device depending on the voltage it required. Pictures of street scenes from the early era when DC was used rather than AC show utility poles having many sets of power lines, each with different voltages (Figure 13.17). An even greater disadvantage of DC was that since the devices such as lamps and motors for which electricity was used required relatively low voltages of perhaps 100 V, and low voltage means high current for a fixed power. Thus, if we transmit high current over the power lines, the i^2R losses would be very large unless the lines were kept relatively short so as to keep the resistance low. As a consequence, power plants could not be very far from customers using DC, and the idea of a very large power plant servicing customers over a wide geographic region (useful based on economy of scale) was out of the question. Not only is it usually less expensive (per megawatt generated) to build and maintain a large power plant than a small one, but also the larger customer base it services means that the variation in load over time would be less and make for still greater efficiency.

Figure 13.17 Etching of scene of New York City streets in 1890. (From Brown, H. C., *Book of Old New York*, 1913.)

WAR OF THE CURRENTS

Thomas Edison may have been a genius and perhaps the most prolific inventor who ever lived, but the war of the currents did no credit to his reputation and may in the end have cost him his commercial empire. It was not simply that he was on the losing side, but Edison's motives were suspect, his understanding of AC flawed, and his tactics in the fight reprehensible. Having pioneered DC power generation and distribution, Edison saw Tesla's new three-phase AC generators (see Chapter 7) as undermining his commercial interests and his stake in patents based on DC. Rather than embracing Tesla who worked for him at the time, Edison's dismissive attitude drove Tesla to George Westinghouse, who had the wisdom and commercial wherewithal to champion the AC cause. During the war of the currents, Edison resorted to some deplorable efforts in his attempt to portray AC as uniquely dangerous by staging public executions of animals. In fact, the electric chair used to execute condemned criminals was invented by an employee of Edison in his further attempt to connect AC current with great danger in the public mind. It is possible that Edison might have succeeded in his efforts, were it not for the successful AC generators built by Westinghouse using Tesla's design that harnessed the power of Niagara Falls in 1896. At the time, Tesla thought this large-scale project would generate enough power to supply the entire eastern United States—and he might have been right, given the limited usage of electricity in the last decade of the ninetieth century. Had Edison won the war of the currents many of today's modern inventions from computers to cell phones to television, with their dependence on such AC circuit, components including capacitors, inductors, diodes, and transistors would have likely been delayed for a long time.

13.2.1.4 High-voltage transmission lines High-voltage transmission lines are normally suspended from steel towers and are above ground, since underground lines are much more expensive, although they are often used in some cases for the lower voltage distribution network from substations. As seen in Figure 13.18, the three-phase lines most commonly used must be suspended from insulators and kept away from the metal structure, so as to avoid any short circuit or flashover and consequent loss of supply. It has been found that dry air begins to breakdown (become conducting) when there is an electric field present of about $E_{max} = 3 \times 10^6$ V/m. For a uniform electric field, the closest separation d between two conductors with a voltage V between them would be $d = V/E_{max}$ before breakdown. In practice, however, this simple formula offers little guidance on the true minimum safe distance needed for power lines, because the electric fields are certainly not uniform, and other complexities such as the presence of sharp edges or points on the supporting structure, and the humidity of the air must also be considered.

For 765 kV power lines (about the highest in general usage), safety requires a minimum spacing of many meters. Failure to maintain adequate clearances for high-voltage power lines can be the result of bad weather, particularly high winds that cause the lines to oscillate or very high temperatures that cause them to sag. Transmission lines are usually not insulated and made of aluminum, which is nearly as good a conductor as copper, but it has lower weight and much lower cost. Typical wire diameters range from about 3.7 mm to 3.2 cm, depending on the maximum amount of power to be carried on the line. Obviously, higher power or longer line lengths require larger diameter lines, so as to avoid large i^2R losses. There is however a point of diminishing returns to increasing the wire diameter so as to cope with larger amounts of power, having to do with the so-called skin effect.

Figure 13.18 High voltage transmission lines.

13.2.1.5 Skin effect In contrast to DC, which flows through the entire cross section of a wire, an AC is more concentrated near the surface of the wire, and the current density tends to exponentially decrease with distance below the wire surface. The skin depth is that depth for which the current has decreased to the fraction $e^{-1} \approx 0.37$ of its surface value. The skin effect occurs because the time-varying magnetic fields generated by an AC cause eddy currents throughout the interior of the wire, and these eddy currents tend to reinforce near the surface and tend to cancel in the interior. It can be shown that when a current having a frequency f (in hertz) passes through a conductor having resistivity ρ and magnetic permeability μ, the skin depth is approximately given by

$$\delta \approx \sqrt{\frac{\rho}{\pi f \mu}}. \tag{13.16}$$

Thus, the skin effect is most important for very high frequencies ($f \gg 60$ Hz) where the current tends to be concentrated on a very thin sheath on the surface of a wire. In the case of a 60 Hz AC through an aluminum wire, the skin depth works out to be 1.10 cm based on Equation 13.16, so the skin effect is therefore only significant for thicker transmission wires.

Even though the skin effect causes the current to decrease as a continuous function of depth, let us simply assume here that it is constant and then suddenly drops to zero at the skin depth, so we imagine that the current travels through a cross section that is an annulus or ring whose outer radius is that of the wire itself r, whose inner radius is $r - \delta$ and whose cross-sectional area is $A = \pi(r^2 - [r - \delta]^2)$. Of course, by extension, when $r < \delta$, we assume that the current passes through the full wire cross section $A = \pi r^2$. Let us use this ring approximation to determine how the fractional power lost due to resistance in a wire varies as a function of its radius. Given the usual formula for resistance $R = \rho L/A$, we can then easily show that the fractional power lost along a line of length L for AC and DC transmission is

$$\left(\frac{p_{\text{Loss}}}{p}\right)_{\text{AC}} = \frac{C}{(r^2 - [r-\delta]^2)}, \tag{13.17}$$

$$\left(\frac{p_{\text{Loss}}}{p}\right)_{\text{DC}} = \frac{C}{r^2}, \tag{13.18}$$

where $C = \rho L p/\pi V^2$ in both cases. Plots of these functions are shown in Figure 13.19. It is clear from the figure that for wires whose radius is small compared to the skin depth, there is virtually no difference in terms of the fractional energy loss between AC and DC transmissions, However, once the wire radius exceeds the skin depth, the gap between AC and DC power losses widens, with the fractional power loss for AC becoming

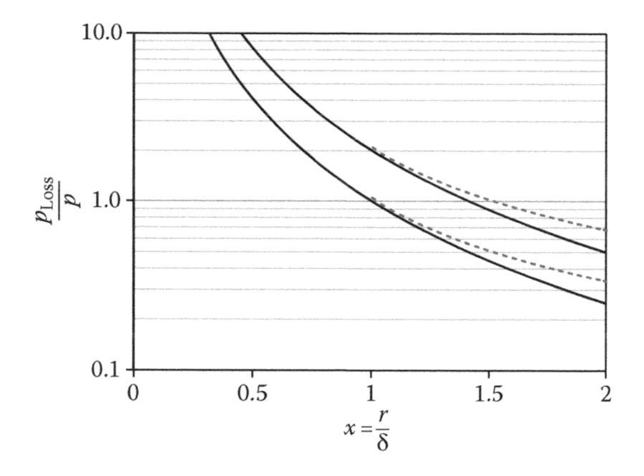

Figure 13.19 Fractional power loss p_{Loss}/p for AC (*dotted curves*) and DC (*solid curves*) versus the quantity $x = r/\delta$. The upper pair of curves is for a C value twice as great as the lower pair. Notice how the AC and DC curves begin to diverge for radii greater than the skin depth, i.e., $x > 1$.

progressively worse. Moreover, given the definition of the quantity $C = \rho Lp/\pi V^2$, the penalty paid for using AC becomes greater for longer lengths L, higher powers p, and lower voltages V. For this reason, there are actually important special applications where DC is the preferred choice over AC for power transmission, even though most power lines do use AC.

13.2.1.6 Direct current power transmission Had Edison known of some simple way to raise and lower DC voltages, the transmission of DC power over long distances would have been much more viable. Nowadays, there is a suitable technology based not on raising or lowering DC voltages, but rather on devices such as thyristors, which are solid-state devices capable of converting AC to DC or the reverse. The low-voltage AC power from a power plant is first raised in voltage using transformers, and then the AC power is passed through a circuit including a stack of thyristors so as to create high-voltage DC (HVDC) before being fed onto a power line. As a result of this technology, the possibility of HVDC transmission lines has become a reality.

As you recall from the previous section, DC will result in less transmission losses when the wire radius exceeds the skin depth, and the advantage over AC becomes especially pronounced for longer length lines and higher amounts of power they might carry. Thus, the longest cost-effective transmission of DC electricity is around 7000 km, while for AC, it is only 4000 km, although most lines fall far short of these values. Nevertheless, DC does tend to be reserved for special applications, so that in the United States perhaps, only about 2% of the miles of lines use DC rather than AC. Some special cases where DC is the preferred way to transmit electrical power include the following:

- Point-to-point bulk power transmission over large distances (no intermediate taps)
- Increasing the power a line can carry without installing any new lines
- Transfer of power between two AC systems that are unsynchronized (no fixed phase relation between them)
- High-power long distance undersea transmission

The last case requires a bit of explanation. Unlike overland transmission, undersea cables must, of course, be insulated, and as a result, there is considerable capacitance between a pair of parallel cables unlike the case where there is just an air gap separating them. The higher capacitance means nothing for DC transmission, but for AC transmission, it causes additional reactive power losses.

Given their lower losses, undersea DC power cables exist between various European nations—one being around 600 km long (between Norway and the Netherlands). An analysis of the cost-effectiveness of DC over AC for transmission lines must take into account not only the relative power losses in the two cases but also the construction costs. For example, DC power lines are less expensive because they require fewer lines than three-phase AC. Offsetting this, however, is the extra hardware they require in order to convert the original AC power to DC before putting it on the line and then converting it back to AC at the other end. As a result of these competing factors, there is usually some breakeven line length above which DC is the less expensive choice.

13.2.1.7 Example 6: Power losses with high-voltage AC and HVDC transmission lines It is desired to carry up to 2000 MW of power from a hydropower plant to a city 1000 km away using 765 kV transmission lines. What minimum-radius wires made of aluminum would be needed to keep line losses under 8%? What would be the power lost if HVDC transmission were instead used over the line? Aluminum has a resistivity $6.2 \times 10^{-8}\ \Omega$ m and a skin depth of 1.1 cm = 0.011 m.

Solution

The value of C in Equation 13.17 may be found from

$$C = \frac{\rho L p}{\pi V^2} = \frac{6.2\times10^{-8}(10^6)(2\times10^9)}{3.14\times(7.65\times10^5)^2} = 6.75\times10^{-5}\ \text{m}.$$

Using Equation 13.17, we may then solve for r:

$$\left(\frac{p_{\text{Loss}}}{p}\right)_{\text{AC}} = 0.08 = \frac{6.75\times10^{-5}}{(r^2 - [r-0.011]^2)},$$

which yields r = 0.044 m = 4.4 cm. The DC line has a larger cross-sectional area through which the current flows and proportionately less resistive losses. From Equations 13.17 and 13.18, the power lost for DC transmission would be less by the factor $r^2/(r^2 - [r - \delta]^2) = 2.29$, or only 3.5%.

DO HIGH-VOLTAGE POWER LINES CAUSE CANCER?

There have been numerous studies on this controversial matter, but a report from the National Academy of Sciences (Committee on the Possible Effects of Electromagnetic Fields on Biologic Systems et al., 1997), which was charged by the US Congress to review these studies, concluded that "there is no credible basis for believing that 2 mG fields are biologically harmful." Moreover, the notion that 2 mG fields are harmful to cellular biology contradicts the most fundamental laws of physics, including the second law of thermodynamics, which calculates the thermal noise level of a cell. The energy of this thermal noise is eight orders of magnitude larger than the energy associated with the external background field from power lines and cannot create mutant strands of deoxyribonucleic acid (Mielczarek and Araujo, 2011).

Finally, in any large community, it is possible to find "cancer clusters," and it is easy to understand that someone whose family member happened to be an unfortunate member of such a cluster, who also lived near a power line, would become convinced that it was the cause of the cancer. However, clusters occur all the time far more often than you might imagine, just based on the laws of probability, and only a careful statistical analysis can reveal whether there may be a real environmental cause for it (Ehrlich, 2003).

13.2.1.8 Problems with the grid The existing grid has organically evolved in a patchwork fashion with a technology that is, in some cases half a century old. There have been recurring calls for upgrading the grid, and one 2012 report from the American Society of Civil Engineers has estimated that $673 billion would be required for the task between now and 2020. The increasing urgency of calls for such upgrading has spurred calls for a smart grid. Note, however, that a smart grid would not be a replacement for the existing grid, but rather a major upgrade to deal with its existing problems. Problems with the existing power grid include the following:

- *Insufficient capacity*: In recent years, the increase in generating capacity has outstripped the increase in transmission line capacity in the United States and many other countries as well. In some areas, there may be considerable resistance to adding more transmission or distribution lines, and this may sometimes derail worthy proposed projects, including a plan to significantly add to the Texas wind turbine capacity. New transmission line capacity need not necessarily require new right of way or expanding their footprint, since technologies exist for increasing the capacity of existing lines. These include dynamic line rating (raising the maximum power level on the line depending on environmental conditions), replacing lines by ones that do not sag as much, and increasing the voltage for the line.

- *Inefficiency*: Currently, the power generated at times of peak demand can be considerably more expensive than power generated at other times, but usually, in most places, customers are charged the same average rate regardless of when they use the power. This flaw is due to the metering system typically used, and it could be remedied if the standard meters were replaced by smart meters that took into account when the power was purchased. It would be even more desirable if customers were aware that power is costing more at certain times and could possibly defer power usage that could easily be deferred. A system of two-way communication between the utility company and appliances in a customer's home or business might even allow selected appliances to be turned off for short periods by the utility company at times of peak demand.
- *Blackouts*: The existing grid is vulnerable to both natural and human-made disruptions and some past blackouts such as the 2003 blackout of the northeastern United States and parts of Canada affecting 55 million people, which arose due to a combination of operator error and a transmission line failing due to a contact with a tree that caused a series of escalating failures that cascaded through the system. Some generating stations were not back on line for as long as 5 days. While the 2003 blackout may have been the largest in US history, smaller power outages affecting at least 50,000 people have been occurring more frequently. During the past two decades, they have risen by 124% according to research at the University of Minnesota.
- *Catastrophic vulnerability*: Many cybersecurity experts worry that a targeted human-made effort to disrupt the grid—either the result of a terrorist attack or a surprise attack by a nation state—could cripple the grid for an extended period. Calculations show that the electromagnetic pulse from a nuclear weapon detonated at high altitude could be a major problem for the grid, and some (of the many) components needing replacement, such as very high-voltage transformers, might take years to replace.
- *Centralized power generation*: When the main components of the grid were designed, the model was one of large power-generating plants connected to the transmission lines, not many small generators, possibly owned by private individuals. The current grid is unsuited to decentralized power generation, especially by renewable sources. Wind and solar present special problems, since the best locations for them are often far from population centers, where there may be few existing transmission lines. One major wind farm project in Texas (with about a quarter of all US wind power) has been stalled due to inadequate transmission line capacity. Similar bottlenecks have prevented the installation of wind farms in a number of other parts of the United States where the wind resource is most abundant.

SOLAR FLARES

At certain times the sun emits prodigious flares whose occurrence tends to be correlated with the sun's 11 year sunspot cycle. Solar flares consist of massive amounts of electromagnetic radiation, especially X-rays and high energy particles (Figure 13.20). They tend to be fairly directional so it would take one directly headed towards Earth to cause a serious problem—for electronics but not living organisms. According to a study by the prestigious National Academy of Sciences:

> A severe space weather event in the U.S. (of the magnitude of one that occurred in 1859) could induce ground currents that would knock out 300 key transformers within about 90 seconds, cutting off the power for more than 130 million people.

The impact of what the report terms a "severe geomagnetic storm scenario" could be as high as $2 trillion—just the first year after the storm. The NAS report puts the recovery time at 4–10 years, and notes that it is questionable whether the United States would ever bounce back (NAS, 2008).

Many other nations, including Japan, China and those in Europe would be affected just as seriously as the United States.

- *One-way information flow*: The present grid in most cases relies on data on power usage to see when additional generating capacity is needed, which is a one-way information flow. Incorporating two-way information flow might involve informing customers about real-time demand and would allow them to directly participate in optimizing the system and reducing costs by choosing to defer power consumption at times of peak demand for appliances such as

Figure 13.20 Charged particles from the sun emitted during a solar flare are shown impinging on the Earth's magnetosphere. The image of the sun is real, but that of the Earth's magnetosphere and the particle streams is an artist's rendition. (Courtesy of NASA, Washington, DC.)

hot water heaters that could easily do so. A smart grid would probably also entail the use of smart meters in residences and businesses that would charge different rates for power consumed at different times and might (with a customer's permission) allow the utility to periodically cycle certain appliances (like hot water heaters) on and off to so as to even out the power load profile during the day.

13.2.1.9 Goals for a smart grid As already noted, a smart grid would merely be an upgrade of the existing grid in light of the current problems with it, and various nations are still in the planning stages concerning just exactly what the priorities and goals might be for a smart grid. In the United States, as of 2011, only a few states have yet implemented some components of a smart grid. It must also be borne in mind that some goals implied by the aforementioned list of problems might actually be in conflict. For example, a grid with two-way communication while more efficient would also be more vulnerable to catastrophic failure due to cyberattack. Even without a mastermind hacker, if all customers had the ability to directly communicate in real time with the power company based on the current price of electricity, some studies have shown that the result might be to crash the grid (Biello, 2011). One could even imagine flash mobs communicating via social media to bring about the crash, thinking that such an outcome might be funny or useful for some looting perhaps.

Obviously, different constituencies (customers, utility companies, states, and federal government regulators) will have very different views on perhaps which goals are most important to pursue. In the United States, the utility companies have a monopoly over the transmission lines and, in return for government regulation, are guaranteed a specific profit based on their costs. This arrangement is understandable given the inefficiency of having multiple sets of long-distance transmission lines. Nevertheless, a monopolistic operation with a fixed rate of return is not one that is likely to foster innovation. Having a new smart grid designed by the utility companies could be a serious mistake—somewhat akin perhaps to what might have happened had new smart phones been invented and designed by the phone company rather than some nimble and innovative computer companies.

Finally, a smart grid also needs to be accompanied by smart policies. For example, utility companies in the United States are required under the 2005 Energy Policy Act to participate in "net metering," which requires them to participate in a program "under which electric energy generated by that electric consumer from an eligible on-site generating facility and delivered to the local distribution facilities may be used to offset electric energy provided by the electric utility to the electric consumer during the applicable billing period." The intent of this law, which encourages greater usage of distributed renewable energy by homeowners and businesses, is excellent. However, its implementation has been left up to the states, and its real impact is negligible in some states.

13.2.2 Summary

We discuss some of the requirements for the transmission and distribution of electricity from power plants to customers and the evolution of the present electric grid. In the early days, a sharp war of the currents pitting AC against DC was fought, with AC winning out because it could more easily be transmitted over long distances with little losses, given the ready means for increasing and decreasing voltage levels. We also note that nowadays, HVDC transmission is also possible, and owing to the skin effect, there are some circumstances under which it is preferable. In a concluding section, we consider problems with the power grid and the need for a major upgrade in the form of a smart grid.

PROBLEMS

1. The Ragone plot (Figure 13.12) is a log–log plot of energy versus power density for various technologies. Prove that on such a plot, the locus of constant discharge times is a straight line having unit slope with an intercept on the energy density axis equal to the log of the discharge time. Copy the figure and show on it such a line with a discharge time equal to 10 h.
2. How many 12 V lead–acid batteries rated at 60 Ah would be needed to store 40 MWh of energy? For how long a time could they discharge 100 kW of power?
3. A hydroelectric plant located in Virginia has a pumped storage capability. There are two reservoirs having equal surface areas of 110 ha separated by 10 m. During operation, the water level in the upper reservoir can vary by 30 m height. How much more energy is stored in the upper reservoir when the height is 30 m than when it is at 0 m? What is the power in the flowing water when it discharges from the upper reservoir to the lower one if the spillway discharge is 850 m^3/s?
4. A 50 MW solar thermal plant uses molten salt technology to generate electricity night and day. If the plant is 20% efficient, and the salt is heated to 300°C, how large must the storage tank be to generate electricity during 8 h without sun if the minimum temperature at which electricity can be produced from the molten salt is 150°C?
5. Consider a flywheel in the shape of a solid uniform disk of radius 0.65 m and mass of 140 kg. How fast would it need to spin to have the energy equivalent of 10 kg of gasoline (3.6 gal), which is about 140 kWh. Could such a flywheel be made of steel, whose tensile strength is $\sigma = 8.5 \times 10^8$ N/m^2?
6. What is the maximum speed that a 0.5 m diameter flywheel made from high-strength steel could spin? Use $\rho = 7850$ kg/m^3 and tensile strength $\sigma = 8.5 \times 10^8$ N/m^2.

7. Show that the maximum volumetric energy density of a flywheel in the shape of a uniform disk is $\sigma/2$.
8. How much energy does a flywheel in the form of a 0.2 m radius disk spinning at 100,000 rpm contain (m = 2 kg)? What is the tangential speed of a point on the rim? How many g centrifugal force acts on a particle on the rim?
9. In a compressed air storage tank, suppose that a million moles of air initially at 1 atm and 300 K is isothermally compressed to 10 bar. How much energy has been stored? How large a cavern would be needed in this case? Hint: You need to use the ideal gas law: $PV = nRT$.
10. Show that a battery connected to a load will deliver maximum power when the resistance of the load equals the internal resistance of the battery and that at that point, the efficiency is 50%.
11. Use the data given in Section 13.1.4 to show that a fuel cell-powered electric car is roughly three times as efficient as one powered by an internal combustion engine.
12. The highest magnitude magnetic field achieved to date (as of 2011) is 91.4 T. Suppose that the interior volume of a coil of length 1 m and diameter 1 m had a uniform field of this magnitude. How much energy would be present?
13. A capacitor is charged to a voltage V and then allowed to discharge through a resistor R. By integration, show that the integral over time of the power dissipated during discharge equals the energy originally stored in the capacitor.
14. Why can there be no way of storing energy that gives a higher energy density per kilogram than antimatter?
15. Consider a pair of protons in a colliding beam accelerator. The protons collide head on with equal kinetic energies K. What is the minimum value of K if a proton antiproton pair is created in the collision? (Their masses can be written as 938.2 MeV/c^2.) If the colliding protons each had 2000 MeV kinetic energy, how many antiprotons could be created in the collision?
16. Prove that the maximum power a battery supplies is when it is connected to a load of resistance $R = r$.
17. Derive Equations 13.17 and 13.18.
18. Find the skin depth for an aluminum wire carrying a 60 Hz current, given that the magnetic permeability of aluminum is 1.0 and its resistivity is 2.7×10^{-8} Ω m. How large would the diameter of a wire need to be before the resistance of the wire to AC is twice that for DC (at 60 Hz)?
19. How much chemical potential energy is stored in a full tank of gasoline? Assume that your tank holds 20 gal of gasoline.
20. What would the weight of a battery be if it possessed the equivalent energy tank of gasoline? Assume a 20 gal tank of gasoline for comparison and assume a battery energy density of 6×10^5 J/kg.

REFERENCES

Biello (2011) What if the smart grid isn't so smart? *Scientific American*, August 7, 2011, http://www.scientificamerican.com/podcast/episode.cfm?id=what-if-the-smart-grid-isnt-so-smar-11-08-07 (Accessed Fall 2011).

Brown, D. (2000) *Angels and Demons*, Washington Square Press, New York.

Brown, H. C. (1913) *Book of Old New York*.

Committee on the Possible Effects of Electromagnetic Fields on Biologic Systems et al. (1997) *Possible Health Effects of Exposure to Residential Electric and Magnetic Fields*, National Academy Press, Washington, DC.

Creutzig, F., A. Papson, L. Schipper, and D. Kammen (2009) Economic and environmental evaluation of compressed-air cars, *Environ. Res. Lett.*, 4(4), 044011.

Ehrlich, R. (2003) *Eight Preposterous Propositions: From the Genetics of Homosexuality to the Benefits of Global Warming*, Princeton University Press, Princeton, NJ, pp. 30–31.

Grossman, J., and A. Kolpak (2011) Azobenzene-functionalized carbon nanotubes as high-energy density solar thermal fuels, *Nanosci. Lett.*, 11(8), 3156–3162.

Mazza, P., and R. Hammerschlag (2005) *Wind-to-Wheel Energy Assessment*, presentation at the Lucerne Fuel Cell Forum, European Fuel Cell Forum, http://www.efcf.com/reports/E18.pdf.

Mielczarek, E. V., and D. C. Araujo (2011) Distant healing and health care, *Skeptical Inquirer*, May/June, 35(3), 40–44.

Climate and Energy

Policy, Politics, and Public Opinion

CONTENTS

14.1 HOW IMPORTANT ARE INTERNATIONAL AGREEMENTS?

Action on reducing CO_2 emissions can take place at a wide range of levels, from the individual to the state and local levels, and finally to the national and the international levels. Individual actions are worth taking and may make you feel virtuous, but their actual impact on the problem of worldwide CO_2 emissions is, of course, miniscule. Worse yet, they may detract you from efforts to promote collective action. Perhaps the best actions that individuals can take are therefore in the area of education—helping to both improve their own and educate others on the nature of the problem. Education needs to include both well-grounded scientific and economic components. Given the way politicians make decisions in a democracy, it is perhaps unavoidable that advocating solutions to reduce CO_2 emissions through national policies, such as an energy tax or cap-and-trade regulations, will be seen through a partisan lens, but the reflexive opposition to such policies on the part of some citizens and politicians need not be a permanent fact of life, and in fact, the political parties in the United States have changed their position on these matters in the past. Until such changes at the national level occur, many useful steps can be taken at the state level, and many states, especially California, have taken actions to foster to varying degrees of energy conservation and renewable energy. As of 2014, a total of 34 states have timetables for generating specific fractions or amounts of their electricity from renewable sources, although in a few, cases the compliance is voluntary.

International agreements to reduce emissions may be even more difficult to achieve than a national consensus, but clearly, global problems such as human-caused climate change induced by greenhouse gases (GHGs) do require a global response. In 2015, representatives of nearly 200 nations met in Paris and unanimously signed an agreement limiting GHGs. The agreement is not legally binding, in large part due to the insistence of the United States, which would otherwise have faced a certain senate rejection. Reactions to this agreement to agree have widely varied between those who hail it as a great achievement to others who believe the planet is now doomed, since the Earth's climate will suffer irreparable harm without steep CO_2 emission cuts beginning immediately. Clearly, the final chapter on the agreement remains to be written. As many commentators have noted, its main significance may be to encourage investors to fund new projects in renewable energy and carbon capture.

Aside from the 2015 agreement, the only other existing international climate agreement in force, as of 2011, is the Kyoto Protocol, which was signed in 1997 by 191 nations, including the United States—the only signatory that never ratified it. In fact, Kyoto was never even submitted for ratification to the US Senate, since its defeat was a foregone conclusion. The treaty called for reducing CO_2 emissions by 37 industrialized countries by an average of 5% below their 1990 levels by the year 2012. Interestingly, the targeted emission cuts widely varied from nation to nation, with, for example, an 8% cut for the United States, no cut for Russia, and an 8% increase allowed for Australia. Although the Kyoto Treaty signatories committed themselves to meeting these targets, the impact of Kyoto in terms of actually curbing carbon emissions has been relatively negligible. For example, the difference between the emissions of the 27 EU nations who have ratified Kyoto and the United States, which has not, has remained relatively constant in the years since the Kyoto Treaty was signed (see Figure 14.1). This suggests that any behavior change specifically related to being a Kyoto-ratifying nation may be minimal. The big news in terms of worldwide CO_2 emissions over the past decade has been the very rapidly growing emissions of the developing Asian nations of China and India. Thus, even if the industrialized nations should meet their 2012 Kyoto targets, worldwide emissions will continue to dramatically increase in the coming years. It was for this reason that Canada, Japan, and Russia have now rejected new Kyoto commitments, with Canada announcing that it would no longer be bound by the treaty.

Binding international agreements having large economic or national security implications are notoriously difficult to achieve. Some optimists point to the successful Montreal Protocol—an international agreement to restrict ozone-destroying chlorofluorocarbons—as a counterexample. However, that treaty involved minimal economic disruption compared

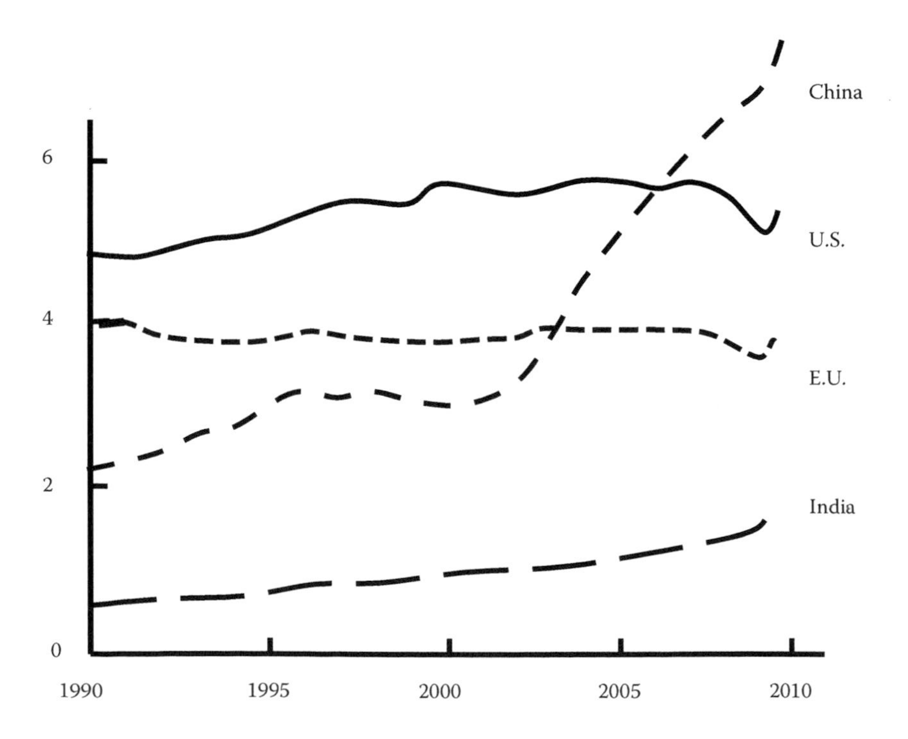

Figure 14.1 Data on CO_2 emissions in Gigatons per year for the top four emitters. The 27 nations making up the European Union have been lumped together as one entity. (From IEA, Data through 2009 is taken from the 2011 edition of the report CO_2 emissions from fuel combustion, by the International Energy Agency. The 2010 data are based on unpublished IEA data, 2009.)

to what would be involved in phasing out fossil fuels. Likewise, the international ban on the usage of dichlorodiphenyltrichloroethane (DDT) is also a poor parallel. That agreement was possible because the economic disruption was minimal, as other mosquito-control methods existed, and the DDT ban exempted malaria-prone areas that lack feasible alternative means of control. It is difficult to think of parallels to a binding international agreement to limit CO_2 emissions strictly within the environmental realm. Perhaps one parallel might be in the security area, such as the nuclear nonproliferation treaty (NPT). The experiences with this treaty suggest a number of challenges that will be faced before and after any future climate change treaty is signed.

Recall that the NPT is a treaty between some states that possess nuclear weapons and some that do not, whereby nonnuclear states agree not to seek the bomb and the nuclear states agree to assist the nonnuclear ones with the peaceful use of nuclear technology. It also requires the nuclear states to work toward general nuclear disarmament over time. One might imagine similar stipulations in a climate treaty, whereby the big industrial nations agree (1) to support the developing nations with technological assistance, (2) to move away from fossil fuels, (3) to adapt to climate change as it occurs, and (4) to reduce their own emissions over time.

There are many useful parallels between a climate change and NPT. As with climate change, the longer a solution is put off to nuclear disarmament, the more difficult a solution becomes—as more nations gain access to nuclear weapons. Additionally, there may be a point of no return in both cases. For nuclear weapons, the possession by certain states (Iran?) could well set off a chain of events that would eventually lead to a rapid arms race in the region or even an actual nuclear war. In the case of climate change, one could at some point be facing an unstable tipping point leading to rapid irreversible climate change.

Furthermore, in both the nuclear and climate situations, the true effectiveness of a legally binding treaty is unclear. Since sovereign nations are always free to withdraw from any treaty, there is the real danger that the treaty will prove to be only a cosmetic solution to a problem. Those nations who find it in their national security interest to seek the bomb will do so regardless of whether they have signed the treaty. In the case of the NPT, there is the supposed benefit that accrues to treaty signers that the "nuclear-haves" will assist the "have-nots" with the peaceful uses of nuclear energy. However, when convenient, this provision is set aside with no consequences for the treaty violator, as in the case of the US nuclear collaboration with India, a nation that has never signed the treaty and then built its own nuclear weapons. Technically, the violation here is that by the United States, not India, which was never bound by the treaty. One could imagine a climate change treaty with stiff penalties for violators, but in that case, if they become too burdensome, the nation would simply withdraw from the treaty. Thus, there is a delicate balance between a penalty being large enough so that it will constrain behavior and being so large that it will induce nations to choose to abrogate the treaty.

Another parallel between a nuclear disarmament treaty and one dealing with climate change is the very different status of some of the parties. In the nuclear case, there are now believed to be eight nations that have nuclear weapons (with many more capable of building them who have simply chosen not to do so). However, the number of weapons in the hands of two of those eight nuclear nations (the United States and Russia) dwarf all the rest combined—even though there are considerable uncertainties in the size of many nations' nuclear arsenals. In a similar vein, while any nation that relies on fossil fuels emits GHGs, two nations (the United States and China) emit 42% of the world's CO_2. Together with number three (India), the trio produces nearly half of the world's GHGs. Thus, just as was the case with limiting nuclear arsenals, where bilateral United States–Russia (or US–USSR) agreements proved to be a fruitful alternative to international agreements between 200 nations, it is possible that the same might hold for agreements on limiting GHGs, which might usefully be pursued between the top two or three emitters.

All nations place extreme value on their sovereignty and will sign an international agreement only when the treaty meets their own interests without imposing severe economic or security costs, and they will abrogate it as soon as these conditions no longer apply. Thus, the key participating countries (especially the largest CO_2 emitters) are likely to agree to a treaty to limit CO_2 emissions only if they believe that without the treaty, the danger to the planet is sufficiently grave that their nation will suffer worse than any sacrifice (monetary or otherwise) they may make in signing the treaty. Whether or not such a treaty actually slows climate change more than it would be slowed by actions taken by nations on their own initiative is an open question. Clearly, the real impetus for action on climate change is likely to hinge on whether future increases in average global temperatures are greater or less than what is now projected and, equally important, how severe those impacts are perceived to be—especially by the public and the governments of the nations that emit the most CO_2.

CAP AND TRADE

One of the governmental tools for controlling GHG emissions involves a system of "cap and trade," whereby once politicians set limits for a nation's permitted level of emissions, the limit is supposed to be enforced by a marketplace in which permits to emit a specific quantity of GHGs are auctioned off. If the emission permits are priced high enough, it becomes more profitable for emissions-producing power companies either to switch to renewable energy sources or perhaps phase out coal plants in favor of natural gas that pollute less. Such a system had been favored by some leaders in the United States but not pursued because of its likely defeat in Congress. In Europe, however, emission trading has been in place for some time, but the system may be having some unfortunate consequences, perhaps because too many permits were issued and the demand for electricity has considerably softened since 2009. As a result, the price for emissions permits has plummeted, and companies have no financial incentive to invest in new technology to reduce their emissions, making the trading system pointless. Interestingly, China, in 2015, announced that it plans to enact a cap and trade policy on its companies in order to limit carbon emissions. Perhaps the Chinese government will be able to pressure companies to comply in ways not possible in the West.

14.2 WHAT ARE THE TOP THREE GHG EMITTERS DOING?

14.2.1 China

The world's leading GHG emitter (China) overtook the United States for that title in 2007, due to its rapid economic expansion. Chinese energy policy is a study in contradictions from an environmental standpoint. It is true, for example, that China invests more on green energy projects such as wind and solar than any other nation (about twice that of the United States), but then it also continues to build an astonishing two to three new coal plants a week—far more than any nation. Moreover, while some of the green technology China builds, such as wind turbines, are used domestically, in other cases, such as solar PV panels, the usage is almost exclusively for export. Nevertheless, China is also the world's largest consumer of solar energy—far more for inexpensive residential solar hot water heaters than solar PV panels.

The nation's heavy reliance on fossil fuels—especially coal, which supplies two-thirds of its electric power—is in furtherance of its goal to establish a nationwide electric grid no later than 2020. As noted in an earlier chapter, quite apart from the contribution of GHG to climate change, the Chinese people are paying a very steep price for their heavy reliance on coal in terms of their health and the state of their polluted environment. Unlike CO_2 emissions that affect the entire world, particulate emissions from coal primarily affect the nation emitting them and its neighbors. China also has plans for the expansion of its nuclear power industry from a very small base of 11 nuclear reactors in 2011. These plans by the government do not seem to have been influenced by Fukushima (even though 61% of Chinese citizens surveyed expressed concern about nuclear power even before Fukushima (Nuclear, 2007). China now aims to increase its nuclear capacity by an astonishing 20-fold by 2030 and 40-fold by 2050—the highest planned expansion of any nation. China should be praised for finally acknowledging that emissions from developing countries need to be subject to limits just as well as those from the industrialized nations. On the other hand, they should not be given too much credit for acknowledging the obvious, especially if they should proceed with their present pace of new coal plant construction, which could result in over 800 new plants by the year 2020, when a treaty might take effect. Moreover, China, as of 2015, does not expect its GHG emissions to reach their peak until 2030. A cynic might note that the only reason China finally agreed to having itself and other developing nations covered by a new treaty is that its massive coal plant expansion would be finished by that time.

It is, therefore, encouraging that in 2007 (and again in 2011), in a rare display of public defiance, 30,000 residents of the seaside town of Haimen mounted a protest against a new coal-fired plant in their area (Wines, 2011). This was the second plant being sited there, and the residents were distressed by the rising cancer rates and destruction of the fishing industry in their area. Even though the protests were squelched by the authorities and will probably not stop the plant from being built, the protests are a welcome sign that the Chinese public may become more assertive in pressing the government on environmental issues.

14.2.2 India

The other Asian giant, India, is currently number three in the world in CO_2 emissions, having just passed Russia for that distinction. The rate of rise in GHGs for India on a percentage basis nearly matches China, but the absolute annual rise is significantly less owing to its smaller base. Nevertheless, the future increases could well be as dramatic as China, given that there is almost no possibility of a one-child-per-family limit being mandated in this democratic nation, and hence, India's population growth is significantly greater than China's. India is expected to be the second-largest energy consumer in the world by 2035. Unlike China, in fact, India is much more dependent on imported energy sources, since it lacks China's abundant coal reserves, and it imports an even larger fraction of its oil than the United States (80%). India's government does recognize the need to shift away from fossil fuels over time, and it has ambitious plans to rely more on renewable and nuclear sources in the future. In fact, it plans to double its nuclear power plant capacity to 37 reactors over the course of the coming decade—second only to China in the rate of expansion. With regard to green energy, India has pledged in 2015 to produce 40% of its energy from renewable sources by 2030 and to reduce CO_2 emissions by a third at that time.

14.2.3 United States

Much has already been discussed in other chapters in relation to energy usage and policy in the United States, so here we just summarize a few highlights. In the past, the nation has probably been more focused on reducing its dependence on imported oil than on weaning itself from fossil fuels or reducing its CO_2 emissions. Although it would be desirable for the United States not to be so heavily dependent on oil imports for economic reasons, the goal of energy independence—or at least energy independence on imports from unstable regions, such as the Middle East—is probably very desirable. Oil, however, is a global commodity, and therefore, the world as a whole remains hostage to sudden major supply disruptions, even if particular nations are less reliant on Middle East oil. Reducing oil consumption through various measures is the only long-term solution, which will become more urgent as global oil supplies fail to keep up with demand. This economic reality may have more urgency in some quarters of the United States than reducing emissions to combat climate change. In fact, the goal of reducing CO_2 emissions has become increasingly partisan with many Republican citizens, candidates, and office holders stating their disbelief in human-caused climate change—see Section 14.4. Other energy-related and environmentally related issues facing sharp partisan divisions include exploiting the nation's newfound natural gas reserves, subsidies for renewable energy, and mandating energy conservation measures—especially when they involve energy taxes. Stymied by political gridlock, the Obama administration has resorted more to executive orders—for example, an EPA rule curbing some non-CO_2 emissions by mandating the installation of scrubbers on coal-fired power plants—a rule that coal companies had lobbied against since 1990. In a 2012 ruling, the EPA further mandated that all new power plants could not emit more than 1000 lb of CO_2/MWh, which effectively prohibits new coal plants without CO_2 sequestration, but not new natural gas power plants. In 2015, under President Obama, the United States expects to cut GHG emissions by 26–28% below 2005 by the year 2025.

Although the nation has taken some modest steps in subsidizing renewable energy sources through tax credits and government-backed loans, even here, the most prominent subsidy (corn ethanol) has been driven more by politics than by a sound consideration of costs and benefits. In the United States, the most promising initiatives in support of renewable energy and conservation have been at the state and local levels in the form of renewable energy credits, which greatly vary from state to state. Nevertheless, some important recent actions have been taken at the federal level as well, including the Energy Independence and Security Act of 2007, which has promoted the goal of energy conservation, including increasing the mandated average gas mileage of new cars—a measure that has been tightened even further under President Obama. As this second edition goes to press, President Trump has withdrawn the United States from the Paris climate accord and has pledged to served the coal industry. Until the hyperpartisanship abates in the United States, it seems likely that major actions at the federal level to promote clean energy are unlikely to succeed, especially in a slow economy when environmental concerns tend to take a backseat to economic ones. As can be seen in Figure 14.2, 2010 marks the first year that US citizens have rated the need for energy production ahead of concerns about the environment. Still, in the 3 years prior to 2012, nonhydro renewable sources in the United States have collectively grown by 55% according to US EIA data. With regard to nuclear

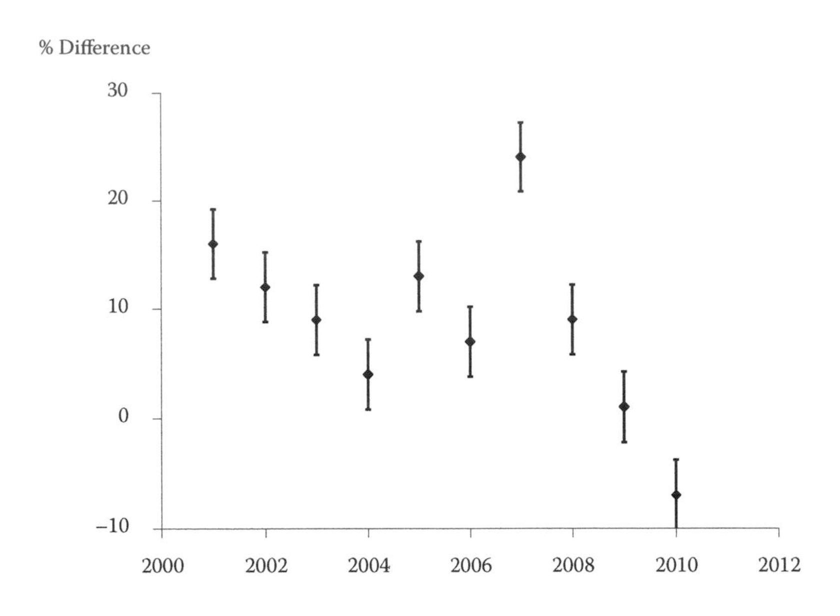

Figure 14.2 Relative degree of concern of U.S. citizens over the priority that should be given to environmental concerns versus energy production. The quantity plotted is the difference between the percentages of survey respondents who identify environmental protection over energy production as having higher priority. Error bars indicate one standard deviation limits based on the polling sample size. The negative value in 2010 marks the first year that energy production was rated as having a higher priority even at the risk of harming the environment "to some extent." Data are based on 10 years of Gallup polling on the question.

power, any expansion that seemed likely before Fukushima is now highly uncertain—especially in light of the new massive natural gas reserves discovered in the last decade.

14.3 HOW MUCH TIME DOES THE WORLD HAVE TO MOVE AWAY FROM FOSSIL FUELS?

The degree of urgency in making the shift away from fossil fuels strongly depends on one's view of the likelihood of the magnitude of future human-caused climate change being catastrophic. Recall that the IPCC, reflecting a consensus view of climate scientists, projected that the temperature increase during the course of the coming century will very likely be somewhere in the range of 1.1–6.4°C and that even if not a single new CO_2 molecule is injected into the atmosphere, the GHGs already there will cause global temperatures to rise about 0.6°C. Readers more familiar with the Fahrenheit scale should recall that increases in Fahrenheit degrees are 1.8 (roughly 2) times as much. The large range in the projected temperature increases are a reflection of both uncertainties in the computer models and uncertainties over future GHG emissions. Obviously, the consequences for life on Earth will clearly be vastly different if the actual increase in temperature over the coming century is closer to one end of the projected range of 1.1–6.4°C or the other. According to the IPCC, an increase over the coming century of more than 2–3°C might be "dangerous."

The IPCC phrase "might be dangerous," of course, has a far less ominous tone than "will be catastrophic," which is preferred by some environmentalists who believe either that (1) it is already too late to take action to halt irreversible climate change or (2) we have at most a few years left to begin to take some drastic steps to cut GHG emissions. This belief is usually coupled with the concept of a tipping point to the global climate, whereby we will either experience a runaway greenhouse effect like the planet Venus or, at best, experience a vastly different climate from the present one that will be essentially irreversible. A major proponent of the tipping point concept has been James Hansen, a NASA climate scientist, who has argued that an increase of only 1°C by 2100 would catastrophic in the end prove. Recall, however, that even if zero new CO_2 is emitted, the projected increase in temperature "in the pipeline" is expected to be 0.6°C by 2100, which explains the claimed need for urgency. In Hansen's view, atmospheric CO_2 concentrations need to be limited to at most 450 ppm, a value that he has subsequently lowered to 350 ppm (even below the present 385 ppm) (Hansen, 2007; Hansen et al., 2008). According to Hansen et al. (2008), "Continued growth of greenhouse gas emissions for just another decade, practically eliminates the possibility of near-term return of atmospheric composition beneath the tipping level for catastrophic effects."

What is the scientific basis for these ominous predictions? Hansen acknowledges that climate models can only offer limited support, since "... it is difficult to prove that models realistically incorporate all feedback processes" (Hansen et al., 2008). He therefore primarily relies on the paleoclimate record from ice core data, which provides both ancient temperatures and GHG levels as functions of time going back more than 400,000 years. Hansen finds that the variations in forcings (primarily GHG levels) correlate almost perfectly with the variations in global temperature over the 400,000 years prior to 1900. The correlation can be used to calculate the expected temperature rise for a doubling of atmospheric CO_2, known as the *climate sensitivity*, and he obtains $\Delta T = 6$°C. Hansen then assumes that a similar climate sensitivity value applies now and argues that the current and future, much higher expected GHG levels will eventually drive the system beyond a tipping point perhaps involving the melting of the Greenland ice sheets.

One flaw in Hansen's reasoning is that GHG levels in paleoclimate data are perhaps more likely to be the result in temperature changes than their cause—since as the oceans are heated, more of their dissolved CO_2 is returned to the atmosphere. Actually, of course, CO_2 is both a result and a cause of changing temperatures, but by different amounts now and during ancient times. Hansen could be right about there being a tipping point and could even be right about an increase of only 1°C by 2100 being catastrophic over the long term, but these claims are merely assertions that have not been proven. An additional reason for skepticism concerning the claim of a tipping point of only 1°C comes from climate data even earlier than what Hansen uses: the late Ordovician period occurring about 450 million years ago. During that era, GHG levels were sky high—perhaps as much as 4000 ppm, or over 10 times (!) the present level. There obviously was no runaway greenhouse then, nor a tipping point—although it should also be stated that those sky-high GHG levels in no way disprove the impact of GHGs on global temperatures. Thus, the basic point about the lack of scientific basis of Hansen's 1°C tipping point remains. There is little question in the opinion of most climate scientists that potentially very serious climate changes are likely to occur over the course of the coming century, but the claim that "catastrophic effects" await us unless we begin immediate action on reducing GHGs is simply an oft-repeated opinion not supported by the science.

Interestingly, Hansen, while he is a reputable scientist, does have a history of making assertions that go beyond what mainstream climate science can support, presumably in order to influence public opinion. In June 1988, he testified to a congressional committee that global warming was underway, and he attributed the abnormally hot weather plaguing our nation to global warming (Hansen, 1988). This claim resulted in much increased media attention to the global warming issue than had previously existed. This connection between the potentially very serious but long-term problem of climate change to regional weather at that time was well beyond what most climate scientists would dare to claim, but it captured the public attention, just as his more recent assertions about tipping points. Scientists who raise alarms about impending environmental threats do serve a useful role in helping to alert the public to underappreciated threats, but when their alarms go beyond what the science can actually prove, they may be seen as scaremongers and actually undermine belief in a real environmental problem by seeming to wish to stampede the public into taking drastic action.

Despite all the reasons for skepticism about a tipping point, just suppose Hansen turns out to be right, and the planet does begin to experience a drastic rise in temperature—what then? The usual response is that by then, it would be too late. However, if all else fails, there are a variety of possible geoengineering solutions, which could reduce the atmospheric CO_2 either through, for example, ocean iron fertilization or alternatively by decreasing the amount of incoming solar radiation. The latter would probably be preferable because it could be done more quickly by, for example, putting space mirrors or many fine reflecting particles into orbit. Most environmentalists are very leery about geoengineering because these methods are unproven and could have unintended side effects. They also worry that if geoengineering might work, it could take attention away from reducing GHGs. Nevertheless, while geoengineering technologies should not be seen as a substitute for reducing GHGs, they should not be ruled out as an important backstop in case global warming, despite everything, does reach catastrophic levels. To reiterate, geoengineering should not be seen as a reason to ignore measures to reduce GHGs, but rather a reason for optimism that the planet will not be doomed to catastrophic climate change if drastic actions are not taken in the next few years to reduce GHG emissions.

14.4 HOW HAS PUBLIC OPINION EVOLVED?

14.4.1 Climate Change

In democratic nations, public opinion plays an important role in shaping government policies—especially through the selection of leaders at various levels. Contrary to the views of some commentators, surveys show that the overall level of concern about global warming or climate change in the United States, while lower than that in some nations (especially those in Latin America and the developed countries in Asia), is comparable to that in many other nations. In fact, the percentage of those surveyed rating the problem as serious or very serious in the United States was 11% higher than for the average of all nations surveyed (53% for United States and 42% for world) based on Gallup surveys conducted in 2010, which represents something of a narrowing of the gap in 2007 (63% for United States; 41% for world) (Newport, 2010). However, in a more recent set of 2013 surveys of publics in 39 countries, the pattern seems to have reversed, with a median of 54% of those surveyed citing climate change as a major threat, compared to 40% in the United States (Pew Research Center, 2013).

Thus, the degree of skepticism has definitely increased in recent years in the United States. For example, according to Gallup Polling data (Newport, 2010), while Americans by a nearly 2-to-1 margin (61% vs. 33%) thought human actions rather than natural causes was the predominant cause of global warming, by 2010, the opinion was nearly equally divided (50% vs. 46%). Moreover, opinion has become increasingly politically polarized over time. As evidence for the rising level of polarization, consider how the poll results have changed over time and how the responses correlate with party affiliation. Consider that in 1998, there was a relatively modest 12% gap between Democrats and Republicans on the question of whether the media exaggerates the danger of global warming (R = 35%; D = 23%). However, by 2009, that gap had become a chasm (R = 66%; D = 22%), with all the shift being on the Republican (and independent) side.

It is important to stress that the question asked in these Gallup polls concerns not the belief in global warming per se, but rather the public's perception of how the issue is handled by the media. In fact, conservatives in the United States routinely attack the mainstream media as being politically biased and often prefer to watch and listen to their own unbiased media. Still a belief that the media exaggerate a danger probably correlates with a belief that the danger is really not so bad. On the other hand, a more subtle interpretation of the results might be that while most of the public is well aware that climate change is a serious long-term threat, they believe that the media are overdramatizing the threat either for sensationalistic reasons or for the promotion a political agenda. In fact, while 63% of US citizens polled in 2010 considered global warming to be a "serious" or "somewhat serious" threat, an even greater percentage

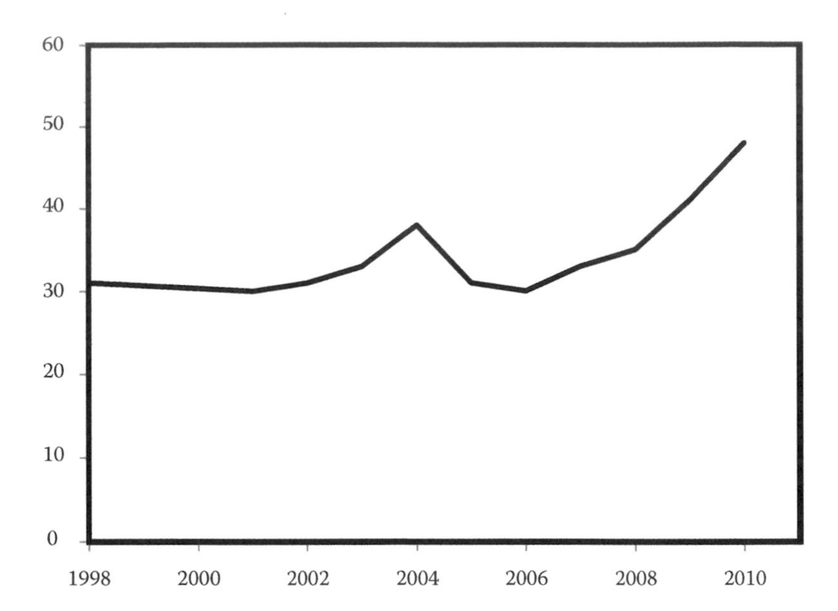

Figure 14.3 Rising percentage of Americans saying that the media exaggerate the seriousness of global warming, according to Gallup Polls. (From Gallup, http://www.gallup.com/poll/126560/Americans-Global-Warming-Concerns-Continue-Drop.aspx, 2010.)

(67%) answered "no" to the question "Will global warming pose a serious threat to you and your way of life in your lifetime"—the highest level since polling on that question began in 1998 (Figure 14.3).

Why the rising levels of skepticism and political polarization in the United States regarding global warming? Is this simply part of a trend wherein many issues have become polarized that had not been previously? It is entirely not. It is likely that some of the polarization has been the result of a highly successful disinformation campaign by global warming deniers and some fossil fuel companies resisting any constraints on CO_2 emissions or energy taxes. There are also many other factors specific to the United States that may contribute to polarization on the need to combat climate change, including the following:

- General distrust of sweeping government intervention.
- Love affair with the automobile and a tradition of suburban living with long commutes and extremely low gasoline taxes.
- Tradition of individual liberty ("If I want to buy cheaper inefficient light bulbs, who is the government to tell me I can't!").
- Rise of cable TV and Internet sites that cater to each political persuasion and position on global warming. The global warming deniers have even created their own Nongovernmental International Panel on Climate Change, whose website asserts that since "we are not predisposed to believe climate change is caused by human greenhouse gas emissions, we are able to look at evidence the Intergovernmental Panel on Climate Change (IPCC) ignores" (NIPCC, 2012).
- Increasing levels of distrust of scientists on the climate change issue in particular—fostered by the global warming deniers and supported by some wealthy individuals.

Finally, some scientists may have themselves contributed to the political polarization by framing a serious long-term environmental threat as being an imminent peril that requires immediate action. Moreover, they appear to be so focused on the environment that they are seen as antibusiness and even antifree enterprise—a fatal stance if they hope to win over an uncommitted middle segment of the population.

14.4.2 Renewable Energy

Although, in past years, strong majorities of the public in the United States favored supporting funding for renewable energy, the level of support (while still a majority) has softened, with the same partisan divide one sees on the global warming issue. In 2006, research in renewable energy was supported by 81% of Democrats and 83% of Republicans. However, in a 2011 survey, the Democratic support was essentially unchanged at 83%, but Republican support had dropped to 53%, according to a Pew poll (2011). To the extent that the reason for renewable energy is strictly couched (or primarily) in terms of avoiding CO_2 emissions, such a connection between the two issues is understandable, but political and budgetary considerations may also be playing an important role.

14.4.3 Nuclear Power

Attitudes toward nuclear power became much more negative worldwide following the Chernobyl accident in 1986, although in the United States, the change occurred right after TMI in 1979, after which there were no new reactors ordered for the next 30 years. Public opposition in the United States toward nuclear power steadily rose after TMI and continued to rise for the next 15 years. According to Louis Harris polls, there were only 19% of the public opposed to nuclear power in 1975, but that number jumped to 62% by 1990. More recent Gallup polling data tell a rather different story. In 2001, the percentages of opposed and in favor were roughly equal, but by 2010, the opposed number was roughly half of those in favor: 62% (favor) versus 33% (opposed)—and then came Fukushima. Obviously, support for nuclear power has dropped somewhat following Fukushima, with considerable variation from country to country. In a survey of residents in 24 nations in the aftermath of Fukushima, support for nuclear power had dropped to 38%—even lower than coal (48%) (Harris, 2011). Still, poll results on nuclear power do tend to be very mixed. For example, based on a March 2011 CBS News poll, while 70% of Americans think that nuclear power plants in the United States are safe, almost an equal percentage (67%) said they were concerned that a major nuclear accident might occur and 30% were "very concerned." There continues to be a strong gender divide on the issue, with a majority of men approving of new nuclear plants and a majority of women opposing, with the gender division even more pronounced than that on the basis of political party.

14.5 BEST WAY FORWARD

Clearly, we must wean ourselves of our addiction to fossil fuels over time, not only because of their contribution to climate change, but for many other reasons—both environmental and economic. There are also the security challenges posed by fossil fuels, and wars have been fought over access to them, as was the case of Japan in World War II and perhaps other wars as well. Finally, there is the peak oil issue, whereby the supply of petroleum is likely to begin to run out (with prices continuing to rise) as drilling occurs in increasingly challenging environments. The longer the transition is delayed, the more difficult it will be, and the greater the harm to the environment in the interim.

Part of the transition away from fossil fuels should include substituting cleaner natural gas for coal when possible. New binary gas electric plants are now even less expensive than coal, so there are both environmental and economic reasons to make the switch. This is clearly feasible for the United States, given the enormous size of the recently discovered shale deposits, although it may prove more difficult for nations such as China. Obviously, there are environmental dangers with the fracking process, but these probably pale next to those posed by coal, and they should be able to be reduced with appropriate government oversight. It makes very little sense for the United States to have its transportation sector become evermore dependent on imported oil from unstable parts of the world, when some of that need could be met by natural gas-fueled vehicles. Over the decades, we may wish to transition away from natural gas as well as coal and oil, particularly if climate change appears to become very severe, but in the meantime, it could be a very useful bridge fuel. Solar, wind, and all the other renewable sources are wonderful and should be aggressively pursued, but natural gas can complement them by providing power at times when the intermittent sources are not.

There is an important need to subsidize some renewable energy sources until they become fully competitive, but it is also a mistake for the government to pick winners and losers among the various alternatives, which can backfire and hurt the public perception of clean energy, as what happened in the United States with the unfortunate loan guarantees given to the solar company Solyndra. A similar story may apply to the strong administration support for fully electric cars, which to date seem to be having a difficult time attracting a large slice of the market, in light of their price tag (even with government rebates) and their limited range compared to hybrid electric vehicles. Support for basic research in support of new technologies (which few companies can afford to do) may be a much more appropriate use of tax dollars that can really pay off in terms of new technology.

The era of renewable energy requiring subsidies may, however, be coming to an end. In one very hopeful development, 2015 was a record year in terms of global investment in green energy, and most strikingly, this occurred at a time when oil, coal, and natural gas were all very cheap. The

BENEFITS OF BASIC ENERGY RESEARCH

Solar cells were first improved and developed in connection with the government-sponsored space program, and space was the first place they were extensively used. The recent vast increase in exploitable natural gas reserves could not have been achieved without several key discoveries made possible by government-sponsored research. The technique of hydraulic fracturing was first developed in the 1970s by DOE. Two other key technologies, for horizontal drilling, and new mapping tools were also made possible with government support. This government agency continues to fund hundreds of energy-related projects in all areas of renewable (and nonrenewable) energy, even though its abolition was advocated by one recent US presidential candidate—at least until he forgot its name.

global leader in clean energy investment in 2015 was China, with $110.5 billion, with the United States in second place at $56 billion. According to Ethan Zindler, an analyst with Bloomberg New Energy Finance, "The technologies have reached an important tipping point in a number of markets in the world." However, before the world congratulates itself too much on this important achievement, it must also be remembered that 2015 was so far the warmest year on record, according to NASA and National Oceanic and Atmospheric Administration data.

Another area where much work needs to be done is energy conservation—the low hanging fruit in terms of cost-effectiveness and where the United States lags behind many other industrialized nations. Many specific measures were identified in previous chapters in the residential, commercial, industrial, and transportation sectors, and they will not be repeated here. The smart grid is also a badly needed improvement (see Chapter 13), not only to help conserve energy, but also to upgrade its security and ability to reconstitute itself in the event of a worldwide catastrophic failure—either one that is human caused or the result of a massive solar flare. Theoretically, consequences of such an event could be the possible deaths of billions of people and having society revert to the Stone Age. The impact of the electric grids being out of commission for a year would almost certainly be worse than that of climate change.

My opinion about nuclear power should not come as any surprise to readers who have gotten this far in the book, as I believe that it should be aggressively pursued in as safe a manner as possible. Nuclear, of course, produces no CO_2 emissions, so it is "clean" in that sense, but it does pose dangers. Nevertheless, as has been pointed out earlier, even taking into account Fukushima and Chernobyl, the actual health and environmental damage done by nuclear pales next to that done by coal. Nuclear is feared by many people on an emotional level, in ways that no other risks are. The fear of nuclear stems, in part, from it being little understood by much of the public, from it being undetectable by our senses, and from misleading images of monsters created by genetic mutations. Part of the fear can also

NEW BREED OF ENVIRONMENTALIST?

Most environmentalists continue to be very leery about nuclear power and generally cast a jaundiced eye toward technology generally—seeing it more as the enemy than the friend of the environment. However, a new breed of environmentalists now have come to realize that we cannot simply turn back the clock and return the planet to some earlier pristine state. The new "technological environmentalists" typified by Stewart (2009) embrace technology (including such heresies as nuclear power, genetic engineering of crops, and even geoengineering). They appreciate that given the changes humans have already made to the planet, we must now take an active role in managing it to achieve a desirable outcome. In Joel Achenbach's (2012) words, writing about the new environmentalism, the time has come for humans to "take the helm" of spaceship Earth.

be attributed to our ability to measure the tiniest level of radiation and the misleading assertion that "no level of radiation is safe." This claim, which has been neither proven nor disproven, simply means that the harm is proportional to the dose, and so very tiny doses may cause a very tiny harm, which can be said about exposures to most risks. At the time of the Fukushima disaster, for example, there was extensive news coverage of the radiation levels in the United States from radiation that had drifted halfway around the globe. Most of the media coverage left the magnitude of the actual risk uncertain, although one TV news reporter did helpfully note that the radiation a US citizen would receive from Fukushima at that time was equivalent to what he or she would get from eating a banana. We clearly need to be able to judge nuclear risks on the same scale as other risks and not expect or demand that nuclear power be absolutely safe with zero risk.

The main problem with nuclear power today is probably one of economics rather than safety. The two issues are, of course, related; however, because given the perceived dangers of nuclear power, investors will demand a higher risk premium in investing into new nuclear plants, which drives up their costs. Given that capital costs are very high for nuclear compared to other ways to generate power, and given that construction times tend to be very long—again partly due to safety concerns, lawsuits, and protests by worried citizens—nuclear has become increasingly noncompetitive with other means of generating electricity, at least in the United States. One possible approach that may greatly reduce the cost of nuclear power, while at the same time enhancing safety, is to emphasize SMRs, which were discussed in Chapter 4.

14.6 SUMMARY

This chapter considers a variety of topics related to energy and climate policy, as well as politics and public opinion. On the subject of international agreements to control human-caused climate change, it discusses the many obstacles to concluding a successful treaty and presents an agnostic view on how successful it will prove to be in controlling emissions. The chapter also discusses public opinion in the United States and around the world and how public opinion has evolved over time. In the United States, there has been rising polarization on climate change, as well as many other environmental and energy-related issues. Finally, the discussion includes the author's personal views on what might be the best way forward.

14.7 SOME CONCLUDING THOUGHTS

As this book is primarily aimed toward undergraduate students who are majoring in science or engineering, it may not be out of place for this educator who is nearing the end of his professional career to offer a few words of advice to such readers. First, as was stated in Chapter 1, I sincerely hope that you will consider a career in the energy field, preferably related to renewable (or nuclear) energy, as well as energy conservation and energy policy. This is a vitally important field for the future; the opportunities promise to be very bright; and the field can certainly use all the talented people who want to make a difference. Your science or engineering degree should prepare you well for such a career, but more than the specific subjects you have taken, the habits of mind that these subjects cultivate is just as important. What are those habits of mind that are especially relevant? I would name three:

1. *Ability to evaluate evidence objectively*: The fields of energy and climate have many complexities and lack the finality of some fields that are relatively settled such as classical physics. You need to be able to evaluate evidence objectively and be prepared to change your mind taking into account all the trade-offs.
2. *Ability to be self-critical*: Always look for flaws in your own beliefs and do not just seek out evidence that supports them. The essence of science is to seek out evidence that disproves a theory, not just evidence that supports it—an all too common pitfall among humans.
3. *Ability to think strategically*: Strategic thinking considers the big picture and the long-term impacts of policy. The practicalities of what works for the present cannot be overlooked, of course, but we must adopt a long-term perspective that combines a balanced consideration of goals in the economic and environmental realms if the world is to survive and prosper. Different cultures have different national characteristics, and the nation of China, for example, is probably noted for its practicality and long-term strategic thinking. Emulating that particular admirable characteristic would be highly desirable for Westerners. Of course, it would be equally important if the Chinese, who have already

emulated to some degree the Western free market economies, were to do the same with regard to its political system. Dr. Steven Chu, physics Nobel laureate and current US secretary of energy, has noted in an interview with *Time Magazine* that what the United States and China do in the coming decade will determine the fate of the world. What do you think about Dr. Chu's assertion?

BECOME A STRATEGIC THINKER: LEARN TO PLAY GO

Some readers may find my last concluding thought a strange one for a book about energy, but it is tied to one of my passions in life, the game of Go, and it does relate to my hope that the reader will develop the habit of thinking more strategically. This final section has been adapted with permission from the US Go Association (Figure 14.4).

Go or Weiqi (pronounced "wei chi") is a fascinating board game that originated in China more than 4000 years ago. The game is played today by millions of people, including thousands in the United States. In Japan, Korea, China, and Taiwan, it is far more popular than chess is in the West. Its popularity in the United States continues to grow, more than 50 years after the founding of the American Go Association. It is said that the rules of Go can be learned in minutes, but that it can take a lifetime to master the game. The rules could not be simpler. Two players alternate in placing black and white stones on a large (19 × 19 line) ruled board, with the aim of surrounding territory. Stones never move and are only removed from the board if they are completely surrounded.

The game rewards patience and balance over aggression and greed; the balance of influence and territory may shift many times in the course of a game, and a strong player must be prepared to be flexible but resolute. Like the Eastern martial arts, Go can teach concentration, balance, and discipline. Each person's style of play reflects their personality and can serve as a medium for self-reflection. Go combines beauty and intellectual challenge. "Good shape" is one of the highest compliments one can pay to a move in the game of Go. The patterns formed by the black and white stones are visually striking and can exercise an almost hypnotic attraction as one "sees" more and more in the constantly evolving positions. The game appeals to many kinds of minds—to musicians and artists, to mathematicians and computer programmers, and to entrepreneurs and options traders. Children learn the game readily and can reach high levels of mastery.

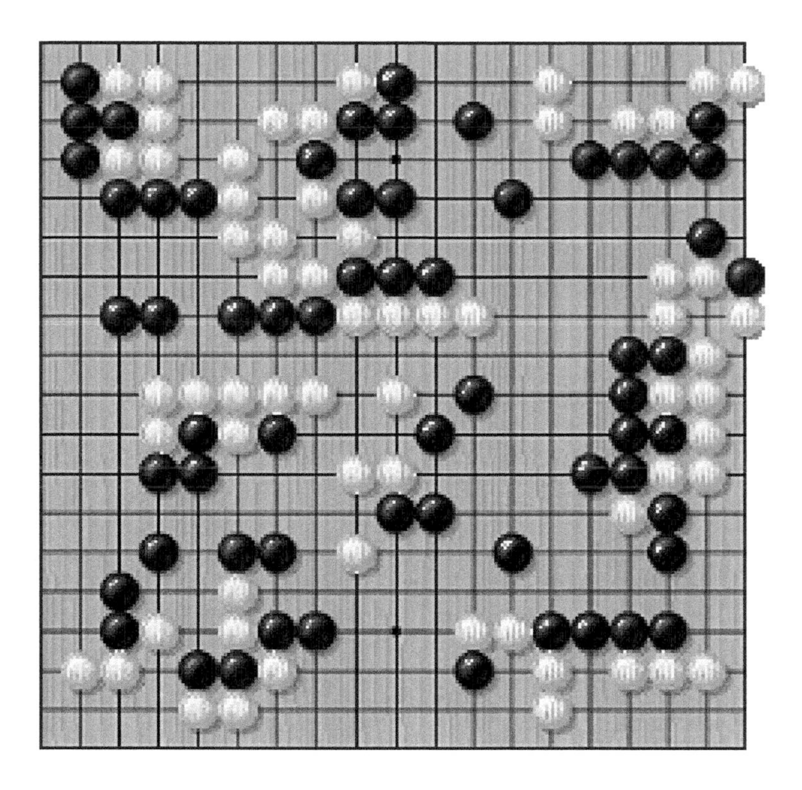

Figure 14.4 Go board with stones for a game in progress. (Image provided by the American Go Association is included with their permission.)

PROBLEMS

1. What do you think is the best way forward in the energy–environment arena?
2. Calculate the amount of carbon dioxide emitted by an automobile in a year. Assume the car gets 20 mpg and travels 10,000 mi in a year. Now estimate the total amount of carbon emitted by all automobiles using the figure just completed and assuming 150 million automobiles in the United States.
3. James Hansen (2012) noted that "if Canada exploits its tar sands, civilization will be at risk." Do you think that statements such as this and others Hansen has made will have the effect of making more members of the public who fall in the two most common categories on the spectrum of global warming opinion, i.e., "concerned" or "cautious," more or less likely to move into the "alarmed" category? Discuss.
4. Imagine if all the coal reserves in the world were burned and 50% of the generated CO_2 is deposited in the atmosphere. Calculate the increase in the concentration of CO_2 in the atmosphere in the world using this scenario. For simplicity, assume that the coal deposits are 100% carbon.
5. In what year would you estimate the CO_2 concentration to have doubled when compared to the 1860 level of CO_2 if the rate of increase were a steady 0.4% per year?

REFERENCES

Achenbach, J. (2012) Spaceship Earth: A new view of environmentalism, *Washington Post*, January 3, 2012.

Hansen, J. (1988) McCright & Dunlap 2000, p. 500.

Hansen, J. (2012) *New York Times*, May 10, 2012.

Hansen, J., M. Sato, R. Ruedy, P. Kharecha, A. Lacis, R. L. Miller, L. Nazarenko et al. (2007) Dangerous human-made interference with climate: A GISS model E study, *Atmos. Chem. Phys.*, 2007(7), 2287–2312.

Hansen, J., M. Sato, P. Kharecha, D. Beerling, R. Berner, V. Masson-Delmotte, M. Pagani, M. Raymo, D. L. Royer, and J. C. Zachos (2008) Target atmospheric CO_2: Where should humanity aim? *Open Atmos. Sci. J.*, 2, 217–231.

Harris Interactive (2011) Recent crisis in Japan has had little impact on Americans, views on nuclear power: Poll, *Harris Interactive*. March 31, 2011, http:www.harrisinteractive.com/NewsRoom/PressReleases/tabrid/446/mid/1506/articleId/774/ctl/ReadCustom%20Default/Default.aspx.

IEA (International Energy Agency) (2009) Data through 2009 are taken from the 2011 edition of the report CO_2 emissions from fuel combustion, by the International Energy Agency. The 2010 data are based on unpublished IEA data.

Newport, F. (2010) Americans' global warming concerns continue to drop, *Gallup*, http://www.gallup.com/poll/126560/Americans-Global-Warming-Concerns-Continue-Drop.aspx.

NIPCC (Nongovernmental International Panel on Climate Change) (2012) http://www.nipccreport.org/about/about.html.

Nuclear (2007) First annual world environment review poll reveals countries want governments to take strong action on climate change, *Global Market Insite*, June 5, 2007.

Pew, P. (2011) http://pewresearch.org/pubs/2129/alternative-energy-solar-technology-nuclear-power-offshore-drilling.

Pew Research Center (2013) Climate change and financial instability seen as top global threats: Survey report. June 4, 2013, http://www.pewglobal.org/2013/06/24/climate-change-and-financial-instability-seen-as-top-global-threats/.

Stewart, B. (2009) *Whole Earth Discipline: An Ecopragmatist Manifesto*, Viking Press, New York.

Wines, M. (2011) *New York Times*, December 23, 2011.

Data Analytics and Risk Assessment

An Overview

CONTENTS

15.1 PROBABILITY CONCEPTS

In 2015, Tyler Vigen published a book with a number of correlations that he called spurious correlations. One so-called spurious correlation was the 99.2% correlation between the federal science budget in support of space, science, and technology versus the number of suicides by hanging, strangulation, and suffocation. There is a very high correlation between these data; however, does that mean that an increase in the size of the federal budget will cause the number of suicides to increase? Alternatively, does it mean that if the number of suicides increases, the federal budget will also increase? Vigen points to many more of these correlations in his book. In fact, Vigen developed an algorithm to search for high correlations between data sets that should have nothing to do with one another. Unfortunately, if you were to conclude that there must therefore be a cause and effect relationship between the two parameters examined, as a high correlation might lead you to do, you would be jumping to a false conclusion. Nonetheless, many in the media are prone to do just that, jump to conclusions. At the heart of the problem in the media and general public is the lack of understanding of probability. In this chapter, we examine the aspects of probability that relate to energy and the environment, especially the risks of certain energy technologies, especially nuclear power. But first, let us examine the basics of probability.

Probability has several definitions. Among practitioners, the two most common definitions are those that take what is called a relative frequency approach and those that take the subjective or axiomatic approach. The relative frequency approach to defining probability involves a sample space and an event which is a member of the sample space. Within this framework, the probability of the outcome of an event, symbolized as P(A), is mathematically defined as

$$\lim_{n\to\infty}(X/n) \to \mathrm{P(A)}. \quad (15.1)$$

The value X/n is the relative frequency of the occurrence of event A.

The more generic definition of *probability* is simply a numeric value, or measure, of the likelihood of event A happening. If event A is certain to occur, then we say that the probability of the occurrence of event A is one. If event A is impossible to occur, then we say that the probability of event A is zero. Thus, mathematically, we have the probability of an event as lying between zero and one or

$$0 \leq P(A) \leq 1. \tag{15.2}$$

In fact, if you consider the set of all events as being the union of sets of the event and the nonoccurrence of the event A, then you mathematically have

$$P(A) + P(\overline{A}) = 1. \tag{15.3}$$

Basically, either event A will occur or it will not occur.

There are more definitions that have to be addressed within our view of probability before we are able to examine the risks associated with the different technologies for generating electricity.

First, we need to consider what is known as conditional probability. That is, given that one event occurs, what is the probability of the second event occurring. For example, if A is the event of a light going on and B is the event of turning on the switch, then the probability of A is zero if the switch is not turned on (not B). Mathematically, we would have

$$P(A|\overline{B}) = 0, \tag{15.4}$$

where $P(A|B)$ means the probability of A given no B.

Now let us examine how mathematical probability defines the intersection of two events as a product rule. The intersection is often referred to using the connector word *and*. Thus, if the probability of A_1 and A_2 is written as $P(A_1A_2)$, we have

$$P(A_1A_2) = P(A_1|A_2)P(A_2) = P(A_2|A_1)P(A_1). \tag{15.5}$$

You may have noticed that if events A_1 and A_2 are mutually exclusive, then the probability of them both occurring is zero, or

$$P(A_1A_2) = 0. \tag{15.6}$$

Also, if there is no connection between two events, we say that these events are independent. Given these conditions, for two events,

$$P(A_1A_2) = P(A_1)\,P(A_2). \tag{15.7}$$

As a simple example, consider how you might calculate the probability of tossing two dice and getting a 1 on the first die and a 2 on the second die.

This can be calculated by applying Equation 15.7 and with the probability of getting a 1 on die one as 1/6 and the probability of getting a 2 on the second die as 1/6, so the probability of getting the combination is 1/36.

Mathematically, generalizing for any number of events, the equation for N independent events would be

$$P(A_1A_2,\ldots,A_N) = P(A_1)P(A_2),\ldots,P(A_N). \tag{15.8}$$

Do not forget that if the events are truly mutually exclusive, you have

$$P(A_1A_2,\ldots,A_N) = 0. \tag{15.9}$$

Just as there is an equivalency in probabilistic mathematics for the English word *and*, there is also a mathematical equivalency for the English word *or*. The statisticians use the addition symbol "+" for the OR operator. Of course, you are not adding two events together; you are adding the probabilities associated with those events. Thus, we have the probability of $P(A_1)$ or $P(A_2)$ mathematically expressed as

$$P(A_1 + A_2) = P(A_1) + P(A_2) - P(A_1A_2). \tag{15.10}$$

In words, this means that the probability of event A_1 or A_2 occurring is merely the sum of the probability of event A_1 occurring plus the probability of A_2 occurring, minus the probability of the two events occurring.

Equation 15.10 above is for two truly independent events. However, what if you have two events, A_1 and A_2, both being conditional on event B. You can show that the following would be true:

$$P(A_1 + A_2|B) = P(A_1|B) + P(A_2|B) - P(A_1A_2|B). \tag{15.11}$$

We previously constructed the case of two independent events. The probability for N events is

$$P(A_1 + A_2 + \ldots + A_N) = \Sigma P(A_N) - \Sigma\Sigma P(A_nA_m) + - \ldots + (-1)^{N-1} P(A_1A_2,\ldots,A_N). \tag{15.12}$$

Such a generalized equation looks messy, does it not? Fortunately, for risk assessment engineers, there is a simplified form. That is, if the probability of any of the independent events is highly infrequent, or rare, then engineers use what is known as the rare-events approximation, and Equation 15.12 is simplified as

$$P(A_1 + A_2 + \ldots + A_N) \approx \sum_{n=1}^{N} P(A_N). \tag{15.13}$$

Equation 15.13 is much simpler and easier to implement in a simulation of an electrical power plant. Another key mathematical concept which is used in risk analysis is referred to as Bayes theorem. In terms of probabilities, the common form of Bayes theorem is represented by the following:

$$P(A_n|B) = \frac{[P(A_n)P(B|A_n)]}{\Sigma P(A_m)P(B|A_m)}. \tag{15.14}$$

In general, the Bayes theorem is the description of the probability of some event based upon certain conditions that are related to the event. One application of the Bayes theorem for nuclear power systems engineers is for the analysis of the risks associated with the transportation of spent nuclear fuel, especially by rail. For example, if you were given data related to the past shipping of spent fuel and then asked to calculate the probability of a single shipment taking place safely, you can derive a probability even without having had a single failure up to that date.

15.2 DATA ANALYTIC CONCEPTS

There are times when the probability of an event has no available data. Engineers then apply what is sometimes called the *principle of insufficient reason*. In this case, engineers assume equal probabilities for each event; that is, they assume that the outcomes are equally probable, just as if you were flipping a coin or throwing dice.

Another data analytic concept is called the frequency of occurrence, which is the number of times a specific event occurs, such as a release of radiation in the transport of radioactive materials. Sometimes, the event of interest, such as the release of radiation, does not occur within the time period where data were provided. In such instances, applications of the Bayes theorem can allow for the calculation of the upper limits of the expected frequency of occurrence.

The word *average* is often used in colloquial discussions as meaning the typical or normal value of some type of measure. However, in risk analysis, there are a number of different averages that can be calculated. The average representing the mean value is derived by summing all values and dividing by the number of values summed. The mean is often symbolized as

$$\bar{x} = \sum_{i=1}^{n} x_i/n. \tag{15.15}$$

Another average statistic which is often confused with the mean is known as the median. Mathematically, the value in the middle of all values listed in magnitude order is the median. Finally, another summary statistic

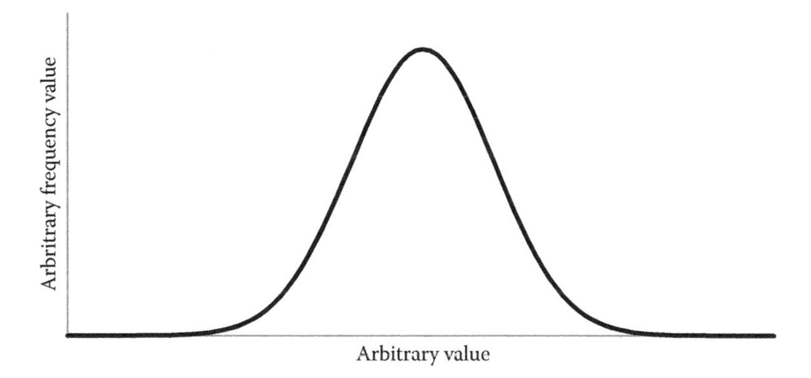

Figure 15.1 Normal distribution curve.

often used is known as the mode. It too is a type of average. In this case, the mode is the value that occurs most often of all ordered values.

As an example, suppose you have data listing the number of radiation safety failure events at a nuclear reactor over a period of a decade, as 2, 3, 5, 7, 2, 3, 3, 4, 2, and 3. With such an example of data for the 10-year period, the mean is 3.4, the median is 4, and the mode is 3. Thus, not all averages are the same, and the reporting should specify exactly which average is being utilized.

It turns out that the mean has another meaning, relative to what is known as the normal distribution, something encountered quite a bit in statistics. First, we should note the standard representation of the normal distribution as thus:

$$f(x)=\frac{1}{\sigma\sqrt{2\pi}}e^{-\frac{1}{2}\left(\frac{x-\mu}{\sigma}\right)^2}. \tag{15.16}$$

Graphically, the normal distribution is depicted as in Figure 15.1.

Using the above notation, μ is the mean of the distribution, and it is distinguishable as the position of a vertical line that bisects the area under the curve, with an equal area to the left and to the right of the mean. Also, for this distribution, σ is the value known as the standard deviation, and σ^2 is known as the variance. The standard deviation is a value that provides the amount of spread of the curve around the mean. The variance is simply the square of the standard deviation.

15.3 FAILURE AND RELIABILITY

Thus far, we have been examining the probability of events largely in terms of a finite number of mutually exclusive events. Engineers have found it useful to study reliability and failures in terms of a function of time, where the outcome is one of an infinite number of mutually exclusive events, allowing for the ease of applying differential equations and integral calculus, in lieu of a large summation of probabilities. Thus engineers use

$$P(X)=\int_{x_{\min}}^{X} p(x)\mathrm{d}x. \tag{15.17}$$

The term $p(x)$ is defined as the probability density or the probability per unit for the specific outcome. Such probability distribution functions allow for the description of failures in terms of being instantaneous or degrading over time.

An example of an instantaneous failure would be the burnout of a filament of an incandescent bulb. Examples of degradation failures include the wearing down of a gear or the rusting of a component. This naturally leads to the view of two major types of systems, those that operate on demand and those that operate continuously. The probability P that a system of operations will work for W_{n-1} operations is

$$\mathrm{P}(W_{n-1})=\mathrm{P}(\mathrm{D}_1\mathrm{D}_2\mathrm{D}_3,\ldots,\mathrm{D}_{n-1}), \tag{15.18}$$

where D_n is the nth demand of event D.

If we assume failures to randomly occur, we have

$$\mathrm{P}(\mathrm{D}_n|W_{n-1})=\mathrm{P}(\mathrm{D}) \tag{15.19}$$

and

$$\mathrm{P}(\bar{\mathrm{D}}_n|\mathrm{W}_{n-1})=\mathrm{P}(\bar{\mathrm{D}}). \tag{15.20}$$

Using Equations 15.18 through 15.20, we can conclude that

$$\mathrm{P}(\mathrm{D}_1\mathrm{D}_2\mathrm{D}_3,\ldots,\mathrm{D}_{n-1}\bar{\mathrm{D}}_n)=\mathrm{P}(\bar{\mathrm{D}})[1-\mathrm{P}(\bar{\mathrm{D}})]^{n-1}. \tag{15.21}$$

Equation 15.21 allows engineers to compute the probability of failure on the nth demand operation of a component given the demand probability failure by the manufacturer and the number of times the component is utilized during the given timeframe.

For components and systems in continuous use, engineers can use the differential form of the failure probabilities mentioned earlier, which can be denoted as

$$f(t)\mathrm{d}t=\lambda(t)\mathrm{d}t[1-F(t)] \tag{15.22}$$

or, more simply,

$$f(t) = \lambda(t)[1 - F(t)]. \tag{15.23}$$

In these equations, $f(t)$ is the failure probability density; $\lambda(t)$ the conditional failure rate (also known as the hazard rate); and $1 - F(t)$ is the probability that the component did not fail prior to time t.

To simplify calculations, if we assume that all failures are random, we can define the hazard rate as

$$\lambda(t) = \lambda. \tag{15.24}$$

The assumption of random failures leads to the often utilized mathematical definition of reliability as being denoted thus:

$$R(t) = e^{-\lambda t}. \tag{15.25}$$

Another basic engineering definition related to the reliability of any system is the mean time to failure (MTTF). The MTTF of any system is defined in relation to the reliability by the following integration:

$$\text{MTTF} = \int_0^{\infty} R(t)\,dt. \tag{15.26}$$

Once again, we can simplify calculations by assuming random failures, which leads to the basic formulation of MTTF being denoted thus:

$$\text{MTTF} = 1/\lambda. \tag{15.27}$$

The engineering model so far developed actually assumes that if you are able to repair or replace failed components, then you should be able to maintain operations of the system for an indefinite period. That of course is not really the case. The reality of the situation led engineers to develop the concept of instantaneous availability. The system instantaneous availability is usually denoted as $A(t)$. Naturally, the probability $1 - A(t)$ is the instantaneous unavailability of a system. It should also be noted that the instantaneous availability of a system has a minimum set by the reliability and, of course, a maximum of one or mathematically

$$R(t) \leq A(t) \leq 1. \tag{15.28}$$

15.4 FROM SIMPLE SYSTEMS TO EVENT AND FAULT TREE ANALYSES

Up to now, we have really only examined how engineers treat the reliability of individual components of systems. We will quickly examine the

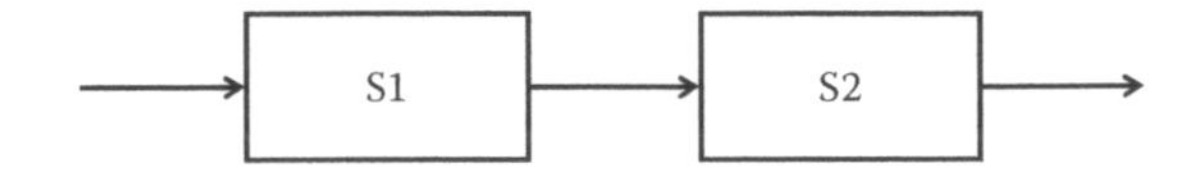

Figure 15.2 Functional components connected in series.

analysis of simple systems before we show how engineers attack the most complex of systems utilizing event and fault tree analyses.

As noted in the previous section, failures that randomly occur allow us to utilize the exponential distribution form of the equation for reliability or

$$R(t) = e^{-\lambda t}. \tag{15.29}$$

A power plant, such as a nuclear power plant (although it really applies to all power plants), can be represented as a series of functional blocks referred to as a block diagram. If our purpose is to determine the reliability of a system, we can then build a reliability block diagram. Such block diagrams appear like electric circuit diagrams, and like electric circuit diagrams, there are two different basic ways to connect the blocks, in a fashion similar to a series circuit or similar to a parallel circuit. In Figure 15.2, we see the series connection of components.

Fundamentally, the reliability of the entire system is the probability product of the components. Thus,

$$R_{\text{system}}(t) = R_1(t)R_2(t). \tag{15.30}$$

A more generalized form of this reliability equation, which assumes no overlap in the components, is as follows:

$$R_{\text{system}}(t) = R_1(t) + R_2(t) - R_1(t)\,\mathrm{R}_2(t). \tag{15.31}$$

It can be shown that for a series connection of n components, the reliability is

$$R_{\text{system}}(t) = \prod_{n=1}^{N} Rn(t). \tag{15.32}$$

Given the configuration as displayed in Figure 15.2, if you were given the operating time for each unit as 200 days and the unit failure rate as 2.5×10^{-4} failures per day, what would be the reliability of the two components in series?

First, we calculate the reliability of each individual component, which is $R(t) = e^{-\lambda t}$ with the hazard rate and time as given, we have the reliability as $e^{-1/20}$ or 0.951. Since the components are identical and we have two in series we have the reliability of the system as 0.951^2 or 0.905.

In Figure 15.3, one finds the parallel connection of components.

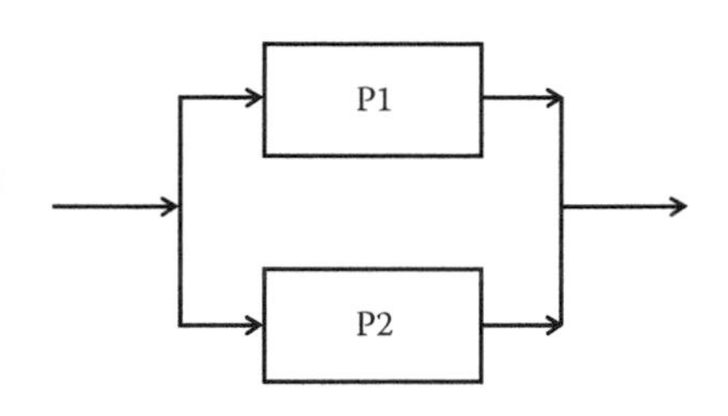

Figure 15.3 Parallel-connected functional components.

It can also be shown that for a parallel connection of n components, the reliability is

$$1 - R_{\text{system}}(t) = \prod_{n=1}^{N} [1 - Rn(t)]. \tag{15.33}$$

Given the parallel configuration as displayed in Figure 15.3, if you were again given the operating time for each unit as 200 days and the unit failure rate as 2.5×10^{-4} failures per day, what would be the reliability of the two components in parallel?

Again, we first calculate the reliability of each individual component, which is $R(t) = e^{-\lambda t}$ with the hazard rate and time as given, and as before, we have the reliability as $e^{-1/20}$ or 0.951. Since the components are identical and we have two in parallel, we have the reliability of the system from Equation 15.33 as $1 - (1 - 0.951)^2$ or 0.998.

Similarly, and especially for identical component subsystems, the MTTF can be calculated for the series and parallel components thusly:

$$\text{MTTF} = (N\lambda)^{-1} \text{ (series)} \tag{15.34}$$

and

$$\text{MTTF} = \sum_{n=1}^{N} (n\lambda)^{-1} \text{ (parallel)}. \tag{15.35}$$

As in electrical circuits, the connections of components and subsystems are rarely simply series or parallel, but a combination of the two. Let us examine an example where there is some simple combination of series and parallel connections of components.

For our combination analysis, let us consider a system which is based upon the following block diagram (Figure 15.4).

For simplicity, we are considering a system which consists of $2n$ identical subsystems. They fail in a random manner, and they have a hazard rate of λ. Provide an equation which would give the reliability and the MTTF of the system.

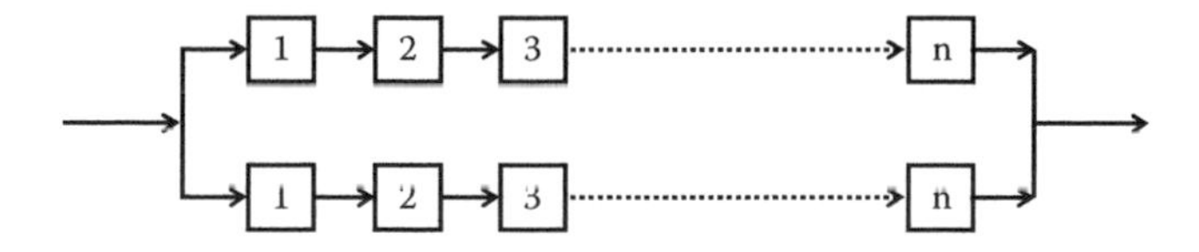

Figure 15.4 Series-connected components in parallel configuration.

$$R_{\text{system}}(t) = [R_1(t)]^n + [R_2(t)]^n - [R_1(t)\,R_2(t)]^n, \tag{15.36}$$

which can be expressed as

$$R_{\text{system}}(t) = e^{-n\lambda t}(2 - e^{-n\lambda t}). \tag{15.37}$$

By the definition of the MTTF, as noted in Equation 15.26, we now have

$$\text{MTTF} = \int_0^\infty R(t)\,\mathrm{d}t, \tag{15.38}$$

which, by applying Equation 15.37, becomes

$$\text{MTTF} = (3/2)(n\lambda)^{-1}. \tag{15.39}$$

In reality, complexities to the combination of components and subsystems quickly add further complexities to the calculations for the reliability of a system. For example, there is the reliability analysis of a system which routinely undergoes maintenance. This typically applies to all systems, as periodic maintenance is the best approach to foregoing a failure. This is similar to why you routinely change the oil in your car, a part of a routine maintenance to prevent catastrophic failure of your automobile.

One of the complexities that have occurred in reality is what we might call common cause failures. For example, consider the disaster at the Fukushima nuclear power plant. Many subsystems failed because of the earthquake and even more failed because of the subsequent tsunami.

In order to probabilistically examine a major system stressor such as an earthquake, you need to examine the occurrence of the earthquakes. As of today, we still cannot predict when an earthquake will occur. However, we can make the assumption that the occurrence has a Poisson distribution which will give us

$$\text{P}_c(r,t) = (e^{-\lambda t})(\lambda_c t)^r / r. \tag{15.40}$$

Using Equation 15.25, we can conclude this equation for the reliability considering a catastrophic earthquake:

$$R_e(t) - e^{-\lambda ct}. \tag{15.41}$$

Using Equation 15.41 as the reliability given a catastrophic earthquake, we can conclude that system reliability can be calculated using

$$R_{\text{sys}} = e^{-\lambda ct} R_{\text{sys}} e'(t). \tag{15.42}$$

In conclusion, the failures that have common causes decrease the system reliability overall. This is not a surprising effect. Nonetheless, the best available data should be used in determining the hazard rate for a system to fail due to a common cause.

We want to emphasize at this point that reliability is merely a probability that a system will perform under certain conditions for a certain period. However, we are assuming that there is no repair state; that is, either a system is operational or it failed. Any complex system may be in a standby mode during which repairs to the system are made. This addresses the true availability of a system, which must consider a periodic maintenance frequency, testing down time, and repair times. This overarching availability of the system is known as the interval availability. It is the fraction of time that a system is operating in any given time interval.

For a much-extended operational or mission time, we can conclude that

$$A_{\mathrm{av}}(\infty) = \lim_{t \to \infty} 1/t \int_{0}^{t} A(t')\mathrm{d}t'. \tag{15.43}$$

The reader is reminded that preventive maintenance is a process performed to restore the system to an optimal operating condition, while repair is a process performed on a failed system, bringing it back to an operational system. Thus, while a system may be completely operational, a portion of the system may be in a state of repair, which means that the primary system may not be 100% operational, but portions of the system are either in failure mode or in a state of preventative maintenance. If we wish to consider the different repair states of the different components, we will be adding to the complexities of the analysis of the reliability of the total system; however, it remains a series of events that are possible to occur during the entire lifetime of the system. Such a complex analysis requires the need for what is called conditional probabilities, wherein a particular unit, which has failed or is undergoing preventive maintenance, is not available to the system for a given time. Such a complex analysis leads to the concept of the mean time to repair (MTTR) a unit. The concept of MTTR is similar to the concept of MTTF we have been previously utilizing in some of the analyses of systems. Another concept related to MTTF is the concept known as the mean time between failures (MTBF), which is merely the sum of the MTTF and the MTTR. In this chapter, we will not be going into the details of these concepts; we merely present them so that the readers are familiar with their descriptions, as they are so common in the engineering analyses of systems.

The calculations needed to analyze the system complexities mentioned earlier are usually handled with what is known as a Markov chain model. Such Markov models allow the engineer to consider the transition probabilities of a component in different states, such as operational, undergoing maintenance, or failed. This kind of Markov model considers the different states independent of all earlier states of the subsystem. The reader is left to investigate, in further details, the intricacies of Markov models. We will now proceed to the main tool of engineers in the analysis of failures of large systems, especially nuclear power plants: fault tree analysis.

Fault tree analysis is the technique utilized by engineers in the analysis of large complex systems, wherein many individual events are interconnected with each other, typically in an electrical or logical manner. Before applying the technique to a complex system, the engineer must understand how the system logically functions, and this will often be accomplished with a logic diagram. The logic diagram for a system is different from a functional block diagram in that the logical relationships between components are explicitly called out in the logic diagram. That is, each possible state of the components will be displayed in a logic diagram, while a block diagram will merely display all states with a single instantiation.

In developing a fault tree, the first step is to establish the so-called top event. For example, the top event may be a circuit break tripping. What can lead to this event? All possible failures that could lead to a circuit breaker tripping off would be indicated in the fault tree analysis. Figure 15.5 is such a fault tree analysis sample.

We find in Figure 15.5 the top failure event, that is, the circuit breaker tripping off, and the contributing events that lead to the top failure event. In this case, the contributing events are joined by the logical OR process, such that the occurrence of any of the contributing events will trigger the circuit breaker to the off position. If we wish, we can now examine the statistics associated with the contributing events that lead to the top failure event. Given enough statistics associated with the events that could lead to the failure of the top event, engineers can estimate the probability of the top failure event.

It is simple for such a simple construct as we have provided, but now imagine a complete logic diagram for every possible contributing failure event! That is a lot of connections and probability calculations. The circuit breaker may only fail once a year, and a trip signal may be coming

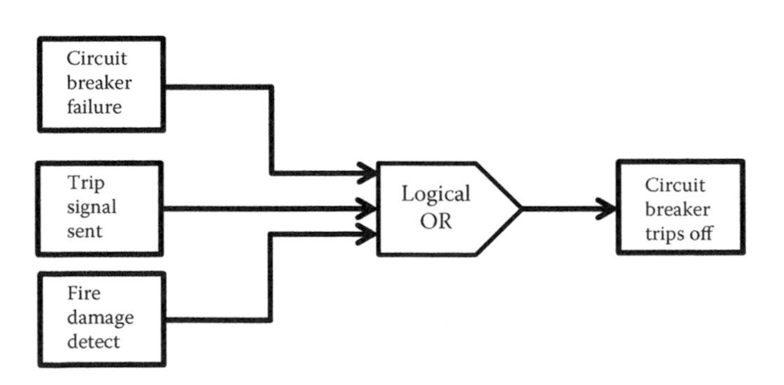

Figure 15.5 Simple fault tree for analysis.

from many different sensors. Finally, a fire, which is a type of common mode failure event may affect not only the circuit breaker of interest, but also many other components as well.

Once some event is established as the initiator, the entire set of possible outcomes, both failure states and success states, can be laid out in a schematic fashion, based upon the building blocks established earlier. Such graphical representations are called event trees. Event trees became extremely popular especially after their demonstrated use and publicity in a report by the NRC in October 1975 called the Reactor Safety Study (RSS). The event trees within the RSS became a cornerstone of engineering reports around the world, addressing failure probabilities.

Basically, any event tree represents the result of a method for identifying possible outcomes given some initial event. The reader should not confuse this with the equally popular decision tree, as the outcomes in an event tree depend only on laws of physics, as opposed to the influence of human control in decision trees. In conclusion, once some event is established as the initiator, the entire set of possible outcomes, both failure states and success states, is laid out in a schematic fashion, known as an event tree.

15.5 CONCEPTS OF RISK

Risk is the potential occurrence of some undesirable event or sequence of events. It can be well defined as the product of the expected frequency F of occurrence and the expected damage D or mathematically thus:

$$R = FD. \tag{15.44}$$

Risk analysis is performed in order to obtain information that may be useful in comparing the risk associated with a number of different events.

Often, risk is confused as a synonym of *hazard*. However, hazard is more properly viewed as a condition with the potential to cause an undesirable event. Another confusing term is *danger*. The term *danger* should more properly be viewed as an exposure to a hazard.

With the concepts mentioned earlier in mind, probabilistic risk assessment was used in the 1975 RSS to break down the frequency of a particular accident sequence into three factors. The first factor is the initiating event. The second factor is the system failure given the initiating event. The third factor is the containment failure given the initiating event and system failure. Mathematically, this can be notated as

$$\begin{aligned} F[\text{release}] = F[\text{initiating event}] \times \text{P}[\text{system failure given the initiating event}] \\ \text{P}[\text{containment failure given the initiating event and system failure}] \end{aligned} \tag{15.45}$$

Do not forget that in the notation mentioned earlier, F is the frequency and P is the probability.

We have now examined how to calculate the risk of some undesirable event using the frequency of occurrence and the damage. The participating authors of the 1975 RSS wished to go further and address the consequences of such risky events. To address the consequences of risky events, the participants developed an equation to address the frequency of property damage given the following: the frequency of radiation release, the probability of radiation transport, and the probability of radiation deposited at any specified location. Mathematically, this was expressed as

$$\begin{aligned} F[\text{property damage at specified location}] &= F[\text{release}] \\ &\times P[\text{radioactivity is transported to specified location}] \\ &\times \mathrm{P}[\text{radioactivity is deposited at specified location}]. \end{aligned} \quad (15.46)$$

In this chapter, we have largely been addressing the risk analyses of electrical power stations, with an emphasis on nuclear reactors. The reason for this emphasis is simple; it is because the American nuclear power industry expended the greatest effort in the mathematical modeling of such complex systems. Nonetheless, it should be noted that while government and industry scientists and engineers did a thorough job of their analyses of electrical power plants, especially nuclear power plants, there were those who maintained a healthy skepticism with respect to all the numbers or, more specifically, the probabilities associated with the analyses.

Soon after the release of the RSS, also known as WASH-1400 report, the American Physical Society (APS) performed an independent review of the RSS. This review had little to criticize regarding the overarching approach of the RSS. However, the one area of the RSS that received the most criticism from the APS was the "values of the probabilities of the various branches." This not only speaks well to the modeling approach of the study, but also brought a focus to the individual probabilities of failure of the individual components.

It should come as no surprise that numerous risk analyses since the RSS have been developed using different approaches. However, it turns out that the alternative studies did not greatly digress from the results of the RSS itself. There were a number of studies specifically related to natural hazards including fires, earthquakes, and tornadoes following the release of the RSS, and these studies highlighted the lack of understanding of risks incurred from such natural disasters.

THREE MILE ISLAND ACCIDENT

The reader should be aware of the fact that while the comprehensive study of nuclear reactor safety was published in 1975, it was in 1979 that the TMI accident took place. The TMI incident was briefly discussed in Chapter 4. Here we wish to question whether the events that took place, which led to the TMI accident, were included as part of the reactor safety study. The TMI accident is a type of loss of coolant accident. The loss of coolant at TMI was due to the failure of a valve to properly reset itself in the pressurizer. Such a failure was indeed part of the RSS. However, the TMI incident was precipitated by the actions or inactions of the personnel on duty at the time. Apparently, the malfunction was picked up by the monitoring system, but it was ignored for too long, and to make matter worse, the operators had shut off of the emergency coolant injection system during the loss of coolant incident. While the malfunction was part of the study itself, the incorrect response of the operators on duty was not considered as part of the study.

15.6 COMPARISON OF RISKS

Risk analysis associated with nuclear reactors attempts to determine risk of a major nuclear disaster. However, it allows engineers to compare the risks associated with all the energy resources from fossil fuels to renewable resources.

There are numerous health and safety risks associated with all energy sources. In 1974, the federal government released a comparison of health and safety risks for a 1 GW power plant generating electricity for the electrical grid. The government compared the generation of electricity using coal, oil, gas, and a nuclear power reactor of the light water reactor type. The total life cycle costs were considered. Thus, coal surfaced as the highest risk source of electrical energy due to the considerations of obtaining the coal from the ground.

Considerations in risk analysis were made for the complete life cycle of the power plant including the material acquisition and construction, the emissions from the plant, the operation and maintenance of the plant, the energy storage requirements, and the transportation of the energy resource from its source.

Once a complete life cycle systems analysis has been performed, only then can we compare the risks among the alternatives for producing energy for consumption by people. Such comparisons are called competing risk assessments. Notice that only the probability and damages are considered in such analyses. Rarely is there no cost accounting and benefit analysis in such assessments. One of the first thorough life cycle systems analyses was performed in 1974, sponsored by the Atomic Energy Commission. These analyses were developed using a risk and cost–benefit analysis comparison for all known alternative modes of production of electricity at that time. These cost–benefit analyses provided the government a means to compare the risks and costs associated with the production of electricity by using coal, oil, natural gas, and nuclear power, specifically a light water reactor.

The results of these comprehensive cost and risk benefit analyses, as published by the Health Physics Society, were somewhat surprising. The production of electricity by the burning of coal had the worst metrics for occupational health and safety and public health and safety. The dominant factor in the occupational health and safety effects was caused by the total number of worker days lost due to black lung disease or coal workers' pneumoconiosis. Injury rates for oil, gas, and nuclear reactors were all about 10 times lower than for the other modes of electricity generation.

Up to the 1970s, it was only fossil fuel generation of electricity, together with nuclear power, that was analyzed to the fullest extent. In 1978 and 1979, Herbert Inhaber addressed the question of the comparison risk analysis between solar generation of electricity and other conventional forms of generation of electricity.

To examine all the risks associated with the production of electricity, you need to include all the aspects of the cycle of the process leading to the generation of electricity. Many people note how solar panels are just sitting there collecting solar energy and not producing any pollution of any kind. However, the solar panels themselves have to be constructed at a factory, and that factory uses energy and natural resources in its construction of the solar panels. In fact, it was demonstrated for that 1978 Inhaber study that solar power panels use a much large amount of construction material, per unit energy, than a standard nuclear power plant. What other components of risk analysis must be included in any cost–benefit analysis? One should at minimum include the cost of the material to produce the system, the emissions caused by the production of the materials, the storage of energy, and the transportation of materials from the factory to the final destination (or house). As it turns out, it was in 1978 that the United States established the US Energy Information Administration (EIA), which, to this date, produces data, including the levelized cost of electricity (LCOE). Data from the US EIA, a portion of the DOE, are available online from the year 2006. By inputting the LCOE data into an Excel spreadsheet, one can perform their own analyses. Here is the data that were published in August of 2016 (in 2015 dollars), comparing the cost of electricity per megawatt-hour for the different resources.

The reader is cautioned about ever comparing data from different studies, unless data were collected in the same exact fashion every year of the

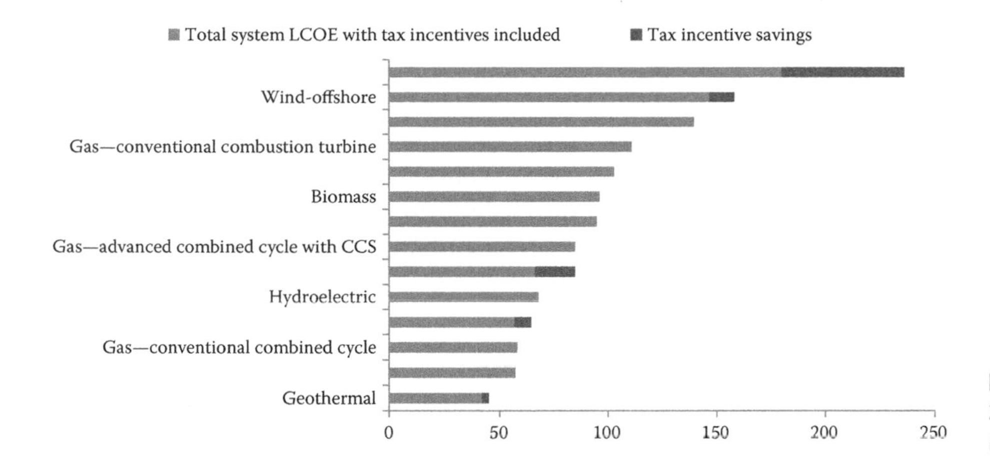

Figure 15.6 US average LCOE in 2015 dollars per megawatt-hour for plants operational in 2022.

study. This is true for all disciplines and something that is often overlooked by the news media of any venue. In this case, the US EIA has been reporting such data in the same manner for the past two decades, so data from the US EIA can be compared over the previous years. It should also be noted that the data graphed here (Figure 15.6) includes a formulation for the forecasting of the construction and operation of electric power plants through the year 2022, so there is the extrapolation of cost data built into these data. Nonetheless, within the study confines, it is apparent that the total life cycle cost of the generation of electricity widely varies from one resource to another.

In the total risk analysis study of any complete system, the analysts need to consider the public health risks associated with the acquisition of the materials. The best example is the acquisition of coal. Power and energy pundits often forget to include the hazards and risks due to the pollution of the process of acquiring the natural resource, as well as the risk to humans in the acquisition of the materials. One objective approach to addressing these risks is to examine the loss of workdays due to illnesses and accidents involved in the removal process (minimally). As early as 1980, risks associated with the acquisition of coal determined that coal had a large loss of workdays due to illnesses such as the black lung or other respiratory illnesses. In fact, initial research into the number of deaths per megawatt-year of energy produced demonstrated that coal was by far (over 10 times the amount) the most deadly resource per megawatt-year. Even when considering the risks of nuclear waste management and nuclear reactor accidents, there was still no comparison as to the least safe method of energy. An analysis of all factors led investigators in the 1980s to conclude that the risk for coal and oil for electricity was most dependent upon the pollution during the production of electricity. However, for the use of natural gas, nuclear power, and hydropower, the largest risks are associated with the acquisition of materials and the construction of the facilities. The need for backup energy, especially with respect to solar energy, was most significant in the costs associated with the generation and use of solar, wind, and ocean energy systems.

Thus far, we have addressed the risks and costs associated with the building, generation, and delivery of energy to the final destination, whether home or office. However, risk analysis, especially what is referred to as complementary risk comparison, is only one part of the total risk analysis picture. The other portion covers what is known as societal risk. This allows engineers to consider the actual number of deaths associated with the generation of energy for people and the calculated number of predicted deaths.

RISKS IN HYDROPOWER

Even the hydropower generation of electricity has some of its own peculiar risks. Dams can fail; for example, the most spectacular failure associated with a hydropower plant is the failure of the dam itself. This was highlighted by the great Johnstown Flood of 1889. This flood was caused by the collapse of the Conewaugh River dam. The dam was 14 mi upstream from the town of Johnstown. When it failed, it killed 2209 people and caused the equivalent damage (in 2016 dollars) of over $450 million.

While radiation risks are often only attached to the risks from the nuclear power generation of electricity, in truth, radiation risks take place in the recovery of all energy resources. Coal mining is most associated with the black lung disease, but coal miners are exposed to higher radiation levels depending upon the precise location of their mine and the mine exposure to higher levels of natural radiation. Flying in an airplane increases your exposure to radiation, and the more you fly, the more exposure you have to cosmic ray radiation risks. Medical X-rays are also a major source of radiation exposure for humans. So we see that risks from radiation have to be considered in all energy resources recovery.

We all associate radiation with the higher risk of cancer, but we forget that higher risks of cancer are associated with many other factors. For example, cancer has also been linked to the production of viable energy by such causes as air pollution produced by the generation of electricity, explosions at power plants, fires at power plants, and even the trash of power plants.

All risks should be addressed in any complete life cycle study of a facility that generates viable energy for consumers, regardless of the location of the risk in the life cycle of the system.

15.7 RISK BENEFITS AND ACCEPTANCE

In the previous section, we have seen that all risks associated with any form of energy generation for consumer needs must be considered before a comparison among different methods can be objectively accomplished. It was 1980 which saw the first publication of the *Annual Review of*

Energy by the US EIA. One of the chapters of the volume dealt with the social costs and the level of risk. The economic costs of the risks from the generation of electricity were couched in terms of economics. This led to the examination of the law of diminishing returns and a development of a rule of thumb, which states that one spends on safety considerations until the increased cost of safety is at the same level as the expenditure of funds. That is the point where expending $1 for safety provides an equal dollar in the reduction of social cost.

By applying the economics mathematicians developed for this cost–benefit analysis, one is led to the following equations:

$$\text{Social cost} = NC(1+i)^x, \tag{15.47}$$

where $x = t$ for $t < 6000$, and $x = 6000$, where $t \geq 6000$. Also, we have N as the number of individuals, C as the cost of one disability day, i as the interest rate, and t as the time in consecutive days of disability. For the record, the National Safety Council considers the loss of 6000 work days for a person to be equivalent to a death or fatality; thus, the significance of the value $t = 6000$.

It is unfortunate that this type of social cost–benefit analysis leads to a mechanism for considering the value of a human life, but this is necessary in order to establish equations whereby such cost comparisons over a life cycle can be examined. Such analyses equate the value of a person's future earnings as the sole value of a human life. There is still a range of values to consider. You may wish to use the average salary of an employee for establishing some baseline. Even with this standard, there is the reality of the difference in the mean and the median income of a worker. Ignoring race and region, according to the US Census Bureau, the mean household income in 2014 was $53,657, while the median household income was $75,738. You should be careful to note which of these averages is being used in the specific study being analyzed. Nevertheless, such a valuation allows for economists and engineers to develop an equivalency in the cost of a job with increased probability of accident. That is, if a job presents more of a risk, then an increase in salary commensurate with the risk would lead to an incentive pay scale for the riskier job. This approach to costs benefit analyses is often known as the personal risk method associated with the value of a human life. It sounds rather cold, but it is an accepted approach for dealing with the cost–benefit analysis of risks associated with the delivery of energy to consumers.

Naturally, we have not exhausted all approaches to cost–benefit analyses, and there are many different factors that we have yet to consider. We may wish to consider the cost of saving lives within our equations. This has to deal with the added costs of preventing human deaths, but death is not avoidable. We all will die sometime; therefore, economists and systems engineers tend to utilize the effective average human lifetime as 65.

This is a typical retirement age and actually not related to the actual average human lifetime of a worker.

Thus, another approach to addressing cost–benefit analyses and considering the frequency of a predetermined set of accidents leads to a cost–benefit ratio for nuclear power plants considering radiation hazards as

$$\text{Cost–benefit ratio} = \mathrm{C}/\sum \mathrm{F}_i\mathbf{R}_i - \sum \mathrm{F}_{i'}\mathbf{R}_{i'}. \tag{15.48}$$

In this equation, C is the cost of safety features in dollars per year, averaged over the years; *F* is the frequency of accidents without the safety features installed; and *R* as the radiological consequences in man–rem of the incident without the safety feature installed. The primed values represent the numbers for the case where the safety features are installed. Once again, this is a cold-hearted approach to the cost–benefit ratio to be sure, but some baselines do have to be established if we are to compare costs across all energy resources.

It was first in 1979 when Cohen and Lee published a table of days reduced in life expectancy for the alternative approaches of producing energy for the consumers. The comparison was based upon the number of days that a person's life was reduced for being in the specified industry of energy generation. For example, for coal, the reduced life expectancy was 11.5 days. For the oil industry, the average days of life expectancy was reduced by only 2.2 days. In the year 1979, the average number of days of life expectancy was reduced by less than 0.02 days. At this point, one may want to consider other life hazards to which these figures could be compared. The average loss of life expectancy due to heart disease is 2100 days. The loss of days of life expectancy due to cigarette smoking for men is 2250 days. Thus, we see that the loss of life expectancy in days for the energy generation sector is very low compared to these more common risks to human longevity.

Today, the National Safety Council, originally chartered by the US Congress in 1953 as a nonprofit, nongovernmental public service organization promoting health and safety for all people within the United States, holds the accident data from which decrements in life expectancy could be calculated. However, doing so for the energy industry is not a simple task, as the manner within which the data are collected and displayed, is not conducive for the particular analysis at hand, which is why it is difficult to uncover the very same data as produced in the 1979 study. Again, this is a common issue with all statistical reporting. If the identical approach is not utilized, the reader cannot properly compare the output results.

15.8 SUMMARY

In this chapter, we have provided the reader with an overview of *probability* and *statistics*. The probability of the outcome of an event is symbolized as P(A). The probability of any event lies between zero and one, and the probability for N events can be approximated as shown by Equation 15.13.

Another key mathematical concept is *Bayes theorem*. It is the description of the probability of some event based upon certain conditions that are related to the event. Applications of Bayes theorem allow for the calculation of the upper limits of the expected frequency of any occurrence.

We then addressed the issue of averages. The *mean* value is derived by summing all values and dividing by the number of values summed. Another type of average is known as the *median*. Mathematically, the value in the middle of all values listed in magnitude order is the median. A third average statistic is known as the *mode*. The mode is the value that occurs most often of all ordered values.

We then examined the so-called *normal distribution*, which is characterized by its mean and its *standard deviation* or σ. Another descriptor associated with the normal distribution is the *variance* or σ^2. The standard deviation is a value that provides the amount of spread of the curve around the mean, while the variance is simply the square of the standard deviation.

We then defined the *hazard rate* or λ of a system. Using the assumption of random failures, we finally derived the concept of *reliability* as shown by Equation 15.25.

Next was the concept of the MTTF, where the MTTF of any system is defined by Equation 15.26.

Looking at larger systems, we noted that the reliability of an entire system is the probability product of the components. For a system of $2n$ identical subsystems, which fail in a random manner and have a hazard rate of λ, the reliability is shown by Equation 15.37.

We examined how a catastrophic event, such as an earthquake, could be assumed to occur with a frequency that approaches a *Poisson distribution*, which led to the equation for the reliability, considering a catastrophe such as an earthquake, as shown by Equation 15.41.

Related to the MTTF are the concepts of the MTTR and the MTBF. The MTBF is simply the sum of the MTTF and the MTTR as shown by Equation 15.49.

We turned to the concept of *risk*, which can be simply represented by Equation 15.44, where F is the expected frequency of occurrence and D is the expected damage.

We examined the modern concept of the LCOE and discovered that solar thermal had the highest cost of producing electricity and geothermal had the lowest cost of electricity generation.

We then defined a concept called *social cost* and represented this concept by Equation 15.47, where $x = t$ for $t < 6000$, and $x = 6000$, where $t \geq 6000$. N is the number of individuals, C is the cost of one disability day, i as the interest rate, and t is the time in consecutive days of disability.

Another approach to addressing cost–benefit analyses while considering the frequency of a predetermined set of accidents leads to a cost–benefit ratio concept for nuclear power plants incorporating radiation hazards as shown by Equation 15.48, where C is the cost of safety features in dollars per year averaged over the years; F is the frequency of accidents without the safety features installed; and R is the radiological consequences in man–rem of the incident without the safety feature installed. The primed values represent the numbers for the case where the safety features are installed, thus the comparison ratio.

PROBLEMS

1. Determine the probability of getting a 3 or a 6 with the toss of a single die.
2. Determine the probability of tossing a die and getting a number that is less than 5.
3. A die is tossed twice. Determine the probability of rolling a 4, 5, or 6 on the first roll and a 1, 2, 3, or 4 on the second roll.
4. Two dice are rolled. What is the probability of not getting a 7 or 11 on the first two rolls of the die?
5. A single die is rolled twice. What is the probability of obtaining the 3 on either of the two rolls of the die?
6. A storage canister has five resistors and four capacitors. An electricians pulls out two items consecutively from the canister. The second item is a capacitor. Determine the probability that the first item pulled out of the canister was also a capacitor.
7. An electronics parts container has 17 identical resistors and 13 identical capacitors. If a technician chooses two items from the container, what is the probability that he/she will pick a pair of resistors or a pair of capacitors?
8. If an engineer has two switches and four bulbs in a container, what is the probability of the engineer choosing a switch and then a bulb out of the container?
9. An engineer has seven different resistors. How many different arrangements are there consisting of just three of the resistors?
10. Given the series configuration as displayed in Figure 15.2, If you were given the operating time for each unit as 100 days and the unit failure rate as 2.5×10^{-14} failures per day, what would be the reliability of the two components in series?
11. Given the parallel configuration as displayed in Figure 15.3, if you were again given the operating time for each unit as 100 days and the unit failure rate as 2.5×10^{-14} failures per day, what would be the reliability of the two components in parallel?

12. An engineer has a container consisting of resistors, capacitors, inductors, and transistors. If the apprentice randomly sticks his/her hand in the container, what is the probability that he/she will grab a resistors? What is the probability that the apprentice will first grab a capacitor and then a transistor? What is the probability that the apprentice will grab four resistors in a row?
13. A power plant has three subsystems. Each subsystem has the probability of a defect as being 0.1, 0.2, and 0.25, respectively. What is the probability that one (only one) of the three subsystems will have a defect if all are used in the power plant?
14. A meteorology company is contracted by a solar power plant company that has a solar power plant in both New York City and Los Angeles. The contractor has data which give the probability of a sunny day in New York City as 0.4 and the probability of a sunny day in Los Angeles as 0.3. The data also show that the probability that it will be a sunny day in both New York City and Los Angeles on the same day? What is the probability that neither New York City nor Los Angeles has a sunny day on the same day?
15. Consider a core monitoring system for a nuclear reactor with only an ionization chamber (IC), a temperature sensor (TS), and a pressure sensor (PS). Let us say that the failure probabilities for these subsystems are as follows: $P(\text{IC}) = 0.02$; $P(\text{TS}) = 0.04$, and $P(\text{PS}) = 0.01$; apply the Bayes theorem for determining the probability that the temperature sensor failed during a reactor operation.
16. Consider the fact that during a 10-year period, there were 4000 spent fuel shipments and not a single incidence of a release of radioactivity. Calculate the probability of radiation release per shipment.
17. Calculate the mean, median, and mode for the following set of numbers: 0, 1, 1, 2, 3, 5, 8, 13, and 21. It turns out that there is a special name for this number sequence, can you tell us what that is?
18. Given a power plant component that fails in a random fashion and a manufacturer demand failure probability of 10^{-14}, with a usage rate of 20 per week, what is the probability that the component will fail within a period of 3 years?
19. The probability that a particular man and a particular woman will be alive in another 20 years is 0.8 for the man and 0.9 for the woman. Calculate the probability that both the man and the woman will be alive in 20 years. Also, calculate the probability that neither the man nor the woman will be alive in 20 years. Finally, calculate the probability that either the man or the woman will be alive in 20 years.
20. What is the reliability of a simple two-part system, whose subsystems are connected in series, if each system is in operation for 200 hours and the subsystem failure rate is 2.5×10^{-14} failures per hour?

21. What is the reliability of a simple two-part system, whose subsystems are connected in parallel, if each system is in operation for 200 hours and the subsystem failure rate is 2.5×10^{-14} failures per hour?
22. An electric power plant is designed to have a reliability of 99%. If each subsystem has a reliability of 60%, how many subsystems must be linked in parallel so that the power plant meets its designed specification?
23. The MTBF of a subsystem of a nuclear power plant is known to follow an exponential probability distribution. If a subsystem has a 90% probability of *not* surviving for more than 30 days, how often would you be expected to replace the subsystem?
24. A thermal solar plant consists of six subsystems. Each subsystem has a MTTF of 8 months. What is the probability that one or two failures occur in a 4-month period of operation?
25. An electric generation power plant using wind has four windmills. Of the four windmills, two are connected in parallel with one another, followed by the remaining two connected in series. The failure probability follows the exponential distribution. Each windmill has a failure rate of 0.01 per day. What is the probability that the power plant will not survive for more than 100 days?

FURTHER READING

Ashley, H., R. L. Rudman, and C. Whipple (eds) (1976) *Energy and the Environment: A Risk Benefit Approach*, Pergamon Press, New York.

Aven, T. (2008) *Risk Analysis: Assessing Uncertainties beyond Expected Values and Probabilities*, John Wiley & Sons, West Sussex.

Brown, R. (ed.) (2012) *30-Second Math: The 50 Most Mind-Expanding Theories in Mathematics, Each Explained in Half a Minute*, Metro Books, New York.

Cohen, B. L., and I.-S. Lee (1979) A catalog of risks, *Health Phys.*, 36, 707.

Frank, M. V. (2008) *Choosing Safety: A Guide to Using Probabilistic Risk Assessment and Decision Analysis in Complex, High-Consequence Systems*, Resources for the Future, Washington, DC.

Gorini, C. A. (2012) *Master Math: Probability*, Course Technology Cengage Learning, Boston, MA.

Huff, D. (1954) *How to Lie With Statistics*, W.W. Norton, New York.

Inhaber, H. (1978) Is solar power riskier than nuclear power. *Trans. Am. Nuclear Soc.*, 30, 11.

Inhaber, H. (1979) Risk with energy from conventional and nonconventional sources, *Science*, 203, 718.

McCormick, N. J. (1981) *Reliability and Risk Analysis: Methods and Nuclear Power Applications*, Academic Press, New York.

Megill, R. E. (1984) *An Introduction to Risk Analysis*, 2nd edn, PennWell Books, Tulsa, OK.

Pishro-Nik, H. (2014) *Introduction to Probability, Statistics, and Random Processes*, Kappa Books, Blue Bell, PA.

Savage, S. L. (2009) *The Flaw of Averages: Why We Underestimate Risk in the Face of Uncertainty*, John Wiley & Sons, Hoboken, NJ.

Spiegel, M. (1975) *Schaum's Outline of Theory and Problems of Probability and Statistics*, McGraw Hill Book Company, New York.

Tetlock, P. E., and D. Gardner (2015) *Superforecasting: The Art and Science of Prediction*, Crown, New York.

Vigen, T. (2015) *Spurious Correlations*, Hachette Books, New York.

Ward, T. L. (ed.) (2007) *Probability and Statistics Exam File*, Kaplan AEC Education, Chicago, IL.

Dynamics of Population

An Overview

16.1 MODELING POPULATION DYNAMICS

In this chapter, we examine some mathematical aspects of population, especially human population. After all, the human population on Earth is a major energy sink. The Earth is not a closed system—it is open to the universe, especially the Sun, which is a source of energy for the Earth and all components of our solar system.

We first address some basic definitions. A population is typically a group of plants, animals, or other such organism. In fact, a population is a group of any objects with some unifying feature. In fact, you may even have a population of stars. Within the context of this volume, we are most interested in the population of human beings. In fact, all populations of living things are indeed consumers of energy of some type. Remember, even plants are consumers of energy, specifically the solar radiant energy.

Often the term *population* is synonymous with the size of the population. In this case, let us choose to represent the size of the human population with the letter N. Determining the size of the human population is simply the numeric count of humans. Unfortunately, the number of human beings on Earth does not remain the same value. Human beings are constantly being born and unfortunately constantly dying. If we wanted to know the human population today, we could simply count the number of human beings on the planet Earth. So we now have our first equation:

$$N = N_t. \tag{16.1}$$

Here N again represents the total count of human beings on Earth, and N_t represents the counted human beings on Earth today. As we said, members of the population *Homo sapiens* do not live forever, so the population changes in time. We may now write the size of the population tomorrow as

$$N_{t+1} = N_t + B - D. \tag{16.2}$$

In this case, B represents the number of births by time $t + 1$, and D represents the number of deaths between t and $t + 1$.

Most population dynamics or ecology textbooks will also note that for a given population, you may increase or decrease the size of the population by having members of the population move into the area under consideration or move out of the study area. For example, if you are studying the size of the population of Chicago, you not only have to consider the

CONTENTS

number of births and deaths, but you must also consider the number of people who move into or out of the city of Chicago. In this chapter, we are most interested in the global human population, as it is the totality of humanity, which consumes the ambient energy, or the energy stored in fossil fuels. Thus, not only because it simplifies the mathematics, but also because we have this global view, we will not include immigration or emigration. And we know of no *Homo sapiens* that would come from somewhere other than Earth. There is simply no evidence for this at all.

Continuing with our mathematics of the human population, we now have the change in the size of the population as simply

$$\Delta N = N_t + B - D. \tag{16.3}$$

In differential form, the equation becomes

$$dN/dt = (b - d)N, \tag{16.4}$$

where b is now the birth rate and d the death rate. As long ago as 1798, the Reverend Thomas Robert Malthus used r to denote the instantaneous rate of increase of the size of population, where

$$b - d = r \tag{16.5}$$

and then

$$dN/dt = rN. \tag{16.6}$$

Equation 16.6 represents the Malthusian model of the growth of population and the constant r, also known as the intrinsic rate of increase, is the determining factor for the population. In the Malthusian model, if $r = 0$, there is simply no population growth. If r is greater than zero, the population will exponentially grow. Of course, if r is less than zero, the population will ultimately go to zero, which also indicates a point of extinction of the particular population.

By integrating the Malthusian model in Equation 16.6, we obtain an equation which allows us to simply plot the population growth based upon an initial value of population and a growth rate using the following resultant:

$$N_t = N_0 e^{rt}. \tag{16.7}$$

In Equation 16.7, N_0 is merely the initial population.

Equation 16.7 is of a form that is easily programmed into a spreadsheet for seeing how it graphically looks. Using an initial population of 1 and an instantaneous rate of 0.1, we can plot the following population curve:

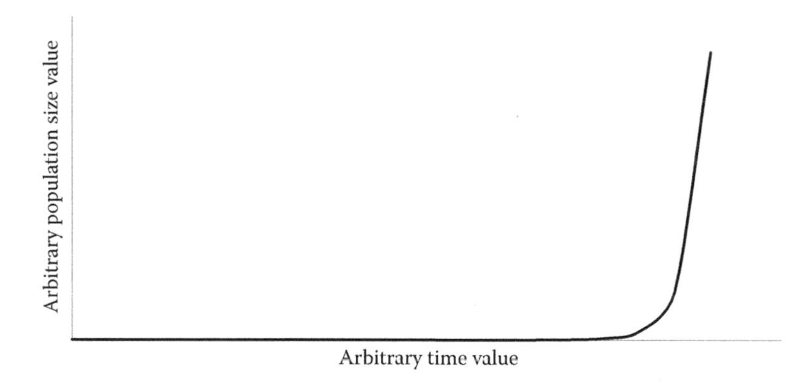

Figure 16.1 Malthusian growth rate curve.

As we can see in Figure 16.1, the growth rate curve of Malthus is exponential in its fundamental nature. Also, it will quickly cause us to expand our y-axis values, pretty much to infinity. Obviously, this is absurd, no matter what population we are dealing with; infinity is not a realistic possible value for the population. After all, even excluding the issue of the amount of energy consumed by the population, the planet is of finite size and cannot handle an infinite number of any population. Before we address the limits to the size of a population, let us perform a little mathematics to determine how long it would take for a Malthusian type population to grow double the size of the population.

To find the time within which the population would double, we set N_t equal to $2N_0$. Substituting this condition into Equation 16.7, we obtain the following equation:

$$2N_0 = N_0 e^{rt_d}. \tag{16.8}$$

In this equation, t_d is the time at which the population has doubled its size. By taking the natural logarithm of both sides, we now have a means by which we can determine the time at which the population size will double. That is,

$$t_d = \ln(2)/r. \tag{16.9}$$

In our example, which led to the development of Figure 16.1, the doubling time is just below 7. It should be noted that the doubling time is not related to the initial size of the population. That is, regardless of the initial population, using the Malthusian model approach, the time to double the size of the population will be 7 time units. But wait a minute; in the initial parameters used to develop Figure 16.1, the initial population was 1. For any sexually reproducing species, you cannot use 1 as an initial population. As in physics, sometimes it is instructive to examine the numbers relative to the physical possibilities. That is a hint for all students reviewing their answers to problems.

Before proceeding to extend our population model, we wish to explicitly state the assumptions that are incorporated within Equation 16.2. First, as noted when we began this chapter, we are examining a population which is closed. That is, there is no immigration into the population nor emigration out of the population. In this case, we are considering only the entire population of the planet and not concerned about any specific geographic location. Another assumption in our preliminary model of human population is the fact that we have incorporated a constant birth and death rate.

A constant birth and death rate leads biologists to imply that the population has no variation in its genetics. Again, this is a simplification of reality to say the least. Not only is there no genetic variation in the population, but the constant birth and death rate also implies that there is no difference in the age or size of the population. Also, the population modeled earlier must be asexual, as we were able to develop a model where the population can grow from a single individual and reproduce as soon as the individual is born. Not exactly the case for most species, let alone *Homo sapiens*. Again, this is part of modeling: gross simplification.

There is one final assumption with this model approach. That is, we are using a calculus-based model which assumes that the population is continuous, not a discrete numeric function, which is actually the case. After all, you cannot have a part of a human being.

There are many ways to add complexity to our basic model derived from Equation 16.2. We will leave it to the reader to explore some of the complexities utilizing such textbooks as Nicholas Gotelli's *A Primer of Ecology* or some similar textbook. For this chapter, we wish to next explore the logistic growth model.

To derive the basic equation for the logistic growth model, we start with the differential growth equation:

$$dN/dt = (b' - d')N, \tag{16.10}$$

where b' is the birth rate per capita and d' is the death rate per capita. There are numerous alternatives for modeling the birth rates and death rates. The simplest linear models are expressed by

$$b' = b - aN \tag{16.11}$$

and

$$d' = d + cN. \tag{16.12}$$

Equation 16.11 assumes that the birth rate will be decreasing, as it would if the population takes into consideration the decrease of resources, including food per person, as births increase. Also, Equation 16.12 assumes that the death rate will be increasing as the number of persons increases, leading to a decrease of resources for the population.

We now leave it to the reader to substitute for b' and d' in Equation 16.10 and use the biologist's definition of carrying capacity as being K, where

$$K = (b - d)/(a + c). \tag{16.13}$$

This leads us to the standard form of the logistic growth equation, which in differential form is

$$dN/dt = rN(1 - N/K). \tag{16.14}$$

Remember that Equation 16.5 allows us to use r for $b - d$.

The reader should note that the logistic growth in Equation 16.14 is similar in nature to the exponential growth in Equation 16.4. To get to a simple algebraic form for forecasting the population, we once again integrate our growth equation. Now the population size can be forecast as a function of time using

$$N_t = K/(1 + [(K - N_0)/N_0)]e^{-rt}. \tag{16.15}$$

The logistic growth curve in Equation 16.15 as a function of time leads to the typical logistic curve as depicted in Figure 16.2.

The parameters used in the generation of Figure 16.2 include a carrying capacity of 1,000,000; a growth rate of 1; and a doubling time of 1.

It is instructive to note the first nine values of Figure 16.2. They are 1, 2, 4, 8, 16, 32, 64, 128, 256, and 512. Note that the population during these

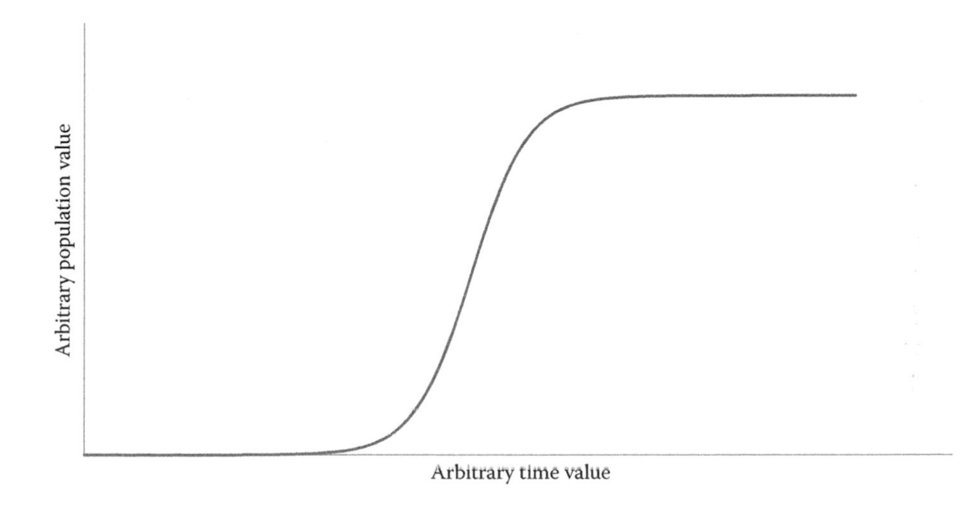

Figure 16.2 Logistic growth rate curve.

first nine values is exponentially increasing. That is the amusing feature of a logistic growth curve. In the early values of the curve, they look exactly like an exponential growth curve. You cannot tell the difference until the later values are calculated and plotted.

Before proceeding, let us emphasize as we did for the exponential population growth model the most important model assumptions in the logistic growth model. First, it must be noted that the carrying capacity utilized in the logistic growth model is a constant. That is, it simply does not vary in time. Second, it is assumed in the logistic growth model that the increase in the population number causes a decrease in the population growth per capita. Such a relationship between the increased population number and the population growth per capita is known as linear density dependence. So while the number of births and deaths is a constant, the birth rate and death rates will change with the increase in population number.

We have now examined two simple models of the size of a population. The first was the simple exponential growth model, and the second, the logistic curve growth model. Both of these excursions into population modeling have been thus far only examining a single population. That is, of course, an oversimplification. There is nowhere on this planet where there is only a single population. In fact, it should be self-evident that no planet could survive with only a single population of any species. So how shall we incorporate other populations into our modeling of a population? Before noting some approaches to this type of modeling, let us first note that our single population models need to be adjusted before they could ever approach modeling the reality of the size of the population.

Populations have members of different ages. Even populations of stars have members of different ages; only with stars is the age well beyond what any living creature exhibits. Not only are the ages of the members of a population different, but not all ages allow for reproduction. Humans, and other animals, are not able to reproduce until they reach puberty. This changes our equations for the population models. In order to bring our models closer to the reality of the population size, we would have to consider not only when a member can begin to reproduce, but also how long any member is able to reproduce. Human females cannot reproduce for the entirety of their lifespans. Of course, there is also a delay from when any animal reproduces to the time that offspring are actually produced. Again, these factors would have to be incorporated in a valid population model.

We have already noted that we were well aware that our models do not consider the emigration or immigration of individual members of the population to another population. This consideration leads to the concept that has been termed *metapopulation* since 1970. A metapopulation is a set of populations that are linked by the ability of members to exit or enter another population. As we noted earlier in this chapter, from the viewpoint of the use of energy resources of the planet Earth, we are most interested in the entirety of the human population. One is inevitably drawn to politics when one considers the human metapopulation and differences in time caused by the movement of members from one population to another.

The concept of a metapopulation leads us to the necessity for developing a population model wherein the populations are in competition with one another for the energy resources.

In the second and third decades of the previous century, Alfred Lotka and Vito Volterra focused on the metapopulation of competing species, and they developed the classical Lotka–Volterran competition model of population growth. The two populations are assumed to behave similar to a logistic growth curve, if their interactions are ignored. Thus, one output curve for such a metapopulation might look like Figure 16.3.

Obviously, once the two logistic growth curves are linked, the model output changes. The simplest linkage for two populations of logistic growth incorporates competition coefficients within the linked logistic curve formulas. We will only express the differential form of this simple Lotka–Volterra formulation, and that is

$$dN_1/dt = r_1N_1\left((K_1 - N_1 - aN_2)/K_1\right) \tag{16.16}$$

$$dN_2/dt = r_2N_2\left((K_2 - N_2 - bN_1)/K_2\right). \tag{16.17}$$

We note the similarity between Equations 16.16 and 16.17 to the original logistic curve formulation expressed in Equation 16.14.

We offer to the readers as an exercise the use of the Lotka–Volterra (Equations 16.16 and 16.17) in a discretized form, within the confines of a spreadsheet. Depending on the initial populations and the values of the

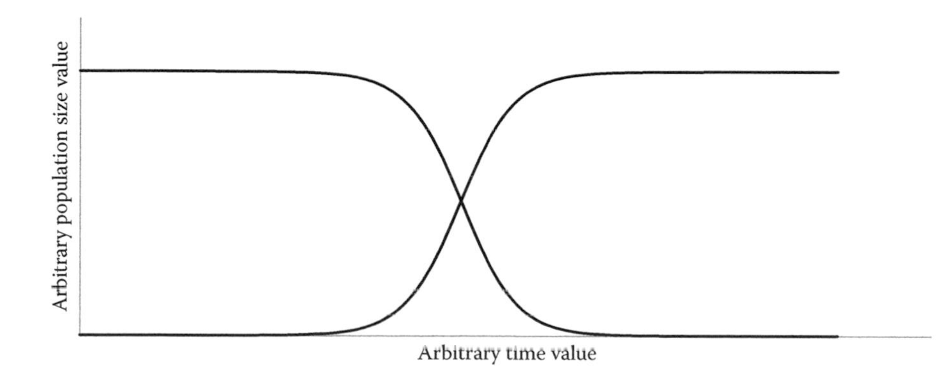

Figure 16.3 Metapopulation growth curve without mathematical linkages.

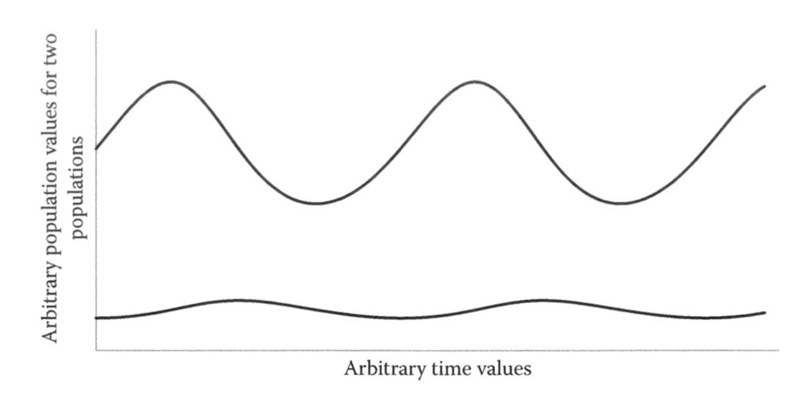

Figure 16.4 Lotka–Volterra population curves.

constants a and b, the reader may develop a Lotka–Volterra plot similar to the one developed for Figure 16.4.

As a hint to the reproduction of the Lotka–Volterra plots earlier, we provide a hint of initial populations of 50 and 8 and competition coefficients of 0.01 and 0.05, respectively.

As with all our simple models, we wish to emphasize the model assumptions for the Lotka–Volterra population growth. For our Lotka–Volterra model, we once again have assumed that there is neither age structure nor any genetic structure in the populations. Also, we assume no migration of any kind in the two populations and no time lags in births and the ability to give birth. All these, of course, will lead to more complex population models.

By the way, ecologists have a term wherein two populations interact strictly as predator and prey. Such a metapopulation is referred to as an *intraguild predation*. There are many other interactions, and they have names as well. The most well known are competition, mutualism, and parasitism.

Although we have only addressed the three most fundamental models of population growth, we hope this has provided an introduction to the models within a population ecologist's repertoire. We hope that this cursory introduction will lead the reader to a more complete investigation of the modeling of populations using one or more of the references at the end of this chapter. At the very least, we trust that the reader has learned the enormous complexities required by such population models and how scientists take increasingly complex steps to alleviate the assumptions made by their current models; much in the way that global change modeling has progressed over the years.

16.2 US POPULATION DYNAMICS AND ENERGY CONSUMPTION

On July 4, 2015, the estimated population in the United States was 321,216,397 as calculated by the US Census Bureau. This number, of course, is constantly changing. If you were so inclined, you could actually watch the estimated US population every second on the Internet. The official US and world population clock maintained by the US Census Bureau can be found at http://www.census.gov/popclock/. Of course, this only shows the instantaneous estimated population. Moreover, if you only know the instantaneous population, you will not know the population from the past nor the future.

The US Constitution provides for the taking of census or the counting the people of the nation, every 10 years. It is Section 2, Article 1 of the US Constitution which specifically states that

> the actual Enumeration shall be made within three Years after the first Meeting of the Congress of the United States, and within every subsequent Term of ten Years, in such Manner as they shall by Law direct.

So how does the Census Bureau estimate the number of people in the United States in the years that do not coincide with the census? They must make an estimate. This estimate is not just a guess. After all, you do have data for the last decadal year. This is the starting point. Then you can use the statistical records and estimate the number of births and deaths. However, Americans are not cemented in place; we do move. Thus, the Census Bureau factors in the data they have for the movement of people within the United States as well as movement of Americans coming into and moving out of the country. This factor is called the migration factor. Thus, the estimated population of the United States in the years between census years incorporates the population base as determined by the latest census and then factors in the births, deaths, and migration of people of the United States.

So we have seen how the US Census Bureau estimates the number of people in the United States for every day, in fact every second of every day of every year. But what about trends over the years? Again, since 1790, the United States has counted the people within its borders. Let us examine the data from the US Census Bureau for the United States from 1790 through 2000 (Figure 16.5).

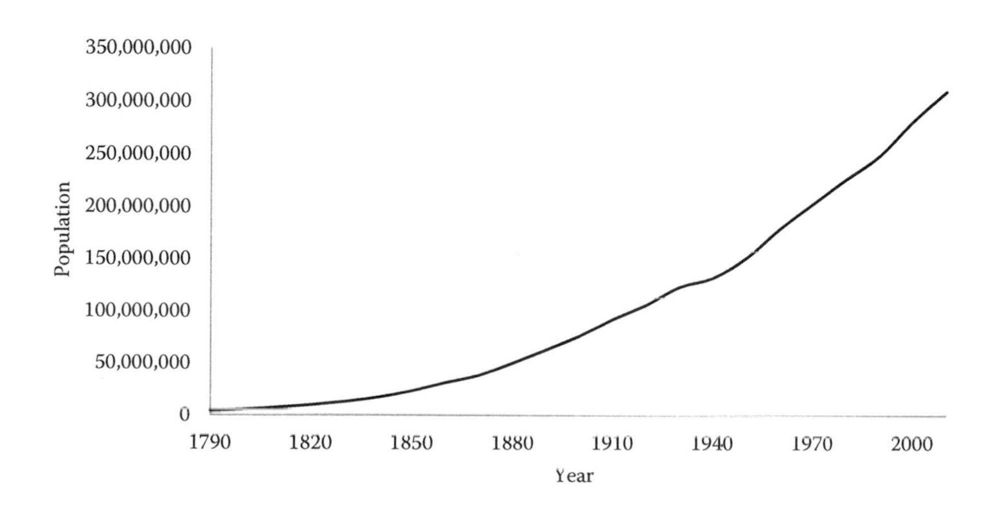

Figure 16.5 US population (1790 to 2010) from US Census Bureau.

One noticeable trend related to the number of people in the United States from 1790 to 2010 is that the number has always increased. In fact, it has not only always increased, but the rate of change has also been increasing itself. Of course, if you know calculus, you know that there is a difference between the rate of change of a value and the rate of change of the rate of change. The first is the first derivative of the function and the second is the second derivative of the function. Thus, while the rate of change of human population has always been positive, the rate of change of the rate of change has also always been positive. But can this continue forever? In fact, just because the trend is this way, does this mean that is how it will be for years to come? Think about events that can change the trends in human population in the United States.

For comparison to the population of the United States since the first census, let us now examine the change in the consumption of energy by the people of the United States since 1790. The US EIA stores and analyzes these data. Here are the data for consumption by energy source since 1790 as generated by these data in our own spreadsheet (Figure 16.6).

Unlike the data for population of the United States, the history of energy consumption in the United States has not always gone up. One can see from these data that there are time periods when energy consumption in the United States actually fell. Now let us examine a scatter plot, which will give us an idea of any correlative relationship between total energy consumption and population size. Again, we display the data from the US EIA in a graph of our own generation from a spreadsheet developed using these data. Figure 16.7 is the resultant scatter plot of the data.

All right, what can you conclude now about the population of the United States versus its power consumption? Fundamentally, as the population increased, so did the power consumption. Does that make sense? Sure, as you have more people in your country, you would expect that power consumption increases. But what could tell us more about the situation? After all, the amount of power or energy consumed by a nation over time

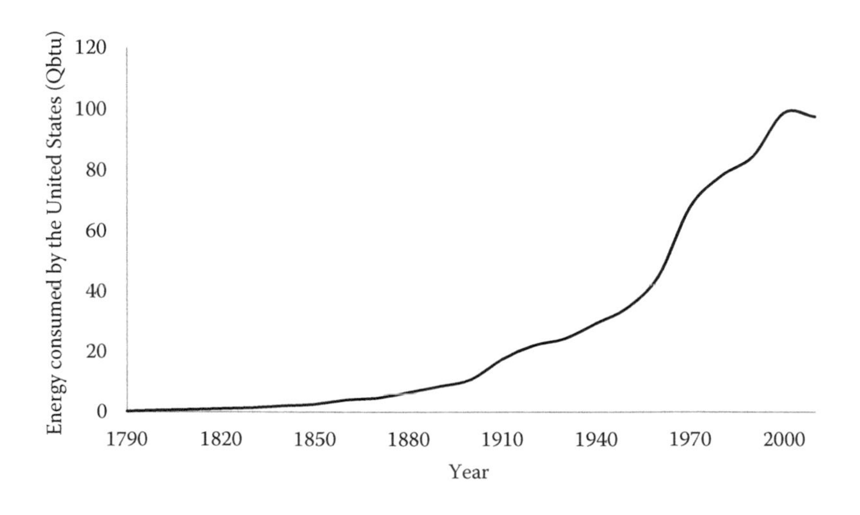

Figure 16.6 US total energy consumption (in quadrillion British thermal units).

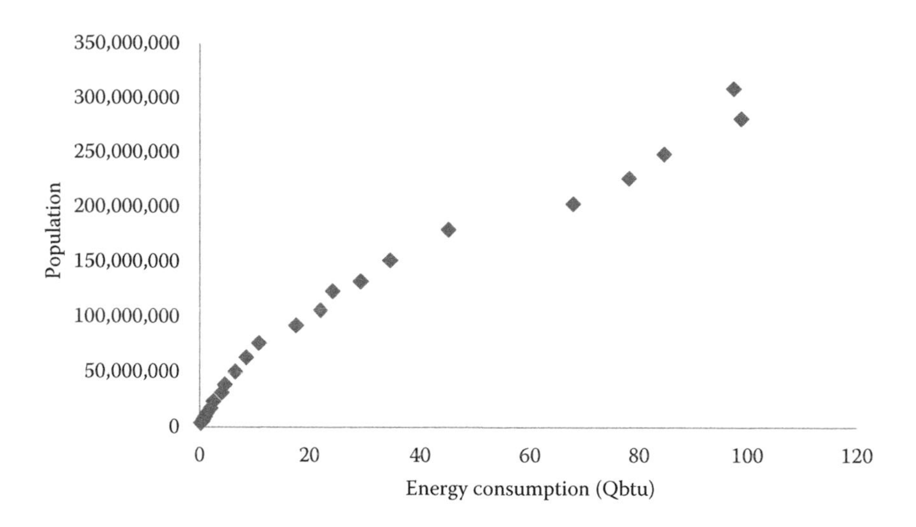

Figure 16.7 Scatter plot of US population vs. total energy consumption.

should be expected not only to increase, but also to increase in the rate of power consumption per capita. Let us explain.

What does per capita mean? Actually, the phrase "per capita" comes from Latin and translates as something like "for each head." It is used to indicate a calculated value which gives an average for each individual person. For example, if there were 100 people attending some event and the total sales during the event was $10,000, we could say that the per capita spending was $100. That is, on the average, each person in attendance spent $100. Now it may be that one person spent $10,000 and all others spent nothing, but the per capita spending would still be $100. So let us once again utilize the available US EIA data in a spreadsheet of our own design. Now we have Figure 16.8 showing us the per capita use of energy in the United States since 1790.

Examining the trend of the energy per capita, one should keep in mind that the largest component of the energy production has changed over the years, but even the oldest carbon-based fuel utilized, namely, wood, is still used today. Nonetheless, while wood was the first primary energy source since the inception of the United States, it has changed. Wood

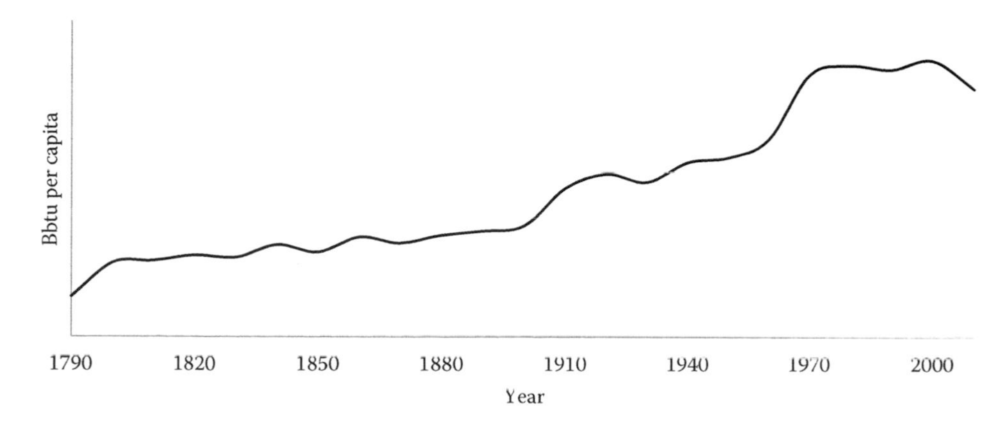

Figure 16.8 Total energy per capita in the United States from 1790 to 2010.

utilization as an energy source peaked around 1875, about two decades after the introduction of coal. Coal, of course, is still utilized as a primary source of energy generation even today, especially in the generation of electricity. Nonetheless, coal energy production per capita peaked in about 1920 and again in 1940. Here it is necessary to recall your American history. You need to remember the effects of the American participation in World War I and the effects of the Great Depression beginning in 1929 on the energy per capita data. Another historical event, that is, World War II, helps explain the rather quick change that occurred in the production of energy and the switch from a more coal-based energy production society to a more petroleum product-based energy production society. Even today, energy production using petroleum products is the dominant source, while the other sources remain a smaller factor in our energy generation.

The most unusual aspect of Figure 16.8 is that the energy per capita utilization has not always increased. In fact, since 1980, the overall energy per capita in the United States has remained somewhat level, with a most recent decrease. The media today tends to give the impression that the trend in the energy per capita has always been increasing. This is simply not true as can be quickly noted by scanning Figure 16.8.

16.3 WORLD POPULATION DYNAMICS AND ENERGY CONSUMPTION

In August of 2016, the estimated population in the world was already over 7.346 billion as estimated by the US Census Bureau. This number, of course, is constantly changing. If you were so inclined, you could actually watch the estimated world population every second just as you can for the US population on the Internet. All you have to do is open your web browser and go to http://www.census.gov/popclock/.

If you were to go online and search for images associated with world population growth, you would find that a majority of those images show some aspect of an exponential growth curve. Using data averaged from six world population estimates, it does look like the world population growth is indeed exponential in nature. Figure 16.9 is our own plot of the estimated world population growth from 10,000 BC to AD 2100, using an average of the most commonly used world population growth models, with their projections out to 2100.

Sometimes, the sample time used makes all the difference in data visualization. Figure 16.9 certainly has all the hallmarks of an exponential growth curve. However, what happens when you limit the world population sample to just the years 1900–2100? If you plot the same data from Figure 16.9 in such a plot, limiting the years displayed, you now get a very different looking curve. This is displayed in Figure 16.10.

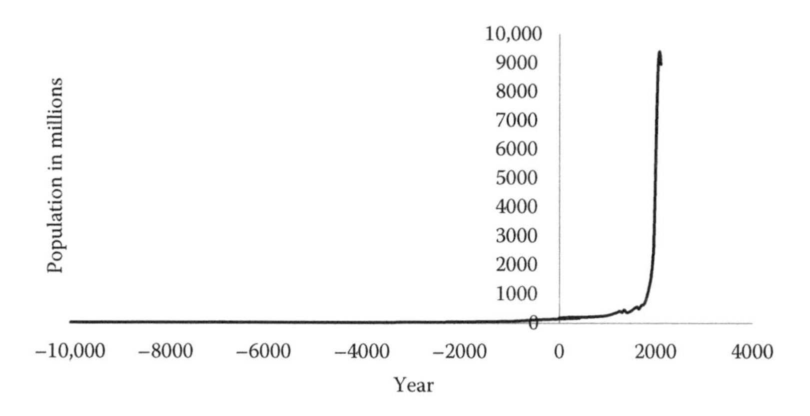

Figure 16.9 World population estimates from 10,000 BC to AD 1950.

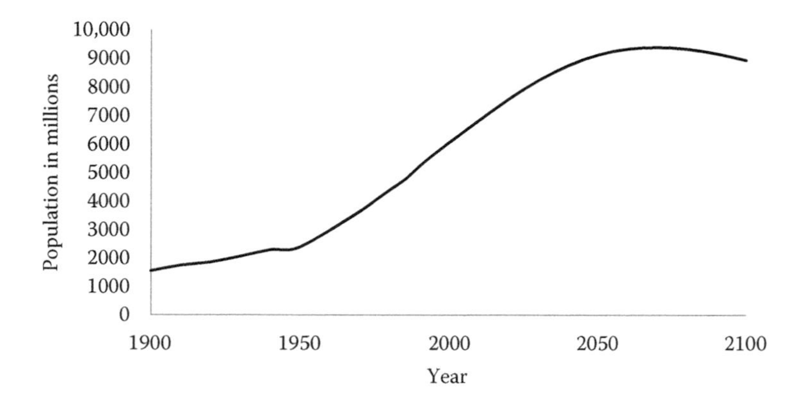

Figure 16.10 World population estimates from 1900 to 2100.

Now we see that world population data from 1900 to 2100 look very similar to a logistic curve as displayed in Figure 16.2. This is very instructive. We have displayed data that look exponential in one representation, while looking very much like a logistic growth model curve in another. So please, be very aware of the time period for which data are displayed, and whether the same data can be displayed in a different manner, giving the appearance of a very different growth model.

As was done for the United States independently, let us now examine world energy consumption over time. Figure 16.11 displays the data for the estimates of world energy consumption since 10,000 BC.

As we can see in Figure 16.11, we again would have no doubts as to the exponential growth of energy consumption in the world over this period. Once again, let us adjust our time period and our time axis and examine the energy growth from 1900 to 2010 for the world. These data are displayed in Figure 16.12.

Once again, now that the time series limits have been adjusted down to a period a little over 100 years, the plotted series for world energy consumption no longer appears like an exponential growth curve as depicted in Figure 16.1, but looks much more like the logistic growth model curve

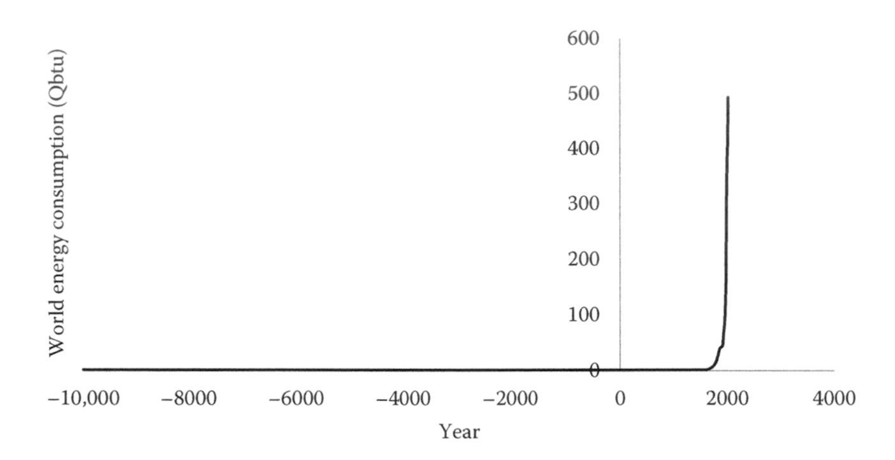

Figure 16.11 World energy consumption from 10,000 BC to AD 2010.

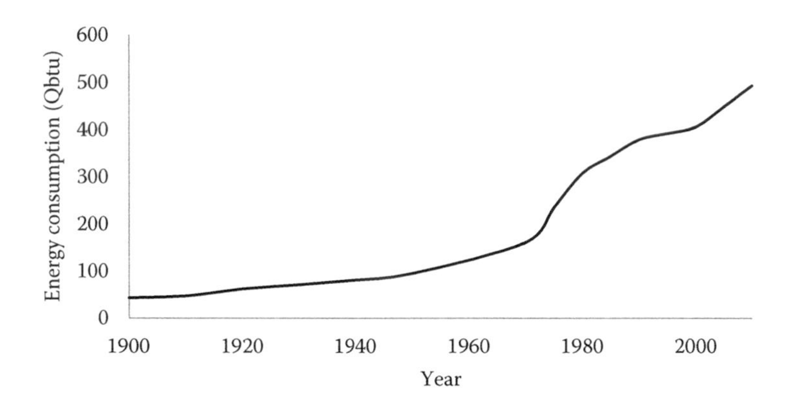

Figure 16.12 World energy consumption from 1900 to 2010.

of Figure 16.2. Remember, the only aspect changed was the time period on the axis displayed.

Finally, let us examine the growth in the world energy consumption per capita. This gives us a look at the changes in the amount of energy consumed per person over all of Earth. These data are displayed in Figure 16.13.

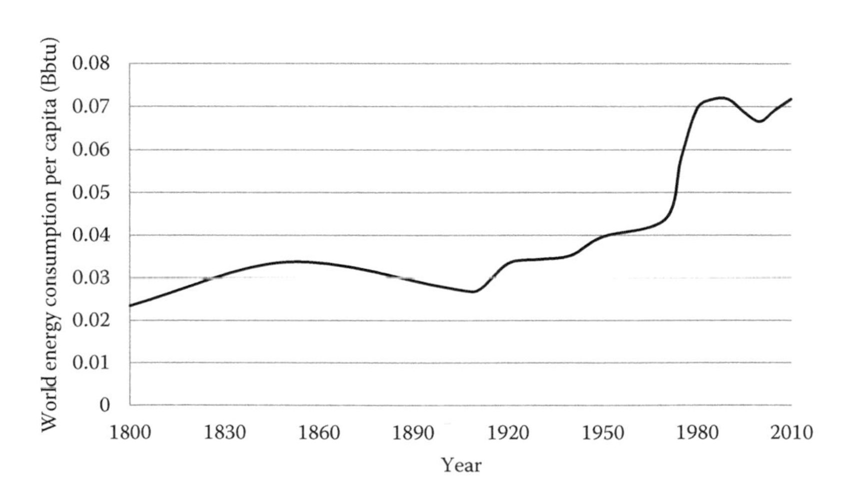

Figure 16.13 World energy consumption per capita.

The reader should compare Figures 16.8 and 16.13, which are data solely for the United States. While there are differences, it is apparent that our modern technology has actually worked to level the amount of energy we consume per person. This is not a conclusion that is commonly assumed in the media.

DELONG AND BURGER STUDY

For a different global perspective of the energy conceptualization, John DeLong of the University of Nebraska at Lincoln, together with Oskar Burger of the University of Kent in Canterbury, United Kingdom, published their results of a study of the scaling of energy use with population in 2015. One word of caution regarding the data displayed by DeLong and Burger is the fact that the authors make use of physical power, rather than physical energy. The reader is reminded of the fact that power is the time rate of change of work, a slightly different concept from energy itself.

16.4 HUMAN POPULATION PROJECTION METHODOLOGIES

The US Census Bureau has a branch within its population division which is wholly responsible for establishing and maintaining the projections for population within the United States specifically and for the population of the world in general. The UN also has a branch responsible for maintaining the modeling of the world's population.

The US Census Bureau utilizes three fundamental principles for their population projection models. The first principle lays out the modeling of populations just as we did in Equation 16.2 in Section 16.1. That is, the projections for populations in the future are based upon population of today and the changes incurred by the principal components of births, deaths, and migration. The second principle underlying the population projections developed by the Census Bureau is that the first principle components stated are driven by the age; sex; race; origin; nativity; and the propensity to reproduce, die, and migrate from one demographic region to another. The third fundamental principle utilized by the US Census Bureau in its projections of population size is a working definition of population universe. This third concept is necessary in order to address the issues of those people not counted by a formal census, the changes in the definition of citizen, or the changes that may occur in the classification of population constituents. The US Census Bureau currently utilizes the definition of a citizen as established in the 1990 US census, with some modifications.

The three principles utilized by the US Census Bureau lead to specific methodologies associated with model projections. For example, the first two principles lead the Census Bureau to apply what is known as the cohort component methodology for projecting population. Using similar notation from Equation 16.2, the cohort component is simply expressed as follows:

$$N_{t+1} = N_t + B - D + M. \tag{16.18}$$

As you can see, the only difference between the cohort component and Equation 16.2 is the incorporation of M for migration. Thus, the cohort component methodology is merely a simplified model which incorporates the demography of migrations of populations.

The second fundamental principle employed by the US Census Bureau requires that the cohort component methodology be applied to each racial and ethnic category as if they were a separate population.

The third fundamental principle employed by the Census Bureau requires a slight modification to the original cohort component methodology. Such a modification is known as the inflation–deflation methodology. The standard cohort component methodology assumes that birth cohorts are only affected by the mortality and migration as they age. If there was a mistake in the year of birth in the population count, this could cause irregularities in the population age distribution in time, advancing the age pattern irregularities and preventing the legitimate comparison of age groups from one year to the next.

In order to properly utilize Equation 16.18 and the cohort component methodology, the numerical projections of births, deaths, and net migration must be modeled as precisely as possible. The derivation of the values for births, deaths, and migration are independent of the instantaneous population itself. The term used by professional census takers to describe this independence is exogenous. If the values used for the births, deaths and, migration are linked to specific values of population, the term applied is endogenous.

Now that we have examined the modeling utilized to project the population of the United States and the world, let us address the uncertainty developed in our models of future population values. In the model development methodology used by the Census Bureau, there are both high and low values generated based upon an assessment of what is viewed as extreme values. So what are the components of the constants utilized in the evaluation of Equation 16.2 by the Census Bureau?

The primary component of the birth parameter in Equation 16.18 that is most influential in the choice of value is the fertility rate. The fertility rate in the United States has been very steady since 1989. The value hovers, just a bit over 2 births for every female. However, there have certainly been changes in this average level during the almost 30-year period. The Census Bureau assumes fertility rates that change over the short-term (<20 years) and over the long-term (>20 years) periods. They also have adjustments based upon the ethnic and geographic differences among all women.

With respect to the death rate, or mortality parameter in Equation 16.18, the Census Bureau examines a number of contributing factors of this parameter. The largest discriminator is gender. Thus, death rates are calculated for both males and females. Another factor which is a major discriminator is the difference in the death rate for different age groups. The Census Bureau breaks up the general population into three major age groups: 0–14, 14–64, and 65 and over. For each age group, a low and a high mortality rate is examined. However, each age group itself is further differentiated by ethnicity, which apparently has a significant difference among the ethnic groups (Hispanic, Caucasian, African American, Native American, etc.).

The final parameter from Equation 16.18 is the migration parameter. The US Census Bureau focuses on the migration in and out of the United States. In order to appropriately link their migration parameter with the different ethnic groups, the Census Bureau examines migration as differentiated by geographic region, always keeping linkages to the different ethnic groups which add to or delete from those being examined for their fertility and their mortality rates. Once again, the Census Bureau develops a model for a high and a low value, ultimately utilizing a middle-level assumption. The margin of uncertainty around this middle-level value for migration is greater than the uncertainty level for the fertility or mortality rates. The Census Bureau discovered that the net migration is a small component of the change in population. Furthermore, the net migration appears to be driven by the movement of US military personnel around the world.

Regardless of all these nuances with respect to the population, specifically of the United States of America, the Census Bureau concludes its methodology and assumptions report with the statement that acknowledges that "projecting the human population continues to be an evolving science."

16.5 SUMMARY

This chapter gives an overview on the study of population dynamics with a focus on human population. In differential form, the equation for *population growth* is shown by Equation 16.4, where b is the birth rate and d the death rate. If r represents the *instantaneous rate of increase*, or $b - d$, we have Equation 16.6.

Equation 16.6 is the classical *Malthusian model* of the growth of population. If $r = 0$, there is simply no population growth. If r is greater than zero, the population will exponentially grow.

The integration of Equation 16.6 leads to another classical representation of population growth (Equation 16.7).

In Equation 16.7, N_0 represents the initial population. From this formulation, we are able to derive Equation 16.9, where t_d is the *population doubling time,* which is the time within which the population doubles its size, a commonly used population growth demarcation.

We then explored the *logistic growth equation,* which in differential form is shown by Equation 16.14.

The integration of Equation 16.14 allows us to represent the logistic growth equation thus shown in Equation 16.15.

We next explored the interactions of populations through the *Lotka–Volterra model,* which in differential form can be represented by Equations 16.16 and 16.17.

We then examined the data from the constitutionally mandated *census,* which takes place every 10 years lead by the US Census Bureau. We also examined how the US Census Bureau estimates the US population for the intervening years between the censuses.

Time series data were examined for the United States since the first census was completed in 1790. We also examined how the US Census Bureau extrapolates the population in the out-years prior to the next census. We looked at data for total *American energy consumption* and then developed a per capita energy consumption value based on data from the US EIA of the DOE. It was discovered that in the last 30 years, the energy consumption per capita has remained nearly level.

After examining data for the United States in terms of population and energy consumption, we next examined the *world population* and *world energy consumption*. We developed infographics which demonstrate the differences between the Unites States and the world in trends for energy consumption. Finally, we examined the methodology and assumptions used in forecasting US and world populations, as emphasized by the US Census Bureau.

PROBLEMS

1. Consider the study of an island human population with no immigration or emigration. In the first year, we count 500 people, and in the second year, we count 800 people. What is the population growth rate?
2. The instantaneous rate increase of some human population center, with no immigration or emigration, is given as 300 humans per year. What is the doubling time for this population?

3. Consider the study of an island human population with no immigration or emigration. What is the doubling time if the instantaneous rate of increase is 0.365 people per year? Assuming that our first census of the island was 50 people, how many years before you would have 310 people on the island?
4. Consider an isolated human population center which has a carrying capacity of 500 and an instantaneous rate of increase of 0.1 people per year. Calculate the maximum possible growth rate for this isolated human population assuming logistic growth.
5. Consider an island nation with a population of 3000. In 1 month, this population has 40 births and 15 deaths. What is the instantaneous rate of increase (r)? Use the calculated instantaneous rate of increase to estimate the population size in 6 months, assuming no migration or emigration and assuming r remains constant.
6. The population of a small town over a 5-year period is 100, 158, 315, 398, and 794. Determine the instantaneous rate of increase (r) for this population, assuming no changes in its value or any migration or emigration.
7. An isolated island population increases in size by 12% every year. Determine the doubling time for the population assuming a constant r and no migration nor emigration.
8. Consider an island whose population is limited by law to no more than 1000. If the instantaneous rate increase is determined to be 0.1 individuals per month, calculate the maximum growth rate for this island population.
9. Consider an island whose population is limited by law to no more than 1000 people. Assume that the initial population is 500 people with $r = 0.005$ per day. Then there is immigration at one time of precisely 600 people. Calculate the instantaneous growth rate (obviously, it must be negative to get the population below its limit) for the population after immigration. Hint: Utilize Equation 15.14.
10. Consider an island nation with a population of 3000. In 1 month, this population has 400 births and 150 deaths. What is the instantaneous rate of increase (r)? Use the calculated instantaneous rate of increase to estimate the population size in 6 months, assuming no migration or emigration and assuming r remains constant.
11. An isolated island population increases in size by 6% every year. Determine the doubling time for the population assuming a constant r and no migration nor emigration. Compare the doubling time for this example to that of problem 7.
12. Consider the total electrical power consumption of the people of the United States. Now calculate the electrical power per capita. Using the Census Bureau historical data and energy usage data from the US EIA, plot the change in electrical power per capita over the years 1950–2010.
13. Calculate the probability that N people pulled at random off the street will have N different birthdays.

14. Calculate the number of people you need to pull off the street at random such that the probability of the people having different birthdays is less than 0.5.
15. Consider the total electrical power consumption of the entire world. Now calculate the electrical power per capita for the world. Using the UN population data and World Bank historical data, plot the change in electrical power per capita in the world over the years 1960–2010.
16. Choose a state you would like to examine. What percentage of the state you have chosen lives in its largest city?
17. Examine the latest US Census Bureau population data. Determine the average of all state populations and develop the average-sized state population. Now determine how much of the population lives within a state more than and less than the average population state.
18. Consider the energy developed by a person exercising on a treadmill. Estimate the total power generated if every person used a treadmill for 1 hour a day and linked it to the power grid.
19. If you assume that each person of the world consumes an average 2500 calories per day, calculate how much electrical energy could be generated by the entire world population if all the food calories consumed each year were put into the generation of electrical energy.

FURTHER READING

Butler, T. (ed.) (2015) *Overdevelopment, Overpopulation, Overshoot*. Goff Books, San Francisco, CA.

Crowther, T. W., H. B. Glick, K. R. Covey, C. Bettigole, D. S. Maynard, S. M. Thomas, J. R. Smith et al. (2015) Mapping tree density at a global scale, *Nature*, 525, 201–205.

DeLong, J. P., and O. Burger (2015) Socio-economic instability and the scaling of energy use with population size, *PLoS ONE*, 10(6), e0130547.

Gotelli, N. (2008) *A Primer of Ecology*, Sinauer Associates, Sunderland, MA.

Hardin, G. (1993) *Living within Limits: Ecology, Economics and Population Taboo*, Oxford University Press, New York.

Khokhar, T., and H. Kashiwase (2015) The future of the world's population in 4 charts, *The World Bank Open Data*. August 5, 2015, https://blogs.worldbank.org/opendata/future-world-s-population-4-charts.

Meadows, D., D. Meadows, J. Randers, and W. W. Behrens (1972) *The Limits to Growth: A Report for the Club of Rome's Project on the Predicament of Mankind*, Signet Books, New York.

Meadows, D., J. Randers, and D. Meadows (2004) *Limits to Growth: The 30-Year Update*, Chelsea Green, White River Junction, VT.

Santos, A. (2009) *How Many Licks? Or How to Estimate Damn Near Anything*, Running Press, Philadelphia, PA.

Vyawahare, M. (2015) Can the planet support 11 billion people, *Scientific American Online*. August 12, 2015, https://www.scientificamerican.com/article/can-the-planet-support-11-billion-people/.

Weinstein, L. (2012) *Guesstimation 2.0: Solving Today's Problems on the Back of a Napkin*, Princeton University Press, Princeton, NJ.

Weinstein, L., and J. Adam (2008) *Guesstimation: Solving the World's Problems on the Back of a Cocktail Napkin*, Princeton University Press, Princeton, NJ.

Answers to Even-Numbered Problems

CHAPTER 1

2. Assuming no down time: 9×10^9 kWh
4. 0.56%
10. 600 km, assuming that the sun shines 12 h out of the day
12. Under the special plan, they would pay 10.97¢/kWh, so they should definitely choose the standard plan.
16. OK

CHAPTER 2

2. A Carnot cycle consists of two isotherms and two adiabatic curves. On a T–S plot, the isotherms are represented by two horizontal lines. The adiabatic curves have $\Delta Q = 0$.

   By the definition of entropy $\Delta S = \Delta Q/T$, we therefore have $\Delta S = 0$ so that the curves are horizontal lines on a T–S plot.
8. The latter is about 46 times the former.
10. No; the first one (since T is higher); $e = e_1 + e_2 + e_3 - e_1e_2 - e_1e_3 - e_2e_3 + e_1e_2e_3$
16. 64%
18. 5×10^9 kg/y

CHAPTER 3

2. 18 events
4. 27.5 MeV
6. 6.58 h; 349 Bq
12. 4.267 MeV; $E_\alpha = 4.19$ MeV; $E_{Th} = 0.08$ MeV
16. Fraction decayed is 1.26×10^{-15}.
20. Scattering from more than one nucleus
22. 13.1 rad
24. Fusion releases about 3.5 times as much per kilogram.
26. 4×10^{12} J

CHAPTER 4

6. 25.2; 29.8; 114.3 (assuming energy loss in each case is half what a 180 scattering predicts)
8. 27.8 K
10. 1360 rad/s
14. 94.9 cm; 32,000 fm^2
16. 1.38 m
18. 702,000 gal/min
20. 1.00 kg
24. 2×10^4 kg/y

CHAPTER 5

4. 8.8 kW
6. 0.367 μmol/min/cm^2
8. You should find around 12 t/acre (within a factor of 10), assuming that wood is about 50% carbon.
10. 3.0 million acres, assuming that dry biomass yields around 8000 Btu/lb
12. The higher compression ratio would increase the 20% efficiency to 22.3%, which is not enough to offset the 38% lower energy density.
16. 1.3×10^{15}, 1%

CHAPTER 6

4. Using a value for $k = 80$ W/m K, and an inner core radius and gradient from Figure 6.3, we find a heat flow of about 1.7 TW out of the inner core, while for the heat flow at the Earth's surface, we find 31 TW, which is about 18 times larger—the difference being due in part to the heat from radioactivity.
6. $c\rho = 3.24 \times 10^6$ J/K m^3; 1.94 EJ/km^3
12. 3.40
14. 3.4×10^8 J

CHAPTER 7

4. Using the cumulative distribution $e^{-v^2/2v_0^2}$ and $v_0 = 2\bar{v}/\sqrt{\pi}$, we find the probabilities indicated in parentheses: $\bar{v} = 4$ m/s(0.8%), $\bar{v} = 5$ m/s(4.6%), and $\bar{v} = 6$ m/s(11.8%).
6. 15.2 m
8. ë = 0.073
10. 1.107 km
12. 1.75×10^5 MWh
14. $v > 12.93$ m/s (Recall that above the rated wind speed for a turbine, the power is kept constant.)
18. 4880 W/m^2

CHAPTER 8

14. 1.32 billion years, unrealistically assuming that the Earth is a sphere of uniform density
16. 98 J

CHAPTER 9

2. About 13.4 doublings or 469 years, according to the rule of 70
4. 202 photons/s
6. 71.0°
10. 0.58 K
12. 1.00, 106.5, 184.9, 123,400. These values do not account for the different atmospheric lifetimes and, therefore, disagree with values found on the web.
20. The 4 days of the year where the equation of time graph of Figure 9.7 crosses the time axis: April 15, June 13, September 1, and December 25
22. $\Delta T = +0.23$ K
28. 1.66×10^{15} W, 5.23×10^{22} J

CHAPTER 10

2. 50 W/m K
6. 40.4°F
8. 125°C above ambient
14. Eightfold
16. 10^8 km

CHAPTER 11

4. 5.0×10^{22}
6. 236 K
10. Power is 11% lower in July due to higher temperature, but 4.0 times higher due to the longer days and higher elevation of the sun. The latter effect is far more important.
16. $0, \frac{L}{6}, \frac{2L}{6}, \frac{3L}{6}, \frac{4L}{6}, \frac{5L}{6}, L$
18. 7.6 A

CHAPTER 12

2. A 1% rise
4. A 30-fold increase in a decade is equivalent to a 40.5% annual increase. By the rule of 70, we therefore have a doubling time of 1.72 years, which means a decade is 5.78 doublings. This is somewhat more of an increase than originally postulated, since five doublings amount to an increase by a factor of 32.
6. e increases by 0.022; r increase by 0.012 would have same effect.
8. 4.0 million m^3; 38.3 million m^3
12. 4.07 mi to save 5¢/gal or 4.13 mi to save 10¢/gal, assuming the car gets 20 mpg and a price per gallon of $3.00. This estimate ignores extra costs including wear and tear on the engine, your time, extra chances of an accident, etc.
14. 11.7

CHAPTER 13

2. 55,556
4. 30 million kg, assuming a specific heat of 1560 J/kg °C
6. 1.73×10^6 rad/s
8. 2.19 MJ; 2093 m/s; 2.24×10^6
12. 2.61 GJ
18. 1.19 mm; 4.06 mm
20. 300 kg

CHAPTER 14

2. 8×10^3 kg CO_2 per year per car or 10^{12} kg/year for all cars in U.S.
4. 230 ppm
16. 0.0183
18. 0.228
20. 0.905
22. 6
24. 43.2%

CHAPTER 15

2. 0.667
4. 0.605
6. 0.375
8. 0.133
10. 0.951
12. 0.25; 0.063; 0.0039
14. 0.12; 0.42

CHAPTER 16

2. 0.0023 years
4. 12.5 persons per month
6. 0.5 persons per year
8. 25 persons per month
10. 0.08 persons per month; 4848 people
14. greater than 23
18. ~500,000,000 kWhr/yr

Appendix A: Useful Physical Constants*

Quantity	Value	Unit
Alpha particle mass	$6.644\ 657\ 230 \times 10^{-27}$	kg
Alpha particle mass energy equivalent	$5.971\ 920\ 097 \times 10^{-10}$	J
Alpha particle mass energy equivalent	3727.379 378	MeV
Atomic mass constant	$1.660\ 539\ 040 \times 10^{-27}$	kg
Atomic mass constant energy equivalent	$1.492\ 418\ 062 \times 10^{-10}$	J
Atomic mass constant energy equivalent	931.494 0954	MeV
Atomic unit of charge	$1.602\ 176\ 6208 \times 10^{-19}$	C
Atomic unit of current	$6.623\ 618\ 183 \times 10^{-3}$	A
Atomic unit of energy	$4.359\ 744\ 650 \times 10^{-18}$	J
Atomic unit of force	$8.238\ 723\ 36 \times 10^{-8}$	N
Atomic unit of length	$0.529\ 177\ 210\ 67 \times 10^{-10}$	m
Atomic unit of mass	$9.109\ 383\ 56 \times 10^{-31}$	kg
Avogadro constant	$6.022\ 140\ 857 \times 10^{23}$	mol^{-1}
Bohr radius	$0.529\ 177\ 210\ 67 \times 10^{-10}$	m
Boltzmann constant	$1.380\ 648\ 52 \times 10^{-23}$	J/K
Characteristic impedance of vacuum	376.730 313 461	Ω
Classical electron radius	$2.817\ 940\ 3227 \times 10^{-15}$	m
Compton wavelength	$2.426\ 310\ 2367 \times 10^{-12}$	m
Deuteron mass	$3.343\ 583\ 719 \times 10^{-27}$	kg
Deuteron mass	2.013 553 212 745	u
Deuteron mass energy equivalent	$3.005\ 063\ 183 \times 10^{-10}$	J
Deuteron mass energy equivalent	1875.612 928	MeV
Deuteron–proton mass ratio	1.999 007 500 87	
Electric constant	$8.854\ 187\ 817 \times 10^{-12}$	F/m
Electron charge to mass quotient	$-1.758\ 820\ 024 \times 10^{11}$	C/kg
Electron mass	$9.109\ 383\ 56 \times 10^{-31}$	kg
Electron mass energy equivalent	$8.187\ 105\ 65 \times 10^{-14}$	J
Electron mass energy equivalent	0.510 998 9461	MeV
Electron mass	$5.485\ 799\ 090\ 70 \times 10^{-4}$	u
Electron-to-alpha particle mass ratio	$1.370\ 933\ 554\ 798 \times 10^{-4}$	
Electron volt	$1.602\ 176\ 6208 \times 10^{-19}$	J
Elementary charge	$1.602\ 176\ 6208 \times 10^{-19}$	C
Faraday constant	96 485.332 89	C/mol
Kilogram–atomic mass unit relationship	$6.022\ 140\ 857 \times 10^{26}$	u
Kilogram–electron volt relationship	$5.609\ 588\ 650 \times 10^{35}$	eV
Magnetic constant	$12.566\ 370\ 614 \times 10^{-7}$	N/A^2
Molar mass constant	1×10^{-3}	kg/mol
Molar mass of carbon-12	12×10^{-3}	kg/mol
Molar volume of ideal gas (273.15 K, 100 kPa)	$22.710\ 947 \times 10^{-3}$	$m^{-3}\ mol^{-1}$
Muon–electron mass ratio	206.768 2826	
Muon mass	$1.883\ 531\ 594 \times 10^{-28}$	kg

* Adapted from http://physics.nist.gov/constants.

Muon mass	0.113 428 9257	u
Muon mass energy equivalent	$1.692\ 833\ 774 \times 10^{-11}$	J
Muon mass energy equivalent	105.658 3745	MeV
Neutron–electron mass ratio	1838.683 661 58	
Neutron mass	$1.674\ 927\ 471 \times 10^{-27}$	kg
Neutron mass	1.008 664 915 88	u
Neutron mass energy equivalent	$1.505\ 349\ 739 \times 10^{-10}$	J
Neutron mass energy equivalent	939.565 4133	MeV
Newtonian constant of gravitation	$6.674\ 08 \times 10^{-11}$	$m^3/kg\ s^2$
Planck constant	$6.626\ 070\ 040 \times 10^{-34}$	J s
Planck constant	$4.135\ 667\ 662 \times 10^{-15}$	eV s
Planck length	$1.616\ 229 \times 10^{-35}$	m
Planck mass	$2.176\ 470 \times 10^{-8}$	kg
Planck mass energy equivalent	$1.220\ 910 \times 10^{19}$	GeV
Planck temperature	$1.416\ 808 \times 10^{32}$	K
Planck time	$5.391\ 16 \times 10^{-44}$	s
Proton charge-to-mass quotient	$9.578\ 833\ 226 \times 10^{7}$	C/kg
Proton Compton wavelength	$1.321\ 409\ 853\ 96 \times 10^{-15}$	m
Proton–electron mass ratio	1836.152 673 89	
Proton mass	$1.672\ 621\ 898 \times 10^{-27}$	kg
Proton mass	1.007 276 466 879	u
Proton mass energy equivalent	$1.503\ 277\ 593 \times 10^{-10}$	J
Proton mass energy equivalent	938.272 0813	MeV
Rydberg constant	10 973 731.568 508	m^{-1}
Speed of light in vacuum	299 792 458	m/s
Standard acceleration of gravity	9.806 65	m/s^2
Standard atmosphere	101 325	Pa
Standard-state pressure	100 000	Pa
Stefan–Boltzmann constant	$5.670\ 367 \times 10^{-8}$	$W/m^2\ K^4$
Tau–electron mass ratio	3477.15	
Tau mass	$3.167\ 47 \times 10^{-27}$	kg
Tau mass	1.907 49	u
Tau mass energy equivalent	$2.846\ 78 \times 10^{-10}$	J
Tau mass energy equivalent	1776.82	MeV
Thomson cross section	$0.665\ 245\ 871\ 58 \times 10^{-28}$	m^2
Unified atomic mass unit	$1.660\ 539\ 040 \times 10^{-27}$	kg
von Klitzing constant	25 812.807 4555	Ω
Weak mixing angle	0.2223	
Wien frequency displacement law constant	$5.878\ 9238 \times 10^{10}$	Hz/K
Wien wavelength displacement law constant	$2.897\ 7729 \times 10^{-3}$	m K

Appendix B: Useful Conversion Factors*

LENGTH

Symbol	When You Know	Multiply by	To Find	Symbol
in	Inches	2.54	Centimeters	cm
ft	Feet	30.48	Centimeters	cm
yd	Yards	0.91	Meters	m
mi	Miles	1.61	Kilometers	km

AREA

Symbol	When You Know	Multiply by	To Find	Symbol
in^2	Square inches	6.45	Square centimeters	cm^2
ft^2	Square feet	0.09	Square meters	m^2
yd^2	Square yards	0.84	Square meters	m^2
mi^2	Square miles	2.59	Square kilometers	km^2
	Acres	0.41	Hectares	Ha

MASS (WEIGHT)

Symbol	When You Know	Multiply by	To Find	Symbol
oz	Ounces	28.35	Grams	g
lb	Pounds	0.45	Kilograms	kg
	Short tons (2000 lb)	0.91	Metric tons	t

* Adapted from http://www.nist.gov/pml/wmd/metric/common-conversion-b.cfm.

VOLUME

Symbol	When You Know	Multiply by	To Find	Symbol
tsp	Teaspoons	4.93	Milliliters	mL
tbsp	Tablespoons	14.79	Milliliters	mL
in^3	Cubic inches	16.39	Milliliters	mL
fl oz	Fluid ounces	29.57	Milliliters	mL
c	Cups	0.24	Liters	L
pt	Pints, liquid	0.47	Liters	L
qt	Quarts, liquid	0.95	Liters	L
gal	Gallons	3.79	Liters	L
ft^3	Cubic feet	0.03	Cubic meters	m^3
yd^3	Cubic yards	0.76	Cubic meters	m^3

TEMPERATURE

Symbol	When You Know	Multiply by	To Find	Symbol
°F	Degrees Fahrenheit	Subtract 32, then multiply by 1.8	Degrees Celsius	°C
°F	Degrees Fahrenheit	Subtract 32, multiply by 1.8, add 273.15	Kelvins	K

Index

A

B

C

D

E

F

G

H

I

Q

R

S